WORKSHOPS IN COMPUTING
Series edited by C. J. van Rijsbergen

C. Rattray (Ed.)

Specification and Verification of Concurrent Systems

Published in collaboration with the
British Computer Society

 Springer-Verlag London Ltd.  BCS

C. Rattray, MSc
Department of Computing Science, University of Stirling,
Stirling FK9 4LA, UK

ISBN 978-3-540-19581-8   ISBN 978-1-4471-3534-0 (eBook)
DOI 10.1007/978-1-4471-3534-0

British Library Cataloguing in Publication Data
Rattray, C. (Charles), *1938–*
Specification and verification of concurrent systems
1. Computer systems
I. Title.
005.42

Library of Congress Cataloging-in-Publication Data
Specification and verification of concurrent systems/C. Rattray, ed.
p. cm.—(Workshops in computing)
"Papers presented at the BCS-FACS Workshop on Specification and Verification of
Concurrent Systems held on 6–8 July 1988, at the University of Stirling, Scotland"—
Pref.
Includes index

1. Parallel processing (Electronic computers)—Congresses.
I. Rattray, C. (Charles). 1938–   . II. British Computer Society.   III. BCS-FACS
Workshop on Specification and Verification of Concurrent Systems (1988: University of
Stirling).   IV. Series.
QA76.58.S64 1990
004'.35—dc20   90–9912
                                                                          CIP

2128/3916–543210   Printed on acid-free paper

Preface

This volume contains papers presented at the BCS-FACS Workshop on Specification and Verification of Concurrent Systems held on 6–8 July 1988, at the University of Stirling, Scotland.

Specification and verification techniques are playing an increasingly important role in the design and production of practical concurrent systems. The wider application of these techniques serves to identify difficult problems that require new approaches to their solution and further developments in specification and verification. The Workshop aimed to capture this interplay by providing a forum for the exchange of the experience of academic and industrial experts in the field.

Presentations included: surveys, original research, practical experience with methods, tools and environments in the following or related areas:

Object-oriented, process, data and logic based models and specification methods for concurrent systems

Verification of concurrent systems

Tools and environments for the analysis of concurrent systems

Applications of specification languages to practical concurrent system design and development.

We should like to thank the invited speakers and all the authors of the papers whose work contributed to making the Workshop such a success. We were particularly pleased with the international response to our call for papers.

Invited Speakers
Pierre America — Philips Research Laboratories
Professor M. Joseph — University of Warwick
David Freestone — British Telecom

Organising Committee
Charles Rattray — Dr Muffy Thomas
Dr Simon Jones — Dr John Cooke
Professor Ken Turner — Derek Coleman
Maurice Naftalin — Dr Peter Scharbach

We would like to acknowledge the financial contribution made by SD-Sysems Designers plc, Camberley, Surrey.

Finally, the work done by Muriel Croll and Jean McInnes is gratefully acknowledged.

June 1989   Charles Rattray

Contents

The Interplay of Theory and Practice in a Parallel Object-Oriented Language

Pierre America
Philips Research Laboratories
Eindhoven, the Netherlands

Abstract

In this paper we give an overview of the techniques that have been used to describe the semantics of the language POOL in a formal way. POOL is a language designed to program a high-performance parallel machine. It was described independently in an operational way and in a denotational way, and these descriptions were subsequently proved to be equivalent. A number of other semantic techniques have also been applied to POOL. Finally we discuss the practical usefulness of this kind of semantic studies.

1   Introduction

At the Philips Research Laboratories in Eindhoven, the Netherlands, currently two research projects are carried out in the area of parallel machine architectures. The DOOM project (Decentralized Object-Oriented Machine) is a subproject of ESPRIT project 415: "Parallel Architectures and Languages for Advanced Information Processing — a VLSI-directed Approach". This ESPRIT project aims at improving the performance of computers in the area of symbolic applications by the use of large-scale parallelism. Several approaches are explored in different subprojects, which are tied together at a disciplinary level by working groups. The DOOM subproject has chosen an object-oriented approach [BJOT86,Odi87]. This subproject is developing a parallel object-oriented programming language POOL in which applications can be written, together with a parallel machine architecture suitable

The work described in this paper was done in the context of ESPRIT project 415: "Parallel Architectures and Languages for Advanced Information Processing — a VLSI-directed Approach".

to execute programs in this language. A number of example applications are developed as well, and a hardware prototype is being built.

Another project, PRISMA (PaRallel Inference and Storage MAchine), is building on the same object-oriented principles as the DOOM project. It aims at developing a system that is able to handle very large amounts of knowledge and data, again using parallelism to reach a high performance. One of the concrete goals of this project is to deliver a prototype relational database machine which can automatically exploit parallelism in evaluating the users' queries [AKO87].

The language POOL (Parallel Object-Oriented Language) used in these projects has been the subject of extensive theoretical studies. In the present paper, we give a survey of the results of these studies, and we shall try to assess their influence on the design and the use of the language.

Section 2 gives an introduction to the language POOL itself. Sections 3 to 6 give an overview of the techniques that have been used to describe the semantics of POOL in a formal way. Finally, in section 7 we shall discuss the practical importance of a formal semantic analysis.

2 An overview of the language POOL

In POOL, a system is described as a collection of *objects*. An object can be thought of as a kind of box, containing some data and having the ability to perform some actions on these data. An object uses *variables* to store its data. A variable contains a *reference* to another object (or, possibly, to the object containing the variable itself). The object's ability to perform operations on its internal data lies in two mechanisms: First, an object can have a set of *methods*, a kind of procedures, which can access and change the values of the variables. (Up to this point, the mechanisms that we have described are generally present in object-oriented languages.) Second, an object has a so-called *body*, a local process that can execute in parallel with the bodies of all the other objects in the system (this is quite specific for POOL).

A very important principle in object-oriented programming is *protection*: The variables of one object are not directly accessible to other objects. In fact, the only way for objects to interact is by sending *messages*. A message is a request to the receiving object to execute one of its methods. The sending object explicitly mentions the receiver and the method name. It can also pass some parameters (again references to objects) to the method.

The sender blocks until the receiver has answered its message. The receiver also explicitly states when it is prepared to answer a message. However, it does not specify the sender but only lists a set of possible method names. As soon as synchronization between sender and receiver takes place, the receiver executed the required method, using the parameters that the sender gave. The method returns a *result* (once again, a reference to an object), which is then passed back to the sender. After that, sender and receiver both continue their own processing in parallel.

Because of the above mechanisms, the only parallelism in the system is caused by the parallel execution of the bodies of the different objects. Inside each object everything happens sequentially and deterministically, so that the object is protected from the parallel and nondeterministic (and therefore "dangerous") outside world. The interesting thing in POOL is that, like in other object-oriented languages, new objects can be *created dynamically* in arbitrary high numbers. In POOL, where as soon as an object is created, its body starts executing, this means that also the degree of parallelism can be increased dynamically. (Objects are never destroyed explicitly; rather, useless objects are removed by a garbage collector working behind the screens).

In order to describe these dynamically evolving systems of objects in a static program, the objects are grouped into *classes*. All the objects in one class (the *instances* of the class) have the same names and types for their variables (of course, each has its own private set of variables) and they execute the same methods and body. In a program, a *class definition* is used to describe this internal structure of the objects. Whenever a new object is to be created, a class is named which serves as a blueprint.

In order to illustrate the language, we give a small example program. It uses Eratosthenes' sieve method to generate an infinite, ascending stream of prime numbers. An object of class Driver is created initially. This in turn creates an object of class Sieve and sends it an infinite stream of messages, containing the natural numbers greater than 1 in ascending order. Each Sieve object takes the first incoming number as its own prime number. This number is sent to the output file. The object also creates another Sieve object, which it remembers in its variable next. Of all the following incoming numbers, only the ones that are not divisible by the local prime number are sent on the next Sieve object in the row.

The program is written in the latest version of POOL, called POOL2 [Ame88a]. Here is the code:

```
IMPL UNIT Sieve
USE File_IO

%% This program uses the sieve method of
%% Erathosthenes to generate prime numbers.

GLOBAL driver := Driver.new ()
%% This object will be created automatically


CLASS Driver
VAR first := Sieve.new ()
BODY FOR i FROM 2 DO first ! input (i) OD
YDOB
END Driver


CLASS Sieve

VAR myprime, current : Int
    next : Sieve

METHOD input (n : Int) : Sieve
BEGIN current := n;
     RESULT SELF
END input

BODY ANSWER (input); %% The first number is prime!
    myprime := current;
    standard_out ! write_Int (myprime, 0) ! new_line ();
    next := Sieve.new();
    DO %% forever
       ANSWER (input);
       IF current // myprime ~= 0
       THEN next ! input (current)
       FI
    OD
YDOB
END Sieve
```

More details about the issues that have played a role in the design of

POOL can be found in [Ame87b].

3 Operational Semantics

A number of different techniques have been used to describe the semantics of POOL in a formal way. In all of these descriptions, a syntactically simplified version of POOL is used. This is more convenient in the semantic definition but not very well readable in concrete programs. There is a straightforward translation from POOL2 or POOL-T (an older version) to this simplified notation, which we shall just call POOL.

The simplest semantic technique is the use of *transition systems* to define an *operational* semantics. This technique has been introduced by Hennessy and Plotkin [HP79,Plo81,Plo83]. It describes the behaviour of a system in terms of sequences of *transitions* between *configurations*. A configuration describes the system at one particular moment during the execution. Apart from a component describing the values of the variables, it typically contains as a component that part of the program that is still to be executed. The possible transitions are described by a *transition relation*, a binary relation between configurations (by having a relation instead of a function, it is possible to model nondeterminism). This transition relation is defined by a number of *axioms* and *rules*. Because of the presence of (the rest of) the program itself in the configurations, it is possible to describe the transition relation in a way that is closely related to the syntactic structure of the language.

The term "operational" can now be understood as follows: The set of configurations defines a (very abstract) model of a machine, and the transition relation describes how this machine operates: each transition corresponds to an action that the machine can perform. The fact that the semantic description follows the syntactic structure of the language so closely (as we shall see below) is a definite advantage of the transition system approach to operational semantics.

In the operational semantics of POOL [ABKR86] uses configurations having four components:

$$Conf = P_{fin}(LStat) \times \Sigma \times Type \times Unit$$

The first component is a finite set of *labelled statements*:

$$\{\langle \alpha_1, s_1 \rangle, \ldots, \langle \alpha_n, s_n \rangle\}$$

Here each α_i is an object name and the corresponding s_i is the statement (or sequence of statements) that the object is about to execute. This models the fact that the objects $\alpha_1, \ldots, \alpha_n$ are executing in parallel. The second component is a *state* $\sigma \in \Sigma$, which records the values of the instance variables and temporary variables of all the objects in the system. The third component is a typing function $\tau \in \textit{Type}$, assigning to each object name the class of which the object is an instance. Finally, the last component is the complete POOL program or *unit*, which is used for looking up the declarations of methods (whenever a message is sent) and bodies (when new objects are created).

The transition relation $\rightarrow$ between configurations is defined by axioms and rules. In general, an axiom describes the essential operation of a certain kind of statement or expression in the language. For example, the axiom describing the assignment statement looks as follows:

$$\left\langle X \cup \{\langle \alpha, x := \beta \rangle\}, \sigma, \tau, U \right\rangle \rightarrow \left\langle X \cup \{\langle \alpha, \beta \rangle\}, \sigma\{\beta/\alpha, x\}, \tau, U \right\rangle$$

Here, X is a set of labelled statements which are not active in this transition, β is another object name, a special case of the expression that can in general appear at the right-hand side of an assignment, and $\sigma\{\beta/\alpha, x\}$ denotes the state that results from changing in the state σ the value of the variable x of the object α into the object name β.

Rules are generally used to describe how to evaluate the components of a composite statement or expression. For example, the following rule describes how the (general) expression at the right-hand side of an assignment is to be evaluated:

$$\frac{\left\langle X \cup \{\langle \alpha, e \rangle\}, \sigma, \tau, U \right\rangle \rightarrow \left\langle X' \cup \{\langle \alpha, e' \rangle\}, \sigma', \tau', U \right\rangle}{\left\langle X \cup \{\langle \alpha, x := e \rangle\}, \sigma, \tau, U \right\rangle \rightarrow \left\langle X' \cup \{\langle \alpha, x := e' \rangle\}, \sigma', \tau', U \right\rangle}$$

According to this rule, *if* the transition above the line is a member of the transition relation, *then* so is the transition below the line. In this way the rule reduces the problem of evaluating the expression in an assignment to evaluating the expression on its own. The latter is described by specific axioms and rules dealing with the several kinds of expressions in the language. Note that as soon as the right-hand side expression has been evaluated completely, so that a concrete object name β results, the assignment axiom above applies and the assignment proper can be performed.

The semantics of a whole program can now be defined as the set of all maximal sequences of configurations $\langle c_1, c_2, c_3, \ldots \rangle$ that satisfy $c_i \rightarrow c_{i+1}$. Each of these sequences represents a possible execution of the program.

4   Denotational semantics

The second form of semantic description that has been used to describe POOL is *denotational* semantics. Whereas operational semantics uses an abstract machine that can perform certain actions, denotational semantics assigns a mathematical value, a "meaning", to each individual language construct. Here, the most important issue is compositionality: the meaning of a composite construct can be described in terms of only the *meanings* of its syntactic constituents.

For sequential languages, it is very natural that the value associated with a statement is a function from states to states: when applied to the state before the execution of the statement, this function delivers the state after the execution. However, for parallel languages, this is no longer appropriate. The first problem is that parallel languages are in general *nondeterministic*: it is no longer possible to predict a single output state for each given input state. In general, there is a *set* of possible output states.

The second problem is that describing the set of possible output states for each input states does not provide enough information to be able to compose a statement in parallel with other statements: information on the intermediate states is also required. This leads us to the concept of *resumptions* (introduced by Plotkin [Plo76]). Instead of delivering the final state after the execution of the statement has completed, we divide the execution of the statement into its atomic (indivisible) parts, and we deliver a pair $\langle \sigma', r \rangle$, where σ' is the state after the execution of the first atomic action and r is the resumption, which describes the execution from this point on. In this way, it is possible to put another statement in parallel with this one: the execution of the second statement can be interleaved with the original one in such a way that between each pair of subsequent atomic actions of the first statement an arbitrary number of atomic actions of the second one can be executed. Each atomic action can inspect the state at the beginning of its execution and possibly modify it.

For a very simple language (not yet having the power of POOL) we get the following equation for the set (the *domain*) in which the values reside that we want to assign to our statements:

$$P \cong \{p_0\} \cup \Big(\Sigma \to \mathcal{P}(\Sigma \times P) \Big). \tag{1}$$

The intended interpretation of this equation is the following: Let us call the elements of the set P *processes* and denote them with letters p, q, and r. Then a process p can either be the terminated process p_0, which cannot perform any action, or it is a function which, when provided with an input state σ, delivers a set X of possible actions. Each element of this set X is a pair $\langle \sigma', q \rangle$, where σ' is the state after this action, and q is a process that describes the rest of the execution.

It is clear that equation (1) cannot be solved in the framework of *sets*, because the cardinality of the right-hand side would always be larger than that of the left-hand side. In contrast to many other workers in the field of denotational semantics of parallelism, who use the framework of complete partial orders (cpo's) to solve this kind of equations (see, e.g., [Plo76]), we have chosen to use the framework of *complete metric spaces*. (Readers unfamiliar with this part of mathematics are referred to standard topology texts like [Dug66,Eng77] or to [BZ82].) The most important reason for this choice is the possibility to uses Banach's fixed point theorem:

> Let M be a complete metric space with distance function d and let $f : M \to M$ be a function that is *contracting*, i.e., there is a real number ϵ with $0 < \epsilon < 1$ such that for all $x, y \in M$ we have $d(f(x), f(y)) < \epsilon . d(x, y)$. Then f has a unique fixed point.

This ensures that whenever we can establish the contractivity of a function we have a *unique* fixed point, whereas in cpo theory mostly we can only guarantee the existence of a *least* fixed point.

Another reason for using complete metric spaces is the naturalness of the power domain construction. Whereas in cpo theory there are several competing definitions (see, e.g., [Plo76,Smy78]) all of which are somewhat hard to understand, in complete metric spaces there is a very natural definition:

> If M is a metric space with distance d, then we define $P(M)$ to be the set of all *closed* subsets of M, provided with the so-called *Hausdorff distance* d_H, which is defined as follows:
>
> $$d_H(X, Y) = \max\left\{\sup_{x \in X}\{d(x, Y)\}, \sup_{y \in Y}\{d(y, X)\}\right\}$$
>
> where $d(x, Z) = \inf_{z \in Z}\{d(x, z)\}$ (with the convention that $\sup \emptyset = 0$ and $\inf \emptyset = 1$).

(A few variations on this definition are sometimes useful, such as taking only the nonempty subsets of M or only the compact ones. The metric is the same in all cases.)

The domain equation that we use for the denotational semantics of POOL (see [ABKR88]) is somewhat more complicated than equation (1), because it also has to accommodate for communication among objects. For POOL, the domain P of processes is defined as follows:

$$P \cong \{p_0\} \cup \left(\Sigma \to P(Step_P)\right)$$

where the set $Step_P$ of *steps* is given by

$$Step_P = (\Sigma \times P) \cup Send_P \cup Answer_P,$$

with

$$Send_P = Obj \times MName \times Obj^* \times (Obj \to P) \times P$$

and

$$Answer_P = Obj \times MName \times \left(Obj^* \to (Obj \to P) \to^1 P\right).$$

The interpretation of these equations (actually, they can be merged into one large equation) is as follows: As in the first example, a process can either terminate directly, or it can take one out of a set of steps, where this set depends on the state. But in addition to internal steps, which are represented by giving the new state plus a resumption, we now also have communication steps. A *send step* gives the destination object, the method name, a sequence of parameters, and *two* resumptions. The first one, the *dependent* resumption, is a function from object names to processes. It describes what should happen after the message has been answered and the result has been returned to the sender. To do that, this function should be applied to the name of the result object, so that it delivers a process that describes the processing of that result in the sending object. The other resumption, also called the *independent* resumption, describes the actions that can take place in parallel with the sending and the processing of the message. These actions do not have to wait until the message has been answered by the destination object. (Note that for a single object the independent resumption will always be p_0, because a sending object cannot do anything before the result has arrived. However, for the correct parallel composition of more objects, the independent resumption is necessary to describe the actions of the objects that are not sending messages.) Finally we have an *answer step*: This consists of the name of the destination object and the method name, plus an even more complicated resumption. This resumption takes as input the sequence of parameters in the message plus the dependent resumption of the sender. Then it returns a process describing the further execution of the receiver and the sender *together*.

Equations like (1) can be solved by a technique explained in [BZ82]: An increasing sequence of metric spaces is constructed, its union is taken and then the metric completion of the union space satisfies the equation. The equation for POOL processes cannot be solved in this way, because the domain variable P occurs at the left-hand side of the arrow in the definition of answer steps. A more general, category-theoretic technique for solving this kind of domain equations has been developed to solve this problem. It is described in [AR88]. Let us only remark here that it is necessary to restrict ourselves to the set of *non-distance-increasing* functions (satisfying $d(f(x), f(y)) \leq d(x, y)$), which is denoted by $\rightarrow^1$ in the above equation.

Let us now give more details about the semantics of statements and expressions. These are described by the following two functions:

$$[\![\ldots]\!]_S : Stat \rightarrow Env \rightarrow AObj \rightarrow Cont_S \rightarrow^1 P$$

$$[\![\ldots]\!]_E : Exp \rightarrow Env \rightarrow AObj \rightarrow Cont_E \rightarrow^1 P.$$

The first argument of each of these function is a statement (from the set *Stat*) or an expression (from *Exp*), respectively. The second argument is an *environment*, which contains the necessary semantic information about the declarations of methods and bodies in the program (for more details, see [ABKR88]). The third argument is the name of the (active) object executing the statement/expression. The last argument is a *continuation*. This certainly deserves some explanation. It seems natural that the semantic function of a statement returns a process describing just the execution of that statement. However, we would get into trouble then, because in defining the semantics of the sequential composition $s_1; s_2$ of two statements we would have to determine the sequential composition of the corresponding processes. This turns out to be impossible, the main source of trouble being the fact that s_1 can create a new object which should execute in parallel not only with the rest of s_1, but also with s_2. In the same way, one would expect that the semantic function for expressions just returns a value (an object name) as its result. This approach, however, would leave us with the problem of describing the possible side-effects of expression evaluation, which can be quite complicated, e.g., involving message sending.

Continuations form the most convenient and elegant solution to these problems. (For a nice introduction to the use of continuations in a sequential setting, see [Gor79].) The semantic function for statements is provided with a continuation, which is just a process ($Cont_S = P$), describing the execution of all the statements following the current one. The semantic function then delivers a process that describes the execution of the current statement plus the following ones. Analogously, the semantic function for expressions is fed with a continuation, which in this case is a function which maps object names to processes ($Cont_E = Obj \rightarrow P$). This function, when applied to the name of the object that is the result of the expression, gives a process describing everything that should happen in the current object after the expression evaluation. Again, the semantic function delivers a process describing the expression evaluation plus the following actions.

Now we are ready to give some examples of clauses that appear in the definition of the semantic functions $[\![\ldots]\!]_S$ and $[\![\ldots]\!]_E$. Let us start with a relatively simple example, the assignment statement:

$$[\![x := e]\!]_S(\gamma)(\alpha)(p) = [\![e]\!]_E(\gamma)(\alpha)(\lambda\beta.\{\langle\sigma',p\rangle\}).$$

This equation says that if the statement $x := e$ is to be executed in an environment γ (recording the effect of the declarations), by the object α, and with continuation p (describing the actions to be performed after this assignment), then first the expression e is to be evaluated, with the same environment γ and by the same object α, but its resulting object is to be fed into an expression continuation $\lambda\beta.\{\langle\sigma',p\rangle\}$ that delivers a process of which the first action is an internal one leading to the new state σ' and having the original continuation p as its resumption. Here, of course, the new state σ' is equal to $\sigma\{\beta/\alpha,x\}$, only different from σ in that the value of the variable x in the object α is now equal to β.

The semantic definition of sequential composition is easy with continuations:

$$[\![s_1; s_2]\!]_S(\gamma)(\alpha)(p) = [\![s_1]\!]_S(\gamma)(\alpha)([\![s_2]\!]_S(\gamma)(\alpha)(p)).$$

Here the process describing the execution of the second statement s_2 just serves as the continuation for the first statement s_1.

As a simple example of a semantic definition of an expression let us take an instance variable:

$$[\![x]\!]_E(\gamma)(\alpha)(f) = \lambda\sigma.\{\langle\sigma, f(\sigma(\alpha)(x))\rangle\}.$$

Evaluating the expression x takes a single step, in which the value $\sigma(\alpha)(x)$ of the variable is looked up in the state σ. The resumption of this first step is obtained by feeding this value into the expression continuation f (which is a function that maps object names into processes).

As a final example of a semantic definition, let us take object creation: The expression $\text{new}(C)$ creates a new object of class C and its value is the name of this object. Its semantics is defined as follows:

$$[\![\text{new}(C)]\!]_E(\gamma)(\alpha)(f) = \lambda\sigma.\{\langle\sigma',\gamma(C)(\beta) \parallel f(\beta)\rangle\}.$$

Here β is a fresh object name, determined from σ in a way that does not really interest us here, and σ' differs from σ only in that the variables of the new object β are properly initialized. We see that execution of this new-expression takes a single step, of which the resumption consists of the parallel composition of the body $\gamma(C)(\beta)$ of the new object with the execution of the creator, where the latter is obtained by applying the expression continuation f to the name of the new object β (which is, after all, the value of the new-expression). The parallel composition operator $\parallel$ is a function in $P \times P \to P$, which can be defined as the unique fixed point of a suitable contracting higher-order function $\Phi_{PC} : (P \times P \to P) \to (P \times P \to P)$ (an application of Banach's fixed point theorem).

From the above few equations it can already be seen how the use of continuations provides an elegant solution to the problems that we have mentioned.

There are a number of further steps necessary before we arrive at the semantics of a complete program. One interesting detail is that in the denotational semantics, sending messages to standard objects is treated in exactly the same way as sending messages to programmer-defined objects. The standard objects themselves (note that there are infinitely many of them!) are represented by a (huge) process p_{ST}, which is able to answer all the messages sent to standard objects and immediately return the correct results. This process p_{ST} is composed in parallel with the process p_U, which describes the execution of the user-defined objects in order to give the process describing the execution of the whole system. From this process it is possible to derive a set of possible execution sequences that resemble the ones that we had with the operational semantics.

5 Equivalence of operational and denotational semantics

Despite the fact that the two forms of semantics described above, the operational and the denotational one, are formulated in widely different frameworks, it turns out that it is possible to establish an important relationship between them:

$$O = abstr \circ D,$$

which in some sense says that the different forms of semantics of POOL are *equivalent*. Here D is the function that assigns a process to a POOL program according to the denotational semantics and O assigns to each program a set of (finite or infinite) sequences of states, which can be extracted from the sequences of configurations obtained from the operational semantics. Finally, *abstr* is an abstraction operator that takes a process and maps it into the set of sequences of states to which the process gives rise. The complete equivalence proof can be found in [Rut88]. In the present section we shall give a rough sketch of this proof, which proceeds in several steps. We apologize for sometimes being somewhat sloppy in our notation.

The first step leads to an operational semantics that delivers a process, instead of a set of sequences of states. For this, it is most convenient to switch to a *labelled* transition system, a transition system in which the transition relation is ternary, and where the extra component gives some more information on the nature of the transition (e.g., whether it is an internal step or a communication action). Now it is possible to define another operational semantics O^* as follows:

$$O^*(X) = \begin{cases} p_0 & \text{if } X \text{ has terminated} \\ \lambda\sigma.\big\{ \langle \sigma', O^*(X') \rangle \mid \langle X, \sigma \rangle \xrightarrow{\text{int}} \langle X', \sigma' \rangle \big\} \cup \dots & \text{otherwise} \end{cases}$$

Here X and X' are finite sets of labelled statements (see section 3), $\langle X, \sigma \rangle$ stands for a configuration containing X and σ, $\xrightarrow{\text{int}}$ stands for a transition that is labelled as an internal one, and the dots $(\dots)$ stand for additional, more complicated terms dealing with communication actions. Note that this definition is a recursive one: O^* also occurs at the right-hand side. However, it can be made into a well-formed definition by taking O^* as the unique fixed point of a suitable contracting higher-order function. Now it is possible to prove that

$$O = abstr \circ O^*,$$

which completes the first step of the equivalence proof.

The second step consists mainly of getting rid of the continuations. Two different techniques have been developed for this. The first technique defines a number of additional semantic operators that allow a denotational (compositional) style of semantic definitions *without* the use of continuations. The resulting semantics can be proved to be equivalent both with O^* and with the denotational semantics *with* continuations. For a language slightly simpler than POOL (instead of methods it has a CSP-like value communication) this approach has been explored in [AB87]. For POOL itself, this has not yet been tried, but it seems feasible. The advantage of this technique is that it provides a clear intuitive idea of the proof, but unfortunately it leads to a large number of tedious calculations (the advantages of using continuations become very clear when one tries to avoid them).

The second technique describes the semantic functions themselves as fixed points of suitable higher-order contractions over different domains, the one with and the other without continuations. By defining mappings between these domains and showing that they commute with the contractions, it can be shown that the two semantic functions are equivalent. This technique has been introduced in [KR87] and it is used in [Rut88] to prove the equivalence of POOL semantics. This technique is more difficult to explain, but it definitely leads to a shorter proof.

The final step consists of dealing with a number of details in the semantic definitions that are not yet solved by the above steps. For example, the standard objects are described by special axioms in the operational semantics, but in the denotational semantics there is a large process p_{ST} to describe them. The problem is that the above two steps only work if in the domain equation for P we take, in the power domain $P(X)$, only the *compact* subsets instead of all the closed ones. (This is because a continuous function maps each compact set into a compact one, which is not necessarily true for closed sets.) However, the process p_{ST} does not fit in this domain. The problem can be solved by proving that if p resides in the "compact" domain then $abstr(p \parallel p_{\text{ST}})$ is compact. (For more details, see [Rut88]).

6 Other forms of semantics for POOL

In addition to the operational and denotational semantics described above, POOL has been the subject of a number of other semantic studies. Let us first mention [Vaa86]. In this paper, the semantics of POOL is defined by means of *process algebra* [BK84,BK85]. This is done as follows: with the help of an attribute grammar, each POOL program is mapped unto a specification of a process in the ACP formalism with a number of additional operators. This specification in turn can be interpreted in each of the different semantic models that exist for ACP, e.g., bisimulation [BBK85] and failure semantics [BKO86].

In addition to describing the semantics of POOL, [Vaa86] also studies a number of related issues. One of them is the implementation of a fair (in a technical sense) communication mechanism with the help of message queues. The analysis in [Vaa86] detected a small error in the language manual. After this had been corrected, the correctness of this implementation could not yet be proved, unfortunately. In bisimulation semantics it can be shown that the process that results from using explicit message queues is different from the process that does not use these queues. However, this difference is due to the strict notion of equivalence in bisimulation semantics: all we are interested in is that the two processes can not be distinguished by observation from outside. This notion is captured by failure semantics (leading to some additional axioms of equality in the formalism), but unfortunately [Vaa86] does not go as far as giving the equivalence proof in this case.

Another semantic technique, which is currently explored for its suitability to describe POOL, uses *graph grammars*. In [JR87], a special type of graph grammars, called *actor grammars*, are used to describe the semantics of actor languages, a more primitive type of concurrent object-oriented languages, where communication is done asynchronously [Agh86,Cli81,Hew77]. In this model, the execution of a program can be seen as a sequence of rewritings of a graph which represents the system. Production rules in the graph grammar describe how these rewritings should take place. In [Lei88] an initial study is made of the viability of such a technique for describing the semantics of POOL.

Finally, we should mention here some work which describes POOL on a different level. In [DD86,DDH87] a description is given of an abstract POOL machine. In contrast to the "abstract machine" employed in the operational semantics described above, this abstract POOL machine is intended to be the first step in a sequence of refinements which ultimately lead to an efficient implementation on real parallel hardware. This abstract POOL machine is described formally in AADL, an Axiomatic Architecture Description Language.

7 Importance of semantic analysis

The most important result of a theoretical analysis of a programming language is a better understanding of the language itself and its basic concepts and constructs. This is also the case with our semantic analysis of POOL. Especially denotational semantics, which by its very nature concentrates on the meaning of each language construct individually, provides a very good feedback on the conceptual integrity of a language design. In our semantic study of POOL we have found no major anomalies. On the one hand, this is a positive result, because it suggests that the initial language design was a fairly good one. On the other hand, the discovery of major flaws in the language design would have clearly indicated the usefulness of formal semantics.

As it is, the total experience with POOL-T (consisting of writing programs and implementing the language, in addition to the formal semantic study), the first version of POOL that was used and studied extensively [Ame85,Ame87b], indicated that nothing should be changed in its basic semantic concepts. Therefore the newer version, POOL2 [Ame88a], definitely retained these concepts. It was even designed in such a way that every new language construct can be described in terms of the original ones [Ame88b].

Our semantic studies are not yet finished. In the future we hope to achieve a better understanding of several new concepts. The most important of these is *inheritance*, a concept that appears in many object-oriented languages (though not yet in POOL). We have the definite feeling that inheritance is not yet understood well enough on a semantic level to warrant its inclusion in POOL [Ame87a]. We hope that this situation will change in the future.

Another important use of formal semantics lies in the verification of the correctness of implementations of the language. Especially an operational form of semantics seems very suitable in this respect. Let us remark first that it seems possible to generate an implementation directly from the axioms and rules that constitute the operational definition of POOL: these axioms and rules can be translated to Horn clauses in a straightforward way and a Horn clause theorem prover could be used to interpret these clauses. Unfortunately it is not possible to use Prolog for this, since the depth-first search strategy used by Prolog is not complete (it does not always find a proof even if one exists). Perhaps some minor adaptations of the Horn clause forms of the axioms and rules could solve the problem. We have not yet tried this. Anyway, it would lead to a very slow implementation.

Formally verifying the correctness of a realistic implementation (preferably on a parallel machine) is a much harder matter, of course. Due to the enormous complexity of such a task we have no hope to achieve this in the near future. However, some of the semantic studies done so far give us some amount of confidence in the chosen implementation model. Let us briefly mention the special treatment of standard objects like integers and booleans (conceptually they are objects, but in the implementation operations on them are not performed by sending messages but by built-in machine instructions). Also the correctness of using message queues to implement a fair communication mechanism, although not yet completely formally established, seems sufficiently corroborated. Finally, we hope that at least some of the refinement steps leading from the abstract POOL machine, specified in AADL, to the actual implementation on the parallel machine, can be verified formally.

A final important use of formal semantic techniques is as a basis for formal methods of program verification. At this moment, some effort is going on to develop a suitable formalism to verify the correctness of POOL programs (see, e.g., [Ame86,Boe86]). We hope to be able to report more on this in the near future.

References

[AB87]   Pierre America and Jaco de Bakker. Designing equivalent semantic models for process creation. In Marisa Venturini Zilli, editor, *Mathematical Models for the Semantics of Parallelism*, 1987, 21–80, Lecture Notes in Computer Science 280, Springer-Verlag.

[ABKR86] Pierre America, Jaco de Bakker, Joost N. Kok, and Jan Rutten. Operational semantics of a parallel object-oriented language. In *Conference Record of the 13th Symposium on Principles of Programming Languages*, St. Petersburg, Florida, January 13–15, 1986, 194–208.

[ABKR88] Pierre America, Jaco de Bakker, Joost N. Kok, and Jan Rutten. *Denotational semantics of a parallel object-oriented language*. ESPRIT Project 415 Document 190, Philips Research Laboratories, Eindhoven, the Netherlands, January 1988. (To appear in *Information and Computation*.)

[Agh86]   Gul Agha. *Actors: A Model of Concurrent Computation in Distributed Systems*. MIT Press, 1986.

[AKO87]   P. M. G. Apers, M. L. Kersten, and A. C. M. Oerlemans. *PRISMA database machine: a distributed, main-memory approach*. PRISMA Project Document 175, Philips Research Laboratories, Eindhoven, the Netherlands, October 1987.

[Ame85]   Pierre America. *Definition of the programming language POOL-T*. ESPRIT Project 415 Document 91, Philips Research Laboratories, Eindhoven, the Netherlands, September 1985.

[Ame86] Pierre America. *A proof theory for a sequential version of POOL*. ES-PRIT Project 415 Document 188, Philips Research Laboratories, Eindhoven, the Netherlands, October 1986.

[Ame87a] Pierre America. Inheritance and subtyping in a parallel object-oriented language. In Jean Bézivin, Jean-Marie Hullot, Pierre Cointe, and Henry Lieberman, editors, *ECOOP'87: European Conference on Object-Oriented Programming*, Paris, France, June 15–17, 1987, 234–242, Lecture Notes in Computer Science 276, Springer-Verlag.

[Ame87b] Pierre America. POOL-T — a parallel object-oriented language. In Akinori Yonezawa and Mario Tokoro, editors, *Object-Oriented Concurrent Programming*, 199–220, MIT Press, 1987.

[Ame88a] Pierre America. *Definition of POOL2, a parallel object-oriented language*. ESPRIT Project 415 Document 364, Philips Research Laboratories, Eindhoven, the Netherlands, April 1988.

[Ame88b] Pierre America. *Rationale for the design of POOL2*. ESPRIT Project 415 Document 393, Philips Research Laboratories, Eindhoven, the Netherlands, April 1988.

[AR88] Pierre America and Jan Rutten. Solving reflexive domain equations in a category of complete metric spaces. In M. Main, A. Melton, M. Mislove, and D. Schmidt, editors, *Mathematical Foundations of Programming Language Semantics*, 1988, 254–288, Lecture Notes in Computer Science 298, Springer-Verlag.

[BBK85] J. C. M. Baeten, J. A. Bergstra, and J. W. Klop. *On the consistency of Koomen's fair abstraction rule*. Report CS-R8511, Centre for Mathematics and Computer Science, Amsterdam, the Netherlands, May 1985.

[BJOT86] W. J. H. J. Bronnenberg, M. D. Janssens, E. A. M. Odijk, and R. A. H. van Twist. The architecture of DOOM. In P. Treleaven and M. Vanneschi, editors, *Future Parallel Computers: An Advanced Course*, Pisa, Italy, June 9–20, 1986, 227–269, Lecture Notes in Computer Science 272, Springer-Verlag.

[BK84] J. A. Bergstra and J. W. Klop. Process algebra for synchronous communication. *Information and Control*, 60:109–137, 1984.

[BK85] J. A. Bergstra and J. W. Klop. Algebra of communicating processes with abstraction. *Theoretical Computer Science*, 37(1):77–121, May 1985.

[BKO86] J. A. Bergstra, J. W. Klop, and E.-R. Olderog. *Failures without chaos: a new process semantics for fair abstraction*. Report CS-R8625, Centre for Mathematics and Computer Science, Amsterdam, the Netherlands, August 1986.

[Boe86] Frank S. de Boer. A proof rule for process creation. In Martin Wirsing, editor, *Formal Description of Programming Concepts III — Proceedings of the Third IFIP WG 2.2 Working Conference*, Gl. Avernæs, Ebberup, Denmark, August 25–28, 1986, 23–50, North-Holland.

[BZ82] J. W. de Bakker and J. I. Zucker. Processes and the denotational semantics of concurrency. *Information and Control*, 54:70–120, 1982.

[Cli81] William Douglas Clinger. *Foundations of actor semantics*. Technical Report 633, Massachusetts Institute of Technology, Artificial Intelligence Laboratory, May 1981.

[DD86] W. Damm and G. Döhmen. *The POOL-machine: a top level specification for a distributed object-oriented machine*. ESPRIT Project 415 Document 1, Lehrstuhl für Informatik, RWTH Aachen, Aachen, West Germany, October 3, 1986.

[DDH87] W. Damm, G. Döhmen, and P. den Haan. Using AADL to specify distributed computer architectures — a case study. In J.W. de Bakker, editor, *Deliverable D3 of the Working Group on Semantics and Proof Techniques*, chapter 1.4, ESPRIT Project 415, Philips Research Laboratories, Eindhoven, the Netherlands, October 1987.

[Dug66] J. Dugundji. *Topology*. Allyn and Bacon, Boston, Massachusetts, 1966.

[Eng77] R. Engelking. *General Topology*. Polish Scientific Publishers, 1977.

[Gor79] Michael J. C. Gordon. *The Denotational Description of Programming Languages: An Introduction*. Springer-Verlag, 1979.

[Hew77] Carl Hewitt. Viewing control structures as patterns of passing messages. *Artificial Intelligence*, 8:323–364, 1977.

[HP79] Matthew Hennessy and Gordon Plotkin. Full abstraction for a simple parallel programming language. In J. Bečvář, editor, *Proceedings of the 8th Symposium on Mathematical Foundations of Computer Science*, 1979, 108–120, Lecture Notes in Computer Science 74.

[JR87] D. Janssens and G. Rozenberg. Basic notions of actor grammars: a graph grammar model for actor computation. In H. Ehrig, M. Nagl, G. Rozenberg, and A. Rosenfeld, editors, *Graph-Grammars and Their Application to Computer Science*, 1987, 280–298, Lecture Notes in Computer Science 291, Springer-Verlag.

[KR87]   Joost N. Kok and Jan J. M. M. Rutten. *Contractions in comparing concurrency semantics*. Report CS-R8755, Centre for Mathematics and Computer Science, Amsterdam, the Netherlands, November 1987.

[Lei88]   George Leih. *Actor graph grammars and POOL2*. PRISMA Project Document 265, University of Leiden, Department of Computer Science, February 1988.

[Odi87]   Eddy A. M. Odijk. The DOOM system and its applications: a survey of ESPRIT 415 subproject A. In J. W. de Bakker, A. J. Nijman, and P. C. Treleaven, editors, *Proceedings of PARLE: Parallel Architectures and Languages Europe. Volume I: Parallel Architectures*, Eindhoven, the Netherlands, June 15–19, 1987, 461–479, Lecture Notes in Computer Science 258, Springer-Verlag.

[Plo76]   Gordon D. Plotkin. A powerdomain construction. *SIAM Journal on Computing*, 5(3):452–487, September 1976.

[Plo81]   Gordon D. Plotkin. *A structural approach to operational semantics*. Report DAIMI FN-19, Aarhus University, Computer Science Department, Aarhus, Denmark, September 1981.

[Plo83]   Gordon D. Plotkin. An operational semantics for CSP. In D. Bjørner, editor, *Formal Description of Programming Concepts II*, 1983, 199–223, North-Holland.

[Rut88]   Jan Rutten. *Semantic equivalence for a parallel object-oriented language*. Report CS-R88XX, Centre for Mathematics and Computer Science, Amsterdam, the Netherlands, 1988. (To appear.)

[Smy78]   Michael B. Smyth. Power domains. *Journal of Computer and System Sciences*, 16:23–36, 1978.

[Vaa86]   Frits W. Vaandrager. *Process algebra semantics for POOL*. Report CS-R8629, Centre for Mathematics and Computer Science, Amsterdam, the Netherlands, August 1986.

Object-Oriented Process Specification

S.A. Schuman, D.H. Pitt and P.J. Byers

Department of Mathematics
University of Surrey
Guildford, Surrey GU2 5XH

1. Introduction

The specification techniques put forward in [Schuman & Pitt 1987] are a variant on the "Z" notation, as promulgated by the Programming Research Group at Oxford [cf. Hayes 1987, Spivey 1988a & 1988b], which was in turn strongly influenced by VDM [Jones 1980, Bjørner & Jones 1982, Jones 1986]. The most obvious difference is our explicit commitment to the familiar "object-oriented" paradigm for structuring and decomposing complex systems. Such methods are already applied quite successfully in current software engineering practice, albeit mostly at the level of various program design languages (PDLs) or system implementation languages (SILs); we are seeking mainly to support and reinforce these same intuitions in the rather more abstract realm of formal specification.

As originally introduced, the emphasis of our approach was on formally specifying and reasoning about the behaviour of user-defined "classes" of abstract objects, under the usual assumption that these will subsequently be instantiated as independent "subsystems" in a great many different contexts. The customary mechanisms for composing such specifications, including multiple inheritance and refinement (by extension and/or restriction), were also provided within that framework. In this paper we turn our attention to consideration of *concurrent systems*, wherein the objects of interest are now to be interpreted as autonomous processes. No modification of our previous notations and conventions is necessary. Instead we simply enlarge here upon their formal interpretation so as to take some of the more fundamental issues of concurrency, *viz.* synchronisation, communication and interference, fully into account.

A particularly important rôle is played in this connection by what is the technically more significant departure from Z or VDM, namely our special rule of "historical inference", whereby only the *minimal* effect (change-of-state) need actually be specified for each possible event. The implications of these conventions in regard to concurrent composition, along with the corresponding conditions for combining independent refinements of such process specifications, are developed in Section 4. We begin, however, with a somewhat tutorial summary of our overall framework in Section 2, which also serves to introduce some very simple examples that will be used throughout the rest of this paper. The behavioural interpretation of process specifications and its underlying model are formulated in Section 3.

2. Classes as Process Specifications

Within this wider perspective, every elementary process is to be viewed as an instance of some specific class of abstract objects. Any such class is introduced in the form of a so-called "state-schema", which gives its name together with a set-theoretic characterisation for the internal state of each individual instance; also specified within this same schema are the conditions which must hold in any initial state. A wholly trivial (but highly typical) example is the class FS (for Finite Set):

```
FS _____
    X: set [S]              -- state components
    n: NAT
  _____
    n = #X                  -- state invariants
  _____
    X' = Ø                  -- initialisation
  _____
```

Thus the state of objects (or processes) belonging to the class FS is described in terms of two distinct components (bound variables), X and n, where X is a *finite* subset of some given (carrier) set **S** and n is (always) equal to the cardinality of X; for every such object, X is initially empty (whence n=0).

A class specification is completed by associating with its given name a number of self-standing "event-schema", which serve to define the various access operations that may be applied to an individual object of that class in order to interrogate and/or update its internal state. These may equally well be interpreted as defining the different events in which a process of that class is possibly prepared to engage when asked to do so (i.e. invoked) by its environment. All such events are taken to be "atomic" in the sense that their occurrences do not overlap in time.

On the class FS above, for example, one would very likely wish to have operations to insert a new element x (of "type" S) into X, as well as to extract some x which is known to be a member of X at the point of invocation:

```
FS.Ins(x) _____          FS.Ext(x) _____

  x: S                       x: S
  _____                   _____

  x ∉ X                      x ∈ X
  ════════                   ════════

  x ∈ X'                     x ∉ X'
```

Such events are characterised in terms of *preconditions* (above the double line) and *postconditions* (below), wherein we adopt the usual convention that dashed variable names denote the corresponding component values *after* each occurrence of that event; parameter names (e.g. the argument x) stand for constant values and so never appear in dashed form. The preconditions specify what must hold (with respect to the parameter and current state component values) in order for the operation to be applicable to an object of that class, whereas the postconditions specify the (explicit) effect of such an application (in terms of the minimal properties that must afterwards hold between dashed and undashed component values). These same conventions are used as part of the state-schema to specify initialisation, which is construed as a pseudo-event (with no precondition) that is implicitly invoked whenever a new object of the class in question is first instantiated.

The events `FS.Ins` and `FS.Ext` introduced above could in fact have been specified somewhat more concisely as follows:

```
FS.Ins(x) _____        FS.Ext(x) _____

    x: ~X                       x: X
_____      _____

    x ∈ X'                      x ∉ X'
```

Exactly the same preconditions have been expressed here entirely in the context of the parameter declarations, which state that x must belong to the complement of X (i.e. the difference set $S \backslash X$) or to X itself (which is a subset of S), respectively. (Textual sections of our schema framework in which nothing needs to be stated are simply omitted.)

Were it intended that these operations should always be applicable, whether the argument x was or was not already a member of X, then one might introduce additional definitions (thereby "overloading" those event names) in order to accommodate the cases where there is no net effect:

```
FS.Ins(x) _____        FS.Ext(x) _____

    x: X                        x: ~X
```

The absence of any postconditions in such schema is taken to mean that though the event may indeed occur, provided its precondition is satisfied, the state of the object to which it is applied remains unchanged. It should be observed that the preconditions of the two separate event schema introduced for each operation are in fact *disjoint* – i.e. either one definition or the other applies at any given invocation (depending on the value of x), so that the behaviour implied by these multiple definitions is fully deterministic in this particular specification.

For the events in question, however, there is no real need to resort to such specification by case-analysis. Instead, the two event schema for each operation could simply be replaced by a single one having a weaker precondition (and hence wider applicability):

If the only intended effect of these events is to ensure that their respective postconditions hold, then there is no need to discriminate between the different cases by means of preconditions (certainly not at the level of a formal specification). Hereafter, in all subsequent examples, we shall assume that just our first definitions of the events Ins(x) and Ext(x) are actually associated with the class FS, so that these operations are only applicable when they will in fact result in a change-of-state.

Alternatively, or in addition, one might wish to introduce operations on FS where an (indeterminate) element y is newly inserted into or extracted from X, with this (internally chosen) value being delivered as a result of the event occurrence:

Note that the pre- and postconditions here are logically identical to those assumed for the events FS.Ins(x) and FS.Ext(x). Conceptually, the distinction is whether the choice of the parameter value is made by the subsystem FS or by its environment at each separate invocation.

A number of other operations might be of interest for this particular class. Consider, for instance, an event to insert two new (and distinct) elements x and y "simultaneously", or another to exchange a new element x for an (arbitrarily chosen) y which is already a member of x:

```
FS.Ins2(x,y) _____          FS.Exch(x→y) _____

   x,y: ~X                       x: ~X ;  y: X


   x ≠ y                         x ∈ X'; y ∉ X'


   x ∈ X'; y ∈ X'
```

(The semi-colon as used above is purely syntactic; it serves only to separate declarations or predicates that would otherwise appear on different lines.)

All of the operations suggested so far involve (at least the possibility of) state change, as indicated by the presence of postconditions within their corresponding event schema. At later stages in the specification process, however, one may want to introduce additional named events solely for purposes of interrogating the internal state of such obects or processes. Operations likely to be of interest for the class FS include a membership test, as either a simple predicate or a boolean function, and a cardinality function:

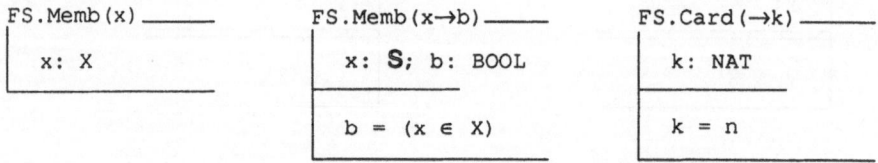

```
FS.Memb(x) _____      FS.Memb(x→b) _____      FS.Card(→k) _____

   x: X                   x: S; b: BOOL            k: NAT


                          b = (x ∈ X)              k = n
```

By analogy with programming practice, adding these definitions may be viewed as a means of "encapsulating" the abstraction embodied in a class specification. One might choose to follow this discipline whenever the class concerned was to be used to specify a component of another class, although there is no obligation to do so; conversely, such queries are not normally introduced when developing a specification by refinement (as discussed below), though this too is a matter of stylistic judgement.

Which events are finally associated with a particular user-defined class will depend upon the projected usage of that abstraction; they are not determined *a priori* by the state characterisation given at the outset. It should be noted that we make no provision for structurally grouping a state-schema and its associated event-schema into some larger syntactic unit, or "module", corresponding to a complete class specification. How such definitions should best be collected together and presented, whether in human-readable documents or machine-readable files, is left open.

Refinement and Inheritance

Much of the pragmatic appeal of the object-oriented approach is owing to its support for subsequent specialisation of any previously defined class of abstract objects, giving rise to one or more *subclasses*. We refer to these latter as *refinements* of the original class specification (wherein the refinement relation over class and subclass names forms a directed acyclic graph in the general case). The apparent power of such methods for structuring the description of a complex system comes from the convention whereby properties associated with a more general class are *inherited* within the definition of any subclass derived therefrom (and hence by all of the objects so described); additional properties, specific only to the subclass in question, will normally be introduced at each separate level of refinement. These capabilities do not provide any extra definitional power *per se*. They are more a matter of methodological convenience, affecting such concerns as modularity and re-use of specifications. Consider, for example, two ways of defining the state schema for a new class FM (for Finite Map) within our framework;

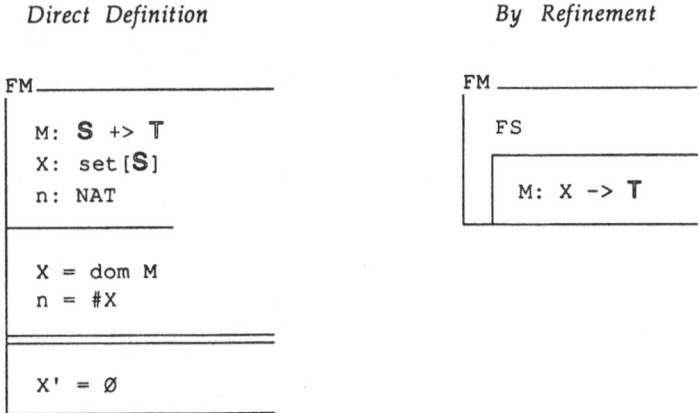

$Direct\ Definition$ $By\ Refinement$

Both schema describe exactly the same state components (and initial conditions). These are just as for FS but with a further component, the mapping M specified as a *partial* function from values of type **S** to values of some other type **T**, wherein the set X now corresponds to the domain of M and n is (as before) the cardinality of X. In the left-hand (direct) definition all properties of FS have been repeated explicitly whereas they are implicitly inherited within the right-hand definition, which characterises FM as a subclass of FS. (Note that because this refinement is effectively "embedded" within the state definition for FS, the constraint that X is the domain of M is here indicated by specifying M to be a *total* function from X to **T**.)

Corresponding facilities are provided for defining the events associated with a class specified by refinement. As an example, suppose the intended application of FM was to support the generation of unique symbolic "references" of type **S** for distinct occcurrences of certain (say textual) information of type **T**. The first operation of interest is likely to be the one for generating a new such reference:

Direct Definition   *By Refinement*

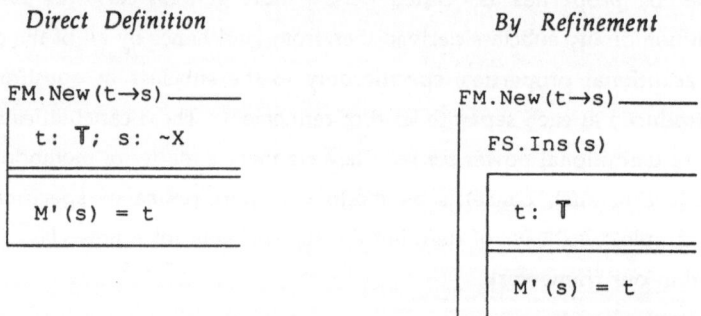

(This refinement might instead have been expressed in terms of the event FS.New(→s) that was suggested previously.) Another operation on FM to nullify (or delete) an outstanding reference would probably be wanted as well:

Direct Definition   *By Refinement*

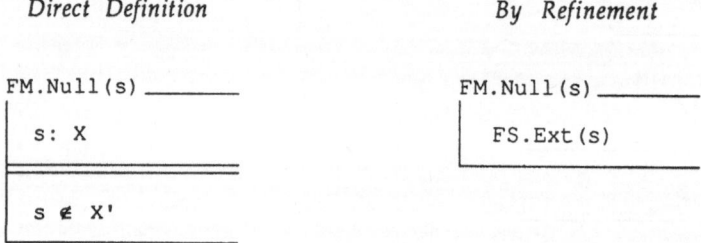

In both cases these definitions by refinement inherit the pre- and postconditions specified for the respective events on the more general class FS. For the operation FM.New, the precondition assumed for FS.Ins ensures that the resultant s is unique; the additional (embedded) conditions relate solely to the component M introduced at the level of the subclass FM. No further conditions (pre or post) need be specified for FM.Null (since deleting an element from the domain of M deletes any association as well), whence its definition is a simple "promotion" (i.e. a renaming) of FS.Ext. Presumably, the intention is also to provide a query operation, which given an s returns the corresponding t.

One might continue in this fashion, deciding at a later stage that unique references should be generated for distinct *values* (as opposed to *occurrences*) of type T, in order to support a "symbol table" or "dictionary" application. This could be expressed as a further refinement of FM by simply specifying a new subclass FI (Finite Injection):

FI _____

  FM

    I: T +> X ; Y: set[T]

    I = M^{-1} ;   Y = cod M

Additional components I (a partial function which is the inverse of M, whence M is "one-one" or injective) and Y (the co-domain of M or the domain of I) are introduced at this level. The associated New and Null operations are then specified as refinements of those same events on FM:

FI.New(t→s) _____

  FM.New(t→s)

    t ∉ Y

FI.Null(s) _____

  FM.Null(s)

The only additional constraint here is that the t supplied as an argument must not already be referred to by some s when the event FS.New is invoked (which suggests that a suitable query operation ought also to be provided for testing this condition).

In the foregoing examples the refinement from FS to FM would intuitively be categorised as an "extension" (in that a wholly independent state component, the mapping M, is introduced without otherwise constraining the values of the original components X and n), whereas the refinement from FM to FI clearly constitutes a "restriction" (since M is therein constrained to be injective by the assertion that its inverse is also functional).

A somewhat more familiar example of the latter category would be an after-the-fact restriction on FS to specify the class BS (for Bounded Set):

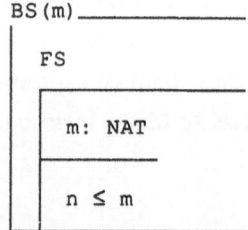

The bound is here introduced as a parameter of the class BS itself (where the parameter value is independently fixed for each separate instantiation), together with an additional invariant which states that the cardinality of X never exceeds m. As for FI, a further precondition is now required on insertion, in order to ensure that this invariant is maintained, but no extra constraints are needed for extraction:

The question which naturally arises at this point is whether one can form the class BI (for Bounded Injection) by simply *combining* the specifications for FI and BS as developed above, making use of what is referred to in object-oriented parlance as "multiple inheritance". The answer is *yes*, although the more formal explanation will be deferred until later sections. For the moment, it is sufficient to compare a possible direct definition with one expressed in terms of such composition.

The following might be the state schema for each of the two different approaches:

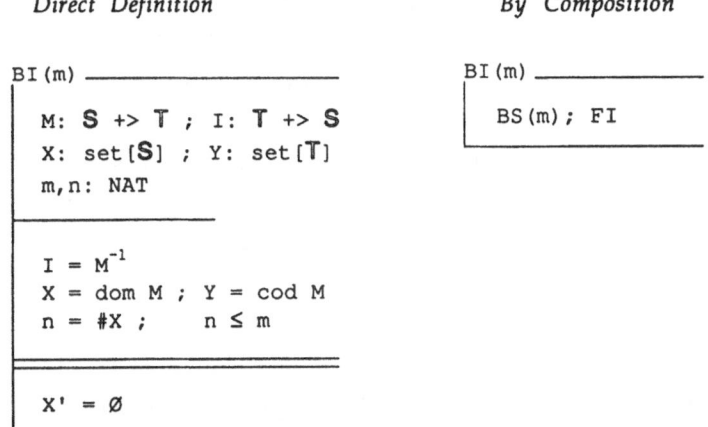

Direct Definition

```
BI (m) _____
    M: S +> T ; I: T +> S
    X: set [S] ; Y: set [T]
    m,n: NAT
    _____
    I = M⁻¹
    X = dom M ; Y = cod M
    n = #X ;      n ≤ m
    _____
    X' = Ø
    _____
```

By Composition

```
BI (m) _____
    BS (m) ; FI
    _____
```

The associated New and Null events for BI would then be specified as follows:

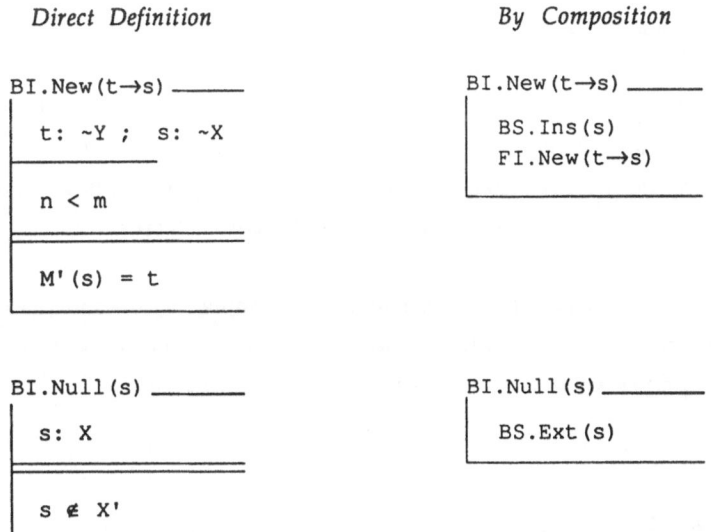

Direct Definition

```
BI.New (t→s) _____
    t: ~Y ;  s: ~X
    _____
    n < m
    _____
    M' (s) = t
    _____
```

```
BI.Null (s) _____
    s: X
    _____
    s ∉ X'
    _____
```

By Composition

```
BI.New (t→s) _____
    BS.Ins (s)
    FI.New (t→s)
    _____
```

```
BI.Null (s) _____
    BS.Ext (s)
    _____
```

The composition here involves a certain degree of subtlety in that BS and FI are both subclasses of FS, whence this definition is effectively "shared" within BI. It should be stressed, however, that such facilities are normally used much more prosaically, to combine classes for which the state specifications are entirely disjoint. As an illustration, the reader is referred to the "Distributed Information System" example in Section 4 of our original [1987] paper.

Schema Validation

The "mission" of formal methods is to bring both the rigour of mathematical expression and the power of mathematical abstraction directly to bear upon the perceived complexities of software design (or systems architecture). Thus all of the various notations proposed in this context have at least one objective in common: to provide a vehicle for stating *precisely* what is intended, *but no more*, such that one can then reason systematically about the resultant (purely symbolic) description. It follows that the first obligation when writing down any would-be specification, however abstract, is to ensure that it is *potentially* meaningful – which is to say that it does not lead to contradictory conclusions within whatever underlying formalism is adopted. This process is referred to as (internal) *validation*. More ambitious aims – i.e. ensuring that a specification meets its external (end-user) requirements – are largely beyond the scope of this paper.

In our object-oriented framework, all such reasoning is founded on ordinary (first-order) logic and, more specifically, on classical (ZF) set-theory [eg. Johnstone 1987]. The different forms of schema presentation that we have superimposed upon this well-known formal system are nothing more than highly stylised conventions for setting out, as predicates or assertions, the particular properties attributed to some other (presumably computer-based) system-of-interest. At any stage in the development of its description, this latter can always be presented as a single class specification.

Formally validating such a specification begins by establishing that the subsystem state, as characterised for the class C under consideration, is *consistent* in the usual sense. The relevant properties are presented in the designated state-schema:

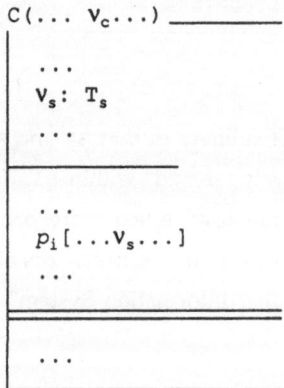

All component names v_s, including any parameter names v_c, are introduced as (mathematical) variables, each of which is declared as belonging to some set-theoretic type (set) T_s. These declarations must be independent of one another; mutual constraints amongst the corresponding values are specified by means of (simple) predicates p_i ranging over those variable names. An overall state invariant is then obtained, as a composite predicate $inv[C]$, by just conjoining all of the individual properties in question:

$$inv[C] \triangleq (...\wedge v_s \in T_s \wedge...) \wedge (...\wedge p_i[...v_s...] \wedge...)$$

Thus the first step in the validation process is to show that it is possible for such objects to exist at all – i.e. that the composite invariant so obtained is satisfiable within the universe of set-theory. This proof obligation can be formulated as a proposition $css[C]$, for consistent subsystem state, by existentially quantifying over the component names appearing in $inv[C]$:

$$css[C] \triangleq \exists ...v_s... \bullet inv[C]$$

So, for our example class FS, it is required to prove the following:

$$css[FS] \triangleq \exists X,n \bullet (X \in set[S] \wedge n \in NAT \wedge n = \#X)$$

In practice, the normal way of discharging such obligations is merely to exhibit a (possibly trivial) "model" wherein the invariants obviously hold. The various types figuring in these propositions are expressed in terms of symbols standing for given sets (herein denoted by letters from a distinguished alphabet $A ... Z$), the names of certain predefined sets (e.g. NAT, the natural numbers) and the usual set constructions over such types – including products, partial or total functions and relations, finite subsets and sequences etc. Thus the state specification may be interpreted as a *structure*, which is some association of actual sets to the given type symbols appearing (directly or indirectly) within this proposition, together with any systematic substitution of values from those sets (or constructed sets) for the existentially quantified variable names. A *model* is then such a structure in which all of the requisite properties are satisfied; the specification is consistent when there is at least one model. (This argument is of course *relative*, based on the assumption that set-theory itself is consistent.)

For example, simple structures corresponding to the class FS might be formed by taking **S** to be NAT as well (with the set $\{0, 1, 2,...\}$ as its carrier). There are then a very large number of possible substitutions for x and n that satisfy all of these invariants, e.g. $[x = \{0, 1\}, n=2]$ etc.; another is $[x=\emptyset, n=0]$, which is the only one that also satisfies the constraints for an initial state of FS. Note, however, that the substitution $[x=\{1\}, n=2]$ violates the explicit invariant, so not all structures are models.

Showing consistency is usually quite straightforward, as long as the state specification is "type-correct" in the first place. This process is further simplified by the fact that we take all of our set constructors and related operators to be "strongly-typed", based on the definitions originally formulated in [Abrial 1982]. This constitutes a more restricted version of ZF set-theory, wherein for instance $x \cup y$ is only defined when x and y are subsets of the same set, say **S**, whence the result also belongs to $\wp[\mathbf{S}]$. The reason for adopting these restrictions is strictly pragmatic, rather than in any way fundamental: our primary purpose is to support the specification of practical computer-based systems, as opposed to the definition of ever more sophisticated mathematical structures. Outside the schema framework, of course, new constructs or notations can always be defined in the traditional fashion of mathematics, exploiting the full power of the underlying formal system; provided that these extensions are logically sound and compatible with our strong-typing conventions, they can (in principle) be used like any other pre-defined types or operators, as and when required.

It is characteristic of the object-oriented approach that the definition of every event associated with a given class (e.g. FS) is logically "embedded" within its corresponding state description, which then stands for some representative instance to be operated upon when that operation is subsequently applied. The actual object, say s1:FS, is designated (as an implicit parameter) in the context of each separate invocation: s1.Ins(x), s1.Card(→k) etc.

Accordingly, the full definition of a query event (without any postconditions) can be visualised as if its explicit declarations and predicates were nested within a textual expansion of the relevant state specification:

As Written   *Expansion*

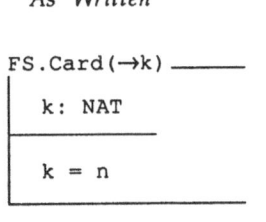

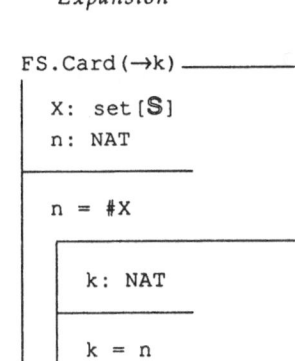

A similar pattern pertains to events involving postconditions, but in this case the explicit properties are further embedded within an entirely separate set of *dashed* state components, together with a dashed version of the invariant, so as to be able to characterise the effect of such operations as a *relation* between undashed and dashed variables:

As written   *Expansion*

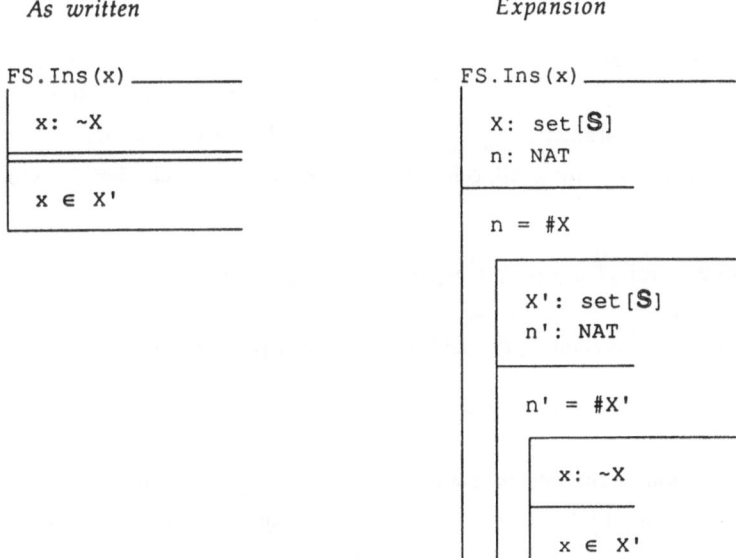

There are two separate aspects to formally validating the definition of each event E associated with some class C, based on properties specified in both the corresponding state-schema and the individual event-schema:

$$\texttt{C.E}\,(\ldots \texttt{v}_a \ldots \rightarrow \ldots \texttt{v}_b \ldots) \underline{\hspace{1.5cm}}$$

$$\ldots$$

$$\texttt{v}_e \colon \texttt{T}_e[\ldots \texttt{v}_s \ldots]$$

$$\ldots$$

$$\ldots$$

$$q_j[\ldots \texttt{v}_s, \texttt{v}_e \ldots]$$

$$\ldots$$

$$\ldots$$

$$r_k[\ldots \texttt{v}'_s, \texttt{v}_s, \texttt{v}_e \ldots]$$

$$\ldots$$

The event variable names v_e, which include any input parameters v_a or output parameters v_b, are again declared as belonging to certain set-theoretic types T_e (but these latter may now depend on the state component names v_s as a consequence of nesting). Mutual constraints amongst values of the event variables and state components are, as before, specified by simple predicates q_j, which are entirely *undashed*. The effect, expressing any resultant change-of-state, is specified by additional *mixed* predicates r_k, which will be considered as relations ranging over the undashed variables v_s and v_e as well as the dashed variable names v'_s (but excluding dashed names of any parameters v_a, v_b or v_c, as these stand for constants). By the same process of simply conjoining all relevant properties, we obtain both a composite predicate *pre*[C.E] for the derived precondition,

$$pre[\text{C}.\text{E}] \triangleq inv[\text{C}] \wedge (\ldots \wedge v_e \in T_e[\ldots v_s \ldots] \wedge \ldots) \wedge (\ldots \wedge q_j[\ldots v_s, v_e \ldots] \wedge \ldots)$$

and also a composite relation *post*[C.E] for the derived postcondition,

$$post[\text{C}.\text{E}] \triangleq inv'[\text{C}] \wedge (\ldots \wedge r_k[\ldots v'_s, v_s, v_e \ldots] \wedge \ldots)$$

Systematic inclusion of the dashed state invariant *inv*'[C] within *post*[C.E] reflects the necessity of ensuring that those properties are maintained after the event, whilst avoiding the need to write explicit postconditions solely for the purpose of restoring such invariants; this is one of the important ways in which our specifications differ from implementations.

The two proof obligations for an event C.E arise directly from these constructions. The first is to show that this event is (ever) *applicable*, which is expressed by the proposition *app*[C.E] as follows:

$$app[\text{C.E}] \triangleq \exists\, ...v_s, v_e... \bullet pre[\text{C.E}]$$

Examples for operations associated with the class FS include:

$$app[\text{FS.Card}] \triangleq \exists\, X, n, k \bullet (X \in set[S] \wedge n \in \text{NAT} \wedge n = \#X \wedge k \in \text{NAT} \wedge k = n)$$

$$app[\text{FS.Ins}] \triangleq \exists\, X, n, x \bullet (X \in set[S] \wedge n \in \text{NAT} \wedge n = \#X \wedge x \in \sim X)$$

Thus these propositions formalise our requirement that the state invariant and the event preconditions must be mutually consistent in order for an event to occur at all. This would normally be established in the same way as for showing the consistency of the state schema – by exhibiting a particular model. Note, however, that the proposition for FS.Card above is satisfied in *all* models (whence this event is always applicable) whereas the one for FS.Ins is only satisfied in models where $\sim X$ is non-empty (which is of course the other precondition for insertion of a new element).

This applicability criterion serves only to ensure that an event *may* occur. It is desirable, however, to guarantee that the event can indeed occur (that is "complete successfully") whenever its derived precondition holds – meaning that there must be a consistent final state for all possible invocations. We say that an event specification is *effective* when this property holds. The corresponding proof obligation is given by the proposition *eff*[C.E]:

$$eff[\text{C.E}] \triangleq \forall\, ...v_s, v_e... \bullet (pre[\text{C.E}] \Rightarrow \exists\, ...v'_s... \bullet post[\text{C.E}]\,)$$

As an example, consider the operation Ins on the class FS:

$$eff[\text{FS.Ins}] \triangleq \forall\, X, n, x \bullet (X \in set[S] \wedge n \in \text{NAT} \wedge n = \#X \wedge x \in \sim X)$$

$$\Rightarrow \exists\, X', n' \bullet (X' \in set[S] \wedge n' \in \text{NAT} \wedge n' = \#X' \wedge x \in X'\,)$$

This proof, unlike those involving simple consistency, would usually have to be carried out formally, although the overall structure of the argument is extremely straightforward.

A subsidiary proof obligation, similar to event effectiveness, is to show that it is always possible (in any model) to establish a consistent initial state for instantiations of some class C. The effect of this pseudo-event is characterised by a composite relation $init[C]$,

$$init[C] \triangleq inv'[C] \wedge (... \wedge r_i[...v'_s,v_s...] \wedge ...)$$

wherein the relations r_i are those explicitly specified in the initialisation part of the state schema for C (in the form of postconditions). Initialisation is applicable if the state specification is consistent. The effectiveness requirement is given by a proposition $uii[C]$, for universal instance initialisation, as follows:

$$uii[C] \triangleq \forall ...v_s,v_e... \bullet (inv[C] \Rightarrow \exists ...v'_s... \bullet init[C.E])$$

For the class FS, this proposition is the following:

$$uii[FS] \triangleq \forall x,n \bullet (x \in set[S] \wedge n \in NAT \wedge n = \#x)$$
$$\Rightarrow \exists x',n' \bullet (x' \in set[S] \wedge n' \in NAT \wedge n' = \#x' \wedge x' = \emptyset)$$

Such proofs are normally trivial except in cases where there are parameters of the class which may figure in the initialisation predicates.

In summary then, static validation for the individual schema constituting a class specification involves the following proof obligations:

1. With respect to the state schema for a given class C:

    (a) showing that the subsystem state is consistent, $css[C]$;

    (b) showing that initialisation is always effective, $uii[C]$;

2. With respect to the event schema for each operation C.E:

    (a) showing the applicability of that event, $app[C.E]$;

    (b) showing the effectiveness of that event, $eff[C.E]$.

One of the major arguments in favour of formal methods is that they provide a suitable basis for carrying out such reasoning at the level of an abstract specification, rather than deferring these same obligations to the rather more complex (and therefore less tractable) domain of some concrete implementation. The validation criteria discussed so far are entirely *local*, in that they pertain to each schema definition in isolation; this aspect will be further developed in Section 4, where schema composition and refinement are considered in relation to concurrency. Validation of the class specification as a whole is covered briefly in the next section.

3. Modelling Process Behaviour

Perhaps the most significant benefit accruing from our commitment to the object-oriented approach is that it then becomes plausible to speak about the "behaviour" of individual instances of a given class – i.e. to consider how (the state of) such objects may evolve over time. This is most easily visualised as a "decision tree". For our simplest example, the class FS, a small part of the initial pattern of behaviour might be depicted as follows:

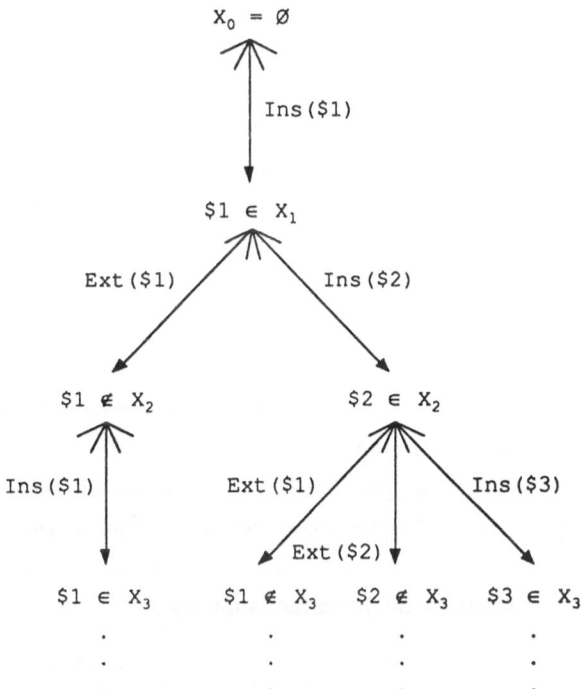

Branching within such a tree portrays *choice*, that is alternative events which may occur at distinct points in some possible "history" (a sequence of nodes along any path starting from the root), where time advances "downwards" in the tree. Branches are labelled by a "denotation" for one particular event, which in general reflects further choice amongst specific argument values admissible at that point (above, these values are assumed to be drawn from some wholly arbitrary set {$1, $2, ...} taken as the type S). The "fan-out" from each node is to suggest other choices (of argument values or of different events that may also be applicable).

The nodes have been annotated with relations corresponding to the postconditions of the immediately preceding event; the root annotation derives from the initialisation relation given in the state schema. Within these annotations, component names are subscripted so as to distinguish different points in time. If one ignores the subscripts, then these and other relations which hold at successive nodes along a given history are to be interpreted as incremental (or cumulative) assertions about the state of an individual object up to that point; note, however, that later assertions may then "override" earlier ones (indicating change-of-state).

As suggested above, it is not possible to apply the operation Ext to an object or process of class FS immediately after initialisaiton, but that process is prepared to engage in an Ins event for any argument value of type S whatsoever. At later times, only values which have not yet been inserted or extracted are admissible for these events – which is precisely what was stated in the pre- and postconditions. Thus the whole of the foregoing characterisation (and its visualisation as a decision tree) is merely an interpretation – albeit a strongly intuitive one – for the purely static description presented in a class specification.

Another, closely related way to characterise the behavioural properties that may be inferred from such a specification is in terms of "traces", following the pioneering work of [Hoare 1985]. Here a trace is a finite sequence of event denotations, which records a possible order of distinct occurrences over time. A set of initial traces for a process belonging to the class FS, including initialisation, denoted FS(), and the labels of all branches shown above, might be the following:

$$
\left\{
\begin{array}{l}
\langle\rangle \\
\langle\text{FS}()\rangle \\
\\
\langle\text{FS}(),\ \text{Ins}(\$1)\rangle \\
\dots \\
\langle\text{FS}(),\ \text{Ins}(\$1),\ \text{Ext}(\$1)\rangle \\
\langle\text{FS}(),\ \text{Ins}(\$1),\ \text{Ins}(\$2)\rangle \\
\dots \\
\\
\langle\text{FS}(),\ \text{Ins}(\$1),\ \text{Ext}(\$1),\ \text{Ins}(\$1)\rangle \\
\langle\text{FS}(),\ \text{Ins}(\$1),\ \text{Ins}(\$2),\ \text{Ext}(\$1)\rangle \\
\langle\text{FS}(),\ \text{Ins}(\$1),\ \text{Ins}(\$2),\ \text{Ext}(\$2)\rangle \\
\langle\text{FS}(),\ \text{Ins}(\$1),\ \text{Ins}(\$2),\ \text{Ins}(\$3)\rangle \\
\dots
\end{array}
\right\}
$$

Allowable sets of traces are *prefix-closed*, meaning that if some trace τ is in the set then so are all of its prefixes (including the empty trace $\langle\rangle$, which stands for nothing having happened yet). Each such trace is just one possible path in the decision tree representation; hence both forms embody the basic principle that the past can only be extended (not altered *ex post facto*).

The form of these dynamic interpretations is essentially the same for classes defined by composition and/or refinement (as the latter are completely static constructions), though the event denotations and node annotations will of course reflect any richer structure that is thereby specified. Consider for example part of the initial behaviour for an object of the subclass BI (where the type T is taken to be simple text strings and the upper bound m is assumed to be initialised to 2).

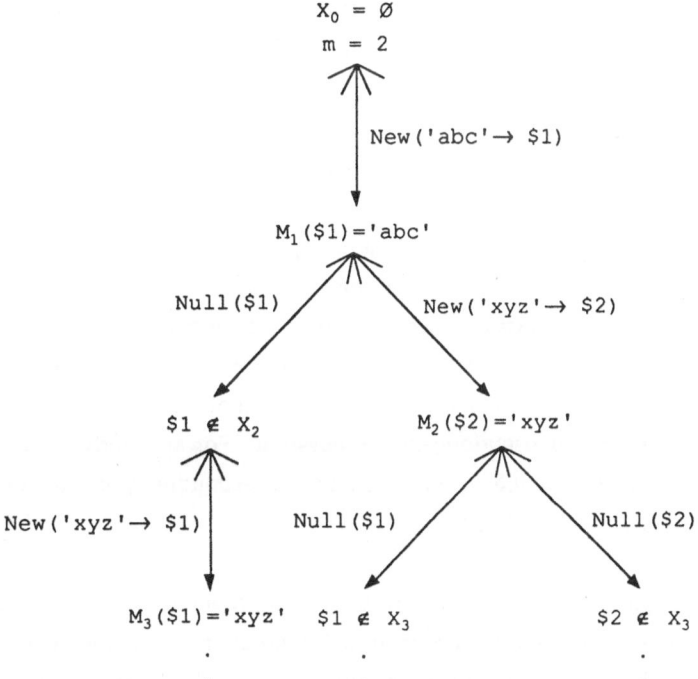

Although this corresponds quite directly to a partial behaviour for the more general class FS, as depicted previously, another insertion event is no longer possible after the first rightward branch above, owing to the bound imposed on the cardinality of X (the domain of M). Similarly, fewer choices will pertain where insertion is applicable because of the more restrictive precondition (and invariant). Again, the essence of this behaviour could equally well be expressed as a prefix-closed set of traces.

It is the very immediacy of these behavioural interpretations that leads to considering the instances of a class of abstract objects as processes, and so to the specification of concurrent systems. Our goal, however, is to be able to reason over such behaviours – that is, to support formal inferences about the dynamic aspects of any subsystem so specified. For this purpose we require a somewhat richer model, one which makes explicit those properties that can be inferred about the implied behaviours.

To this end we first assume a model for (our strongly typed) set theory and extend its language with names to denote distinct elements in the universe of that model, at the same time adding appropriate axioms for those names to the underlying theory. All predicates and relations in the definitions which follow are taken to be over this extended language, call it L. We further extend L by the names of all given sets, state variables (dashed and undashed) and parameters as introduced within the specification for some class C of interest, obtaining thereby a language $L^+[C]$ which is specific to that class. This language serves as a basis for our behavioural model. In what follows we shall assume as well that the specification for C satisfies all of the static validation criteria introduced in the previous section.

Taking some association for the given set names appearing within the specification of C, a *denotation* $\delta C . E$ for an event E on that class is just its signature with any systematic substitution of element names from those sets in place of the corresponding parameter names; an *initial denotation* δC is just the class signature after a similar substitution for any instantiation parameters. Such denotations stand for individual *occurrences* of the event (or pseudo-event) in question. For any derived predicate or relation $p[C...]$ we shall write $p[\delta C...]$ to denote the corresponding predicate or relation with the substitutions implicit in $\delta C...$.

As already indicated we define a *trace* to be a finite sequence of such denotations, which is either empty or begins with an initial denotation. A *history* is then a pair (τ,H), where τ is a trace comprising a possible sequence of event occurrences for a process of class C and H is a set of (undashed) predicates closed under logical inference: H embodies all properties which can be (syntactically) inferred about the state of that process after it has engaged in the particular sequence denoted by τ.

We are interested in obtaining, from a given history (τ_0, H_0) and a denotation $\delta c...$ for some event which is applicable at that point, a new history (τ_1, H_1) such that $\tau_1 = \tau_0 {}^\wedge \langle \delta c... \rangle$ (concatenation of the new denotation onto τ_0) and H_1 contains all of the predicates (again expressed in undashed form) which can be historically inferred from both H_0 and the derived relation $r[\delta c...]$, where r is either *init* or *post*.

Hence the overall structure of our required model for the behaviour implied by a class specification takes the form of a set of histories, *behav*[c], which is defined inductively as follows:

(i) for any class c,

$(\langle \rangle, \{true\}) \in$ *behav*[c];

(ii) for any initial denotation δc,

$(\langle \delta c \rangle, H_1) \in$ *behav*[c]

if H_1 is historically inferred from {true} and *init*[δc];

(iii) for any event denotation $\delta c . E$ and non-empty trace τ_0,

$(\tau_0 {}^\wedge \langle \delta c . E \rangle, H_1) \in$ *behav*[c]

if $(\tau_0, H_0) \in$ *behav*[c]

and *pre*[$\delta c . E$] is consistent with H_0

and H_1 is historically inferred from H_0 and *post*[$\delta c . E$];

(iv) There are no other members of *behav*[c].

The set of trace components included in such a model is clearly prefix-closed since any trace τ is only extended by a single event denotation when all of its prefixes are already present within *behav*[c]. Every such occurrence (except instantiation) is of course subject to conditions which ensure applicability of the event and admissibility of its parameter values at that point, but these involve nothing more than consistency in the usual sense. Thus the only aspect of this model which is in any way novel is exactly how "H_1 is historically inferred from H_0 and $r[\delta c...]$".

This special rule of inference, which is the heart of our whole approach, will be formulated in the next subsection after development of the necessary background and definitions.

Historical Inference

Heretofore, in all examples, we have made systematic use of our distinctive conventions where only the minimal effect of an event need be specified within its explicit postcondition. There is, however, no obligation to do so in all cases. It may sometimes be entirely appropriate to give a "stronger" specification. Consider an alternative (and probably more familiar) formulation of the finite set abstraction, which class we shall call SS (for Simple Set):

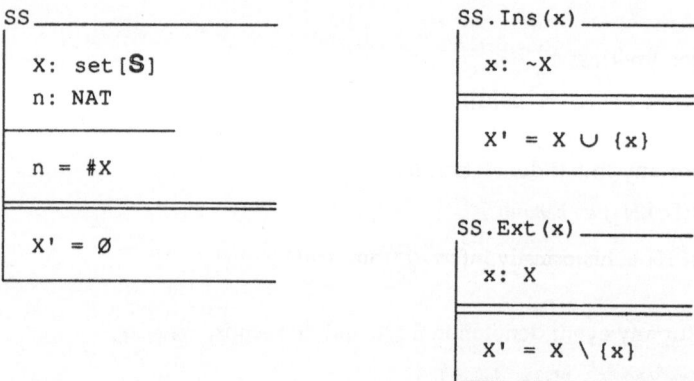

This is identical to the specification of the class FS except for the way in which the postconditions of these two events have been expressed:

	for FS	for SS
insertion	$x \in X'$	$X' = X \cup \{x\}$
extraction	$x \notin X'$	$X' = X \setminus \{x\}$

Both would seem to specify exactly the same (net) effect, namely inclusion or exclusion of x with X, but they are still *different* specifications. (Intuitively, the events for FS are characterised in a "pointwise" fashion whereas those for SS correspond to "copy-rewrite" definitions.)

Traditionally, the primary reason for writing postconditions in the latter form has been the need to carry forward "historical" state component values (e.g. to ensure that the subset $X \setminus \{x\}$ is still included within X' after these events). This is clearly guaranteed by the specification of SS above, as can be seen from again portraying a part of its initial behaviour:

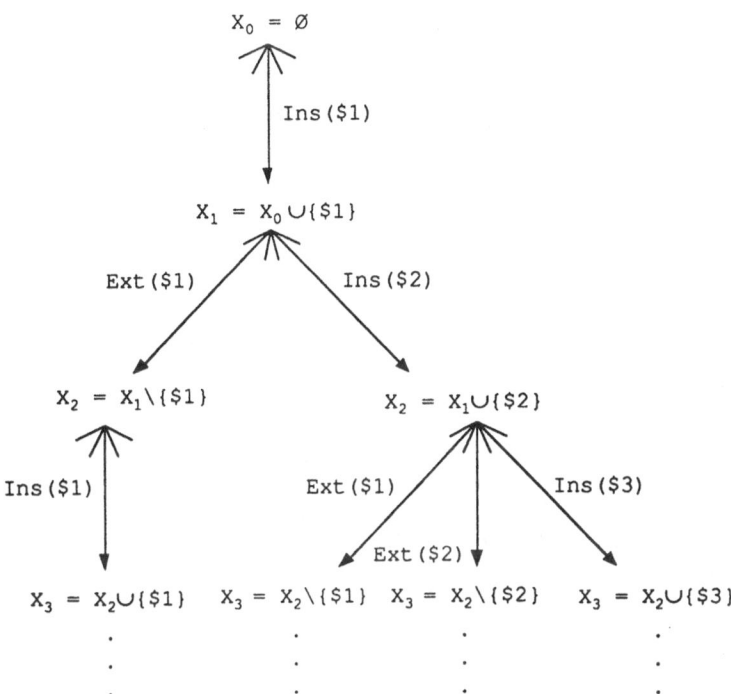

This is to be compared with the behaviour depicted earlier for FS, which differs only in the form of the node annotations. We nonetheless intend, *in the absence of other constraints,* that the behaviour implied by these two different specifications should be equivalent in the sense that any implementation of one is also an implementaion of the other. But we have also taken the position that the stronger postconditions of SS are tantamount to "over-specification" when the only intended effect of these events is to insert or extract a single value x. This was the original motivation behind introducing our special rule of inference, whereby propagation of the appropriate historical values is achieved by implicitly *augmenting* the postconditions for each individual event. Thus we would argue that the best way of saying "the rest stays unchanged" is to leave it unsaid!

This strategy is therefore dependent upon identifying suitable "neutral relations", which simply assert that certain aspects of the state remain unchanged, in order to be able to carry out the required augmentation.

Examples. For the class FS the following equality relations are all neutral in the intended sense. We observe that they may each be written in a "normal form", which is given in parentheses afterwards:

$X = X'$ $\qquad\qquad$ ($\equiv \{\alpha \in X \mid \text{true}\} = \{\alpha \in X' \mid \text{true}\}$)

$X \cap \{x\} = X' \cap \{x\}$ $\qquad$ ($\equiv \{\alpha \in X \mid \alpha = x\} = \{\alpha \in X' \mid \alpha = x\}$)

$X \setminus \{x, y\} = X' \setminus \{x, y\}$ $\quad$ ($\equiv \{\alpha \in X \mid \alpha \neq x \wedge \alpha \neq y\} = \{\alpha \in X' \mid \alpha \neq x \wedge \alpha \neq y\}$)

true $\qquad\qquad\qquad$ ($\equiv \{\alpha \in X \mid \text{false}\} = \{\alpha \in X' \mid \text{false}\}$)

$x \in X \Rightarrow X = X'$ $\qquad$ ($\equiv \{\alpha \in X \mid x \in X\} = \{\alpha \in X' \mid x \in X\}$)

For the class FM, we might also have the following neutral relations:

$M = M'$ $\qquad\qquad$ ($\equiv \{(x,y) \in M \mid \text{true}\} = \{(x,y) \in M' \mid \text{true}\}$)

M restricted to X

$= M'$ restricted to X $\quad$ ($\equiv \{(x,y) \in M \mid x \in X\} = \{(x,y) \in M' \mid x \in X\}$)

M restricted to $X \setminus \{s\}$

$= M'$ restricted to $X \setminus \{s\}$ $\quad$ ($\equiv \{(x,y) \in M \mid x \neq s\} = \{(x,y) \in M' \mid x \neq s\}$)

The normal forms shown with the above examples are in fact the ones of interest. Given a subset S of the set of state variables of a class specification, var[C], we define an *S-neutral relation* to be a conjunction of equality relations,

$$\{\alpha \in v_s \mid p_s(\alpha)\} = \{\alpha \in v'_s \mid p_s(\alpha)\}$$

where (...v_s...) are the variables from S and, for each s, $p_s(\alpha)$ is a formula with one free variable α over the language $\mathcal{L}^+[C]$. Notice that:

(i) each S-neutral relation is reflexive between pre- and post-states;

(ii) the set of all S-neutral relations is closed under conjunction, since

$$\{\alpha \in v_s \mid p_s(\alpha)\} = \{\alpha \in v'_s \mid p_s(\alpha)\} \wedge \{\alpha \in v_s \mid q_s(\alpha)\} = \{\alpha \in v'_s \mid q_s(\alpha)\}$$

$$\Leftrightarrow \{\alpha \in v_s \mid p_s(\alpha) \vee q_s(\alpha)\} = \{\alpha \in v'_s \mid p_s(\alpha) \vee q_s(\alpha)\}$$

The neutral relations with which we augment a particular event specification are drawn from the set of those which are consistent with the predicates of that event:

Definition. Let C.E be an event and S a set of state variables; then an S-neutral relation n is termed *S-neutral with respect to* C.E if, for every assignment of undashed variables satisfying *pre*[C.E], there is a corresponding assignment of dashed variables satisfying $(n \wedge post[C.E])$. The set of all such n is denoted *neutral*[C.E](S).

This set of relations is not necessarily closed under conjunction. For example, consider a new event Opt on the class FS, corresponding to a non-deterministic choice between two different argument values:

```
FS.Opt (x,y) _____

   x,y: ~X
   _____

   x ≠ y
   _____

   x ∈ X' ∨ y ∈ X'
   _____
```

Taking $n_1 \triangleq (X \backslash \{x\} = X' \backslash \{x\})$ and $n_2 \triangleq (X \backslash \{y\} = X' \backslash \{y\})$ then both n_1 and n_2 are are contained in *neutral*[FS.Opt]({X}) but $(n_1 \wedge n_2)$ is not. This means that not all neutral relations are mutually compatible. To avoid such "clashes" there is a need to restrict further the choice of neutral relations, as follows.

Definition. Let C.E be an event and S a set of state variables; then an S-neutral relation n is termed *S-central with respect to* C.E if, for all relations $m \in$ *neutral*[C.E](S), the relation $(n \wedge m) \in$ *neutral*[C.E](S). The set of all such n is denoted *central*[C.E](S). Similar definitions, but involving *init*[C] rather than *post*[C.E] pertain for instantiation.

Proposition. For any set S, the set *central*[C.E](S) is closed under conjunction. With regard to the example suggested just above, the relation $(X \backslash \{x, y\} = X' \backslash \{x, y\})$ is in *central*[FS.Opt]({X}).

This means that relations in *neutral*[C.E](var[C]), for example, are not only consistent with the event specification C.E but are mutually consistent with any other such consistent neutral relations. In making inferences about occurrences of events we will augment the event specification with relations from *neutral*[C.E](var[C]), since these are guaranteed to be compatible with any notion of "minimal change". Consider for example, the specification of the exchange event on the class FS given earlier:

```
FS.Exch(x→y) _____
    x: ~X ;  y: X
   ─────────────────
    x ∈ X'; y ∉ X'
```

Then the relations n'=n, X'\{x,y}=X\{x,y} are both in *central*[FS.Exch](var[FS]). Thus on an occurrence of this event we will infer that n does not change and neither does the presence or otherwise of elements other than x or y in X.

However, by considering other examples, it becomes clear that this choice of neutral relations still does not allow us to infer everything that we would like about the effect of an event. Recall the example of the insert event on the class FS:

```
FS.Ins(x) _____
    x: ~X
   ─────────────────
    x ∈ X'
```

The intention of this specification is to insert x into X, leaving X\{x} unchanged. The only variable of apparent concern is X and the relation X'\{x}=X\{x} is an element of *central*[FS.Ins]({X}). The difficulty lies in the fact that the class specification involved two state variables, X and n.

Suppose that we wished to apply the event Ins and that there was already an element in X, y say. Then whilst satisfying the explicit postcondition x∈X' it would be possible to keep n the same, simply by deleting y and inserting x. This means that the relation n>0 ⇒ n'=n is an element of *neutral*[FS.Ins](var[FS]). As already observed, X'\{x}=X\{x} is also an element of *neutral*[FS.Ins](var[FS]) but these two cannot be simultaneously satisfied so their conjunction (n>0 ⇒ n'=n) ∧ (X'\{x}=X\{x}) is *not* in *neutral*[FS.Ins](var[FS]). Therefore neither can be elements of *central*[FS.Ins](var[FS]); in particular, we are not at present able to augment FS.Ins with X'\{x}=X\{x}. The conflict arises in the presence of the second variable n which is of no concern to the event except that it is linked to x by the invariant in the class specification.

The purpose of the invariant is to mutually constrain the variables of the class; if it were omitted the relations $n'=n$ and $x' \setminus \{x\}=x \setminus \{x\}$ would both be elements of *central*[FS . Ins](var[FS]). It is for this reason that the notion of *primary* and *subsidiary* variables is introduced – the distinction is that a primary variable is one which the event is genuinely concerned with changing (such as x in the example), whereas a subsidiary variable (such as n) would not be affected by the event were it not necessary to maintain the invariant.

Definition. A state variable, $v_s \in$ var[C] is *subsidiary* to an event C . E if the relation $(v_s=v'_s)$ is an element of *central*[C . E](var[C]) when the explicit invariant is omitted from the specification of the class C. A state variable is *primary* to C . E if it is not subsidiary. The set of all primary variables is denoted prim[C . E].

As the principal concern of an event is to effect minimal change to its primary variables, we may safely augment the event specification with relations from *central*[C . E](prim[C . E]). Thus we will augment our event with relations of the form $n_1 \wedge n_2$ where $n_1 \in$ *central*[C . E](prim[C . E]), consistent with minimal change to the primary variables, and $n_2 \in$ *central*[C . E](var[C . E]) then being consistent with minimal "adjustment" to preserve the invariant.

Definition. The set of relations of the form $n_1 \wedge n_2$ where $n_1 \in$ *central*[C . E](prim[C . E]) and $n_2 \in$ *central*[C . E](var[C . E]) is termed *central*[C . E].

The required rule of historical inference can now be formulated, using these central relations to carry forward (into H_1) those properties holding in some previous history H_0 that are not explicitly "interfered with" by the event as specified.

Definition. Let H_0 and H_1 be sets of undashed predicates closed under logical inference, and r[δc...] be the derived relation (*init* or *post*) obtained for a particular occurrence of some event on the class C. A predicate q is included in H_1 precisely when it can be inferred from predicates holding before the event and from r itself, *together with* appropriate central relations:

$$q \in H_1 \Leftrightarrow \exists p \in H_0, n \in central[\text{C}...] \bullet (p \wedge r[\delta\text{C}...] \wedge n[\delta\text{C}...]) \Rightarrow q'$$

where q' is just the dashed form of q. We then say "H_1 is historically inferred from H_0 and r[δc...]". It may, of course, be possible to infer that predicates hold for a given state without using central relations. This is illustrated immediately below.

Let r be a mixed relation; then r^n_m stands for the relation obtained by replacing every occurrence of a dashed variable by its counterpart dashed n times, and by replacing every occurrence of an undashed variable by its counterpart dashed m times. Thus for any such r, we have $r = r^1_0$.

For any event C.E, we define the *weakest postcondition* for that event, $wpc[C.E]$, to be

$$wpc[C.E] \triangleq \exists \ldots v'_s, v_e \ldots \bullet (\, pre[C.E]^0_1 \wedge post[C.E]^0_1 \,)$$

Proposition. For any event denotation, δC.E, and history $(\tau^\wedge\langle\delta$C.E$\rangle,H)$, $wpc[C.E] \in H$. We also have $wpc[C]$ for initialisation, where its definition involves $inv[C]^0_1 \wedge init[C]^0_1$.

In general, however, the rule of inference will enable us to infer more about the state if in addition we use predicates holding before the event and central relations. The examples which follow show the usage of such central relations to augment the relations derived from different forms of event specification.

Proposition. For any event C.E, the relation

$$post[C.E]^0_0 \Rightarrow (\ldots \wedge (v_s = v'_s) \wedge \ldots) \text{ is in } central[C.E].$$

That is, for any event, if the pre-state enables the event to complete with no change to any variable then after the event we can infer exactly the same about the state as we could before. In particular, if there is no explicit postcondition then

$$(\ldots \wedge (v_s = v'_s) \wedge \ldots) \text{ is in } central[C.E].$$

Thus the event is interpreted as having no effect on the state, which justifies our earlier use of the term "query" to describe such events.

Proposition. For any event C.E, if there is a var[C]-neutral relation n such that $post[C.E] \Rightarrow n$ then n is in $central[C.E]$.

Such relations may be thought of as being *explicitly* central. Consider, for example, the class SS introduced earlier where the postcondition for insertion was specified in the form $x' = x \cup \{x\}$ which states explicitly that $x \setminus \{x\}$ should remain unchanged. As this form of the postcondition does in fact imply the neutral relation

$$m \triangleq (x \setminus \{x\} = x' \setminus \{x\}),$$

then m is indeed central to SS.Ins.

This is to be contrasted with the class FS where the postcondition for insertion was specified in the form $x \in X'$, which certainly does not imply m. However, m is nonetheless an element of *central*[FS.Ins](prim[FS]) as is shown by the following example (which illustrates derivation of a central relation that is not explicitly central). Here, the relations of *entral*[FS.Ins](prim[FS]) are shown to be all the {x}-neutral relations that can be inferred from the conjunction of m, above, and the relation $x \in X$ $\Rightarrow X = X'$:

Proposition. Let $c \triangleq \{\alpha \in X \mid x \in X \vee \alpha \neq x\} = \{\alpha \in X' \mid x \in X \vee \alpha \neq x\}$. Then for every $n \in$ *neutral*[FS.Ins]({x}), we have $c \Rightarrow n$.

Proof. (i)   The relation c is in *neutral*[FS.Ins]({x}) (Easily verified).

(ii)   Suppose $n \in$ *neutral*[FS.Ins]({x}), so $n \triangleq \{\alpha \in X \mid p(\alpha)\} = \{\alpha \in X' \mid p(\alpha)\}$.

then $\forall x \bullet \exists x' \bullet x \in X' \wedge \{\alpha \in X \mid p(\alpha)\} = \{\alpha \in X' \mid p(\alpha)\}$

$\Rightarrow$ $\forall x \bullet \exists x' \bullet x \in X' \wedge \{\alpha \in X \cap \{x\} \mid p(\alpha)\} = \{\alpha \in X' \cap \{x\} \mid p(\alpha)\}$

$\Rightarrow$ $\forall x \bullet \{\alpha \in X \cap \{x\} \mid p(\alpha)\} = \{\alpha \in \{x\} \mid p(\alpha)\}$

$\Rightarrow$ $x \in X \vee \neg p(x)$

$\Rightarrow$ $p(\alpha) \Rightarrow x \in X \vee \alpha \neq x$

$\Rightarrow$ $c \Rightarrow n$.

Any relation $c \in$ *neutral*[C.E](S) which implies all other elements of *neutral*[C.E](S) is said to be *strongly S-central* to C.E. It can be shown that if c is strongly S-central, then it is S-central. In particular, c above is {x}-central to FS.Ins. The existence of such a strongly S-central relation implies that *central*[C.E](S) = *neutral* [C.E](S), in which case *neutral*[C.E](S) is closed under conjunction. The set *neutral*[FS.Opt]({x}) mentioned in an earlier example did not have this property, which illustrates that not all events have a strongly S-central relation.

In our original formulation, neutral relations were simply all those which did not force a change of state. This gave rise to certain anomalies. Consider the specification of a class DS (for Double Set), with an event that always increases the cardinality of one of its state components:

```
DS ─────────────────        DS.Exp ─────────────
┌─────────────────────      ┌──────────────────────
│  X,Y: set [S]             │   #X' > #X
├─────────────────────      └──────────────────────
│  X' = Y' = ∅
└─────────────────────
```

We would certainly expect $(Y'=Y)$ to be central to this event, as Y is not affected by the explicit postcondition at all. Previously, the relation $(Y' = ∅ \vee X = X')$ would have been termed neutral, and indeed would be in the set $neutral$[DS.Exp]$(\{X,Y\})$. This is undesirable since, in the presence of such neutrals, $(Y=Y')$ is no longer central.

Other criteria for neutrality have been investigated, but all gave rise to counter-intuitive effects. For example, the relation $(X = Y) \Leftrightarrow (X' = Y')$ is an equivalence relation on states. If it were to be considered neutral then it would be neutral with respect to the event DS.Exp, simply by changing Y to X', as would $(Y=Y')$ by leaving Y alone. But clearly these two are not mutually consistent, so neither can be central. In particular, $(Y=Y')$ would again not be central.

Class Validation

The static proof obligations outlined in the previous section pertain to the various (state or event) schema comprising the specification for a particular class, but they apply to each separate schema definition *in isolation*. These criteria are concerned with essentially syntactic aspects of that specification, relating solely to consistency; the apparently semantic constraints imposed upon the individual events, namely applicability and effectiveness, serve only to ensure that *some* behaviour is always implied. For this reason such properties are said to be mandatory – i.e. they must hold for every class. We shall now take a brief look at the spectrum of "dynamic", that is to say behavioural, properties which may be of interest. Unlike our static criteria, however, none of the dynamic properties suggested here can be considered mandatory. It is rather a matter of what behaviour is (or is not) desired for a given class. Thus we are concerned more with overall characteristics of such behavioural requirements, and with how the corresponding verification conditions would then be formulated. These latter are necessarily expressed with respect to a *complete* class specification. We therefore assume that the (finite) set of event definitions that are associated with any class C, denoted $\mathcal{E}$[C], is somehow established from the context in which its specification is presented.

When speaking of concurrent systems it is customary to distinguish between"safety" as opposed to "liveness" properties. The former are analogous to the verification conditions for partial correctness in sequential programming (which serve to show that they satisfy their specifications), whereas the latter are related to the (additional) termination conditions needed to prove total correctness within that domain.

We are of course dealing here with the specifications themselves, so correctness as such is not an issue in the same sense as for implementations. However, concerns about *safety* of process specifications are nonetheless a matter of verifying that their implied behaviour always remains within some acceptable limits – i.e. that they do not violate any (possibly non-functional) requirements which, for whatever reason, have not already been specified explicitly. Not surprisingly, various techniques used in proving the correctness of sequential programs are still directly applicable, *viz.*
- *invariants* (e.g. the value of some variable x is unchanged within a given scope) or
- *variants* (e.g. the variable x always decreases in value within a certain loop body).

Every (dynamic) safety property can be cast as one or the other of these notions; the corresponding proposition would then be formulated as a relation, say r, between dashed and undashed variables. The general verification condition to show that r holds (as either a variant or an invariant) for a particular event C . E is as follows:

$$\exists\, n \in central[\text{C.E}] \bullet (pre[\text{C.E}] \wedge post[\text{C.E}] \wedge n) \Rightarrow r$$

(In some cases this condition can be simplified, in particular where the implication can be proven without involving central predicates.) If it is required that r should hold over the entire behaviour specified for the class C, then this same condition must be satisfied for all $E \in \mathcal{E}[\text{C}]$. In many situations, however, the safety requirement expressed by r is relevant only to some designated subset $\{... E_i, E_j ...\} \subset \mathcal{E}[\text{C}]$; in principle, one might then go on to prove complementary properties in regard to the other events.

By their very nature, all *liveness* concerns can be expressed in the same general form:

"starting from any state satisfying some proposition p, it is always possible
to reach some (later) state in which another proposition, say q, will hold."

Such properties may be formalised, with respect to the specification for a given class C, by the following verification condition:

> **for every** history $(\tau_0, H_0) \in behav[C]$ such that $p \in H_0$,
> **there exists** a history $(\tau_1, H_1) \in behav[C]$ such that:
> (i) τ_0 is a prefix of τ_1 and (ii) $\#\tau_1 > \#\tau_0$ and (iii) $q \in H_1$

A wide variety of more specialised properties can then be formulated just by making appropriate substitutions for the fixed predicates p and q as they appear within this overall condition. For example taking $p \equiv \mathtt{true}$ is to enquire about the whole of the behaviour specified for C, whereas taking $p \equiv wpc[C...]$ is to ask about the potential behaviours following any occurrence of the designated event (which might be, for certain properties, instantiation of the class); this could be extended to encompass all behaviours after some particular set of alternative events $\{... E_i, E_j ...\} \subset \mathcal{E}[C]$ simply by taking $p \equiv (... \vee wpc[C.E_i] \vee wpc[C.E_j] \vee ...)$. Similarly, taking $q \equiv \mathtt{true}$ is merely to posit the *existence* of possible successor states whereas to take $q \equiv pre[C.E]$ is to require that the event C.E will always be "enabled" at some (future) point; selected subsets may again be of interest in this connection, but in such cases one might take either $q \equiv (... \vee pre[C.E_i] \vee pre[C.E_j] \vee ...)$ or $q \equiv (... \wedge pre[C.E_i] \wedge pre[C.E_j] \wedge ...)$ in order to assert, respectively, that *any* or *all* of the events in question must become enabled.

Although the different forms for the predicates p and q discussed here are certainly the most commonly occurring ones, no restrictions whatsoever are placed upon the structure of these propositions – thus the full range of liveness properties, as identified above, can in principle be formulated and verified. This basic approach is especially useful for investigating conditions that are highly specific to a particular class, when the properties of interest are closely related to the domain of application. However, there do exist a rather limited number of properties that are considered to be of universal interest in the context of concurrent systems. For these cases at least, it is sometimes more convenient in practice to make use of *sufficient conditions* that involve just the individual event specifications, if only to avoid the complexities attendant in reasoning over the behaviours themselves.

The most obvious example is of course showing *absence of deadlock* for the given class of processes – which corresponds to substituting $p \equiv q \equiv \mathtt{true}$ within our overall verification condition. A sufficient condition for such purposes might be:

>**for every** event $\mathtt{C.E_i} \in \mathcal{E}[\mathtt{C}]$,
>>**there exists** an event $\mathtt{C.E_j} \in \mathcal{E}[\mathtt{C}]$ such that $wpc[\mathtt{C.E_i}] \Rightarrow pre[\mathtt{C.E_j}]$

It must be emphasised, however, that this property is by no means desirable in all cases. One may also conceive of classes for which it is specifically intended that each process should instead be able to reach some characterisable "final state" (where the actual conditions for termination would normally depend upon the context of their subsequent instantiations). In these situations one could well wish to verify what is essentially the opposite property: namely that certain events always lead to expected final states, after which no further events associated with that class are applicable.

A related property, one which is quite frequently a matter of pragmatic concern, is to show that transient subsystem states, as characterised by a suitable proposition r, are always *recoverable*. A trivial example on the class $\mathtt{FS}$ is that any state in which the event $\mathtt{Ins(x)}$ is applicable can always be restored by applying $\mathtt{Ext(x)}$ immediately afterwards. A special case that is often of interest is when r corresponds to $wpc[\mathtt{C}]$, i.e. to possible initial states for the class $\mathtt{C}$, which must hold if the specified processes are meant to be *restartable*. The condition for verifying such properties in general is obtained by taking $p \equiv q \equiv r$, but again one would be more likely to have recourse to sufficient conditions in the first instance. Any property of this sort implies that some aspect of the class behaviour is *cyclic* in nature, which is certainly not desired in all cases. Indeed, a major proof obligation for many specifications is to show that the system as a whole always *makes progress*, even though many of its constituent subsystems do behave cyclically – as with almost every class of "server" process, for example. It follows that, for such classes, conditions concerned with (global) progress cannot be established in isolation. Their verification will depend upon assumptions about the actual environment in which those processes are intended to operate: for instance, that the number of "requests" to be serviced is at least countable [cf. Abrial & Schuman 1979]. One approach is to introduce a suitable refinement for the class specification under consideration, which embodies just those properties that are necessary to carry out the required proof; in effect, this acts as a "surrogate" for all environments. Then one need only validate that these "enabling" assumptions are consistent in each separate context of use. Proof strategies of this sort are greatly facilitated by object-oriented composition, as discussed in the next section.

There is another set of concerns, sometimes referred to as *responsiveness* as opposed to liveness, which also involves reasoning about the dynamic properties implied by a class specification. One example of interest is establishing whether its behaviour is *deterministic*, in the sense that those processes always offer their environment only one possible choice as to the next event (even though that event may itself be highly non-deterministic, as in the case of an operation like FS.Opt(x,y) mentioned in the previous subsection). A sufficient condition for deterministic behaviour is that the preconditions of all events in $\mathcal{E}[c]$ are *disjoint*. In most situations, however, the desired behaviour is exactly the opposite. Such requirements are illustrated by the traditional client/server paradigm, where the intention is that the server processes should offer some range of alternatives to their environment whenever possible – precisely so that the overall behaviour (i.e. the particular choices made at each stage) will ultimately be responsive to the demand originating with individual clients, in order to maximise system "throughput".

A closely related concern is the (often misunderstood) concept of *fairness*. Again, this is most easily illustrated in the context of a (multiple) client/(single) server configuration. One might well wish to impose the requirement that every such server should be "fair" in the sense that it will never favour any particular client over others or, more typically, that the choice amongst alternative events is not biased. For example, on the class FS, it may be required that neither insertion nor extraction is given priority. But this is exactly what is meant by *non-deterministic* choice, which is how its behaviour is in fact specified. Some (additional) degree of determinism will, of necessity, be introduced by any possible implementation, since there is no such thing as a "fair arbiter". However, each of those implementations would (and indeed should) be regarded as "satisfying" the original specification, in that they constitute consistent refinements where the indeterminacy is (partially or totally) resolved. Such resolutions can of course be analysed formally to determine whether they are acceptable, but then the criteria involved are no longer concerned with fairness *per se*. What must be emphasised is that it is nevertherless possible to reason effectively over non-deterministic behaviours. For instance, it can be shown that, for any process of the class FS, the number of extraction events before it is then obliged to engage in an insertion is *bounded* (by n, the cardinality of x); the reverse cannot be said of FS, but it will apply in the (more realistic) case of the class BS (where an upper bound m is imposed on the cardinality of x). Assertions of this sort provide the basis for analysing "overtaking" and/or "starvation" properties, which are the key to reasoning about whether the implied behaviour will be sufficiently responsive.

4. Combining Process Specifications

Almost all essential issues involving concurrency arise in the context of *combining* independently specified processes in order to construct the definition of a new one. It is a hallmark of the various different "calculi" for concurrent systems, originating with CCS [Milner 1980] and CSP [Hoare 1985], that the meaning of any such process is defined *compositionally* – i.e. in terms of the meanings of its constituent processes; this is always subject to additional constraints, which serve to ensure that the resultant process definition is also meaningful. In extending the interpretation of our object-oriented specifications so as to encompass concurrent composition, the basic combinator is still *logical conjunction*, which is the most natural (and least "operational") notion of concurrency; the constraints in question are the same ones that have already been introduced, namely those which are needed to preserve both applicability and effectiveness.

In the preceding section we modelled the behaviour of every process, whether elementary or constructed, in terms of derived predicates and relations, but without taking into account the various "layers" of composition and/or refinement that may have been involved in their specification. We now wish to consider these derivations in a way that strictly respects any such (static) compositional structure, so as to reinforce the strong intuitions evoked by the phrase "combining processes".

When one speaks of composition, however, it is necessary to distinguish from the outset two quite different ways in which such combinations might be defined in our object-oriented framework. The distinction is between *instantiation* of previously specified classes, say A and B, to define named components within some larger class C, as opposed to multiple *inheritance* within the specification of C as a new subclass of both A and B:

    Instantiation   *Inheritance*

```
C _____              C _____
 |                              |
 |  a1: A                       |  A; B
 |  b1: B                       |  _____
 |  . . .                       | |
 |                              | |  . . .
 |_____              |_|
```

In the case of instantiation above every process of class C comprises two independent subprocesses, a1 and b1, which are instances of the class A and the class B, respectively. One might say the same in regard to inheritance of A and B within the subclass C, even though such "instances" are not named, but this only applies when A and B are wholly independent classes (i.e. when their component names are entirely *disjoint* to any level of generalisation upwards in the subclass hierarchy); otherwise, when the inherited component names intersect at some level, those names are taken to identify the *same* component (for which all of the properties, as specified in both A and B, must then be mutually consistent). The real distinction between these two approaches can therefore be seen in terms of how the composite "state-space" is constituted: in the case of separate instantiations, the resultant state is effectively a *disjoint union* of all constituent instances (after systematic renaming of their individual components, such that they are made completely distinct); in the context of a subclass definition, by contrast, the resultant state is to be construed as the ordinary union of any inherited component names, constrained by the *conjunction* of all corresponding properties.

Thus instantiations of one class within another are most easily thought of as "combining instances", whereas defining subclasses by means of (multiple) inheritance is perhaps more appropriately viewed as "composing specifications". No matter which way such a constructed process definition is expressed, however, it is entirely suitable to speak of concurrent composition when one comes to definition of their associated operations since this will in general involve conjoining the pre- and postconditions of previously defined events. Assuming an event F on the class A and an event G on the class B, a new event E on the class C might then be specified in either approach as follows:

Instantiation	*Inheritance*

```
    C.E(...) _____          C.E(...) _____
   |                            |
   |  a1.F(...)                 |   A.F(...)
   |  b1.G(...)                 |   B.G(...)
   |  ...                       |    _____
   |_____            |   |
                                |   |   ...
                                |___|_____
```

Both express *synchronisation* in the usual sense, *viz.* the event E is only complete when (the occurrences of) F and G are also complete.

The concept of *communication*, in the sense of "data-flow", is then readily encompassed within such synchronised event specifications:

Whilst the communication constraints expressed above may seem to imply *sequential* composition, this is not necessarily the case: true simultaneity or indeed any arbitrary *interleaving* is permissible providing there is agreement on identically named parameter values. The above forms may be viewed as simple abbreviations:

$$
\begin{array}{|l}
\texttt{C.E}\,(\alpha{\to}\gamma)\,\underline{\hspace{2cm}} \\[4pt]
\quad \texttt{a1.F}\,(\alpha{\to}\beta_1) \\
\quad \texttt{b1.G}\,(\beta_2{\to}\gamma) \\
\quad \ldots \\[4pt]
\quad \boxed{\beta_1{=}\beta_2}
\end{array}
\qquad
\begin{array}{|l}
\texttt{C.E}\,(\alpha{\to}\gamma)\,\underline{\hspace{2cm}} \\[4pt]
\quad \texttt{A.F}\,(\alpha{\to}\beta_1) \\
\quad \texttt{B.G}\,(\beta_2{\to}\gamma) \\[4pt]
\quad \boxed{\beta_1{=}\beta_2} \\
\quad \ldots
\end{array}
$$

When constructing a process definition by instantiation, it is of course reasonable to introduce more than one distinct instance of the same class. New events might then be specified in terms of occurrences of different (or even the same) operations on these two independent components:

$$
\begin{array}{|l}
\texttt{D}\,\underline{\hspace{2cm}} \\[4pt]
\quad \texttt{a1,a2; A} \\
\quad \ldots
\end{array}
\quad
\begin{array}{|l}
\texttt{D.E1}\,(\ldots)\,\underline{\hspace{1.5cm}} \\[4pt]
\quad \texttt{a1.F1}\,(\ldots) \\
\quad \texttt{a2.F2}\,(\ldots) \\
\quad \ldots
\end{array}
\quad
\begin{array}{|l}
\texttt{D.E2}\,(\ldots)\,\underline{\hspace{1.5cm}} \\[4pt]
\quad \texttt{a1.F}\,(\ldots) \\
\quad \texttt{a2.F}\,(\ldots) \\
\quad \ldots
\end{array}
$$

Because the underlying state components are entirely disjoint, the apparent concurrency is still nothing more than interleaving in the final analysis. This is the fundamental assumption made in most approaches to process composition, which has the effect of forcing all interactions between processes to be specified in terms of (synchronous) communication.

In the case of subclass definitions, however, properties of a more general class can only be inherited once, so it is redundant to combine the same class name with itself:

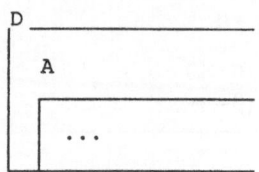
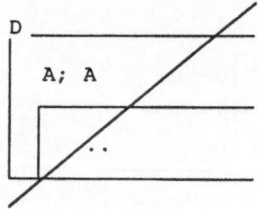

One may nonetheless choose to make use of concurrent composition when specifying the events associated with such a subclass, though there is no need to mention the same event twice unless different parameters are involved:

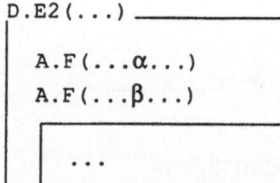

Suppose for example that the various additional operations suggested for the class F S were to be introduced as refinements, on a subclass F S1, instead of directly extending F S. One might begin by explicitly promoting the basic insertion and extraction operations:

```
FS1 _____        FS1.Ins(x) _____        FS1.Ext(x) _____
   ┌──────────────            ┌──────────────             ┌──────────────
   │  FS                      │  FS.Ins(x)                │  FS.Ext(x)
   └──────────────            └──────────────             └──────────────
```

Such promotions could be used as well to specify the "indeterminate selection" events that were discussed earlier (by simply "reversing" the direction of each argument):

```
      FS1.New (→y) _____             FS1.Any (→y) _____
         ┌──────────────                 ┌──────────────
         │  FS.Ins(y)                    │  FS.Ext(y)
         └──────────────                 └──────────────
```

The argument for concurrent composition arises very naturally when one comes to specifying the double-insertion and exchange operations that were also proposed:

```
FS1.Ins2(x,y) _____          FS1.Exch(x→y) _____

    FS.Ins(x)                       FS.Ins(x)
    FS.Ins(y)                       FS.Ext(y)

        x ≠ y
```

These examples raise, in its simplest possible form, the issue of *interference* – since the two events involved in each of the above specifications are quite obviously operating on the same (inherited) state component, namely the set X. (Notice, however, that the preconditions ensure these occurrences are concerned with different element values).

Here is precisely where our principle, that over-specification should be avoided, comes into its own. Had one attempted to specify the same composition of events on a subclass of SS (with its stronger postconditions) rather than FS, then these refinements would not even satisfy our static validation criteria:

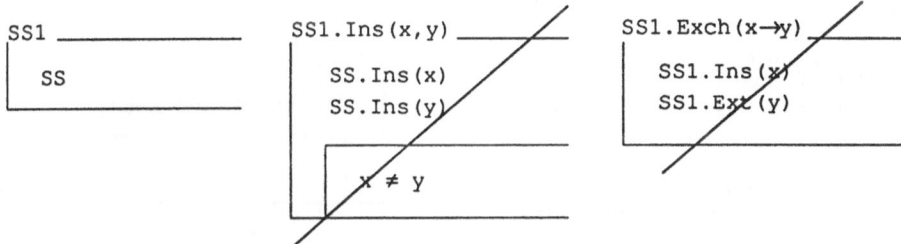

This comes about because the conjoined postconditions of these occurrences are manifestly INCONSISTENT when $x \neq y$ (as imposed by the preconditions):

$$\text{for} \quad \text{SS1.Ins2}(x,y): \quad (X' = X \cup \{x\}) \wedge (X' = X \cup \{y\})$$

$$\text{for} \quad \text{SS1.Exch}(x \rightarrow y): \quad (X' = X \cup \{x\}) \wedge (X' = X \setminus \{y\})$$

No such conflict appears when composing the weaker postconditions of FS:

$$\text{for} \quad \text{FS1.Ins2}(x,y): \quad (x \in X') \wedge (y \in X')$$

$$\text{for} \quad \text{FS1.Exch}(x \rightarrow y): \quad (x \in X') \wedge (y \notin X')$$

The fact that our implicit augmentation is *context-dependent* means that more specifications can be composed concurrently!

This problem of inteference does not arise when combining separate process *instances,* which is exactly why such an approach is much to be preferred in most circumstances. By way of illustration, a variant of the class F I specified earlier might better be formulated in two steps, first introducing a class FP (for Finite Product) and then refining that definition within a subclass FB (for Finite Bijection):

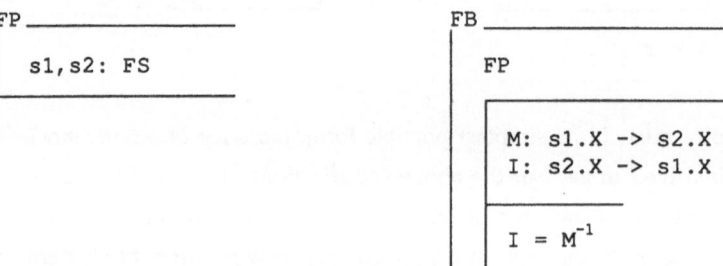

FP _____
|
| s1,s2: FS
|_____

FB _____
|
| FP
| _____
| |
| | M: s1.X -> s2.X
| | I: s2.X -> s1.X
| |
| | I = M^{-1}
| |_____

The renaming effected as part of the instantiations in FP means that s1.X and s2.X are distinct state components. The mappings M and I, specified as further components within the refinement FB, are therein constrained to be a total bijection between those two sets. The associated operations are then most naturally defined as follows:

FB.New(x,y) _____
|
| s1.Ins(x); s2.Ins(y)
| _____
| |
| | M'(x) = y
| |_____

FB.Null(x,y) _____
|
| s1.Ext(x); s2.Ext(y)
| _____
| |
| | M'(x) = y
| |_____

Both specifications involve concurrent composition of the same event, albeit operating on different components. Additional pre- and postconditions, for Null and New respectively, are needed in order to preserve the state invariants of FB, but the two occurrences of Ins or Ext are otherwise entirely independent.

However desirable it may be from a methodological point of view to minimise cases of interference in the course of specifying (or implementing!) concurrent systems, we would nevertheless argue that the issue cannot simply be side-stepped by precluding such specifications altogether (e.g. [He 1988]). Rather, the need is for a compositional framework wherein the proof rules allow potential interference to be reasoned about *statically.*

Concurrent Composition

Given two events $C.E_1$ and $C.E_2$, we can construct the event $C.E_3$ which is simply the logical conjunction of these two components, that is:

$$pre[C.E_3] \triangleq pre[C.E_1] \wedge pre[C.E_2]$$

$$post[C.E_3] \triangleq post[C.E_1] \wedge post[C.E_2]$$

We say that $C.E_1$ and $C.E_2$ may be *composed concurrently* if $C.E_3$ is both applicable and effective. This composition may be extended to any number of events irrespective of the order and nesting structure of their presentation, as conjunction is both associative and commutative.

We now investigate sufficient conditions for concurrent composition, and for the compatibility to be preserved through refinement. We define a relation $core[C.E]$, which preserves those components of the state which are necessary for the satisfaction of the postcondition of the event $C.E$.

A structure for a class C, as already stated, is an assignment of the state variables $(...v_s...)$ over a model of set-theory (with assignment within this to the given set names). We may obtain a structure for the whole of $\mathcal{L}^+[C]$ from two class structures σ_0 and σ_1 by assigning the undashed variables in accord with the assignments in σ_0 and the dashed variables in accord with those in σ_1 (provided that they are *compatible* in the sense that the underlying models of set-theory and the assignment of the given sets within this are the same.) If the resulting structure is a model of a relation r, we say r *holds from σ_0 to σ_1*, written $\sigma_0, \sigma_1 \models r$.

The relation *core* for an event $C.E$ is then defined as follows:

$$core[C.E] \triangleq \forall ...v_s''... \bullet [post[C.E]_2^0 \Rightarrow post[C.E]_2^1]$$

An example for the event $FS.Ins$, with postcondition $x \in X'$, is

$$core[FS.Ins(x)] \triangleq \forall X'' \bullet (x \in X \Rightarrow x \in X')$$

which is equivalent to $(x \in X \Rightarrow x \in X')$.

If *core*[C...] holds from σ_1 to σ_2, then whenever *post*[C...] holds from σ_0 to σ_1, we have *post*[C...] holding from σ_0 to σ_2.

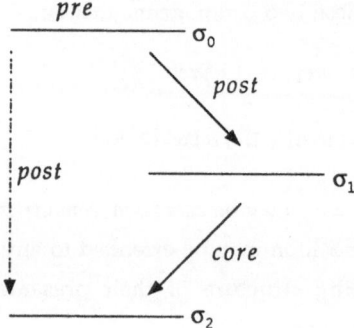

The relation *core*[C.E] embodies those portions of the state which the event C.E is liable to "re-write".

If these state components are disjoint for two events, we would expect to be able to compose those events concurrently. Two relations q and r are said to be *complementary* if for any two (compatible) states σ_0 and σ_1, there is a state σ_2 such that q holds from σ_0 to σ_2, and r holds from σ_1 to σ_2.

This may be illustrated with respect to the specification for FS.Ins2(x,y). The relations

$$core[\text{FS.Ins}(x)] \triangleq (x \in X \Rightarrow x \in X')$$

$$core[\text{FS.Ins}(y)] \triangleq (y \in X \Rightarrow y \in X')$$

are complementary (when $x \neq y$) since for any two assignments of the set X, we can construct an assignment in which $x \in X$ if this is the case in the first assignment, and $y \in X$ if this is the case in the second assignment.

Proposition. If, for two events C.E_1 and C.E_2, *pre*[C.E_1] $\wedge$ *pre*[C.E_2] is consistent and *core*[C.E_1] and *core*[C.E_2] are complementary, then C.E_1 and C.E_2 may be composed concurrently.

Proof.

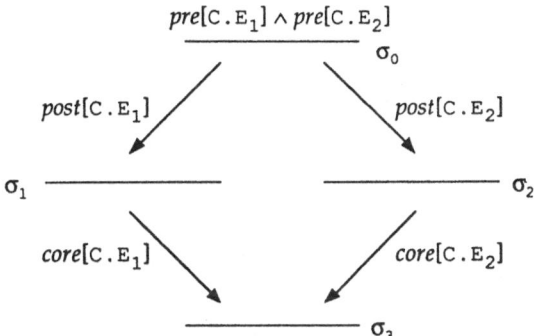

For any initial state σ_0 which satisfies $pre[\text{C}.\text{E}_1] \wedge pre[\text{C}.\text{E}_2]$, there exist states σ_1 and σ_2 such that $\sigma_0, \sigma_i \models post[\text{C}.\text{E}_i]$ (i=1,2) by effectiveness of $\text{C}.\text{E}_i$. Since $core[\text{C}.\text{E}_1]$ and $core[\text{C}.\text{E}_2]$ are complementary, there exists a state σ_3 such that $\sigma_i, \sigma_3 \models post[\text{C}.\text{E}_i]$ (i=1,2), whence, from the definition of $core[\text{C}...]$, $\sigma_0, \sigma_3 \models post[\text{C}.\text{E}_i]$ (i=1,2).

A relation q is a *complement* of relation r if for any two (compatible) states σ_0 and σ_1, there is a unique state σ_2 such that $\sigma_0, \sigma_2 \models r$, and $\sigma_1, \sigma_2 \models q$. For example, the two relations

$$q \triangleq \{\alpha \in X \mid \alpha=x\} = \{\alpha \in X' \mid \alpha=x\}$$

$$r \triangleq \{\alpha \in X \mid \alpha \neq x\} = \{\alpha \in X' \mid \alpha \neq x\}$$

are complements of one another.

Proposition. A neutral predicate n, where

$$n \triangleq ... \wedge \{\alpha \in v_s \mid p_s(\alpha)\} = \{\alpha \in v'_s \mid p_s(\alpha)\} \wedge ...$$

in which neither v_s or v'_s occur in $p_s(\alpha)$ is called an *element of state*. In such cases

$$n^c \triangleq ... \wedge \{\alpha \in v_s \mid \neg p_s(\alpha)\} = \{\alpha \in v'_s \mid \neg p_s(\alpha)\} \wedge ...$$

is always a complement of n.

An immediate corollary, which guarantees concurrent composition is the following:

Corollary. If, for two events $C.E_1$ and $C.E_2$, there exists an element of state n, such that

$$n \Rightarrow core[C.E_1] \land n^c \Rightarrow core[C.E_2]$$

then $C.E_1$ and $C.E_2$ may be composed concurrently.

We can now formulate 'local' rules, whereby classes (and their associated events) may be refined (implemented!) independently whilst preserving the ability to compose them concurrently. Suppose two events $C.E_1$ and $C.E_2$ can be 'separated' by an element of state n in the sense of the corollary above, that is

$$n \Rightarrow core[C.E_1] \land n^c \Rightarrow core[C.E_2]$$

Then not only can they be composed concurrently, but if $C.F_1$ and $C.F_2$ are refinements of $C.E_1$ and $C.E_2$ respectively, and if $C.F_1$ and $C.F_2$ can still be separated by n, then they may also be composed concurrently. Thus $C.E_i$ may be refined independently, provided we ensure that the relevant n still implies the *core* of any such refinement.

Example. Consider the following events on class FS:

$$FS.Opt(x,y) \triangleq x \in X' \lor y \in X'$$

$$FS.Ins(z) \triangleq z \in X', \text{ where } z \notin \{x,y\}.$$

Then
$$(X \cap \{x,y\} = X' \cap \{x,y\}) \Rightarrow core[FS.Opt(x,y)]$$

$$(X \cap \{z\} = X' \cap \{z\}) \Rightarrow core[FS.Ins(z)]$$

Thus these events are separated by

$$n \triangleq (X \cap \{x,y\} = X' \cap \{x,y\})$$

$$n^c \triangleq (X \setminus \{x,y\} = X' \setminus \{x,y\})$$

since the latter implies the element of state $m \triangleq (X \cap \{z\} = X' \cap \{z\})$.

We may refine these events by resolving the non-determinacy of FS.Opt to yield

$$\text{FS.Ins(x)} \triangleq x \in X'$$

and by strengthening FS.Ins(z) to give

$$\text{FS.Ins2(z,w)} \triangleq z \in X' \wedge w \in X', \text{where } w \notin \{x, y, z\}.$$

In each case we have ensured the *core* of the refinement is within the limits described above. That is,

$$n \Rightarrow core[\text{FS.Ins(x)}] \quad \wedge \quad n^c \Rightarrow core[\text{FS.Ins2(z,w)}]$$

and so they may still be combined concurrently.

However, if we had instead separated FS.Ins(z) and FS.Opt(x,y) using m and m^c above, the refinements would have violated the condition. Thus it may be necessary to anticipate subsequent refinements by carefully circumscribing the degree of interference that is or is not permitted by a specification.

The *core* of an event locally determines the extent to which that event may interfere with the state, *and thus the conditions under which it might be interfered with by other events*. During refinement, the extent of such interference may be reduced as in the case of FS.Ins, or increased, as seen in FS.Ins2. As observed above, if we wish to do independent refinement it may be necessary to delimit the possible extent of interference at the outset.

This approach may be compared with that of [Owicki & Gries 1976]. If we consider the events FS.Ins(x) and FS.Ins(y), it would be possible to find implementations with non-interfering proofs of correctness for each. Such implementations would also be provably correct with respect to the specifications of SS.Ins(x) and SS.Ins(y), but these proofs will always (necessarily) interfere since it will never be possible to find a correct *joint* implementation (in the case $x \neq y$). If, in refinement as above, we preserve mutual consistency, then we also preserve the possibility of obtaining implementations with non-interfering proofs, which will then be compatible.

Our approach is also clearly related to the "Rely/Guarantee" conditions proposed in [Jones 1981, 1983a & 1983b], although the precise nature of this relationship is extremely elusive. However, our formulation appears to give insight into overcoming certain difficulties with such conditions, as raised by [Woodcock & Dickinson 1988]. In particular, the fact that all of the necessary relations are derived, rather than being introduced as additional assertions, provides a methodological advantage. More importantly, our proof-rules are natural and follow directly from taking composition to be logical conjunction. This applies equally to our sufficient conditions for independent refinement.

5. In Conclusion

This paper constitutes the sequel that was envisaged at the end of our previously published work on object-oriented subsystem specification [1987]. Part of the purpose here has been to make explicit all of the associated proof obligations and verification conditions, in order to emphasise the potential for reasoning formally about both static and dynamic properties of any system specified within our overall framework; in the course of formulating the various rules involved, we have also clarified (and considerably improved upon) our original definitions of the so-called "central relations" which underlie the key concept in our whole approach, namely the notion of historical inference. However, the main objective has been to underpin the interpretation of class specifications in terms of process behaviours, so that the widely appreciated (practical) advantages of an object-oriented structuring discipline can then be carried over into the domain of formally specifying concurrent systems. We have described what is now almost a complete calculus for composition and refinement of such classes, regarded as process specifications. It will be observed that no facilities for sequential composition have been provided, and indeed there is no intention to do so! The only missing capability that is normally associated with concurrency is "hiding" – wherein certain events are permitted to occur without any (possibility of) participation by their environment. Any such facilitiy leads directly to the issue of "divergence", which in turn strongly affects the system "responsiveness" properties that were discussed in passing under the heading of class validation. Accordingly, we have simply deferred introducing the corresponding mechanisms until their implications can be more thoroughly investigated. This is now underway in the context of our continuing work on reasoning over process behaviours.

Acknowledgements. Research on object-oriented formal specification at the University of Surrey has been funded in part by the Science and Engineering Research Council under Grant No. GR/F 1448.2 in conjunction with the Alvey FORSITE Project (SE/065). We are grateful to our colleagues within this group for several constructive discussions about the work described here; the same applies in regard to those of our collaborators from other organisations who have shown a sustained interest both in the development of this particular approach and in its subsequent application.

References

1.   J-R. Abrial [1982]: "Theoretical Foundations to Formal Programming", Unpublished Lecture Notes, Programming Research Group (Oxford).

2.   J-R. Abrial and S.A. Schuman [1979]: "Non-Deterministic System Specification", LNCS 70 (Springer-Verlag) pp. 34-51.

3.   D. Bjørner and C.B. Jones [1982]: *Formal Specification and Software Development* (Prentice-Hall International).

4.   I. Hayes (ed.) [1987]: *Specification Case Studies* (Prentice-Hall International).

5.   He Jifeng [1988]: "Process Refinement", *Proc. Workshop on Refinement* (University of York).

6.   C.A.R. Hoare [1985]: *Communicating Sequential Processes* (Prentice-Hall International).

7.   P.T. Johnstone [1987]: *Notes on Logic and Set Theory* (Cambridge University Press).

8.   C.B. Jones [1980]: *Software Development: A Rigorous Approach* (Prentice-Hall International).

9.   C.B. Jones [1981]: "Development Methods for Computer Programs Including a Notion of Interference", D.Phil. Thesis, Technical Monograph PRG-25 (Oxford University).

10. C.B. Jones [1983a]: "Specification and Design of (Parallel) Programs",
    in Mason (ed.), *Information Processing 83* (North Holland).

11. C.B. Jones [1983b]: "Tentative Steps Towards a Development Method for
    Interfering Programs", *ACM Trans. on Programming Languages and Systems*,
    5(4) pp. 576-619.

12. C.B. Jones [1986]: *Systematic Software Development Using VDM*
    (Prentice-Hall International).

13. R. Milner [1980]: *A Calculus of Communicating Systems*, LNCS 92
    (Springer-Verlag).

14. S. S. Owicki and D. Gries [1976]: "Verifying Properties of Parallel Programs:
    an Axiomatic Approach", *Comm. ACM*, 19(5) pp. 279-285.

15. S.A. Schuman and D.H. Pitt [1987]: "Object-Oriented Subsystem Specification",
    in Meertens (ed.), *Program Specification and Transformation* (North Holland)
    pp. 313-342.

16. J. M. Spivey [1988a]: *Understanding Z: A Specification Language and its Formal
    Semantics* (Cambridge University Press).

17. J. M. Spivey [1988b]: *The Z Notation. A Reference Manual*
    (Prentice-Hall International).

18. J.C.P. Woodcock and B. Dickinson [1988]: "Using VDM with Rely and
    Guarantee Conditions: Experiences From a Real Project",
    Proc. VDM Symposium 88, LNCS 328 (Springer-Verlag).

FORMAL OBJECT ORIENTED SPECIFICATION OF DISTRIBUTED SYSTEMS

Elspeth Cusack
Formal Methods Group
Research and Technology/Information Services Standards Division
British Telecom
St Vincent House, Ipswich, IP1 1UX

INTRODUCTION

Ideas from the field of object oriented programming are currently exciting interest as an intuitively attractive approach to the problem of the design of distributed systems. In particular, there are initiatives within the International Organisation for Standardisation (ISO) and the International Telephone and Telegraph Consultative Committee (CCITT) [1, 2].

There is as yet no generic theory of objects, although all object oriented approaches to design share common elements. In particular, they depend on the concept of an *object* as a "black box" with prescribed interfaces and the exploitation of similarities or relationships between objects or classes of objects. The latter is usually termed *inheritance* in the literature. The central problem of object oriented design is to express a system specification or architecture in terms of a collection of cooperating objects, which may or may not have a separate physical existence (decomposition). Designers also wish to determine which objects may be substituted for which others without altering system behaviour for the worse.

Objects can be modelled as generalized order sorted algebras, allowing definitions of types, subtypes and inheritance to be formalised [3]. However, communications system engineering demands the less abstract, event-based approach provided by specification languages such as CSP [4] and the ISO standard formal description technique LOTOS [5]. The absence of formal event-based definitions of basic object oriented concepts makes their incorporation into a convincing specification and design methodology difficult: this paper represents an attempt to rectify the situation. It is further motivated by the observation that many of the ideas arising naturally when distributed systems are considered already exist as formal concepts in CSP and LOTOS. The failures model of CSP makes it a suitable research tool at this early stage of our work and it is assumed that the reader is familiar with it. Translation of the ideas in this paper into LOTOS for a wider engineering audience is planned for the future.

Similar research is being undertaken within the Alvey project Advanced Systems Networked Architecture (ANSA). ANSA has developed a theory called "object physics" [6]. The relationship with the work presented here is discussed at the end of the paper.

This paper begins by showing how *objects, instances* and *classes* can be described as communicating processes or families of processes. Next, a definition of the *refinement* of a communicating process is introduced which generalises the partial order on processes defined in [4, Chapter 3] and serves the same function as *subclassing* or *strict inheritance*.

We continue with a discussion of how a system specification can usefully be expressed as a collection of component processes (objects) operating concurrently according to a prescribed configuration. The method depends on the simultaneous consideration of two levels of abstraction, rather than one level as is usually the case in CSP. One level deals with system events and the other with object events. This leads to the formulation of "design rules" dealing with the central design problems of *decomposition* and *substitution* (Theorems 2 and 3).

In addition to clarifying our ideas about objects, the work reported here is directed at making the accepted benefits of formal specification (analysability, possibility of correctness preserving transformation and so on) available to designers of open distributed systems. The notation follows Hoare's book [4] supplemented according to the glossary at the end of the paper.

OBJECTS AS COMMUNICATING PROCESSES

Object oriented design relies on abstraction and hiding. Classically [7] an object consists of private memory and a set of operations. The set of messages to which an object can respond is called its interface with the rest of the system. The only way to interact with an object is through its interface – this means that its private memory can be manipulated only by its own operations. A class is a set of objects which all represent the same system component. The individual objects described by the class are called instances.

These ideas are precisely captured by the concept of a process in CSP [4]. A process P is the behaviour pattern of some entity, described in terms of its alphabet, αP, the set of externally–observable events in which the entity is logically equipped to engage. Communication between processes is represented as their simultaneous participation in the same event. Processes can be constructed to be history–sensitive, modelling the idea of private memory.

Example

Let $P = x \rightarrow Q$, and $Q = y \rightarrow P$, with $\alpha P = \alpha Q = \{x, y\}$.

Let R be the process ($x \rightarrow Q \mid y \rightarrow P$). Then the behaviour of R is determined by the environment's choice of x or y as initial event. In other words, R has a private memory storing one bit of information.

Whenever f is a one–one function mapping αP into another set of symbols, let $f(P)$ denote the process derived by changing the symbols of αP in accordance with f. "Change of symbol" thus induces an equivalence relation on communicating processes. The equivalence class containing P is simply the set of all processes which can be obtained from P by change of symbol. (The choice of P as class representative is of course arbitrary.) This definition of a *class of processes* captures the sense of the informal definition introduced above.

Example

Let $A = a \rightarrow (b \rightarrow A \sqcap c \rightarrow A)$ with $\alpha A = \{a, b, c, d\}$. Define $f(\alpha A)$ to map b and c to x and y respectively leaving a and d unchanged. Then
$f(A) = a \rightarrow (x \rightarrow f(A) \sqcap y \rightarrow f(A))$ and $\alpha f(A) = \{a, x, y, d\}$. The processes A and $f(A)$ are members of the same class.

Cooperation and communication between processes is expressed using the CSP concurrent composition combinator $\parallel$ [4, Chapter 2]. For example, if C is a collection of processes then the system formed by all the processes in C running concurrently is
$\parallel \{ X \mid X \in C \}$. Theorem 3 of the paper shows how a given system specification can be decomposed in this object oriented fashion.

The substitution problem is best considered in the light of Hoare's partial order $\sqsubseteq$ on processes with a given alphabet. To quote from [4, Chapter 3.9] , "$P \sqsubseteq Q$ means that Q is equal to P or better in the sense that it is less likely to diverge and less likely to fail. Q is more predictable and more controllable than P, because if Q can do something undesirable P can do it too; and if Q can refuse to do something desirable, P can also refuse".

Substitution Problem

What processes can replace a given process P in a collection C? In other words, given P what restrictions on Q ensure that

$$\| \{ X \mid X \in (C - \{P\}) \cup \{Q\} \} \sqsupseteq \| \{ X \mid X \in C \} ?$$

Theorem 2 states the solution to the substitution problem.

SUBCLASSING, INHERITANCE AND REFINEMENT

In object oriented programming, class A is said to be a *subclass* of class B when all instances of class A are also instances of class B. Class A may also be said to *strictly inherit* from class B. (To quote from [8], "Strict inheritance requires descendants to be behaviorally compatible with ancestors while non–strict inheritance allows operations of ancestors to be arbitrarily redefined and captures the notion of 'similarity' rather than 'behavioral compatibility' ".) Designers usually wish to be able to use an instance of a subclass to replace an instance of a class without changing system behaviour for the worse. Subclassing therefore captures ideas of specialisation and resolution of nondeterminism, and is the typical approach to the substitution problem in object oriented programming.

In this paper we express subclassing in terms of the related concept of *refinement*. We define the refinement of a CSP process as follows:

Process Q is said to refine process P precisely when the following conditions hold:

1       $\alpha P \subseteq \alpha Q$

2       $(tr\ Q) \lceil \alpha P \subseteq tr\ P$

3   *If (s, X) is a failure of Q with $\alpha Q - \alpha P \subseteq X$, then $(s \lceil \alpha P, X \cap \alpha P)$ is a failure of P.*

4   *If s is a divergence of Q then $s \lceil \alpha P$ is a divergence of P.*

(In general we will not wish to use or refine processes which diverge, so the sets of divergences in condition 4 will be empty).

Refinement is transitive and reduces to $\sqsubseteq$ in the case that $\alpha P = \alpha Q$. (Both assertions can be immediately derived from the definition). It follows that deterministic processes can only be refined by processes with larger alphabets, corresponding to the idea of specialisation. For example, if P and Q have disjoint alphabets, then both processes are refined by P $\|$ Q.

We observe from this simple example that a process may refine more than one other process. In other words, in a strict inheritance hierarchy a class may have more than one ancestor. This is a special case of *multiple inheritance*, an appealing design concept which has proved very difficult to define in the general case of non–strict inheritance [8].

Nondeterministic processes can also be refined by the resolution of nondeterminism. For example, if $\alpha P = \alpha Q$ then it is always the case that P [] Q, P and Q refine P $\sqcap$ Q.

Examples

1   Consider the processes $P = x \rightarrow P$ with $\alpha P = \{x\}$; $Q = x \rightarrow y \rightarrow Q$,
$R = (x \rightarrow R \mid y \rightarrow R)$ and $S = y \rightarrow x \rightarrow S$ with $\alpha Q = \alpha R = \alpha S = \{x, y\}$. Then
Q, R and S refine P.

2   Consider the processes $C = (a \sqcap b) \rightarrow C$, $A = a \rightarrow A$ and $B = b \rightarrow B$ with
$\alpha A = \alpha B = \alpha C = \{a, b\}$.

Then A [] B refines A $\sqcap$ B which in turn refines C.   A and B each refine A $\sqcap$ B.
Neither A nor B refines   A [] B since, for example, (< >, {b}) is a failure   of A
but not of A [] B. Thus condition 3 of the   definition of refinement is not
fulfilled.

Our interest in refinement is provoked by the following result (proved in the Appendix)
which answers the substitution problem in the special case that $\alpha Q = \alpha P$.

Theorem 1

*If Q refines P then whenever R is a process satisfying $\alpha Q \cap \alpha R = \alpha P \cap \alpha R$,
Q $\parallel$ R   refines P $\parallel$ R.*

It is not difficult to find an example showing that the syntactic restrictions of the
theorem cannot be weakened:

Example

*Let $P = x \rightarrow P$ and $R = x \rightarrow y \rightarrow R$ with $\alpha P = \{x\}$ and $\alpha R = \{x, y\}$. Then P is
refined by $Q = y \rightarrow x \rightarrow Q$ with $\alpha Q = \{x, y\}$. However, Q $\parallel$ R = STOP, which does
not refine P $\parallel$ R.*

If Q refines P, then an observer monitoring events in αP is unable to distinguish P from
Q. Refinement thus reflects an intuitive idea of "realisation" or "abstract implementation".
It also suggests a CSP interpretation of the concept of a subclass : we can think of class
Q (the equivalence class containing Q) as a subclass of class P whenever there exists a
change of symbol mapping f on αP with the property that Q refines f(P). An observer
monitoring events in $\alpha f(P)$ is unable to distinguish Q from f(P) and in this sense Q is an
instance of class P. In other words, strict inheritance is precisely the composition of
change of symbol and refinement.

Refinement is also the concept we need to formulate the decomposition problem for CSP
processes.

Decomposition problem

*Find a collection of processes C with the property that $\parallel \{ X \mid X \in C \}$ refines a
given process S.*

The collection of processes C = {S} is always a trivial solution of the decomposition
problem. In fact, the substitution problem is now revealed as a special case of the
decomposition problem. However, in general we are interested in "realisations" or "abstract
implementations" of S composed of more than one process and with an alphabet strictly
larger than αS, so we will continue to treat substitution separately.

COMMUNICATION AND CONFIGURATION

This section explores the ways in which a given collection of processes can be configured or "connected" to make a system. The aim is to gain insight into the substitution and decomposition problems.

The *configuration* of any collection of processes C' with disjoint alphabets is introduced. The configuration identifies distinct events in these alphabets and gives rise to a change of symbol mapping for each process alphabet. More precisely, let $A = \cup \{ \alpha X \mid X \in C' \}$. A configuration of C' is any subset D of $A \times A$ which is subdiagonal in the sense that each element of A appears at most once as a component of an element of D.

Each such configuration D induces a change of symbol mapping f(D, X) on the alphabet of each process X in C' :

for $x \in \alpha X$, $f(D, X)(x) = d$ if x is a component of some d in D and $f(D, X)(x) = x$ otherwise.

We can now define the system arising when the processes in C' are configured according to D to be

$$\|_D \{ X \mid X \in C' \} = \| \{ f(D, X)(X) \mid X \in C' \}.$$

Notice that the processes in C' are instances – so long as alphabets are disjoint, C' may contain more than one member of any class.

Example

Let *X, Y and Z be processes with disjoint alphabets defined as follows and set* $C' = \{X, Y, Z\}$.

$X = x \to y \to X$ $\qquad \alpha X = \{x, y\}$

$Y = (a \to Y \mid b \to Y)$ $\qquad \alpha Y = \{a, b, c\}$

$Z = (d \sqcap e) \to Z$ $\qquad \alpha Z = \{d, e\}$

Then $D = \{ (x, a), (d, b), (e, c) \}$ is a configuration of C'. Renaming the members of D as $r = (x, a)$, $s = (d, b)$ and $t = (e, c)$ and writing f(X) for f(D, X)(X) and so on, we obtain

$f(X) = r \to y \to f(X)$ $\qquad\qquad \alpha f(X) = \{r, y\}$

$f(Y) = (r \to f(Y) \mid s \to f(Y))$ $\qquad \alpha f(Y) = \{r, s\}$

$f(Z) = (s \sqcap t) \to f(Z)$ $\qquad\qquad \alpha f(Z) = \{s, t\}.$

It follows that $X \|_D Y \|_D Z = f(X) \| f(Y) \| f(Z).$

We notice that in general it will be possible to so configure a collection of objects in more than one way, yielding different systems. Since objects are "black boxes" communicating only by prescribed messages to each other we wish to ignore all events in the alphabet of only one process in C'. The next section shows how this is achieved.

OBJECT ORIENTED CSP

The notion that objects communicate by sending messages to each other is captured by restricting the alphabet of a system to consist only of special events known as *communications*. ("A communication is an event that is described by a pair c.v where c is the name of the channel on which the communication takes place and v is the value of the message which passes" [4, Chapter 4].) The set of possible values of v is called the channel alphabet αc. We follow Hoare's convention that channels are used for communication in only one direction and between only two processes. We call the set of channel names the *system events*. The system event level can be thought of as the objects' view of the system. At this level of abstraction, then, process alphabets necessarily have events in common.

However, consider the system $T = \| \{ X \mid X \in C \}$. The alphabet of each process in C can be considered independently as a set of input and output "ports" – that is, the channels linking system objects can be "unconnected". We define an *object event* to be such an input or output "port" – in other words, the preparedness to transmit or receive some message. Precisely which other object is involved in the communication is irrelevant. The object events of a process P therefore determine the possible communications in which P can engage, and can be thought of as the designer's view of the process.

Let C' denote the collection of processes obtained in this way from those in C. The processes in C' can be considered to be precisely the processes in C but viewed at a lower level of abstraction which ensures that (object event) alphabets are disjoint. We now investigate the recovery of T from C' using the concept of configuration introduced in the previous section. In other words, we are going to think of C' as a "box of components" which we are going to configure to obtain T.

As usual in CSP, x! denotes an output "port" and y? denotes an input "port". Following the definition of channel alphabet, the alphabet of an object event is defined to be the set of messages which the process is equipped to transmit or receive at that "port". In constructing the required configuration we can only identify an output "port" x! with an input "port" y? where $\alpha x! \subseteq \alpha y?$ and x! and y? are in the alphabets of different processes.

For each process X in C' let $\alpha_{out} X = \{ \alpha x! \mid x! \in \alpha X \}$ and let

$\alpha_{in} X = \{ \alpha y? \mid y? \in \alpha X \}$ (so $\alpha_{out} X$ and $\alpha_{in} X$ partition αX).

Set $A_{out} = \cup \{ \alpha_{out} X \mid X \in C' \}$, $A_{in} = \cup \{ \alpha_{in} X \mid X \in C' \}$, and

$A =$
$\{ (x!, y?) \in A_{out} \times A_{in} \mid \alpha x! \subseteq \alpha y?,$ x! and y? belong to different process alphabets$\}$.

Any subdiagonal subset of A is necessarily also subdiagonal in $A_{out} \times A_{in}$, and is thus a configuration of C'. Such a configuration identifies object events, thereby creating channels and determining the system events.

Configuration law

Using the notation introduced in the previous section, we have demonstrated that whenever we are given a specification expressed as $T = \| \{ X \mid X_{,} \in C \}$ for some collection of processes C, a collection of processes C' and a configuration D can always be found such that

$C = \{ f(D, X)(X) \mid X \in C' \}$, $T = \|_D \{ X \mid X \in C' \}$ *and* $\alpha T = D.$

It is immediate that $\alpha T = D$, since no events can happen on "unconnected channels" . They are ignored at the level of abstraction of system events, and represent merely design possibilities that the designer has chosen not to exploit.

We can now state the general solution to the substitution problem. The result is proved in the Appendix and ensures that a process P can be replaced (subject to a simple alphabet restriction) by its refinement Q. The refinement takes place at object event level, leaving the system events unchanged. In other words, we have established that any object P can be replaced by another object Q inheriting from the class of P. The theorem's syntactic conditions on Q are to ensure that αQ – aP consists of "new events" (compare the conditions in Theorem 1) . The configuration law implies that the theorem is generally applicable.

Theorem 2

Assume that $\| \{ X \mid X \in C \} = \|_D \{ X \mid X \in C' \}$ for some collection C' of processes with disjoint alphabets and configuration D. Suppose that P is a process in C and let P' be the corresponding process in C'. So $P = f(D, P')(P')$. Write $\alpha C = \cup \{ \alpha X \mid X \in C \}$ and let Q' be any process refining P' with the property that $\alpha Q' \cap \alpha C \subseteq \alpha P \cap \alpha(C - \{P\})$ and $\alpha Q' \cap \alpha(C' - \{P'\}) = \varphi$.

Then a change of symbol mapping $f(D, Q')$ can be defined on $\alpha Q'$ which ensures that D is also a configuration of $C' - \{P'\} \cup \{Q'\}$. Setting $Q = f(D, Q')(Q')$ it follows that at system event level $\| \{ X \mid X \in (C - \{P\}) \cup \{Q\} \} \sqsupseteq \| \{ X \mid X \in C\}$.

In order to decompose a system we need to be able to introduce new system events. The final theorem guarantees that this can be done subject to certain syntactic constraints. The result depends on the CSP notion of hiding [4, Chapter 3] and is proved in the Appendix. It shows how decomposition and refinement at object event level can be "carried through" to system event level.

Theorem 3

Suppose that $\| \{ X \mid X \in C \}$ refines a system S and that we can express some component process U in C as $U = V \| W \setminus \alpha V \cap \alpha W$. Setting $R = \| \{ X \mid X \in (C - \{U\}) \cup \{V, W\} \}$ and $T = \| \{ X \mid X \in C \}$ then R refines S with $\alpha T \subseteq \alpha R$.

Theorem 3 can be simply generalised to the case where process U is expressed as more than two processes running concurrently.

WORKED EXAMPLE USING DESIGN RULES

The design rules incorporated in Theorems 2 and 3 can be used to expand a given (monolithic) system specification by substitution and decomposition. The designer is free to choose which technique to apply at any stage. The techniques themselves incorporate design freedom, namely the choice of processes V and W in Theorem 3 and the choice of Q' in Theorem 2. The following worked esample shows how the techniques can be used.

We wish to design a system which will accept any two natural numbers and return their sum. The system specification is

$S = in?(x, y):N^2 \to out!(x+y) \to S.$

In order to apply Theorem 3 we set $U = S$, $C = \{U\}$,

$$V = in?(x, y):N^2 \rightarrow VS(x, y, 0)$$

where

$$VS(m, n, i) = if \; n=i \; then \; out!m \rightarrow V$$
$$else \; win!m \rightarrow wout?p:N \rightarrow win!i \rightarrow wout?j:N \rightarrow VS(p, n, j)$$

and $W = win?a:N \rightarrow wout!(a+1) \rightarrow W$.

The alphabets are $\alpha V = \{ in, out, win, wout\}$ and $\alpha W = \{ win, wout\}$.

Then $U = V \parallel W \setminus\{win, wout\}$ so Theorem 3 ensures that $V\parallel W$ refines S.

Design freedom is exploited in this example to incorporate an "off the shelf" component, the natural number incrementer W.

The configuration law ensures that we can find processes V' and W' with disjoint alphabets and a configuration D such that $V\parallel W = V'\parallel_D W'$. Let f^{-1} be the change of symbol mapping on αW defined by $f^{-1}(win) = yin$ and $f^{-1}(wout) = yout$ and set $W' = f^{-1}(W)$. Let g^{-1} be the change of symbol mapping on V defined by $g^{-1}(win) = vin$ and $g^{-1}(wout) = vout$ and set $V' = g^{-1}(V)$. (In other words, $f(D, W') = f$ and $f(D, V') = g$.)

Suppose now that W is no longer available and that instead we are offered a more complicated component which will increment or square a given number:

$$Y' = (\; yin?a:N \rightarrow yout!(a+1) \rightarrow Y' \; | \; sqin?a:N \rightarrow yout!a^2 \rightarrow Y' \;).$$

Since Y' refines W' with $\alpha Y' \cap (\alpha V \cup \alpha W) = \varphi \subseteq \alpha W \cap \alpha V$ and $\alpha Y' \cap \alpha V' = \varphi$, the conditions of Theorem 2 are satisfied. We can therefore use Y' in place of W' to implement S. In other words, $V\parallel Y \sqsupseteq V\parallel W$.

OPEN QUESTIONS AND RELATED WORK

The identification of configurations giving rise to undesirable system behaviour such as deadlock or livelock is outside the scope of the paper, and demands further study. So too does the determination of constructive rules for the refinement of a given process – this clearly relates to the problem of context equations [9].

It is intended that the ideas in this paper will be translated into LOTOS. In order to exploit the flexibility of LOTOS to the full, a definition of an object will be required capable of consistent expression in both LOTOS process algebra and the data typing sublanguage ACT–ONE. The lack of partial functions in ACT–ONE make this a nontrivial problem.

The "object physics" developed by the ANSA project [6] is an empirical approach which abstracts away from any particular existing formal language. An object is characterised by the events in its alphabet, so that we can identify an object P with its alphabet αP. ANSA points out that the behaviour of P could be described in one of several formal languages. A "binding" of objects P and Q is defined to be any subset B of $P \times Q$. If B is subdiagonal (in the sense of this paper) then the binding is a special sort called a "connection". Thus a connection of two objects is syntactically similar to the concept called a configuration in this paper. To quote from [6], "Connections are convenient forms of bindings and bindings in a system are often restricted to this one form".

However, ANSA bindings differ semantically from our configurations in not treating pairs as events in which two objects participate simultaneously. Instead the events are linked but not identified. A binding of P and Q may therefore contain both (p, q) and (p, r) for events p ϵ αP and q, r ϵ αQ. This permits some non-determinism to be captured at object description level without relying on a particular specification language – a facility which our use of CSP renders unnecessary.

CONCLUSIONS

Formal event-based definitions have been given of the basic concepts of object, class, instance, and subclass which correspond to informal definitions from object oriented programming, and demonstrate the behaviour expected. We have derived a way of decomposing a system or architectural specification into a collection of cooperating objects or communicating processes, and set out the conditions under which one object may be substituted for another without changing system behaviour for the worse.

The paper demonstrates that it is possible to use a process algebra (CSP) in a natural and integrated way within the context of object oriented specification of distributed systems, making available the accepted benefits of formal specification languages. In object oriented CSP both inheritance (relating objects in the "component catalogue") and decomposition (with substitution as a special case) are shown to have the same semantics, derived from the concept of refinement of a CSP process. This insight gives the design techniques formulated in the paper coherence, clarity and elegance.

Acknowledgements David Freestone is thanked for many helpful comments and suggestions. The paper is published with the permission of British Telecom's Director of Research and Technology.

REFERENCES

1   ISO/IEC JTC1 SC21 N1547, *Proposal for new work item on Open Distributed Processing*, 1987

2   CCITT SGVII, COM VII-R26(A), Annex 5, *Qx/y Framework for CCITT Distributed Applications*, 1987

3   K B Bruce and P Wegner, *An algebraic model of subtype and inheritance*, Technical Report No CS-87-21, Brown University, Providence, Rhode Island, August 1987

4   C A R Hoare, *Communicating Sequential Processes*, Prentice-Hall, 1985

5   ISO DIS 8807, *LOTOS – a formal description technique based on the temporal ordering of observational behaviour*, July 1987

6   *ANSA Reference Manual* Issue 00.03, Part IV, Cambridge, June 1987

7   A Goldberg and D Robson, *Smalltalk-80 : The language and its implementation*, Addison-Wesley, 1983 (reprinted 1985)

8   P Wegner, *Dimensions of object-based language design*, Proceedings of "Object oriented programming, systems, languages and applications '87", October 1987

9   M T Norris, M W Shields and J Ganeri, *A theoretical basis for the construction of interactive systems*, British Telecom Technology Journal, Vol 5 No 2, April 1987

APPENDIX

Proof of Theorem 1

The proof is obtained by checking that each of the four conditions defining refinement are fulfilled by Q ‖ R and P ‖ R.

1     $\alpha(P \parallel R) = \alpha P \cup \alpha R \subseteq \alpha Q \cup \alpha R = \alpha(Q \parallel R)$.

2   (tr Q ‖ R) ⌈ $\alpha P \cup \alpha R$

      = {t ∈ $(\alpha Q \cup \alpha R)^*$ | t⌈αQ ∈ tr Q , t⌈αR ∈ tr R }⌈ $\alpha P \cup \alpha R$

      ⊆ {t ∈ $(\alpha P \cup \alpha R)^*$ | t⌈αP ∈ (tr Q)⌈αP and t⌈αR ∈ tr R}

      ⊆ tr P ‖ R.

4   Suppose that s is a divergence of Q ‖ R. Then s⌈ $\alpha P \cup \alpha R$ ∈ $(\alpha P \cdot \cup \alpha R)^*$ and by [4, Chapter 3.8 L8] either s⌈αP can be written as (s⌈αQ)⌈αP where s⌈αQ is a divergence of Q and s⌈αR ∈ tr R (case 1), or s⌈αR is a divergence of R and s⌈αP ∈ (tr Q)⌈αP (case 2).

In the first case, the fact that Q refines P implies that s⌈αP is a divergence of P and that s⌈αR ⊆ tr R. We are led to conclude that in both cases s⌈ $\alpha P \cup \alpha R$ is a divergence of P ‖ R.

3   Suppose that (s, X) is a failure of Q ‖ R with $\alpha Q - \alpha P \subseteq X$. If s is a divergence of Q ‖ R then we have already shown that s⌈ $\alpha P \cup \alpha R$ is a divergence of P ‖ R. It follows from [4, Chapter 3.9 D10] that (s⌈ $\alpha P \cup \alpha R$, X ∩ $(\alpha P \cup \alpha R)$) is a failure of P ‖ R.

If s is not a divergence of Q ‖ R [4, Chapter 3.9 D10] allows us to write X = Y ∪ Z, where (s⌈αQ, Y) is a failure of Q and (s⌈αR, Z) is a failure of R. Since $(\alpha Q - \alpha P) \cap \alpha R$ is empty, it must be the case that $\alpha Q - \alpha P \subseteq Y$. Since Q refines P, we conclude that (s⌈αP, Y ∩ αP) is a failure of P. This implies (again, using [4, Chapter 3.9 D10]) that (s⌈ $\alpha P \cup \alpha R$, Z ∪ (Y ∩ αP) is a failure of P ‖ R.

It remains to show that Z ∪ (Y ∩ αP) = X ∩ $(\alpha P \cup \alpha R)$. This follows immediately from the observations that $\alpha Q \cap (\alpha P \cup \alpha R) = \alpha P$ and that Z ⊆ αR, since
  X ∩ $(\alpha P \cup \alpha R)$
= (Y ∪ Z) ∩ $(\alpha P \cup \alpha R)$
= ($((\alpha Q - \alpha P) \cup (Y \cap \alpha P) \cup Z)$ ∩ $(\alpha P \cup \alpha R)$.

Proof of Theorem 2

Let C be a collection of processes and assume that
‖ { X | X ∈ C } = ‖$_D$ { X | X ∈ C' } for a second collection C' of processes with disjoint alphabets and a configuration D. Suppose that P ∈ C and let P' be the corresponding process in C'. So P = f(D, P')(P'). Write $\alpha C = \cup$ { αX | X ∈ C } and let Q' be any process. refining P' with the property that $\alpha Q' \cup \alpha C \subseteq \alpha P \cap \alpha(C - \{P\})$ and $\alpha Q' \cap \alpha(C' - \{P'\}) = \varphi$. It is immediate that the alphabets of processes in (C' - {P'}) ∪ {Q'} are disjoint. This implies that if we can find a suitable change of symbol mapping on Q' we can regard D as a configuration„ of (C' - {P'}) ∪ {Q'}.

A mapping f(D, Q') on $\alpha Q'$ with the required properties can be defined by

f(D, Q')(a) = f(D, P')(a) if a ∈ $\alpha P'$, f(D, Q')(a) = a otherwise.

If we set Q = f(D, Q')(Q') it follows that $\alpha Q - \alpha P = \alpha Q' - \alpha P'$.

Since Q' refines P' it must be the case that Q refines P. We also notice that

$\alpha Q \cap \alpha(C - \{P\})$

$= \quad [(\alpha Q - \alpha P) \cap \alpha(C - \{P\})] \cup [\alpha P \cap \alpha(C - \{P\})]$

$\subseteq \quad (\alpha Q' \cap \alpha C) \cup (\alpha P \cap \alpha(C - \{P\}))$

$= \quad \alpha P \cap \alpha(C - \{P\}).$

It follows from Theorem 1 that

$Q \parallel (\parallel \{ X \mid X \in C - \{P\} \})$

refines $P \parallel (\parallel \{ X \mid X \in C - \{P\} \}).$

Rearranging these two expressions establishes that

$\parallel \{ X \mid X \in (C - \{P\}) \cup \{Q\} \}$ refines $\parallel \{ X \mid X \in C\}.$

Since the alphabet of each process is D, the result follows.

Proof of Theorem 3

The proof depends on the following observation: the definition of refinement implies that whenever $A \subseteq \alpha Q$ and Q possesses no unbounded traces in A^*, then Q refines Q\A. (This is a straightforward consequence of the definition and [4, Chapter 3.9 D14].)

Suppose that $T = \parallel \{X \mid X \in C\}$ refines S, and that U is a process in C which can be written $U = (V \parallel W) \setminus \alpha V \cap \alpha W$. Thus $V \parallel W$ refines U, and Theorem 2 ensures that it can be substituted for U in C. In other words,
$R = \parallel \{ X \mid X \in (C - \{U\}) \cup \{V \parallel W\} \} = \parallel \{ X \mid X \in (C - \{U\}) \cup \{V, W\} \}$
refines T. The transitivity of refinement ensures that R refines S as required.

By definition $\alpha V \cap \alpha W \cup \alpha C = \varphi$, where $\alpha C = \cup \{ \alpha X \mid X \in C\}$ and there exists (by the configuration law) a configuration D and a collection of processes C' with disjoint alphabets such that $C = \{ f(D, X)(X) \mid X \in C' \}$. It remains to show that $\alpha T \subseteq \alpha R$, and to do this we must find a configuration underlying R and containing D as a subset.

Let U' be the process in C' corresponding to U. So $U = f(D, U')(U')$. Define change of symbol mappings g^{-1} and h^{-1} on V and W as follows:

for each $a \in \alpha V \cap \alpha U$, $g^{-1}(a) = f^{-1}(D, U')(a)$;

for each $b \in \alpha W \cap \alpha U$, $h^{-1}(b) = f^{-1}(D, U')(b)$;

for each $c \in \alpha V \cap \alpha W$, $g^{-1}(c) = h^{-1}(c) = c.$

Set $V' = g^{-1}(V)$ and $W' = h^{-1}(W)$. Then $U' = V' \parallel W' \setminus \alpha V' \cap \alpha W'.$

Although $\alpha V'$ and $\alpha W'$ have no events in common with $\alpha(C' - \{U'\})$, $\alpha V' \cap \alpha W'$ is non-empty. The next step is therefore to find a configuration E and processes V" and W" such that $V' \parallel W' = V" \parallel_E W"$. Thus the alphabets of V" and W" are necessarily disjoint and the associated change of symbol mappings are f(E, V") on $\alpha V"$ and f(E, W") on $\alpha W"$. Clearly $D \cap E = \varphi$, D is subdiagonal in $\alpha C' \times \alpha C'$ and E is subdiagonal in $\alpha V" \times \alpha W"$. Thus $D \cup E$ must be subdiagonal in $\alpha((C' - \{U'\}) \cup \{V", W"\}) \times \alpha((C' - \{U'\}) \cup \{V", W"\}).$

We now define a change of symbol mapping related to D ∪ E for each process X in C' − {U'} ∪ {V", W"} as follows:

f(D ∪ E, V") = g(f(E, V"))

f(D ∪ E, W") = h(f(E, W"))

f(D ∪ E, X) = f(D, X) for X ε C' − {U'}.

It is immediate that f(D ∪ E, V")(V") = g(V') = V and

$$f(D ∪ E, W")(W") = h(W') = W.$$

Consequently

$$\|_{D ∪ E} \{ X \mid X ε (C' − \{U'\}) ∪ \{V", W"\} \}$$

$$= \| \{ f(D ∪ E, X)(X) \mid X ε (C' − \{U'\}) ∪ \{V", W"\} \}$$

$$= \| \{ f(D, X)(X) \mid X ε C' − \{U'\} \} \| (V\|W)$$

$$= \| \{ X \mid X ε C − \{U\} ∪ \{V, W\} \}.$$

$$= R$$

and αR = D ∪ E ⊇ D = αT.

GLOSSARY OF SYMBOLS

The paper adheres to the set–theoretic and CSP notation set out in Hoare's book [4] with the following alterations and additions:

φ	the empty set
tr P	set of traces of process P
$\parallel \{ X \mid X \in C \}$	concurrent composition of processes in a set C
$\parallel_D \{ X \mid X \in C \}$	concurrent composition according to a configuration D, where C is a set of processes with disjoint alphabets
αC	$\cup \{ \alpha X \mid X \in C \}$ for a set of processes C
$T \upharpoonright A$	$\{ t \upharpoonright A \mid t \in T \}$ for a set of traces T and a set of events A

The Design and Development of Ada Real-Time Embedded Systems

Robert G Clark
Department of Computing Science
University of Stirling
Scotland FK9 4LA

Abstract

This paper describes how a combination of CSP and the *me too* method of software design can be used to formalise the early stages of the object-oriented development of embedded systems including those with time constraints. Using the example of the watchdog timer, we show how a specification can be developed, exercised in a prototyping environment and then transformed into an outline concurrent Ada program.

Introduction

In this paper we describe an approach to the design and development of embedded real-time systems which is based on a combination of the *me too* method of software design [1,2,3] and CSP [4]. The end product is an Ada program, but the approach can be applied to any language which supports the notion of *communicating concurrent objects*.

The method is an example of the *operational approach* to software development [5] where an executable specification of a complete system is constructed in terms of *problem-oriented* structures. An implementation of the system is then created by transforming the *problem-oriented* structures into *implementation-oriented* structures while preserving the system's external behaviour.

The main criticism of the operational approach is that an *executable specification* has many of the characteristics of a high-level design and premature design decisions may be taken. Although this can be a danger, we agree with Swartout and Balzar [6] that specification and design are intertwined and one cannot be considered in the absence of the other.

A major advantage of the operational approach is that it enables a prototype of the intended system to be produced at an early stage in the project. This prototype can be demonstrated to the clients who drew up the original requirements and can be used as a basis for discussion as to whether or not the specification describes the intended system.

The *me too* Method

The *me too* method of software design was developed to deal with sequential systems. Like VDM [7], it is model-oriented although it is simpler and leads to executable specifications. It has three steps; the model step, the specify step and the prototype step.

In the model step, an informal model of a real-world situation is built by identifying abstract objects and their associated operations. Although the functionality of each of the operations is given at this stage, the effect of the operations is described informally in English.

Me too is both a method and a purely functional language [8]. In the specify step, each object in the model is given a formal representation in terms of sets, sequences, relations, maps, records and tuples. Operations on each object are defined constructively in the *me too* language in terms of the mathematical operations available on the object's formal representation. This gives a simple and clearly defined semantics for the objects and their operations. Each *me too* object definition corresponds to an *abstract data type* or, in Smalltalk terminology [9], an object *class*.

As the *me too* language is executable, a specification written in *me too* is executable and can be exercised as a prototype of the system being specified. Feedback from the prototype step can then be used to improve the original informal model either by correcting errors, by remedying omissions or by exploring alternative designs.

Specification Modules

Objects in a *me too* specification typically form a hierarchy with higher level objects having lower level objects as components. The association of object definitions with modules forms the basis of our extension to the *me too* method so that it can deal with the specification and design of large software systems [10]. Each module has a clearly defined interface which shows the information imported and exported by the module.

The syntax of a specification module is

specification of ⟨module name⟩
imports from ⟨module name list⟩
objects
    ⟨object names and visible representations⟩
exported operations
    ⟨functionality of visible operations⟩
object representations
    ⟨hidden object representations⟩
operations
    ⟨functionality of internal operations⟩
definitions
    ⟨operation definitions⟩
end ⟨module name⟩

with each of the component parts being optional.

The importance of specification modules is that they support information hiding, reusability, incremental software development, the division of labour and the localisation of design and code changes which result from modifications to a system's design or specification.

Example Specification Modules

Let us now look at a simple example. Suppose that we want to catalogue the books in our personal library. If we describe a book by its title and authors, we get the following *me too* definition for a book.

specification of books
objects
    title = seq(atom)

```
      author = seq(atom)
      book
exported operations
      mk-book : title × set(author) −> book
      the-title : book −> title
      authors : book −> set(author)
object representations
      book = tuple(title, set(author))
definitions
      mk-book(ti, aus) == (ti, aus);
      the-title(b) == first(b);
      authors(b) == second(b);
end books
```

Both the name and the representation of *title* and *author* are exported by this module, but while the name *book* is exported, its representation is hidden. We can only access *book* objects through the exported operations.

In the *me too* language the symbols { } are used in set construction, [ ] in sequence construction and ( ) in the construction of tuples. Function application also uses round brackets. Elements in a tuple are selected by the functions first, second etc, while the set operations $\cup$ and $\in$ are written as **union** and **member**. In this paper we have written operators in bold, but they are not typed in any special way in the version presented to the computer.

The definition of the catalogue uses the information exported by the *books* module.

```
specification of catalogue
imports from books
objects
      catalog
exported operations
      emptycat : −> catalog
      addbook : catalog × book −> catalog
      titlesby : catalog × author −> set(title)
      incat : catalog × title −> Boolean
object representations
      catalog = set(book)
definitions
      emptycat() == {};
      addbook(cat, b) == cat union {b};
      titlesby(cat, au) == {the-title(bk) | bk <− cat; au member authors(bk)};
      incat(cat, ti) == ti member {the-title(bk) | bk <− cat};
end catalogue
```

Other *catalog* operations would normally be declared, but they have been omitted for reasons of space. The expression

$$\{c \mid a <− s; b\}$$

in which c will be an expression involving a, should be read as "the set of c such that a is drawn from the set s and the expression b is true".

Exercising the Prototype

The characteristics of the overall state of a sequential system, together with operations which act on the state, can be described by an abstract data type which we shall refer to as *state-obj*. Exercising a *me too* prototype then involves the user invoking operations with the functionality

$$\text{state-obj} \times p_1 \times p_2 \times \ldots p_n \rightarrow \text{state-obj}$$

with each invocation causing a transition from an old to a new *instance* of *state-obj*. When, for example, the state object is *catalog*, each invocation of *addbook* will cause a transition from an old to a new instance of *catalog*.

Concurrent Systems

A concurrent system can be considered to be a set of sequential systems (processes) which communicate with one another. An alternative view is that a concurrent system is a set of objects in the Smalltalk sense (ie object instances) which may exist in parallel and which communicate with one another by passing messages. Each object instance has a local state together with operations which act on that state.

Our approach is a combination of these two views [11,12]. A *me too* description of a concurrent system is constructed from *me too* descriptions of each of the constituent sequential systems. It has an overall state object which has the state objects of each of the sequential systems as a component. Operations on the overall state object (*state-obj*) are defined to model both external operations (ie operations initiated by the system environment) and interprocess communication operations.

In a system of communicating objects, we must be able to show how the objects (ie the processes) interact with one another. As the language CSP [4] has been widely used to describe communicating sequential processes, we use a subset of CSP to show process interaction. *Me too* operations on the overall state correspond to *events* in CSP while each sequential *me too* state object corresponds to a CSP *process*. Communication is synchronised and is point-to-point, ie only two processes are involved in a communication. If asynchronous communication is required, a buffer process must be inserted between the two processes.

We therefore view a concurrent system as a set of communicating objects which engage in *events*. A similar combination of CSP and *me too* has been used in the specification of human-computer interaction [13].

The Model Step

In the model step, we identify abstract objects and their associated operations and describe them informally as before. This is done for each of the component sequential systems. We also identify which operations initiate events and which operations are either called by other objects or called from the environment.

When two processes A and B synchronise by sharing in an event, one process (say A) initiates the interaction while the other (B) is the passive responder. A pair of *me too* operations corresponds to each synchronisation event. The initiating process (object) has an operation which calls an operation exported by the responding process. If the name of the operation in the initiating process A is *op*, then the name of the operation exported by process B will be *B-op*. In the CSP description, the name of the event on which processes A and B synchronise is called *op*.

The Specify Step

The specify step is in two parts. In the CSP part, we construct a CSP description of the overall system. The purpose of the CSP description is to show how the *me too* operations interact with one another.

The *concurrent* specification module which describes the overall state object has *inner specification modules* to describe each of the sequential state objects. Each inner specification module has the same syntax as an ordinary specification module. As *me too* is a functional language, the overall state will commonly be passed as a parameter and returned as the result of a function call. Modularity is maintained by only allowing a component of the overall state object (ie a sequential state object) to be interrogated or modified by operations defined in its own inner specification module.

Me too operations which correspond to CSP events may have Boolean guards. Such operations will only take place when they are scheduled in the CSP description and when the guard evaluates to true. Boolean guards are not part of the standard *me too* method.

CSP/*me too* Prototyping System

A joint CSP/*me too* prototyping system has been developed from a CSP simulator devised by Peter Henderson [14]. The prototyping system uses the CSP description to control the order in which events may occur while the *me too* description is used to describe the changes which will occur in the system state when the operation corresponding to an event is executed. The user exercising the prototype acts as the system scheduler. The complete system is implemented in *me too*.

As we are interested in modelling real-time systems, the prototyping system supports a real time clock and the abstract data types *duration* and *time*. The function *cu-time* returns the current time while *at-or-past(ti)* returns true when *ti* is at or past the current time, otherwise it returns false.

Estimated "worst-case" execution times may be associated with operations and processes may be allocated to different processors. This enables a prototype to be used to examine the circumstances under which a system can meet imposed time constraints and to determine what will happen when it does not.

We now look at the different stages in more detail through a simple example.

The Watchdog Timer

The example which we shall use is that of a *watchdog timer* [15]. The function of a *watchdog timer* is to monitor a process, which we shall call MP, running on another processor. Once the *watchdog timer* has been enabled, it must be reset periodically by a signal from the *watchdog timer handler* process which runs on the same processor as MP. The *watchdog timer handler* initially enables the *watchdog timer* with a fixed expiry time *t* and subsequently resets the timer within an interval of less than that time. If the timer is not reset within the predetermined time due, for example, to not having been sent a reset signal, the *watchdog timer* expires and sends a halt signal to the processor running MP.

There are two main reasons why a reset signal may not be sent in time. One is that there is a malfunction and the other is that MP is putting such a high load on the system that the *watchdog timer handler* process cannot be scheduled. The purpose of the watchdog timer is therefore to detect either an error or a system overload and to close down the monitored system before too much harm is done. The watchdog timer system is shown in Figure 1.

The Development Method

We are primarily interested in developing software for the *watchdog timer handler* and the monitored process. Embedded systems do not however exist in isolation, but within an environment of hardware, other software and users. The CSP/*me too* approach enables us to build a model of a complete system, not just the embedded software.

In the *watchdog* example, the *watchdog timer* can be considered to be part of the environment in which the handler software operates while the monitored process is likely to interact with the environment and this interaction can be described by one or more environment processes.

It is up to the system specifier to decide which, if any, of the environment processes are to be explicitly modelled. It is important to build a model of the *watchdog timer* process, but, as no details have been given about the monitored process, its interaction with the environment cannot usefully be described at this stage.

The Model Step

We identify three main objects (processes); the *watchdog timer* (*wd*), the *watchdog timer handler* (*wdh*) and the monitored process (*mp*). We consider each in turn.

There are three operations which can affect the state of the *watchdog timer* plus a synchronisation operation (*halt*) which is initiated by the *watchdog timer*. They are given below with their functionality and an informal description of their effect.

> wd-enable : duration × state-obj −> state-obj
>> On receipt of an *enable* signal, enable the *watchdog timer* with a given expiry interval, ie record the time within which it must be reset.
>
> wd-reset : state-obj −> state-obj
>> On receipt of a *reset* signal, reset the *watchdog timer*, ie update the time within which it must next be reset.
>
> time-out : state-obj −> state-obj
>> Initiate the sending of a *halt* signal to the monitored process if some fixed time has elapsed since the *watchdog timer* was last reset.
>
> halt : state-obj −> state-obj
>> Send a *halt* signal to the monitored process.

The *watchdog timer handler* has two operations which affect its state plus two synchronisation operations which it initiates.

> wdh-start : duration × duration × state-obj −> state-obj
>> On receipt of a *start* signal, initialise the *watchdog timer handler* with its wake-up interval and with the *watchdog timer* expiry interval and then initiate the sending of an *enable* signal to the *watchdog timer*.
>
> wake-wdh : state-obj −> state-obj
>> An operation which, after some fixed time has elapsed, wakes the *watchdog timer handler* so that it can send a *reset* signal to the *watchdog timer*.
>
> enable : state-obj −> state-obj

Send an *enable* signal to the *watchdog timer*.
reset : state-obj —> state-obj
　　　Send a *reset* signal to the *watchdog timer*.

To keep the system as simple as possible we have given no details of the system being monitored. It can be thought of as having two operations.

mp-work : state-obj —> state-obj
　　　Carry out the next stage in the work of the monitored process.
mp-halt : state-obj —> state-obj
　　　On receipt of a *halt* signal, halt the monitored process.

Operations *wdh-start* and *mp-work* are external operations called by the environment. If the model was extended to include objects which describe the interaction of *mp* and *wdh* with the environment, these operations would correspond to synchronisation events. As we shall see later, both kinds of operation lead to the generation of the same outline Ada code.

The CSP Specify Step

A CSP description is given for each of the objects (processes) identified in the previous stage. It shows the order in which the events take place and how two processes can synchronise with one another by engaging in the same event.
　　　Our model of the watchdog timer system has three processes running in parallel.

WATCH = WDH ∥ WD ∥ MP

Once process WDH has been initiated it synchronises with WD on an *enable* event and then moves to state WDH-1 where it waits to be woken up before again synchronising with WD, this time on a *reset* event. It then returns to state WDH-1.

WDH = (wdh-start —> (enable —> WDH-1))
WDH-1 = (wake-wdh —> (reset —> WDH-1))

After being enabled, WD moves to state WD-1 and waits for either a *reset* or a *time-out* event to occur. If WD synchronises with WDH on a *reset* event, WD returns to state WD-1 while, if a *time-out* event occurs, a *halt* signal is sent to terminate MP. Process WD then terminates.

WD = (enable —> WD-1)
WD-1 = ((reset —> WD-1) [] (time-out —> (halt —> SKIP)))

Process MP continues to take part in event *mp-work* until it is interrupted by a *halt* event.

MP = (mp-work —> MP) ↑ (halt —> SKIP)

The *me too* Specify Step

The three *me too* objects (*wdh*, *wd* and *mp*) are now given a formal representation. The *watchdog timer handler* state has three components; the *watchdog timer* expiry interval, the interval at which the handler should be woken up and the time after which *wdh* may next be woken up. The *watchdog timer* state is the time after which a *time-out* event may occur together with the expiry interval. The state of the monitored process can either have the value RUNNING or the value HALTED.

A *me too* specification of a concurrent system consists of a single *concurrent specification module* together with any ordinary specification modules which may be needed to define abstract data types used by the system. The concurrent specification module defines the overall state object which is represented by a record with a component from each of the sequential systems. Each sequential system is then described by an *inner* specification module declared within the concurrent specification module.

If the sequential state objects are $a, b, c, \ldots$ then the overall state object is defined as

$$\text{state-obj} = \text{record}(\text{a-st} : \text{a},$$
$$\text{b-st} : \text{b},$$
$$\text{c-st} : \text{c}, \ldots)$$

The definition of this record in *me too* causes the automatic generation of functions to access components (*a-st, b-st, c-st, ...*) and to update components (*upd-a-st, upd-b-st, upd-c-st, ...*) of the record. Parameterless operations *init-a, init-b, init-c, ...* must be defined in the inner specification modules to initialise each of the components of the overall state object.

Although the overall state object is available globally, information hiding is maintained by only allowing operations such as *a-st* and *upd-a-st* to be used within the inner specification module of object *a*. Operations defined in an inner specification module can therefore only directly access or change that module's component of the overall state. To access or change components of another module, operations exported by the other module must be called. Communication between objects is therefore by message passing. Two objects cannot share state information.

A pair of *me too* operations corresponds to each synchronisation event. For example, corresponding to the CSP event *reset* we have the operations *reset* in WDH and *wd-reset* in WD. The operation in the initiating process (*reset* in WDH) calls the operation (*wd-reset*) exported by the responding process WD.

Operations which are called by the environment or from another object are listed under the heading **exported operations**. These operations always have the object name as a prefix. Other operations, including all those with guards which involve time, are listed under the heading **operations**.

A *me too* description deals with time, but not with the passage of time, ie it describes a snapshot of the system. The current time is obtained by calling the constant function *cu-time*. Different snapshots may then be obtained by redefining *cu-time*. In the *me too* description of the watchdog system given below, it should be noted that the Boolean guards ensure that *wake-wdh* and *time-out* cannot be scheduled until the current time has reached some pre-determined value.

concurrent specification of WATCH
objects
$$\text{state-obj} = \text{record}(\text{mp-st} : \text{mp},$$
$$\text{wdh-st} : \text{wdh},$$
$$\text{wd-st} : \text{wd})$$

specification of MP
objects

```
      mp
exported operations
   mp-work : state-obj -> state-obj
   mp-halt : state-obj -> state-obj
object representations
   mp = {"RUNNING", "HALTED"}
operations
   init-mp : -> mp
definitions
   init-mp() == "HALTED";
   mp-work(st) == upd-mp-st(st, "RUNNING");
   mp-halt(st) == upd-mp-st(st, "HALTED");
end MP
```

specification of WDH
imports from WD
objects
   wdh
exported operations
   wdh-start : duration × duration × state-obj -> state-obj
object representations
   wdh = tuple(duration, duration, time)
operations
   init-wdh : -> wdh
   wake-wdh : state-obj -> state-obj
   enable : state-obj -> state-obj
   reset : state-obj -> state-obj
definitions

```
   init-wdh() == (0, 0, cu-time());
   wdh-start(wd-intrvl, wdh-intrvl, st) ==
            upd-wdh-st(st, (wd-intrvl, wdh-intrvl, cu-time() + wdh-intrvl));
   wake-wdh(st) when at-or-past(third(wdh-st(st))) ==
            let wd-intrvl == first(wdh-st(st)),
                wdh-intrvl == second(wdh-st(st)) in
            upd-wdh-st(st, (wd-intrvl, wdh-intrvl, cu-time() + wdh-intrvl));
   enable(st) == let wd-intrvl == first(wdh-st(st)) in
            wd-enable(wd-intrvl, st);
   reset(st) == wd-reset(st);
end WDH
```

specification of WD
imports from MP
objects
   wd
exported operations
   wd-enable : duration × state-obj -> state-obj
   wd-reset : state-obj -> state-obj

object representations
   wd = tuple(duration, time)
operations
   init-wd : $->$ wd
   time-out : state-obj $->$ state-obj
   halt : state-obj $->$ state-obj
definitions
   init-wd() == (0, cu-time());
   wd-enable(tim, st) == upd-wd-st(st, (tim, cu-time() + tim));
   wd-reset(st) == **let** expiry-intrvl == first(wd-st(st)) **in**
              upd-wd-st(st, (expiry-intrvl, cu-time() + expiry-intrvl));
   time-out(st) **when** at-or-past(second(wd-st(st)))
              == st;
   halt(st) == mp-halt(st);
end WD
end WATCH

Local definitions are introduced by the reserved word let. The effect of

   let n_1 == e_1, n_2 == e_2, ... n_k == e_k **in**
     expression

is to return the value of *expression* evaluated in a context enriched by binding the names n_i to values e_i. Boolean guards are prefixed by the reserved word **when**. An operation with a Boolean guard can only be scheduled when the expression following **when** evaluates to true.

Exercise the CSP/*me too* Prototype

Prototyping is an important part of the *me too* method. Exercising a CSP/*me too* prototype consists of performing a series of state $->$ state transitions. These are initiated by the human user calling an operation which corresponds to one of the CSP events. In a concurrent system, events associated with different processes may be interleaved and the exact order in which they occur can affect the result. To model this, a CSP description splits each process cycle into separate parts with the CSP events indicating where a process may interact with another process or where suspension of the process and the scheduling of another process could affect the result.

    Each CSP event corresponds to a *me too* operation. The *me too* operations that are available are determined by a combination of the CSP description and the values of the guards.

    The prototyping system maintains a real time clock which is initialised to zero. The current time can be determined by calling the function *cu-time*. Time may be advanced between transitions which has the effect of redefining *cu-time*.

    The prototyping system can operate in either manual or automatic mode. Manual mode is used before execution times have been associated with operations or processes have been allocated to processors and is described below. In automatic mode the prototyping system only requests user intervention when a choice has to be made or when the *me too* operation requires a parameter.

At each stage in exercising the prototype in manual mode the user is shown the current system state together with a list of the operations which may be scheduled, ie a list of the events which may be scheduled according to the CSP description and which correspond to *me too* operations with true (ie open) guards. For example, at the start of the *watchdog* system we will get the message

```
current state:
    mp -> HALTED
    wdh -> ( 0 0 0 )
    wd -> ( 0 0 )
time = 0:     pick one of { wdh-start mp-work }
```

At each stage we chose which event is to be scheduled and give any necessary parameters excluding the system state which is supplied automatically. If we respond with

```
(wdh-start 300 200)
```

we will get the response

```
current state:
    mp -> HALTED
    wdh -> ( 300 200 200 )
    wd -> ( 0 0 )
time = 0:     pick one of { enable mp-work }
```

Event *enable* is a synchronisation event and if it is chosen we step through both the WD and WDH processes and get the response

```
current state:
    mp -> HALTED
    wdh -> ( 300 200 200 )
    wd -> ( 300 300)
time = 0:     next event is mp-work
```

Events *wake-wdh* and *time-out* cannot be offered because they have closed guards. Event *reset* is not offered as it is a synchronisation event and we have not reached the correct position in WDH. If the real time clock is advanced by 200 or more time units then the guard for *wake-wdh* will become open. The real time clock is advanced by responding with a number instead of an event name. If we type 105 then 105 time units are added to the real time clock and we get the message

```
time = 105:     next event is mp-work
```

If the real time clock is then advanced by a further 100 time units, the set of available events will change. The system will respond with

```
time = 205:     pick one of { wake-wdh mp-work }
```

If the *wake-wdh* event is now chosen, the system will be ready to offer a *reset* synchronisation event.

```
current state:
      mp -> HALTED
      wdh -> ( 300 200 405 )
      wd -> ( 300 300 )
 time = 205:      pick one of { reset mp-work }
```

The time after which the next *wake-wdh* event can occur has been advanced by 200 time units from the current time.

Using this version of the prototype, we can explore what will happen when the events are scheduled in various orders including, for example, what will happen if a *reset* event is not scheduled in time and a *time-out* event occurs.

Exercising the prototype is likely to show up shortcomings in the CSP and *me too* descriptions. This requires a return to stage one and a complete iteration through the different stages.

Refinement

An initial design of a large system is likely to have omitted a lot of detail. Once we are satisfied with an initial CSP/*me too* prototype, we are in a position to produce a more detailed design. The detailed design should be a refinement of the earlier one. We could, for example, extend the watchdog example by giving details of the process being monitored [12]. The *watchdog timer* and the *watchdog timer handler* parts of the system can of course be re-used with different monitored systems.

The CSP/*me too* approach allows the components, or groups of components, in a large system to be prototyped independently before being integrated into a complete system. In such a prototype, each group of components will have the rest of the system defined as part of its environment.

If, when we consider the system in more detail or integrate system components, we find that the existing design cannot be preserved, then we must return and modify the high level design. The final result will be a series of designs, each more detailed than the previous one. They will all describe exactly the same overall architecture and we will be able to trace features in a detailed design back to where the design decisions were made in a higher level design. Each of the designs will have been prototyped.

Create Outline Tasking Model

Once we are satisfied that a prototype is a suitable model of the intended system, it can be used as the starting point for the design of an efficient implementation in an imperative language. It is proposed that a practical method of producing reliable Ada programs is to start with a *me too* specification which has already been tested in a prototyping environment. This approach can be viewed as formalising the early stages in the method of object-oriented design which is widely used in the Ada community [16] for the design of both sequential and concurrent systems.

A functional specification with a global state is to be converted into a series of packages, each of which encapsulates a local hidden state instance. In place of a local state object defined in an inner specification module, a type and an instance of the type are declared within each package body. The object instances are the local hidden states of each of the communicating objects. If the local state object is *a*, the instance will be called *a_st* and references to *a-st(st)* in the *me too* definition are replaced by references to *a_st*. Each operation exported by an inner specification module becomes an operation which is exported by the corresponding Ada package.

Concurrency is achieved by the package body encapsulating a task. The operations exported by the Ada package act as a procedural interface to the hidden task entries. The reasons for using a procedural interface rather than accessing tasks directly are summarised by Welch [17]. If the interface procedures are declared to be *inline*, there will be no run-time overhead.

The resulting Ada programs have no shared variables as all process communication is by message passing. Tasks may not be nested and as each task is encapsulated in a package each process is implemented as a self-contained module. The approach therefore rules out some of the more dangerous Ada constructs.

Transformation rules have been developed [12] and implemented in a translator [18] which takes a CSP/*me too* specification and produces a set of outline Ada packages. The *me too* part of the specification determines the object representations and the functionality of the operations while the structure of the task body is controlled by the CSP description.

The definition of a *me too* operation remains as a comment in an Ada function or accept body and acts as a specification of the code which must then be implemented manually. As we have a set of generic packages [19] which implement sets, sequences, relations and maps, producing an Ada prototype is straightforward. We take the view that human beings are good at writing ten line function bodies. It is with the organisation of large software systems that they need help. The next step is reification [7] where the Ada prototype is converted into an efficient Ada program. This will have the same structure as the prototype, but the type representations and the implementation of the function bodies may be changed. A similar approach can be used in the transformation of VDM specifications of sequential systems into Ada [20,21].

The outline Ada code produced from object WDH is shown below.

```
with wd;
package wdh is
    procedure start(wd_intrvl : duration; wdh_intrvl : duration);
    pragma inline(start);
end wdh;
package body wdh is
    type wdh is
        record
            first : duration; second : duration; third : time;
        end record;
    function init_wdh return wdh;
    task tt is
        entry start(wd_intrvl : duration; wdh_intrvl : duration);
    end tt;
    task body tt is
        wdh_st : wdh := init_wdh;
    begin
        accept start(wd_intrvl : duration; wdh_intrvl : duration) do
            --wdh_st' <= (wd_intrvl, wdh_intrvl, cu_time() + wdh_intrvl)
```

```
      end start;
      −−let wd_intrvl == first(wdh_st) in
            −− wd.enable(wd_intrvl);
      loop
         −− delay third(wdh_st) − clock;
            −−wdh_st' <= let wd_intrvl == first(wdh_st),
                −−wdh_intrvl == second(wdh_st) in
            −− (wd_intrvl, wdh_intrvl, cu_time() + wdh_intrvl).
         −−wd.reset;
      end loop;
   end tt;
   procedure start(wd_intrvl : duration; wdh_intrvl : duration) is
   begin
      tt.start(wd_intrvl, wdh_intrvl);
   end start;
   function init_wdh return wdh is
   begin
      −− return (0, 0, cu_time())
   end init_wdh;
end wdh;
```

It is normally the case that the requirements of a system will continue to evolve once the system is in use and that the costs of maintenance will outweigh the initial development costs. Any change in requirements should be reflected by a change in the specification followed by a re-iteration of the development cycle. When a system has been developed using the CSP/*me too* method, this will not require a re-implementation of the complete system for, as the specification consists of a set of self-contained specification modules, the effect of changes will be localised.

By continuously updating the specification we will always have a single up to date document which describes the current system no matter how many modifications may have been made to the original requirements.

Related Work

Several other systems have been, or are being, developed to specify embedded systems although they do not all deal with both data structuring and process interaction. Three which do are the PAISLey system [22] which, like the CSP/*me too* approach uses a functional language to describe computation within a process, the RAISE project [23] which is developing a new language by combining VDM and CSP, and the approach proposed by Nielson and Shumate [24] which, like CSP/*me too* uses a combination of processes and object-oriented design. We are applying the CSP/*me too* method to the problem described in [24] to compare the relative advantages of the two approaches.

Summary

We have described how *me too* and CSP can complement one another in the development of the design of a real-time embedded system. A concurrent system is considered to be a set of objects which communicate and synchronise with one another by sending messages. These interactions correspond to CSP events.

Once the objects and their interactions (ie messages, events or operations) have been identified, we give a *me too* description of the objects and their associated operations. The CSP and the *me too* descriptions give two complementary views of the overall system.

The combined CSP/*me too* description can be exercised as a prototype. We can assign execution times to events and their associated operations and allocate processes to processors. By prototyping such a system we can examine its behaviour under differing workloads. Once we are satisfied with its performance we can use transformation rules to automatically convert the functional prototype into an outline concurrent Ada program.

This research was carried out within ESPRIT Project 937, Descartes (Debugging and Specification of Ada Real-Time Embedded Systems).

References

[1] P. Henderson, "Functional Programming, Formal Specification and Rapid Prototyping". *IEEE Transactions on Software Engineering 12*, 241-250, 1986.

[2] P. Henderson and C. Minkowitz, "The *me too* method of software design". *ICL Technical Journal 5*, 64-95, 1986.

[3] H. Alexander and V. Jones, *Software Design and Prototyping Using me too*. Prentice-Hall, (*to be published*).

[4] C. A. R. Hoare, *Communicating Sequential Processes*, Prentice-Hall, 1985.

[5] P. Zave, "The Operational versus the Conventional Approach to Software Development". *CACM 27*, 104-118, 1984.

[6] W. Swartout and R. Balzer, "On the Inevitable Intertwining of Specification and Implementation". *CACM 25*, 438-440, 1982.

[7] C. B. Jones, *Systematic Software Development using VDM*. Prentice-Hall, 1986.

[8] S. M. Bennett, C. Minkowitz and J. S. Rowles, "*me too* Reference Manual".*STC Technology Ltd and Stirling University*, 1988.

[9] A. Goldberg and D. Robson, *Smalltalk-80: The Language and its Implementation*. Addison-Wesley, 1983.

[10] R. G. Clark, "Ada Programs from *me too* specifications". *Descartes Technical Report D3-3*, 1987.

[11] R. G. Clark, "Designing Concurrent Objects". *International Workshop on Real-Time Ada Issues, ACM Ada Letters VII no 6*, 107-109, 1987.

[12] R. G. Clark, "Extension to *me too* to Incorporate Real Time".*Descartes Technical Report D3-6-2*, 1988.

[13] H. Alexander, *Formally-Based Tools and Techniques for Human-Computer Dialogues*, Ellis Horwood, 1987.

[14] P. Henderson, *Private Communication.*

[15] P. Leo, E. Zijlstra and U. Zwart, "Paradigms of Real Time Systems". *Descartes Technical Report D5-1*, 1987.

[16] G. Booch, "Object-oriented Development". *IEEE Transactions on Software Engineering 12*, 211-221, 1986.

[17] P. H. Welch, "A Structured Technique for Concurrent Systems Design in Ada". *Proceedings of the Ada-Europe Conference 1986*, 261-272, Cambridge University Press, 1986.

[18] K. McPherson, *MSc Dissertation, University of Stirling*, 1988.

[19] R. G. Clark, "Report on the Set of Generic Packages". *Descartes Technical Report D3-1-2*, 1987.

[20] C. Chedgey, S. Kearney and H-J. Kugler, "Developing Ada Software Using VDM in an Object-Oriented Framework". *EUUG Conference Proceedings, Dublin* 41-58, 1987.

[21] M. I. Jackson, "Developing Ada Programs Using the Vienna Development Method (VDM)". *Software Practice and Experience 15*, 305-318, 1985.

[22] P. Zave and W. Schell, "Salient Features of an Executable Specification Language and its Environment". *IEEE Transactions on Software Engineering 12*, 312-325, 1986.

[23] E. Meiling and C. W. George, "The RAISE Language and Method". *Esprit '86: Results and Achievements*, 607-617, Elsevier, 1987.

[24] K. W. Nielsen and K. Shumate, "Designing Large Real-Time Systems with Ada". *CACM 30*, 695-715, 1987.

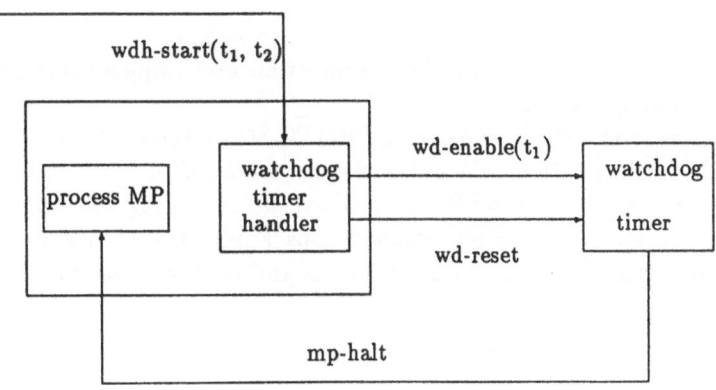

Figure 1: The watchdog timer

Protocol Analysis and Implementation using NPNs and SDL

Kenneth R. Parker   Rainer A. Berger
Kong E. Cheng

CSIRO Division of Information Technology
55 Barry Street, Carlton, Victoria 3053, Australia

May 1988

1   Introduction

Communications protocols are examples of practical concurrent systems. Formal Description Techniques (FDTs) are being used for the specification of communication protocols to achieve greater accuracy and efficiency in their design and implementation [22]. These specifications can be analysed to verify accuracy and can be used as a rigorous basis for implementation.

In this paper, we will discuss our experience using Numerical Petri Nets (NPNs) and the CCITT Specification and Description Language (SDL) to formally specify, design and implement protocols. NPNs are a type of high-level Petri net which contain inhibitor arcs [25,26,27,28]. SDL is a standardized language widely used for the specification and implementation of telecommunications protocols.

We will describe the PROTEAN and PROTGEN systems which aid the analysis of NPN specifications. We also discuss the MELBA system which automatically converts high level SDL specifications into programming language code for a variety of telecommunications applications. We discuss the advantages and limitations of our approach and outline areas for future research.

2   Numerical Petri Nets

Petri nets have been recognized as useful for representing concurrent systems, and communication protocols in particular [11]. NPNs are a type of Petri net which includes tokens with attributes and powerful input conditions, transition conditions and transition operations. This allows many transitions and places to be folded into one. Because of this higher information density, NPNs are more useful than Petri nets for practical communication protocols.

As well as the intrinsic properties of the NPN formalism, the success achieved in using NPNs has relied on the availability of analysis tools. In particular, the PROTEAN emulation and analysis system, developed by Telecom Australia [5,6], enables the automated analysis of NPN specifications. This has led to the specification and analysis of important protocols, such as the OSI Transport Protocol Class 0 [1,2,3].

3   The PROTEAN NPN Analysis System

The PROTEAN (PROTocol Emulation and ANalysis) system for Numerical Petri Nets (NPNs) aids in the analysis of NPN specifications. It has many different capabilities, including single stepping through NPN nets and automatically generating the reachability graph (RG). It can detect livelocks, deadlocks and cycles in the reachability graph.

PROTEAN can also focus on particular parts of the NPN which are of interest. For example, the LANGUAGE program reduces the reachability graph to show only those transitions the user has specified and the SCE-NARIO program highlights sequences of interest. The PATH program determines which nodes cannot reach a specified node. PROTEAN has previously been used to analyse the OSI Transport Protocol Class 0 [2].

The principal limitation of PROTEAN, and reachability analysis in general, is the occurrence of the well known state explosion problem. However this also limits many other approaches to the modelling of discrete systems. Our approach is to try to develop tools and techniques to reduce the impact of this problem.

4   NPN Manipulation by PROTGEN

Inspite of its usefulness, PROTEAN still has important deficiencies[1]. Firstly it requires the tedious input of NPNs in a precise textual form, which is error prone and time consuming, rather than allowing input in a graphical form. Another limitation is the restriction of PROTEAN to a subset of the NPN formalism.

The PROTGEN system [23] was developed to overcome these limitations of PROTEAN and to allow manipulation of NPN specifications. It allows direct graphical input of NPNs via the GENIE diagram editor [14,15], followed by the translation of the graphical specification into PROTEAN input format. Manipulation of an NPN is then performed—this includes the automatic expansion of high-level NPN constructs into a simpler form acceptable

---

[1]This section is a revised version of part of [24].

to PROTEAN. Finally, PROTGEN allows the automatic duplication (and interlinking) of a net for testing the communication between two entities.

The GENIE diagram editor was developed by CSIRO and Telecom Australia Research Laboratories to allow entry and editing of connected graphs, such as Petri Nets. It runs on a high resolution interactive graphics workstation. The system has many powerful facilities for editing diagrams, such as zooming in and out to focus on the desired level of detail. The NPN files from GENIE are easily converted to PostScript[2] and printed on a laser printer, producing diagrams with fine detail. We have tailored the GENIE "shapes" file to allow for a wide range of sizes for Transition boxes and other text.

Once a net has been entered and edited on GENIE, the file needs to be changed into PROTEAN format. The TRANSLATE program extracts all the information in the GENIE format file, including the graphical positioning information, text and comments, then writes out a new file which can be read by PROTEAN. The TRANSLATE program can easily be modified to reflect any changes to the PROTEAN format.

The output files created by TRANSLATE may still incorporate NPN constructs which PROTEAN cannot yet handle. The CONVERT program simplifies and expands NPN forms automatically—avoiding tedious and error prone manual conversion steps. This results in a change from the advanced NPN forms which are entered, into greatly expanded nets executable by PROTEAN. These expanded nets may be difficult to understand and create manually, simply because of their size and fine detail. This expansion process may be compared to the compilation of a high level programming language into a machine language.

To allow handling of the intricate mechanisms found in practical communications protocols the features catered for by CONVERT are numerous and varied. CONVERT provides the following facilities:

- A Macro expansion facility allows transition conditions or operations that are used several times to be defined only once. This decreases the size of the specification and makes it more readable and maintainable.

- IF statement constructs in a transition operation cause that transition to be expanded into multiple transitions. For example, a transition whose transition operation has n sequential IF statements is expanded into 2^n separate transitions. Nested IF statements are also handled correctly.

- "Wild card" token attributes (represented by underscores) are handled by introducing new consistent-substitution variables (which are local to one transition). Use of such attributes avoids the explicit introduction of unnecessary variables.

---

[2]PostScript is a trademark of Adobe Systems, Inc.

- CONVERT optionally creates a file of P-variables and place definitions, which are defined globally for the net. This allows multiple NPN net files to be conveniently loaded into PROTEAN.

- Modular arithmetic constructs are expanded by inserting extra transitions. Modular arithmetic is particularly useful for sequence numbers in protocol data units.

- CONVERT allows patterns to be used to specify related sets of tokens. These are expanded by the use of net loops. For example, in the Transport Protocol specification <ALL,J> stands for all tokens relating to the Jth Transport Connection (see example below).

- Arrays of P-variables are expanded by introducing a new place containing one token for each array element. Arrays are very useful for complex specifications, such as protocols that allow multiple connection instances.

We will now illustrate some of the manipulation performed by CONVERT using simplified versions of constructs which occur in the NPN specification of the OSI Transport Protocol. Figures 1 and 3 show the simplified transitions and Figures 2 and 4 show the corresponding expanded transitions resulting from processing by CONVERT.

The IF statement in Figure 1 is expanded as shown in Figure 2. IF statements in NPNs are based on Dijkstra's Guarded Command Language [12]. If more than one guard is true then one of the true guards should be chosen nondeterministically. This is the case in Figure 2, due to the nondeterministic nature of Petri Nets, if the upper two transitions are both enabled. The net is not well defined if none of the guards is true. The bottom transition in Figure 2 checks for this condition. In practice, the structure of the net guarantees that at least one guard is true whenever the transition is enabled (thus the bottom transition only serves to help debug the net).

The transition in Figure 3 requires the deletion of all tokens representing Network Service Data Units related to the Jth network connection, represented by <ALL,J>. Figure 4 shows the expanded transition and the following features of this expansion should be noted:

- The places in Figure 4 are "queueing places" in which tokens are ordered by the time of their arrival. "O=" is used to select the oldest token from the input place and "o:" is used to delete the oldest token.

- The variable LOOP is set so that the transitions of the expanded net will fire in sequence to completion, while other transitions cannot fire. The other transitions of the specification net have a transition condition of LOOP=0 automatically inserted, and are therefore disabled while the tokens related to the Jth network connection are being removed.

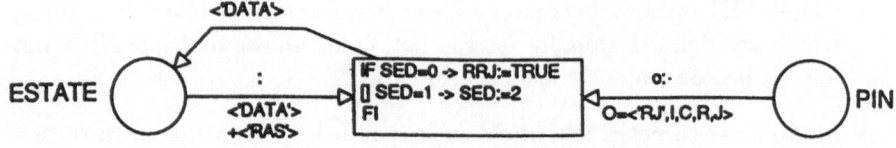

Figure 1: A simple IF-statement

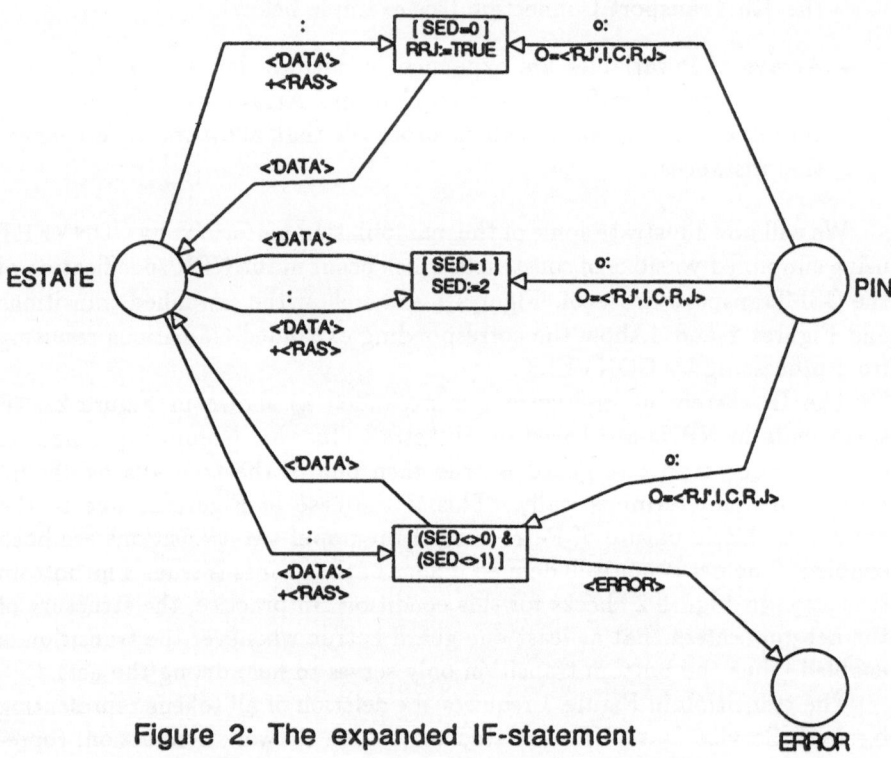

Figure 2: The expanded IF-statement

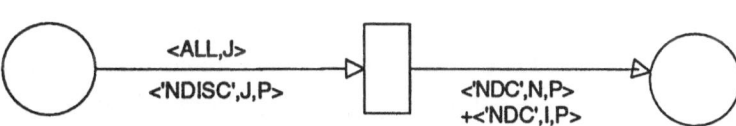

Figure 3: A simple transition

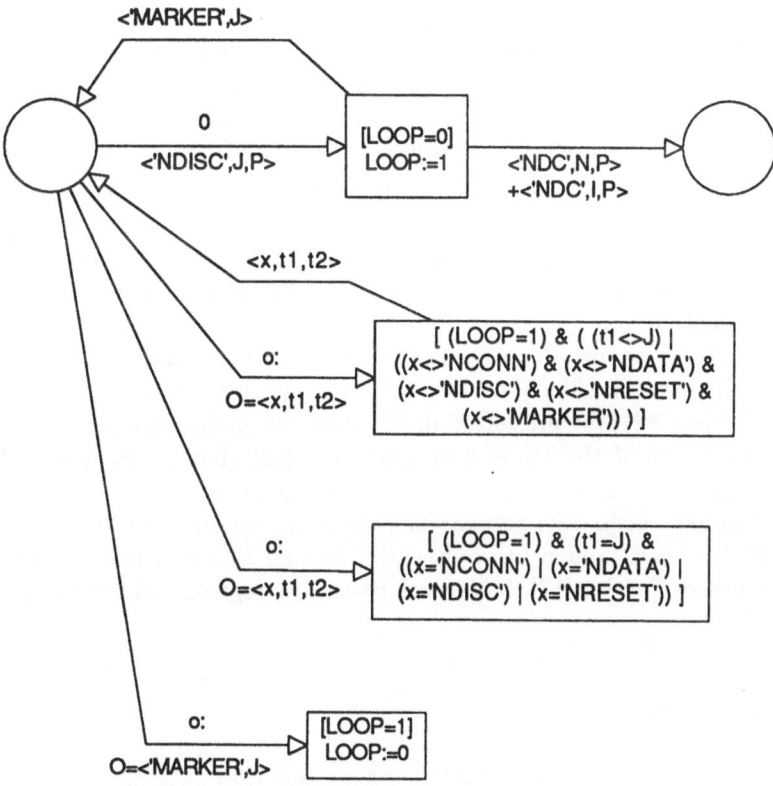

Figure 4: The expanded transition

- The attribute t2 in the tokens can represent a variable number of actual attributes. PROTEAN restricts each place to contain only one type of token. CONVERT automatically adds zeros, as extra attributes, to all the tokens where necessary, so that they all have the same number of attributes.

- The expansion is performed using a data file, which contains definitions of the patterns to be expanded and the structure of the replacement transitions.

For analysis of communication protocols it is essential to incorporate initiator and receiver copies of the protocol within the one net. This enables the analysis of realistic interchanges of primitives and data units to be simulated and analysed. PROTGEN incorporates the DUP program to automatically duplicate the net, with different labels for place names and transitions, and linking of the input and output queues for the two sides.

Used in conjunction, PROTEAN and PROTGEN constitute a powerful tool for the specification and analysis of communication protocols using Numerical Petri Nets. Other graphical interface packages have been developed since GENIE, which have advantages over it in some respects (e.g. Design[3]). PROTGEN can be adapted to use these simply by modifying the TRANSLATE program.

The OSI Transport Protocol Class 3 is being used as a test case for the PROTGEN system. The complete NPN specification of the OSI Transport Protocol Class 3 [18] is presented in [4]. PROTGEN has enabled us to enter each component of the Class 3 specification and then input it into PROTEAN. The general methodology of analysis used is the same as described in [2]. We have been able to analyse part of the specification for the occurrence of deadlocks and other behaviour, as well as verifying it against the OSI Transport Service [17]. This work is still in progress and will be reported more fully later.

5 MELBA

MELBA [10,20] is a software development system that automatically converts system specifications into compilable high level language code. The input specification is written in the CCITT Specification and Description Language (SDL) [8] and the code generated is in the CCITT High Level Language (CHILL) [9]. The MELBA system was developed at the Royal Melbourne Institute of Technology (RMIT) under contract to Telecom Australia.

SDL is a standard international language designed by the CCITT. SDL is based on flowcharts and state transition diagrams. It is intended for specifying and describing the SPC (Stored Program Control) logic of telecommunications applications, such as message and packet switching nodes and

---

[3]Design is a trademark of Meta Software, Cambridge MA, USA

telephone exchanges. In addition to providing a graphical means of specifying telecommunication software, SDL can be used as a software design and documentation tool. The graphical representation (SDL/GR) can be presented in an equivalent textual (program) representation form (SDL/PR). SDL provides three major functions for specifying and describing software systems:

- Structuring concepts

- Description of the behaviour of systems

- Specification of abstract data types

CHILL is a standardized programming language based on Pascal, PL/I and Algol-68. It was developed by CCITT for programming SPC telephone exchanges. However, it is general enough to be used for other applications. It provides extensive concurrency and modularity constructs with capabilities similar to the programming language Ada.

Since SDL and CHILL are both standard languages developed by CCITT, they are or will be familiar to the majority of telecommunications engineers. If current trends continue, system development and documentation will commonly be performed using SDL while the system itself will be programmed using CHILL. A system such as MELBA for automatically converting SDL specifications to CHILL code will therefore greatly improve software productivity and increase system reliability.

As the specification is presented in a readily comprehended graphical form, the system specification (with added comments) can become the documentation of the system. Since the code is generated directly from the specification, the documentation and system specification always match the final code produced. In this way, MELBA provides a coherent software development system which enables software development and documentation to proceed in parallel.

MELBA also reduces the maintenance cost of operational software. System modification is usually performed by entering design level changes, avoiding the need to understand and alter detailed code to produce a new system. Thus, MELBA is also suited to rapid prototyping by allowing the user to work at the design level, removing the tedious task of translating design specification into high level language code.

6   MELBA Methodology

The overall MELBA methodology [7] supports two main concepts: *decomposition* and *abstraction*. Decomposition provides for the partitioning of a design, and results in a hierarchical structure of modules, with well-defined interfaces between modules at different or the same levels (see Figure 5).

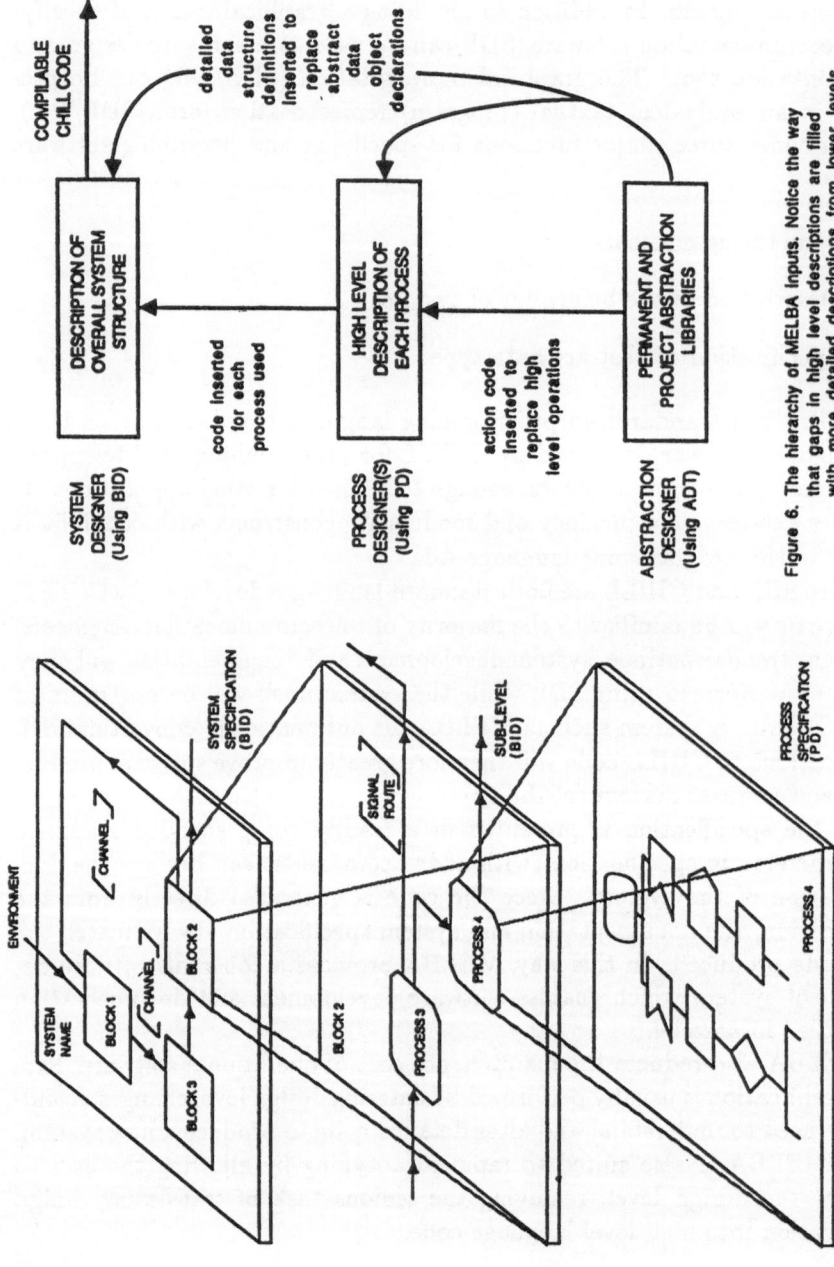

COMPILABLE CHILL CODE

detailed data structure definitions inserted to replace abstract data object declarations

DESCRIPTION OF OVERALL SYSTEM STRUCTURE

SYSTEM DESIGNER (Using BID)

code inserted for each process used

HIGH LEVEL DESCRIPTION OF EACH PROCESS

PROCESS DESIGNER(S) (Using PD)

action code inserted to replace high level operations

PERMANENT AND PROJECT ABSTRACTION LIBRARIES

ABSTRACTION DESIGNER (Using ADT)

Figure 6. The hierarchy of MELBA inputs. Notice the way that gaps in high level descriptions are filled with more detailed descriptions from lower levels. Items are inserted textually to produce the complete CHILL program.

SYSTEM SPECIFICATION (BID)

CHANNEL

SIGNAL ROUTE

SUB-LEVEL (BID)

ENVIRONMENT

SYSTEM NAME

BLOCK 1

BLOCK 2

CHANNEL

BLOCK 3

BLOCK 2

PROCESS 3

PROCESS 4

PROCESS 4

PROCESS SPECIFICATION (PD)

Figure 5. BID/PD hierarchy for system structuring.

Abstraction [21] enables the data aspects of the design and the flow of control to be treated independently. With the data abstraction approach in MELBA, the aim is to allow the use of data objects without having to know the details of their implementation. The representation details of data objects are hidden from the user to preserve representation independence. Hence, any changes in the representation have no effect on the behaviour of the data objects, and thus no effect on the system design.

In large projects, it is essential to specify the work performed by various team members. As such, a methodology is adopted for users of the MELBA system, in which the development of a target system is under the supervision of a 'system designer' who designs and controls the overall system structure under development (see Figure 6). The task of detailed system design is allocated to one or more 'process designer(s)'. To allow the system and process designers to work on the design level, an additional designer, the 'abstraction designer' is required to provide primitive operations used in the specifications.

In addition to establishing a well defined interface among the team members, MELBA also provides a sound communication means between them. Good communication is essential to ensure a smooth flow of accurate information to all team members. For instance, the process designer can use the algebraic technique [8,16] to specify the behaviour of abstract types required; these are then implemented, according to that specification, on the MELBA system, with human assistance from an abstraction designer.

6.1   Block Interaction and Process Diagrams

The system designer can define the structure of a proposed system in a series of hierarchical diagrams consisting of blocks or processes By using SDL Block Interaction Diagrams (BID) (see Figure 7). Each block can be decomposed into further blocks or processes. In MELBA, a process is an atomic entity which cannot be decomposed into subprocesses; its functions are described using SDL Process Diagrams (PD) by the process designer.

Communication between processes is shown in the BID by indicating all signals that are sent to/from each process. These must correspond to signals used in the PD for the particular process. Two processes in the same block can communicate directly, but if a process needs to send a signal to a process in another block, the signal must go via a channel. A channel can be viewed as a uni-directional pipe used to carry signals from one block to another.

The following example illustrates the development of an 'Automatic Call Distributor' (ACD) using various features of the MELBA system. The ACD is a call queue processor which places calls into a queue and connects 'music' to the callers while waiting. The calls are served by a variable number of operators according to a strict first-in-first-out (FIFO) order. When the queue is full, any incoming calls will receive the 'busy tone'. Figure 7 shows the overall system structure of the ACD, including channels and signals com-

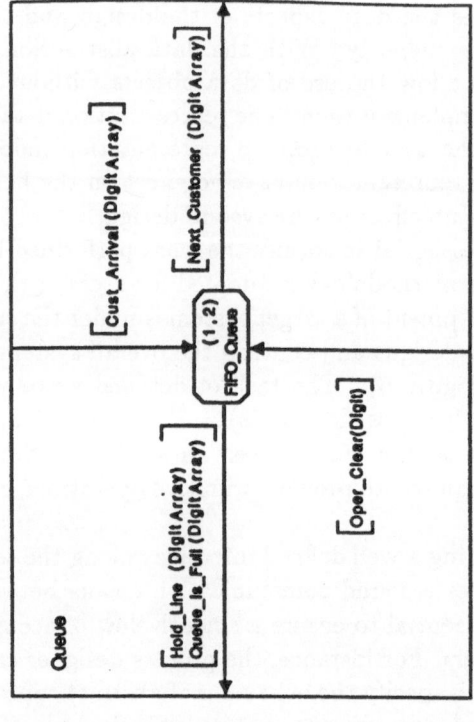

Figure 8. The above diagram describes details of the Queue block in Figure 7. The only process in the block is 'FIFO_Queue'. The first integer value on the left indicates the number of process instances instantiated during target system initialisation (one), while the other integer value indicates the maximum number of similar process instances that can exist simultaneously within the target system (two).

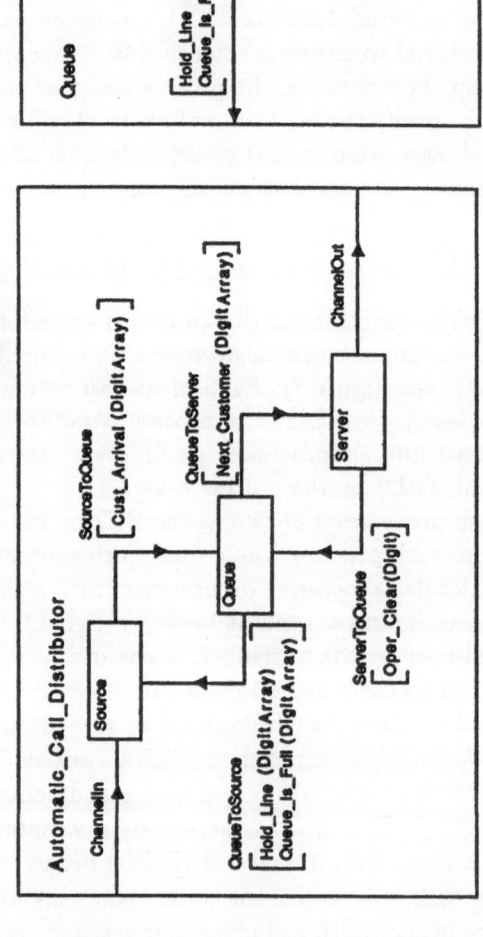

Figure 7. The BID shows the overall design of the system. A complete system would include the details of the Source and Server blocks.

111

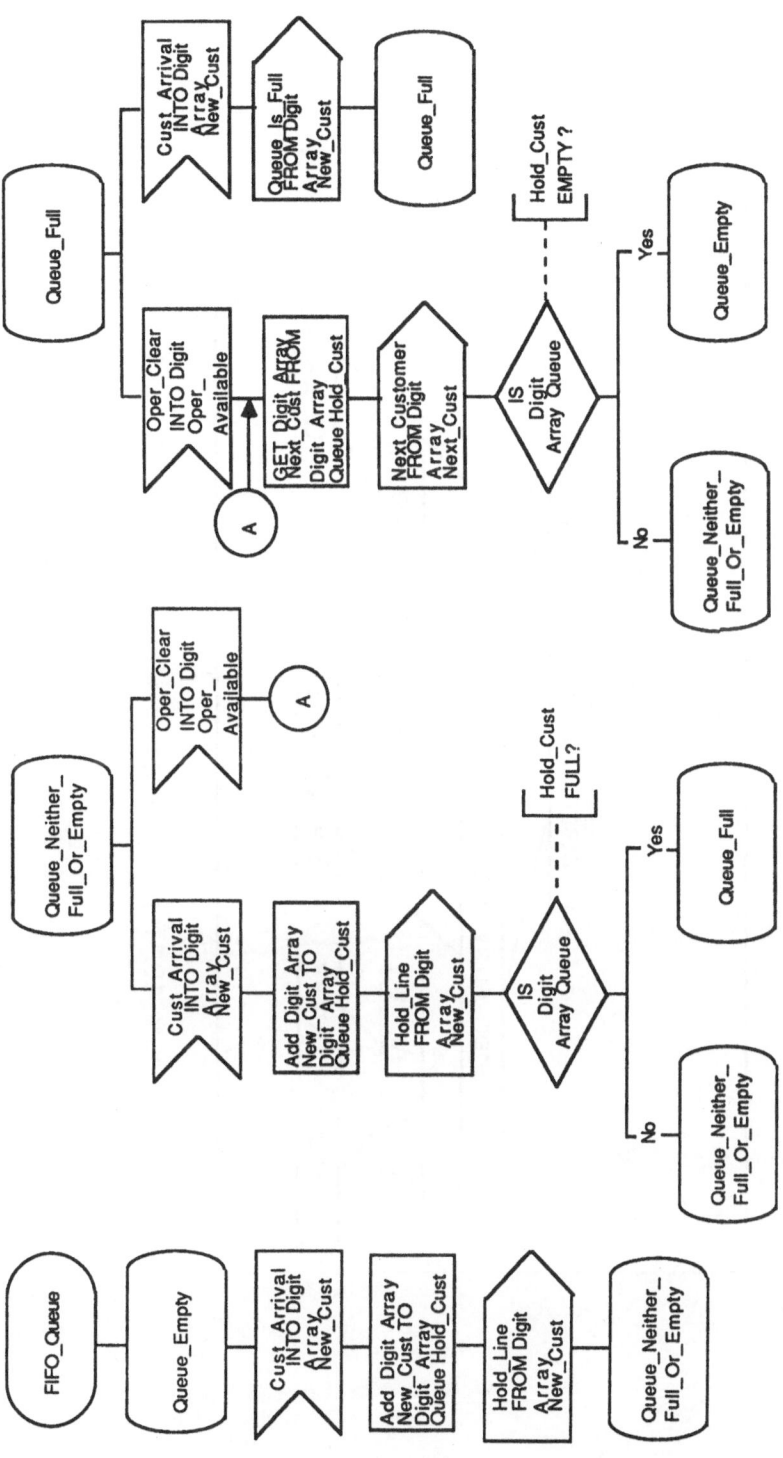

Figure 9. Process Diagram for FIFO_Queue process.

112

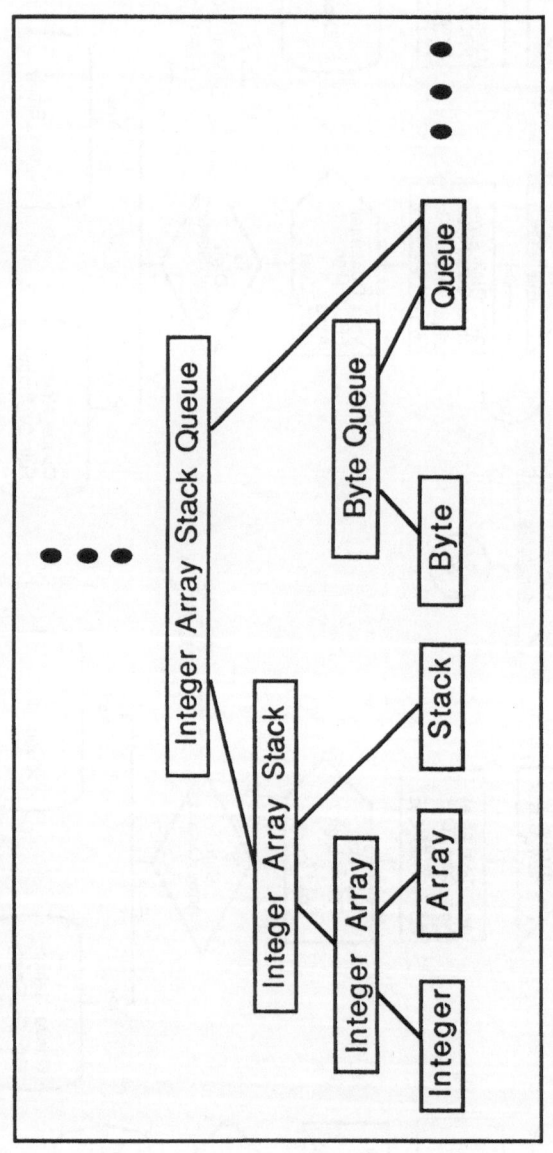

Figure 10. Type parameterisation concept in MELBA.

municating between various blocks. The Source block handles the physical aspect such as the connection and disconnection of 'music' and 'busy tone'. The Server block handles the connection of the call to the operator. The block of interest is the Queue block which consists of a FIFO_Queue process (Figure 8) that distributes the calls in the FIFO queue and generates and accepts signals to and from the Source and Server blocks. For instance, the signal Cust_Arrival, which carries an array of digits (Digit Array), is sent from the Source block to the FIFO_Queue process inside the Queue block via the channel, SourceToQueue.

The PD is based primarily on state transition diagrams and includes the concept of signals which allows communication and synchronization between concurrently executing processes. Each PD is used to describe a single concurrent process. A detailed description of the FIFO_Queue process is shown in Figure 9.

6.2   Abstract Data Types

MELBA incorporates the data abstraction concept into the methodology to help bridge the gap between the specification (SDL) and the implementation (CHILL) level. Implementation details can be obtained where necessary from a library of software modules. This approach is used since it relieves the process and system designers from the burden of defining implementation details (e.g. data structures), something that is not normally done using BIDs and PDs. To increase reliability and stop this library from becoming a random collection of arbitrarily designed routines, MELBA adopts a very strict approach to the way the library entries are stored. All information must be stored as specific, well-designed abstract data types. Further, the full advantages of the abstract type concept are used by allowing types to be parameterised. To the BID and PD user these types appear to be recognised by the system and can be used as if they were built-in types in a typical programming language. Their parameterisation allows them to act as building blocks for creating an even wider range of types as shown in Figure 10. Strong typing is adhered to within the abstract data type system. In addition, this approach allows for software reusability and avoids the necessity to recreate abstract data types that already exist.

Although it is hoped that a thorough library of abstractions can be constructed, it would be unrealistic to assume that this library, no matter how large, will be suitable for all MELBA users' needs. To overcome this, it is possible for an abstraction designer to create a separate temporary library for any particular project under development containing any special types required.

One of the most important features of the system from the user's point of view is the self-documenting text employed in SDL. The user works solely with high-level operations defined in the abstractions library. The syntax for these operations has been defined so that it looks similar to English (see

Figure 9). With the use of meaningful identifiers for the local variables, the process designer can draw SDL diagrams containing text that reads like natural language and is fully self documenting. This conforms with the SDL's main aim of being a specification rather than programming language.

Also, since the users work with types and operations available from the library, they do not need to be aware of the target language and in fact do not even need to know which language is being generated. Target language independence and self-documenting diagrams were both given high priority during design of the MELBA system. Part of the CHILL code generated for the ACD example is shown in Figure 11.

7   Future Work

We are continuing with development of tools for protocol analysis and implementation based on NPNs and SDL. Some of the directions for this work are indicated below.

There is a need for software to aid in the interpretation of the large amounts of data produced by PROTEAN. During our analysis of the OSI Transport Protocol we often wanted to perform operations unavailable in the PROTEAN system, on the reachability graphs generated by PROTEAN. We are currently developing an interactive program that can be integrated with PROTEAN. The user will be able to execute queries on the RG, simplify a RG by removing transitions or P-variables that are not of interest and view the new RG using PROTEAN's Graph module. The program will accept incomplete RGs, such as those that result when PROTEAN ends prematurely (e.g. if it runs out of memory). Such an incomplete RG cannot be used for verifying a protocol specification, however it could aid in determining the cause of the state explosion. We intend the simplification process to be iterative and under interactive control.

The state explosion problem also dictates that more attention must be focused on the initial generation of the reachability graph. A pre-processing program which will take an NPN and convert it into a reduced, but equivalent, net has the potential to reduce the extent of the state explosion. By a reduced, equivalent net, we mean one leading to a smaller reachability graph, but still exhibiting the same behaviour (in the features of interest). This reduction is difficult to do with any great generality, and the program will need to be able to be told which features of the NPN are to be regarded as important, and which are not.

There is a need for software to enable automated implementation from an NPN specification, to complement the analysis tools. Telecom Australia is sponsoring some work in this area and some success in the automated implementation of OSI protocols has already been achieved.

The MELBA system is being further developed by researchers at RMIT. MELBA is being extended to enable it to produce code in a number of other

```
      .
      .
Automatic_Call_Distributor :
MODULE
      .
      .
  SIGNAL Cust_Arrival = (Digit_Array) TO Process_FIFO_Queue;
  SIGNAL Oper_Clear = (Digit) TO Process_FIFO_Queue;

  GRANT
    Process_FIFO_Queue;

  Process_FIFO_Queue :
  PROCESS ( );

     DCL Digit_Array_Queue_Hold_Cust Digit_Array_Queue :=
         [.Front : 1, .Rear : 1, .No_Items : 0,
         .Queue_Array : [ (0, 10 - 1): *]];
     DCL Digit_Array_New_Cust Digit_Array :=
         [(1, 10) : [.Occupied : TRUE, .Contents : *]];
     DCL Digit_Array_Next_Cust Digit_Array :=
         [(1, 10) : [.Occupied : TRUE, .Contents : *]];
     DCL Digit_Oper_Available Digit;

  ABSTRACT_TYPE_Digit_Array_Queue:
  MODULE

     NEWMODE Digit_Array_Queue=
        STRUCT (Front, Rear, No_Items INT,
              Queue_Array ARRAY (0 : 10 - 1) Digit_Array);

     ADD_Digit_Array_TO_Digit_Array_Queue:
     PROC (New_Item Digit_Array IN,
       The_Queue Digit_Array_Queue INOUT);

       DO WITH The_Queue;
         IF No_Items = 10 THEN
           CAUSE QueueFull;
         FI;
         Rear := (Rear + 1) MOD 10;
         Queue_Array (Rear) := New_Item;
         No_Items := No_Items + 1;
       OD;
     END;

     IS_Digit_Array_Queue_FULL:
     PROC (The_Queue Digit_Array_Queue IN,
       Result Answer INOUT) (Answer);

       IF The_Queue.No_Items = 10 THEN
         RETURN Yes;
       ELSE
         RETURN No;
       FI;
     END;

     IS_Digit_Array_Queue_EMPTY:
     PROC (The_Queue Digit_Array_Queue IN,
       Result Answer INOUT) (Answer);

       IF The_Queue.No_Items = 0 THEN
         RETURN Yes;
       ELSE
         RETURN No;
       FI;
     END;

     GET_Digit_Array_FROM_Digit_Array_Queue:
     PROC (Front_Item Digit_Array INOUT,
       The_Queue Digit_Array_Queue INOUT);

       /* implements both the delete and front operations */
       DO WITH The_Queue;
         IF No_Items = 0 THEN
           CAUSE QueueEmpty;
```

```
        FI;
        Front := (Front + 1) MOD 10;
        Front_Item := Queue_Array  (Front);
        No_Items := No_Items - 1;
    OD;   ,

   END;

   SEIZE
      Digit_Array,
      Answer,
      STORE_Digit_IN_Digit_Array_AT_Nat_No,
      RETRIEVE_Digit_FROM_Digit_Array_AT_Nat_No,
      No,
      Yes;
   GRANT
      Digit_Array_Queue,
      ADD_Digit_Array_TO_Digit_Array_Queue,
      IS_Digit_Array_Queue_FULL,
      IS_Digit_Array_Queue_EMPTY,
      GET_Digit_Array_FROM_Digit_Array_Queue;
END;

Queue_Empty :
RECEIVE CASE
   (Cust_Arrival IN Digit_Array_New_Cust):
      CALL ADD_Digit_Array_TO_Digit_Array_Queue
         (Digit_Array_New_Cust,Digit_Array_Queue_Hold_Cust);
      SEND Hold_Line (Digit_Array_New_Cust);
      GOTO Queue_Neither_Full_Or_Empty;
ESAC;

Queue_Neither_Full_Or_Empty :
RECEIVE CASE
   (Oper_Clear IN Digit_Oper_Available):
      GOTO A;
   (Cust_Arrival IN Digit_Array_New_Cust):
      CALL ADD_Digit_Array_TO_Digit_Array_Queue
         (Digit_Array_New_Cust,Digit_Array_Queue_Hold_Cust);
      SEND Hold_Line (Digit_Array_New_Cust);
      CASE IS_Digit_Array_Queue_FULL (Digit_Array_Queue_Hold_Cust) OF
         (No):
            GOTO Queue_Neither_Full_Or_Empty;
         (Yes):
            GOTO Queue_Full;
      ESAC;
ESAC;

Queue_Full :
RECEIVE CASE
   (Oper_Clear IN Digit_Oper_Available):
      A :
      CALL GET_Digit_Array_FROM_Digit_Array_Queue
         (Digit_Array_Next_Cust,Digit_Array_Queue_Hold_Cust);
      SEND Next_Customer (Digit_Array_Next_Cust);
      CASE IS_Digit_Array_Queue_EMPTY (Digit_Array_Queue_Hold_Cust) OF
         (No):
            GOTO Queue_Neither_Full_Or_Empty;
         (Yes):
            GOTO Queue_Empty;
      ESAC;
   (Cust_Arrival IN Digit_Array_New_Cust):
      SEND Queue_Is_Full (Digit_Array_New_Cust);
      GOTO Queue_Full;
ESAC;

END Process_FIFO_Queue;
      .
      .
```

Figure 11. Segment of the CHILL code generated for the Automatic_Call_Distr
ibutor. Notice the insertion of abstract types from the library.

programming languages, and in particular the C language.

We are involved in the development of techniques for the analysis of SDL specifications in collaboration with RMIT. SDL has proved to be a popular language for specifying telecommunications protocols, but it is hampered by a lack of analysis tools. The aim of the work is to generate reachability graphs from SDL specifications, for analysis purposes, possibly via conversion of SDL to NPNs. This will enable us to use our existing tools for reachability analysis to analyse SDL specifications.

We are investigating the application of the techniques discussed here to the specification and analysis of other OSI protocols, such as the FTAM protocol [19], which is significantly more complex than the Transport Protocol. To handle such complexity we are engaged in the development of more powerful analysis tools.

8   Summary and Conclusions

Our PROTGEN software complements the PROTEAN system to give us a system which allows the specification and analysis of protocols using NPNs with a high degree of automation. This automation extends from the input of NPN nets into a diagram editor, through manipulation into a simplified lower level form and input into a package for emulation and reachability analysis. This automation will be extended by the development of tools for automated implementation.

The state explosion problem remains the prime limitation to our analysis techniques, particularly in applying them to practical cases, such as complex OSI protocols. Nevertheless by dividing an NPN specification into suitable parts, the techniques have been successfully used for such protocols. There is considerable potential for further development of these techniques.

We have discussed MELBA as an automated system that assists various stages of the software development cycle using SDL and CHILL. The approach integrates various areas of software engineering, including automatic programming, to provide a working software development support system. It also emphasizes the importance of following international standards if the tools are to be widely accepted and used. The system is currently being used in a number of applications and is available as a commercial product.

One facility lacking in MELBA is the ability to analyse input specifications. The generation of reachability graphs from SDL will overcome this limitation.

We conclude that NPNs are a powerful formalism for protocol analysis because the process can be automated using PROTEAN and PROTGEN. This will be further enhanced by tools for automated implementation currently being developed. We also conclude that MELBA makes SDL a powerful formalism for automated implementation. This power will also be enhanced by developments underway to build an SDL reachability graph generator, as that would enable SDL analysis using PROTEAN.

ACKNOWLEDGEMENTS

We acknowledge the assistance and sponsorship of Telecom Australia Research Laboratories, in particular for the use of their PROTEAN software and sponsorship of the MELBA project. We also acknowledge the researchers of the MELBA system, in particular L.N. Jackson and R.S.V. Pascoe at the Centre for Advanced Technology in Telecommunications in RMIT. We acknowledge the contribution of G. Freeman as principal developer of the GENIE diagram editor and for his advice. We also thank M. Bearman for helpful comments and for use of the Transport Protocol specification.

References

[1] M. Y. Bearman, M. C. Wilbur-Ham and J. Billington, "A Formal Specification of the OSI Class 0 Transport Protocol using NPNs", Telecom Australia Research Laboratories Report No. 7736 (October 1984)

[2] M. Y. Bearman, M. C. Wilbur-Ham and J. Billington, "Analysis of the OSI Class 0 Transport Protocol using NPNs", Telecom Australia Research Laboratories Report No. 7737 (November 1984)

[3] M. Y. Bearman, "Formal Specification of the Open Systems Interconnection Transport Protocol Class 2, using NPNs", CSIRONET Technical Report No. 25 (1986)

[4] M. Y. Bearman, K. R. Parker and R. A. Berger, "A Formal Specification of the OSI Class 3 Transport Protocol using NPNs", CSIRO Division of Information Technology, Technical Report TR-ED-88-02 (February 1988)

[5] J. Billington, M. C. Wilbur-Ham and M. Y. Bearman, "Automated Protocol Verification", in *Protocol Specification, Testing, and Verification, V* (ed. M. Diaz), North-Holland, pp. 59–70 (1986)

[6] J. Billington, G. R. Wheeler and M. C. Wilbur-Ham, "PROTEAN: A High-Level Petri Net Tool for the Specification and Verification of Communication Protocols", *IEEE Transactions on Software Engineering*, Vol. 14, No. 3, pp. 301–316 (March 1988)

[7] G. J. Cain, K. E. Cheng, C. J. Fidge, L. N. Jackson and R.S.V. Pascoe, "Computer-Aided CHILL Code Generation Report No. 12", RMIT Research and Development Memorandum No. 112034M, contract report to Telecom Australia Research Laboratories (Feb 1984)

[8] CCITT, "Specification and Description Language (SDL) — Recommendations Z.100" (1986)

[9] CCITT, "CHILL Language Definition — Recommendation Z.200", (May 1984)

[10] K. E. Cheng, R.S.V. Pascoe, T. S. Choong, L. N. Jackson and G. J. Cain, "An Integrated Approach to Automated Telecommunications Software Development", *The First Pan Pacific Computer Conference*, Melbourne (Sep 1985)

[11] J. P. Courtiat, J. M. Ayache and B. Algayres, "Petri Nets Are Good For Protocols", *ACM SIGCOMM'84* , pp. 66–74 (June 1984)

[12] E. W. Dijkstra, "A Discipline of Programming", Prentice Hall (1976)

[13] C. J. Fidge and R.S.V. Pascoe, "A Comparison of The Concurrency Constructs and Module Facilities of CHILL and Ada", *Australian Computer Journal*, Vol. 15, No.1 (Feb 1983)

[14] T. G. Freeman, "The Diagram Editor GENIE", *AUSGRAPH 86 — Fourth Australasian Conference on Computer Graphics*, Sydney, Australia, pp. 15–19 (July 1986)

[15] G. Freeman, "User Guide for the Diagram Editor GENIE", CSIRO Division of Information Technology, Canberra (May 1986)

[16] J. V. Guttag and J. J. Horning, "The Algebraic Specification of Abstract Data Types", *Acta Informatica*, Vol. 10 (1978)

[17] ISO 8072, "Information Processing Systems — Open Systems Interconnection — Transport Service Definition" (1986)

[18] ISO 8073, "Information Processing Systems — Open Systems Interconnection — Connection Oriented Transport Protocol Specification" (1986)

[19] ISO DIS 8571, "Information Processing Systems — Open Systems Interconnection — File Transfer, Access and Management" (1986)

[20] L. N. Jackson, C. J. Fidge, R. S. V. P. Pascoe and P. H. Gerrand, "Computer-Aided Program Generation from System Specifications: Design Experience from the MELBA Project", *International Switching Symposium*, Florence (May 1984)

[21] B. Liskov and S. Zilles, "An Introduction to Formal Specifications of Data Abstractions", *Current Trends in Programming Methodology*, edited by R. T. Yeh, Vol. 1, pp. 1–32, Prentice-Hall, New Jersey (1977)

[22] K. R. Parker, "Formal Description Techniques in the Development of Open Systems Interconnection Standards", *Australian Software Engineering Conference*, Canberra, Australia, pp. 267–280 (May 1988)

[23] K. R. Parker and R. A. Berger, "The PROTGEN NPN Graphical Input and Conversion System", CSIRO Division of Information Technology, Technical Report TR-ED-88-01 (February 1988)

[24] K. R. Parker, R. A. Berger and M. Y. Bearman, "Analysis of Numerical Petri Net Specifications of OSI Protocols", *Australian Software Engineering Conference*, Canberra, Australia, pp. 193–209 (May 1988)

[25] F. J. W. Symons, "Modelling and Analysis of Communication Protocols using Numerical Petri Nets", PhD. Thesis, University of Essex (May 1978)

[26] F. J. W. Symons, "Representation, Analysis and Verification of Communication Protocols", Telecom Australia Research Laboratories Report No. 7380 (1980)

[27] G. R. Wheeler, "Numerical Petri Nets—A Definition", Telecom Australia Research Laboratories Report No. 7780 (April 1985)

[28] M. C. Wilbur-Ham, "Numerical Petri Nets—A Guide", Telecom Australia Research Laboratories Report No. 7791 (May 1985)

A TOOL FOR THE PERFORMANCE ANALYSIS OF CONCURRENT SYSTEMS

Vincenza Carchiolo, Albert Faro, Michele Malgeri
Instituto di Informatica e Telecomunicazioni
Facoltà di Ingegneria - Università di Catania
Viala A. Doria 6, 95125 CATANIA (Italy)
tel + 39 95 339449 - TELEX 970255 UNIVICT I

ABSTRACT Formal Description Techniques have been successfully experimented with for specifying and verifying concurrent systems. Generally, Formal Description Techniques are not able to specify time and probability, and are unsuitable for proving that a given system works according to desired performance requirements. For these reasons, this paper presents a language, named ELLIPSe, based on SCCS proposed by Milner providing the designer with a formal framework for reasoning about time and probability of parallel systems. An automated tool is also presented to perform the performance analysis of real systems.

1 INTRODUCTION

The increasing complexity of discrete concurrent systems makes it very difficult to handle their design and even more so the performance analysis of their behaviour.

The use of Formal Description Techniques (FDTs) for aiding the designers of computer networks has been successfully investigated. These techniques allow the designers to easily and unambiguously describe the system and to avoid such unwanted situations as deadlock or livelock. Generally, FDTs are not suitable for proving that the system under study works according to acceptable performances. This is because FDTs are not able to describe both time and probability requirements.

ELLIPSe (Extended Language LIPS) is an extension of language LIPS (Language for Interacting Parallel Systems [CAR1]. It is a FDT based on MILNER's Synchronous Calculus for Communicating Systems (SCCS) [MIL] and it provides us with the following facilities:

* time is explicitly described;

* asynchronous and synchronous behaviours can be formally specified;

* probability can be taken into account;

* formal reasoning for verification is supported.

For these reasons the ELLIPSe language has been chosen as a base for a tool aiming at analyzing the performance of concurrent systems.

The specification of a real system (e. g. communicating protocols, control processes) often becomes quite large so it is hard to analyze any critical case only by hand. On the contrary, an automatic tool seems well suitable for helping the designers deal with specification details. The tool proposed in this paper falls into the class of the symbolic simulators, i. e. it expands the system specification, allowing the designers to follow any path through the derivation tree and to evaluate probability and time constraints associated with these paths.

The paper begins with a brief overview of the behavioural FDT ELLIPSe. Section 3 deals with the performance analysis of a system by means of the analysis of its ELLIPSe specification and points out how the tool aids the user to do this. Sectin 4 sketches the tool architecture, pointing out the subdivision into functional modules. Section 5 describes each of these modules more in more detail and lays emphasis on their functionalities and on the algorithm used. Finally, some conclusions and items for further work are discussed.

2 ELLIPSE

ELLIPSe is a FDT belonging to the class of behavioural languages. It is based on Milner's SCCS [MIL] and on LIPS [CAR].

ELLIPSe models the system to be specified as a black box able to communicate through some access points (viz. the gates) with its environment.

The observer of this black box is provided with a discrete time clock. An ELLIPSe specification consists in describing the observation of the interactions (or their absence) between the black box and the environment in each time unit. To manage the probability, any interaction (or its absence) has an associated coefficient measuring the probability that the interaction (or its absence) can be observed.

The interactions and their absence are referred to as actions. More precisely, the interaction is called "observable action" and its absence is called "idle action".

GATE, ACTION NAME AND ACTION

A gate is characterized by the point where the interaction takes place and the direction of this interaction. Then, the gates can be split in two sets: the set of input gates and the set of output gates. Two gates are said to be complementary if they have the same interaction point and opposite directions.

In the following we refer to gates with the lowercase letters g, h, k and we recognize the output gate by the countermark "$-$" (e. g. g is an input gate and $\bar{g}$ is its complementary output gate).

The action-name set can be derived from the set of gates as an abelian group, where the action-name of the idle action, named 1, represents the null element.

In the following we refer to action-names with the lowercase letters a, b, c.

An action consists of a pair of action-names and real number, ranging over $[0, 1]$, that represents the probability that the interaction takes place. In the following we refer to probability with the letter w. The syntax of an action is the following:

$$a\{w\}$$

The action is atomic in the sense that it cannot be divided in time, therefore, each action has duration of one time unit. An action can consist of the simultaneous occurrence of more than one action. The simultaneous occurrence of

$$g\{w1\} \quad \text{and} \quad h\{w2\}$$

gives rise to the action with action-name $g \cdot h$ and probability $w1 \times w2$, that is

$$gh\{w1 \times w2\}$$

Two actions taking place simultaneously at two complementary gates give rise to an idle action (e. g. $g\{w1\}$ and $\bar{g}\{w2\}$ give rise to the action $1\{w1 \times w2\}$).

When the probability of an action is omitted, it is assumed equal to 1.

PROCESS EXPRESSION

The syntax of the ELLIPSe process expressions is given in Table 1. We refer to process expressions with the set of the capital letters.

In the following we give an informal semantics of the ELLIPSe constructs. The formal semantics can be given in terms of inference rules. An inference rule is a pair of premisses and conclusions. The premisses and conclusions are given by means of the operator $-a->$. $P - a -> P'$ means that P can execute action a and transforms itself into P'.

Note that an ELLIPSe specification can be easily mapped onto a stochastic timed-tree. A tree branch is an action and its projection on a time axis is one unit. A node with a name represents a process. An example of a stochastic timed-tree is shown in Figure 1.

Inaction
Process 0 cannot execute any action, its duration in time is 0 time units.

Time
The behaviour of time is an infinite number of actions 1.

Action guard
$a : P$ executes action a at the first time unit and transforms itself into P. The exeuction of action a is observed with the probability w associated with a.

Summation
$P1 + P2$ behaves like either $P1$ with probability 0.5 or $P2$ with probability 0.5.
$P1 +_w P2$ behaves like either $P1$ with probability w or $P2$ with probability $1 - w$.
Figure 3 shows an example of summation.
Note that the summation operator is not associative and commutative.
$\sum_I P_i$, where $I = \{1, 2, \ldots N\}$, denotes the choice of alternative behaviours P_i with probability $1/N$.

Produce (x)
$P1 \times P2$ denotes the synchronous evolution of the processe $P1$ and $P2$. If $P1$ can execute action a (associated with the probability $\{wb\}$), process $P1 \times P2$ may execute action ab (with probability $\{wa \times wb\}$) and becomes $P1' \times P2'$ at the next time unit. Note that $P \times 0 = 0$.

Restriction
The behaviour of $P \backslash G$ is obtained from the behaviour of P by pruning away the subtrees that start with an action that has a factor belonging to set G. The probability connected with the branch cut away is distributed over the other branches rooted at the same node. This preserves both the sum of the probability equal to 1 and the relative probabilities among uncut actions.

For example, we have the process P of Figure 4, where the four actions $a\{wa\}$ (where $wa = 0.25$), $b\{wb\}$ (where $wb = 0.05$), $c\{wc\}$ (where $wc = 0.5$) and $d\{wd\}$ (where $wd = 0.2$) can be observed. The process $P \backslash c$ can execute the actions $a\{w'a\}, b\{w'b\}$ and $d\{w'd\}$ where

$$w'a = wa/(wa + wb + wd) = 0.5$$
$$w'b = wb/(wa + wb + wd) = 0.1$$
$$w'd = wd/(wa + wb + wd) = 0.4$$

Note that $w'a + w'b + w'd = 1$

Relabelling
The process $P[S]$ is obtained from P by relabelling the gate names in P according to the morphism S.

Behaviour identifiers
A name p can be associated with a process expression P with the syntax $p := P$. This facility allows us to specify recursion.

Delay
Four delay operators are defined.

$\delta_n P$ can wait n time units before behaving like P. P may fire at any time unit with probability 0.5. It is a shorthand for

$$\delta_n P := 1 : \delta_{n-1} P + P$$

$$\delta_0 P := P$$

$\delta_{n \backslash w} P$ can wait n time units before behaving like P. P may fire at any time unit with probabiilty w. It is a shorthand for

$$\delta_{n \backslash w} P := 1 : \delta_{n-1 \backslash w} P +_w P$$

$$\delta_{0 \backslash w} P := P$$

δP can wait an arbitrary time before behaving like P. P may fire at any time unit with probability 0.5. It is a shorthand for

$$\delta P := 1 : \delta P + P$$

$\delta_{\backslash w} P$ can wait an arbitrary time before behaving like P. P may fire at any time unit with probability w. It is a shorthand for

$$\delta_{\backslash w} P := 1 : \delta_{\backslash w} P +_{\backslash w} P$$

Exponentiation

$1^n : P$ behaves like P after n time units.

Timeout

$P1\ \theta_{n\backslash w}\ P2$ behaves during the first n time units either like $P1$ or idle; when the n time units have expired it must behave like $P2$. $P1$ may fire at any time unit with probability w.

Observation equivalence

We do not wish to distinguish processes which have observation equivalent stochastic-timed trees. P is observation equivalent to Q (written $P \approx Q$) iff there exists a bisimulation relation $\mathcal{E}$ with $(P, Q) \in \mathcal{E}$. $(P, Q) \in \mathcal{E}$ iff, for all actions a,

(i) if $P - a- > P'$ then (there exists $Q' : Q - a- > Q'$ and $(P', Q') \in \mathcal{E}$)

(ii) if $Q - a- > Q'$ then (there exists $P : P - a- > P', (P', Q') \in \mathcal{E}$)

Probability

We have associated a probability with an action; in other words, we have defined the probability that a process P transforms itself into P' by executing an action a as the probability associated with this action, i. e.

if $P - g\{w_a\}- > P', \pi_g(P, P') = w_a$

Now, we generalize this concept by introducing the probability that a process P transforms itself into P' through a sequence of actions s, as follows:

if $P - s- > P'$, that is

$P - g_1\{w_1\}- > P_1, P_1 - g_2\{w_2\}- > P_2, \ldots, P_{N-1} - g_N\{w_N\}- > P'$

$\pi_s(P, P') = w_1 \cdot w_2 \cdot \ldots \cdot w_N$.

3   TOOL FACILITIES

During the design of this tool we focussed our attention only on some indices we considered significant in order to evaluate the main characteristics of the system under study.

The idea underlying this simulator/analyzer is to provide the system designer with a tool giving him the behaviour of the process both in terms of a sequence of events, and also in terms of time and probability figures. In this way, a designer gets all the information concerning the specification, guaranteeing not only that the system behaves correctly but also that its time performances are within the requirements.

The designer may choose to investigate the correctness of the sequence of events, before investigating the time constraints, or to stress both characteristics together.

Moreover, the designer may use the tool to calculate the probability indices which provide him with further information about the performance of the process.

The indices the tool is able to calculate are the following:

- minimum time: time needed to traverse a path from a starting process (P_s) to a final process (P_f). We refer to it as $T_{\min}(P_s, P_f)$

- probability that P_s becomes P_f along a path s

- average time (T_m): the average time a starting process (P_s) needs to transform itself into a final process (P_f). It is equal to

$$T_M = \sum_{inI} \pi_i \cdot T_i$$

where I is the set of all paths between P_s and P_f,
   π_i is the probability of following the i-th path,
   T_i is the time needed to follow the i-th path.
If only one path exists between P_s and P_f then $T_M = T_{min}$

- throughput ($T_{e,t}$): average number of the occurrences of a given event e in the time interval t.

$$T_{e,t} = (\sum_{inI} \Pi_i \cdot T_{ie,t})/N$$

where I is the set of all paths between P_s and P_f,
   π_i is the probability of following the i-th path,
   $T_{ie,T}$ is the throughput of the i-th path,
   N is the number of the paths of set I.

Average time is often more significant than T_{min} because it takes into account the probability associated with each path. For example, if T_{min} is associated with a path having a very low probability, then it does not influence T_M.

Moreover, the tool allows the designer to perform sensitivity analysis to investigate the critical points of the system, i. e. the parameters that have major impact on the performances of a process. In fact, often very little modification of the system parameters heavily influences system behaviour. Therefore the designer may modify the project in order to obtain a better system behaviour.

4   TOOL ARCHITECTURE

As discussed in the previous section we can identify two phases in computing the performance requirements of a system specified using ELLIPSe; that is, a symbolic simulation of the specification and an evaluation of the results.

The structure of the tool mirrors this procedure. The tool consists mainly of four modules dealing with difference functions (see Figure 5); a manager module(viz. Control Module), a data base module (viz. Specification Data Base) and two functional modules. The first of the functional modules (viz. Symbolic Simulator) is concerned with the simulation phase, and the second module (named Analyzer Module) is concerned with the analysis of the results.

Control Module (CM) manages the cooperation among the other three modules. It drives the modules in order to collect the data needed to calculate the performance indices requested by the user as described below.

Specification Data Base (SDB) is a data base containing the ELLIPSe specification of the system to be analyzed.

Symbolic Simulator (SS) is able to simulate, step by step, the behaviour of the specified system under the strict control of CM.

To calculate the performance indices, CM receives from the user the appropriate input and passes on the specification to be simulated, from SDB to SS. CM interacts with SS choosing the paths that permit the calculation of the desired indices; it evaluates step by step the results of SS and, depending on these, issues the next command. Moreover, CM oversees the end of the simulation. The simulation results, in terms of times and probabilities, are delivered from CM to AM which is able to calculate the requested indices.

Analyzer Module (AM) receives from CM information about the performance indices requested by the user. Then it provides CM with some additional parameters to perform the appropriate simulation and it waits for the simulation response in order to calculate the requested indices.

The tool is implemented in the UNIX environment. The languages used are the logic programming language PROLOG [CLO] and the C language. The analyzer module is implemented in C. The language chosen for implementing SS is PROLOG. This allows us to easily map the ELLIPSe operator semantics onto PROLOG predicates and to facilitate the management of the complex backtracing operations needed by the simulation algorithm. CM is involved in both symbolic and numerical aspects and therefore it is implemented by means of both PROLOG and C blocks.

5 FUNCTIONALITIES OF THE MODULES

This section is devoted to discussing more deeply CM, SS and AM, and points out the internal structure and strategies needed to perform their goals.

5.1 Control Module

CM is the engine of the tool. It is structured in three submodules as depicted in Figure 6. Driver Module (DM) is the kernel of the Control Module and it is the unique module that exchanges information with SS and AM. It is implemented in C.

DM has the following main functions:

a. manages the operation concerned with reading from and writing to the data base, and performs syntactic analysis of the processes to be written into SDB;

b. handles the errors and is capable of recovering from some of them. These errors are divided into three classes:
   - simulator and verifier errors (e. g. overflow);
   - specification errors (syntax error);
   - other errors.

c. analyses the user inputs (or commands);

d. asks the analyzer for the additional parameters needed to calculate the performance indices requested by the user;

e. starts the simulation;

f. controls step by step the simulation results, collects data and continues with the next step of the simulation;

g. halts the simulation;

h. provides the analyzer with the data collected during the simulation;

i. notifies the user of the requested performance indices calculated by the analyzer.

The most interesting function among these is f. Function f can be performed according to both the indices requested by the user and the simulation parameters defined by the analyzer.

In the following we give the algorithm to calculate $T_{\min}, \pi, T_M$ and $T_{e,t}$. The computation of sensitivity is made using the same algorithm with different input parameters given by the Analyzer Module.

$T_{\min}$

The inputs of the procedure are:

☐ starting process P_s

☐ final process P_f

☐ $T_{\max}$ is the time associated with the maximum depth of the stochastic timed-tree reachable during the simulation. It allows us to obtain a more efficient algorithm.

The outputs of the procedure are:

☐ $T_{\min}$

☐ flag, to signal if the simulation is stopped for a memory overflow

☐ path-num: number of the crossed paths before the simulation halt.

An outline of the algorithm to calculate $T_{\min}$ follows.

1. $T_{min} = T_{max}$

2. $P1 = P_s$

3. if there exists a non-visited subtree of $P1$, then the leftmost subtree of $P1$ is simulated in order to obtain the resulting process Q and the time, T, that P_s needs to transform itself into Q

4. if $Q \approx nil$ or $T > T_{min}$ or $Q \approx P_f$

   then { if $Q \approx P_f$ then $T_{min} = T$
   backtrack and goto step 3
   }
   else $P1 = Q$; goto step 3.

π

The inputs of the procedure are:

☐ starting process P_s

☐ final process P_f

☐ π_{min} is the minimum probability to be taken into account in order to calculate $\pi_s(P_s, P_f)$

The outputs of the procedure are:

☐ array of π, where each element of the array contains the probability to reach P_f from P_s through a given sequence of actions;

☐ flag;

☐ path-num.

An outline of the algorithm to calculate π follows

1. $P1 = P_s$

2. if there exists a non-visited subtree of $P1$, then the leftmost subtree of $P1$ is simulated in order to obtain the resulting process Q and the probability π to reach Q from P_s.

3. if $Q \approx nil$ or $\pi < \pi_{min}$ or $Q \approx P_f$

   then {if $Q \approx P_f$ then put π in the array of results
   backtrack and goto step 2
   }
   else $P1 = Q$; goto step 2.

T_M

The inputs of the procedure are:

☐ starting process P_s

☐ final process P_f

☐ T_{max}

☐ π_{min}

The outputs of the procedure are

☐ array of π and T, where each element of the array contains the probability π to reach P_f from P_s through a given sequence of actions and the time T needed to pass from P_s to P_f;

☐ flag;

☐ path-num.

An outline of the algorithm to calculate T_M follows.

1. $P1 = P_s$

2. if there exists a non-visited subtree of $P1$, then the leftmost subtree of $P1$ is simulated in order to obtain the resulting process Q, the probability π and the time T to reach Q from P_s.

3. if $Q \approx nil$ or $\pi < \pi_{\min}$ or $T < T_{\max}$ or $Q \approx P_f$

> then {if $Q \approx P_f$ then put the pair π, T in the array
> backtrack and goto step 2
> }
> else $P1 = Q$; gotostep 2.

$T_{e,t}$

The inputs of the procedure are:

☐ starting process P_s

☐ final process P_f

☐ $\pi_{\min}$

☐ $T_{\max}$

The outputs of the procedure are:

☐ array of π, T and K; where each element of the array contains the probability π that P_s transforms itself into P_f through a given sequence of actions, the time T needed to perform this transformation and the number K of actions e performed in this transformation;

☐ flag;

☐ path-num.

An outline of the algorithm to calculate $T_{t,e}$ follows.

1. $P1 = P_s$

2. if there exists a non-visited substree of $P1$, then the leftmost subtree of $P1$ is simulated in order to obtain the resulting process Q, the time T and the probability π to reach Q from P_s.

3. if $Q \approx nil$ or $\pi < \pi_{\min}$ or $T < T_{\max}$ or $Q \approx P_f$

> then {if $Q \approx P_f$ then put π, T and K in the array
> backtrack and goto step 2
> }
> else $P1 = Q$; goto step 2

User Interface (UI) makes both inputs of data and the interpretation of results simple. Performance analysis gives as result a number of indices in graphical format, the meanings of which are almost clear; for this reason, a great deal of effort has been devoted to obtaining a significant representation of the results.

UI is also able to show simultaneously the results of different simulations in order to compare the impact of a given parameter on the bahaviour of the system under study. The designer often needs to debug the specification of a process so the tool provides it with "trace" facilities, i. e. the tool is able to show the process behaviour event by event. The input part of UI deals with (1) processes and (2) performances. The former has the capability of reading process expressions and of storing them in the SDB. The latter is devoted to receiving information about the performance indices to be calculated. Some menus and forms to be completed aid the user in avoiding errors or misunderstandings. Verifier Module (VM) is an observation equivalence verifier. It proves the existence of a bisimulation relation between two processes. VM is used to check if a simulation must be halted; that is, if the final behaviour has been reached. It implements the relation operator $\approx$, between two processes, used in the algorithms introduced above.

VM is implemented in PROLOG and uses an algorithm similar to the one presented in [CAR2]. This algorithm is based on the bisimulation definition (see Section 2). It is recursive and symmetric, as required by the bisimulation definition. It aims at building a list containing the process pairs that are in a bisimulation relation. An outline description of the algorithm is shown in the following:

1. read pair $< P, Q >$

2. for all $P' : P - a - > P'$ check if there exists a $Q' : A - a - > Q'$; if Q' does not exist the claimed bisimulation is disproved and the algorithm is halted; if Q' exists the algorithm starts again with the pair $< P', Q' >$

3. for all $Q' : Q - a - > Q'$ check if there exist a $P' : P - a - > P'$; if Q' does not exist the claimed bisimulation is disproved and the algorithm is halted; if P' exists the algorithm starts again with the pair $< P', Q' >$.

5.2 Symbolic Simulator

This module is implemented in PROLOG. It makes use of the same concepts and facilities of the simulators proposed in [PAP] and [CAR3].

SS mainly consists of a part concerning the ELLIPSe operator declaration and a part concerning the definition of the ELLIPSe operator semantics.

The declarative part uses the built-in PROLOG predicate **op** to define the ELLIPSe operator symbols and their properties (precedence and associativity). The syntax of actions and processes is given in Table 2 in a BNF form.

The syntax is the same as the one presented in Table 1 except for:

- unary operator δP has a prefix and the symbol used is **delta1** (i.e., δP is replaced by **delta1** P)

-binary operators have an infix form, the symbols used are the same as the respective ELLIPSe constructs except for:

□ exponentiation ($1^n : P$ is replaced by P **ex** N)

□ delay operator ($\delta_n P$ is replaced by N **delta2** P)

□ delay operator ($\delta_{\backslash w} P$ is replaced by PI **delta3** P).

-ternary operators have a parenthesis form and the symbols used are the following:

□ summation (i.e., $P +_w Q$ is replaced by **plus**$(P, PI < Q)$)

□ delay operator ($\delta_{n \backslash w} P$ is replaced by **delta4** $(P < N < PI)$)

-quaternary operator θ has a parenthesis form and the symbol used is **theta** (that is, $P\theta_{n \backslash w} Q$ is replaced by **theta**(P, N, PI, QQ)).
The semantics definition part consists of a predicate, viz. der, that implements (by means of Horn Clauses) the ELLIPSe operator semantics given in terms of inference rules. Each inference rule is given in the form

conclusion:- premiss1, premiss2, ... premissN.

der is a PROLOG predicate having as parameters the starting process (P), the action (A), and the resulting process (Q); that is, der(P,A,Q) represents the parenthesis form of the infix operator -a->. Table 3 shows the predicate der.
The following notes on some predicates are useful in understanding Table 3:

□ action (A) is true if A is an action whose syntax is in Table 2.

□ nofactor(A,G) holds if the action A has no factor belonging to the set G.

□ rel(S,A,AS) implements the relabelling function.

□ clock(T) represents the global clock. It maintains the current value of time by adding to T a time-unit when an action takes place; if a backtracking happens, then clock recalculates the correct time.

pro(_,PI) maintains the current value of probability by forming the product between PI and the probability of the action taking place at the current time unit. It manages the backtracking operation by the recalculation of the correct probability. When the first parameter of this predicate is equal to 1 this means it is used in the restriction operator. In this case, predicate pro must visit the brothers of the current action in order to recalculate the probabilities associated with them as introduced in Section 2.

5.3 Analyzer Module

Analyser Module is devoted to providing the control module with the simulation parameters. It has an internal data base in which all the possible values of the simulation parameters and the formulas to compute the performance indices are stored.
In order to calculate T_{min}, AM reads from the data base a value of T_{max} and provides CM with it. On the basis of the value of T_{min}, flag and path-num obtained by the simulator, AM decides either to accept the obtained T_{min} (and provides CM with it) or to make another simulation attempt with a new value of T_{max}.

In order to calculate π, AM reads from the data base a value of π_{min} and provides CM with it. If flag is equal to false, AM finds the maximum of the values of π, i.e. the maximum probability to reach P_f from P_s (and provides CM with it). If flag is equal to true, AM decides either to accept this value of the array π and finds the maximum of the values of π or makes another simulation attempt with a major value of the parameter π_{min}.

In order to calculate T_M, AM reads from the data base the values of T_{max} and π_{min} and provides CM with them. If flag is equal to false, AM calculates T_M by using the formula given in Section 3, on the basis of the array of T and π, and it provides CM with T_M. If flag is equal to true, AM decides either to accept the values obtained or to make another simulation attempt with a major value of the parameter π_{min}.

In order to calculate $T_{e,t}$, AM reads from the data base the values of T_{max} and π_{min} and provides CM with them. If flag is equal to false, AM calculates $T_{e,t}$ by using the formula given in Section 3, on the basis of the array of T, π and K, and it provides CM with $T_{e,t}$. If flag is equal to true, AM decides either to accept the value obtained or to make another simulation attempt with a major value of the parameter π_{min}.

The algorithm used by *AM* to calculate sensitivity is more complex. In fact, in this case, *AM* must generate a set of parameters for which the performance indices have to be calculated and provides *CM* with the set of points to draw the sensitivity curve.

6 Conclusions

This paper describes how a tool to compute some performance indices works and gives some general information about its implementation.

The application of this tool to study OSI communication protocol performance is planned. For further study is the extension of the Analyser Module in order to allow the tool to compute other performance indices. Moreover, we have in mind to make the tool more flexible by introducing an intelligent module. In this way, it is possible to automatically perform more powerful analysis.

REFERENCES

[CAR1] V. Carchiolo et al, "ECCS and LIPS, two languages for OSI systems specification and verification", Internal Report, Instituto di Informatica e Telecomunicazioni, Catania, Italy, 1987.

[CAR2] V. Carchiolo and A. Faro, "BIP system", Proc. of Ninth International Conference on Computer Communication, Tel Aviv, Israel, 1988

[CAR3] V. Carchiolo, "A Simulator Tool for the Behavioural Timed Language LIPS", Proc. of Nineteenth Annual Modelling and Simulation Conference, Pittsburgh, USA, 1988

[CLO] W. F. Clocksin and C. S. Mellish, "Programming in Prolog", Springer-Verlag, Berlin, Heidelberg, New York, 1981

[MIL] R. Milner, "Calculi for Synchrony and Asynchrony", Theor. Comp. Science, Vol.25, July 1983

[PAP] Guiseppe Pappalardo, "Experiences with a Verification and Simulation Tool for Behavioural Language", Proc. of VII IFIP on Protocol Specification, Texting and

Operator	Syntax	Remark
Inaction	0	
Time	time	
Action guard	$a : P$	
Summation	$P1 + P2$	
	$P1 +_w P2$	w is a probability $\in [0,1]$
	$\sum_I P_i$	$I = \{1, 2, \ldots N\}$
Product(x)	$P1 \times P2$	
Restriction	$P\backslash G$	G is a set of gates
Relabelling	$P[S]$	S is a relabelling
Behaviour identifiers	$p := P$	p is an identifier
Delay	$\delta_n P$	
	$\delta_{n\backslash w} P$	
	δP	
	$\delta_{\backslash w} P$	
Exponentiation	$1^{n\backslash w} : P$	
Timeout	$P1 \; \theta_n \; P2$	

Table 1: ELLIPSE SYNTAX

Figure 1: The stochastic timed-tree of process P

Verification, North-Holland, 1987

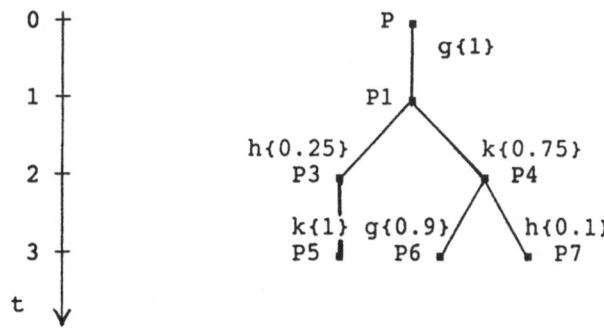

Figure 2: The stochastic timed-tree of process $a : P$

<process> : :=	0|time| <action>:<process> | <process> + process> |
	plus(<process>, <prob>, <process>)
	<process>*<process> -<process> \ <gate set>
	<process>{relabelling}| <processex<num>—
	delta1 <process> | <num> delta2 <process> |
	<prob> delta3 <process> |delta4 (<process>,<num>,<prob>)
	theta (<process>,<action>,<prob>,<process>)|
	<proc_identifier> := <process>
<num> : :=	integer
<gate_set> : :=	<gate_name>, {<gate_name>}*
<proc_identifier> : :=	<identifier>
<action> : := '	<gate_name>? | <gate_name>^ |1| <action> · <action>
<gate_name>: :=	<identifier>
<relabelling> : :=	<gate_set>/<gate_set>
<identifier> : :=	PROLOG atom

Note that prob is a real, ranging over [0,1]

Table 2:

Operator	Inference Rules
INACTION	der(0, _,_):- fail.
TIME	der(time, 1, time):- clock(T).
ACTION GUARD	der(A : P, A, P):- action(A), clock(T), pro(X).
SUMMATION	der(P + Q,A, P1) :- pro(0,0.5), der(P,A,P1).
	der(P + Q,A,Q1) :- pro(0,0.5), der(Q,A,Q1).
	der(plus(P,P1,Q),A,Q1) :- pro(0,PI), der(P,A,P1).
	der(plus(P,P1,Q),A,Q1) :- pro(0, 1-P1), der(Q,A,Q1).
PRODUCT	der(P*Q,A · B,P1*Q1) :- der(P,A,P1), der(Q,B,Q1).
RESTRICTION	der(P\G,A,P1\G) :- pro(1,X), der(P,A,P1), nofactor(A,G).
RELABELLING	der(P{S},AS¡P,1{S}) :- der(P,A,P1), rel(S,A,AS).
EXPONENTIATION	der(N ex P, 1, N-1 ex P) :- N ==\ 0.
	der(N ex P, A, P1) :- N = 0, der(P, A, P1).
DELAY	der(delta1 P, 1, delta P) :- pro(0,0.5), clock(T).
	der(delta1 P, A, P1) :- pro(0,0.5), der(P,A,P1).
	der(N delta2 P,1, N-1 ndelta P) :- N ==\0, pro(0,0.5), clock(T)
	der(N delta2 P, A, P1) :- der(P,A,P1).
	der(p delta3 P, 1, p delta3 P) :- pro(0,p), clock(T).
	der(p delta3 P, A, P1) :- pro(0,1-p), der(P,A,P1).
	der (delta4(P,p,N),1, N-1 ndelta P) :- N ==\0, pro(0,p), clock(T)
	der(delta4(P,p,N), A, P1) :- der(P,A,P1).
TIMEOUT	der(theta(P,N,Q),1, theta(P,N-1,Q) :- N ==\0, pro(0, 0.5), clock(T).
	der(theta(P,Q,N), A, P1) :- N ==\0, pro(0,0.5), der(P,A,,P1).
	der(theta(P,Q,N), A, Q1) :- N = 0, pro(0,0.5), der(Q,A,Q1).
BEHAVIOUR	der(PIden, A, P1) :- PIden:= P, der(P,A,.P1)

Table 3:

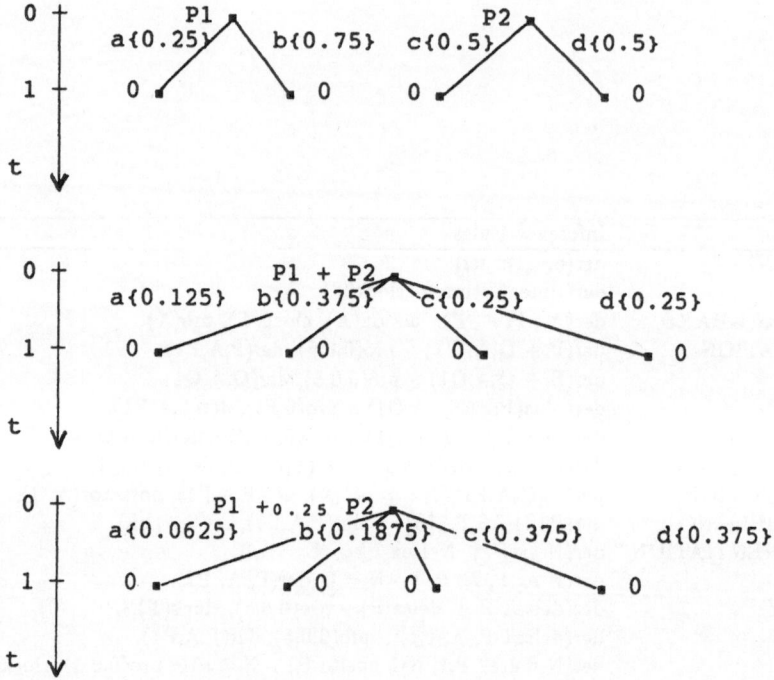

Figure 3: The stochastic timed-tree of summation

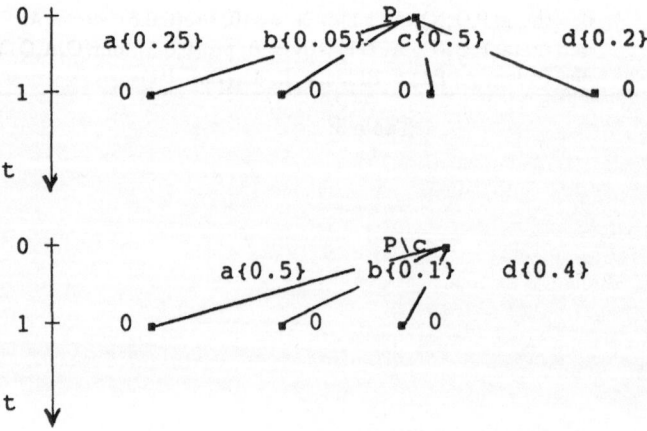

Figure 4: Stochastic-timed tree of restriction

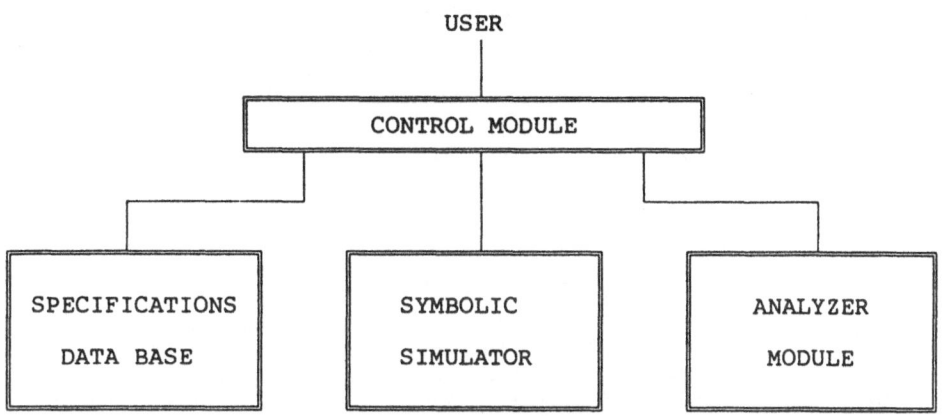

Figure 5: Tool Architecture

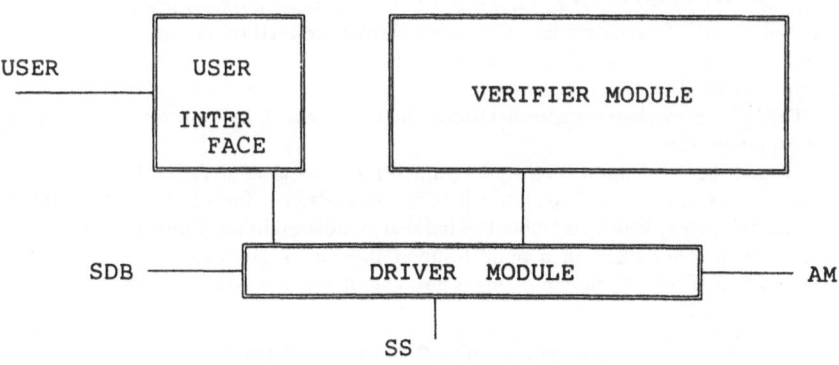

Figure 6: Structure of CM

Winston

A Tool for Hierarchical Design and Simulation of Concurrent Systems

Jawahar Malhotra[†]   Scott A. Smolka[‡]   Alessandro Giacalone[‡]   Robert Shapiro[†]

May, 1988

Abstract

Winston is an interactive environment that offers hierarchical editing, analysis, and simulation of concurrent systems. Winston is built using the *design/OA*™ Development System and hence inherits all the graphics capabilities of design. Winston supports the development of *hierarchically structured networks of processes*, in which networks of communicating processes may be decomposed recursively into subnetworks. A system specification created using Winston can be studied through multi-level simulation, and by analysis tools which include efficient procedures for deciding Milner's strong and weak observational equivalence between Finite State Processes.

1   Introduction

We describe *Winston*, an interactive environment that offers hierarchical editing, analysis, and simulation of concurrent systems. Winston is a joint effort between SUNY at Stony Brook and Meta Software Corporation of Cambridge, Mass. In particular, Winston is an application built using the *design/OA*™ Development System [OA88], a sophisticated environment for creation and manipulation of graphical descriptions of complex systems.

Winston supports the development of *hierarchically structured networks of processes*: networks may consist of collections of subnetworks, each of which may be further decomposed recursively into subnetworks. The basic object of Winston is the process. This structuring capability is useful in system design. The level of detail increases gradually as one progresses deeper and deeper into the hierarchy. This allows an abstract specification to be transformed, by a process of refinement, into a concrete representation.

The model of distributed computation underlying Winston is that of Finite State Processes (FSPs). This model was studied in [KS83] and [KS88] and is derived from Milner's CCS [Mil80] [Mil84]. In this model, FSPs, which resemble the familiar nondeterministic finite state automata, may communicate with other FSPs in a point-to-point fashion: a given message type may be exchanged by exactly two FSPs. Communication is also unbuffered, involving a handshake protocol between the two participants.

† Meta Software Corporation, 150 Cambridge Park Drive, Cambridge, MA 02140, USA
‡ Dept. of Computer Science, SUNY at Stony Brook, Stony Brook, NY 11794, USA. The research of Professors Smolka and Giacalone was supported by the National Science Foundation under Grant No. CCR–8704309.
design/OA and *design/2.0* are trademarks of Meta Software Corporation, Cambridge, Mass.

1.1 Running Example – The Alternating Bit Protocol

To illustrate the capabilities of Winston we will use the well-known Alternating Bit (AB) Protocol as it appears in [Par85]. The following nicely written description of the protocol, also from [Par85], is intended to provide the reader with an intuitive understanding of the protocol's behavior.

> The protocol specification (cf. Figure 1) consists of a *sender* and a *receiver*, communicating over two media *M1* and *M2* (in a lower layer) that may lose or corrupt (but not reorder) messages. The operation of the protocol is as follows. The *sender* accepts a message to be transmitted on the channel *send*. It adds a one bit sequence number to the message (starting with 0 for the first message) and transmits it through the medium *M1*. The sender then awaits an acknowledgement with the same sequence number on the medium *M2*. After reception of the correct acknowledgement the procedure is repeated: a new message can be accepted for transmission. This time the sequence number is inverted. If the sender receives an acknowledgement with wrong sequence number, or corrupt acknowledgement, or no acknowledgement at all within a specified time, the sender assumes medium failure and retransmits the message on *M1* (with the same sequence number). Retransmissions are repeated until a correct acknowledgement arrives.
>
> The *receiver* acknowledges all messages from *M1* by transmitting an acknowledgement on *M2* with the same sequence number as the message. Each message with a sequence different from that of the immediately preceding one is reported on the channel *rec*. In addition, if a corrupt message arrives, medium failure is assumed and the last acknowledgement is retransmitted.

The structure of the rest of the paper is as follows. The model of concurrent computation underlying the Winston environment is presented in Section 2. The graphical editing capabilities of Winston are the topic of Section 3. Section 4 describes the multi-level simulation facilities of Winston and Section 5 the verification facilities. Finally, the availability of Winston is discussed in Section 6.

2 Winston's Model of Concurrent Computation

As mentioned above, Winston supports the design and analysis of hierarchically structured networks of FSPs. In this section, we formally define such networks. We do this by first considering "flat" networks, and then lifting our definitions to the hierarchical case, where networks can be decomposed recursively into subnetworks

The finite state process, the basic building block of our model, closely resembles the familiar nondeterministic finite state automaton.

Definition 1 A *Finite State Process* (FSP) is a quadruple (K, Σ, Δ, p_0) where

> K is a finite set of *states*,
> Σ is the FSP's *alphabet*, a finite set of *observable* actions,
> $\Delta \subseteq K \times (\Sigma \cup \{\tau\}) \times K$ is the *transition relation*,
> where τ is the *unobservable action*,
> $p_0 \in K$ is the *initial state*.　　　　　　　　　　　　　　　　■

An FSP can be represented as a labeled directed graph (labeled transition system) whose nodes are the states, and for each $(p, \lambda, p') \in \Delta$, there is an arc from state p to state p' labeled by λ. This in fact *is* the graphical representation used by Winston, and editing facilities are provided for relatively painless construction of such diagrams (Section 3). Examples of FSPs taken from the AB protocol example appear in Figures 2 and 3.

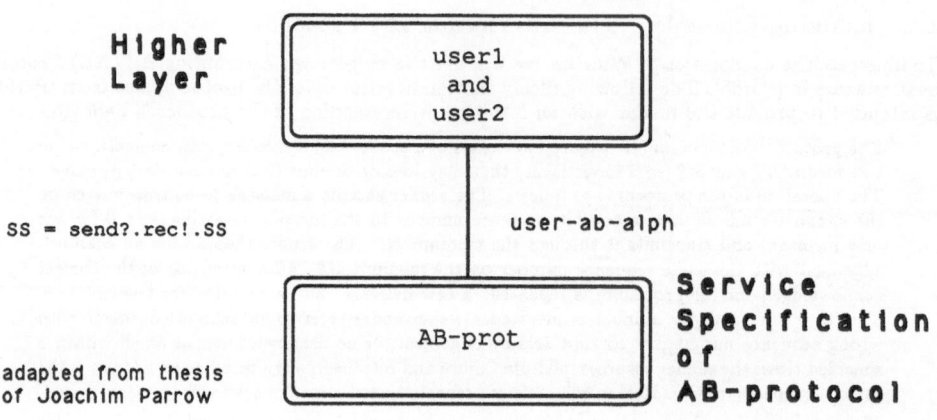

Higher Layer

user1 and user2

SS = send?.rec!.SS

user-ab-alph

AB-prot

adapted from thesis of Joachim Parrow

Service Specification of AB-protocol

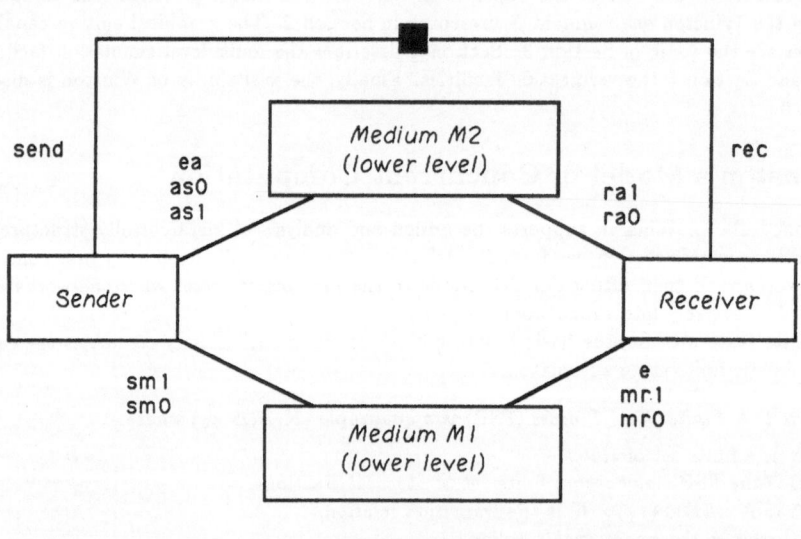

send

ea
as0
as1

Medium M2 (lower level)

ra1
ra0

rec

Sender

Receiver

sm1
sm0

Medium M1 (lower level)

e
mr1
mr0

Architecture of AB-protocol specification

Figure 1: Specification of the Alternating Bit Protocol

Process S :

S = send?.S'
S' = smOl.(as1?.S' + ea?.S'
        + asO?.send?.S")
S" = sm1l.(asO?.S' + ea?.S'
        + tau.S' + as1?.S)

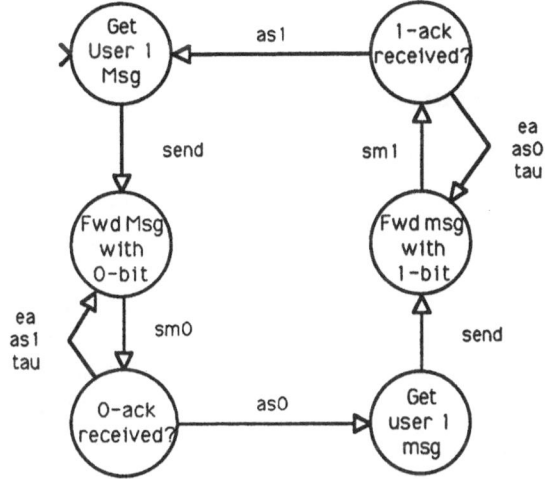

Process R :

R = e?.ra1l.R + mr1?.ra1l.R
    + mrO.recl.R'
R' = raO.(mrO?.R' + e?.R'
        + mr1?.recl.ra1l.R)

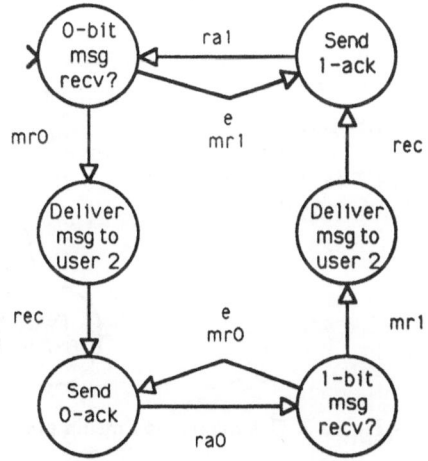

Figure 2: *Sender* and *Receiver* processes of the Alternating Bit Protocol

Medium M1 :

$$M1 = \quad sm0?.(mr0l.M1 + el.M1$$
$$+ tau.M1)$$
$$+ sm1?.(mr1l.M1 + el.M1$$
$$+ tau.M1)$$

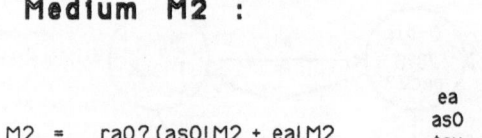

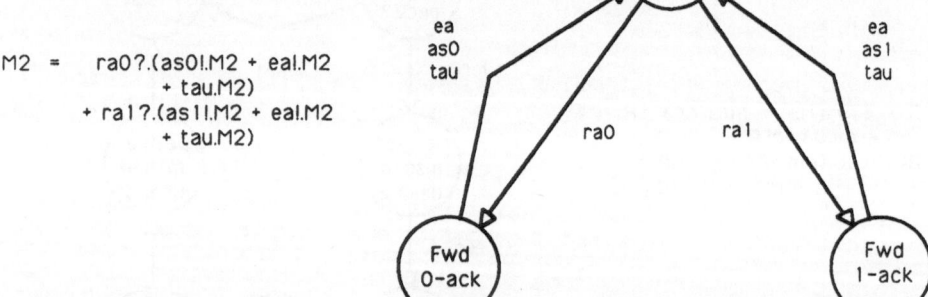

Medium M2 :

$$M2 = \quad ra0?.(as0l.M2 + eal.M2$$
$$+ tau.M2)$$
$$+ ra1?.(as1l.M2 + eal.M2$$
$$+ tau.M2)$$

Figure 3: Communication media processes of the Alternating Bit Protocol

Intuitively, the meaning of an action of an FSP is the exchange of a message with another FSP. For example, if $x \in \Sigma_1 \cap \Sigma_2$ then x is a message that FSP_1 could exchange with FSP_2. As will be clear from the definition of composition below, the message exchange is in the form of a "handshake" between the two processes; no distinction is made between *send* and *receive*. A message can only be exchanged between two processes. The meaning of τ is a step inside the FSP invisible to the outside world.

2.1 Flat Networks of Communicating Processes

Definition 2 A *(flat) network N of processes* is a set of $m \geq 1$ FSPs $P_1, P_2, \ldots, P_m$ where, if we let P_i denote $< K_i, \Sigma_i, \Delta_i, p_i >$, then

the K_i's are mutually disjoint sets of states, $1 \leq i \leq m$, and

each $x \in \bigcup_i \Sigma_i$ belongs to *exactly two* process' sets of actions. ∎

A network is a closed system of communicating processes. Since each action symbol belongs to exactly two processes, we can describe the potential to communicate using a labeled undirected graph C_N (the *communication structure* of N). The nodes of C_N correspond to processes in N and there is an edge $\{i, j\}$ between nodes i and j iff $\Sigma_i \cap \Sigma_j \neq \emptyset$. The label of the edge $\{i, j\}$ is $\Sigma_i \cap \Sigma_j$ (i.e., process P_i can communicate with process P_j using any $x \in \Sigma_i \cap \Sigma_j$). We refer to the edges of C_N as *channels*.

We can now describe the interaction of processes in a network using the operation of *composition* ($\|$).

Definition 3 Let N be a (flat) network of FSPs and $P_i = < K_i, \Sigma_i, \Delta_i, p_i >$, $P_j = < K_j, \Sigma_j, \Delta_j, p_j >$ two distinct processes in N. Let

$$P_i \| P_j = < K_i \times K_j, (\Sigma_i \cup \Sigma_j), \Delta, (p_i, p_j) >$$

where the new transition relation Δ is defined as follows:

if $(q_i, \lambda, r_i) \in \Delta_i$ and $\lambda \in (\Sigma_i \cup \tau) - \Sigma_j$ then
$((q_i, q_j), \lambda, (r_i, q_j)) \in \Delta$ for all $q_j \in K_j$,

if $(q_j, \mu, r_j) \in \Delta_j$ and $\mu \in (\Sigma_j \cup \tau) - \Sigma_i$ then
$((q_i, q_j), \mu, (q_i, r_j)) \in \Delta$ for all $q_i \in K_i$,

if $(q_i, \lambda, r_i) \in \Delta_i$, $(q_j, \mu, r_j) \in \Delta_j$, and $\lambda = \mu \in \Sigma_i \cap \Sigma_j$ then
$((q_i, q_j), \tau, (r_i, r_j)) \in \Delta$.

There are no other elements of Δ. ∎

Let N be the network of FSPs $P_1, P_2, \ldots, P_m$. Intuitively, the transitions of $P_1 \| P_2$ are either the moves of P_1 with respect to $P_3, \ldots, P_m$, or moves of P_2 with respect to $P_3, \ldots, P_m$, or internal τ-moves occurring whenever P_1 and P_2 can "handshake."

The composition operator is associative and commutative, and hence its arity can be extended arbitrarily. In fact, the simulation facility of Winston (Section 4) is based on the operation of composition. Simulation via composition is illustrated in Figure 5 for the alternating bit protocol.

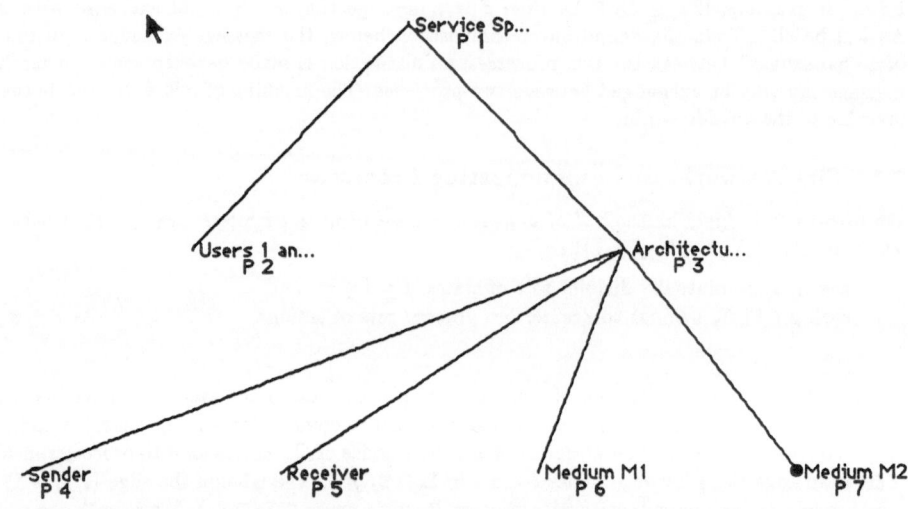

Figure 4: Page hierarchy of the Alternating Bit Protocol

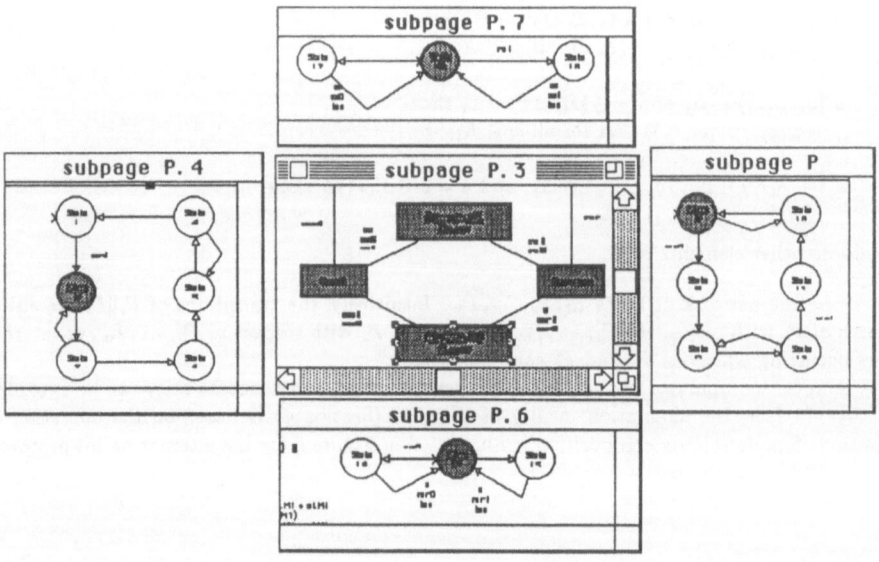

Figure 5: Multi-level simulation of the Alternating Bit Protocol – state of the system after *user1* has transmitted a message to *Sender*

2.2 Hierarchically Structured Networks of Processes

We now describe how the above model of FSP networks can be extended to hierarchically structured networks. The presentation is at an intuitive level, although the formal details are not difficult to provide. We proceed by way of analogy with the Unix or any similarly tree-structured file system. In such systems, one first defines a root directory which may then be decomposed into subdirectories. This process continues recursively until the "leaves" (regular files) of the file system are reached. The organizational benefits of such a structured file system are well accepted and understood.

In Winston, we similarly have a root network consisting of one or more subnetworks, which in turn may be decomposed recursively into subnetworks. Now the leaves of the system are the processes!

The definition of the communication structure of a tree-structured network is somewhat more involved than in the flat case. First, we must generalize the notion of communication alphabet from process to network. This is easy: the alphabet of a network N is simply the union of the alphabets of the processes that constitute the leaves of N. Now, like before, the communication structure of N will be represented as a labeled undirected graph C_N, where the nodes correspond to the networks contained in N, and the edges correspond to the channels interconnecting these subnetworks.

Consider two subnetworks of N, say $SN1$ and $SN2$, having alphabets Σ_1 and Σ_2 respectively. The alphabet of the channel between $SN1$ and $SN2$ will naturally be $\Sigma_1 \cap \Sigma_2$. Let us now look inside $SN1$, a level of refinement greater in the description of N. When doing so, we must remember the symbols $SN1$ shared with $SN2$ as they will appear in the alphabets of the subnetworks of $SN1$. We do this by introducing a new type of node into C_{SN1} called an *external port*. This node will be incident to one or more channels of $SN1$, which thereby reflects another level of detail as to how $SN1$ communicates with $SN2$. Note that an eternal port node can never share an edge with another external port node.

To illustrate, consider the AB protocol example (Figure 1). The root network consists of subnetworks *user1 and user2* and *AB-prot*, which are interconnected by the channel with alphabet *user-ab-alph* = {*send, rec*}. If we look inside *AB-prot*, we find four processes, two of which (*Sender* and *Receiver*) are connected to *user1 and user2* by way of an external port (represented graphically as a small black box). The transition diagrams of the four processes of *AB-prot* are given in Figures 2 and 3.

3 Editing in Winston

The Winston environment is really a restricted version of the *design/2.0*™ drawing tool environment [des88]. Specifications of networks created using Winston are called *diagrams*, and can be saved in and read from files.

Corresponding to our model of computation, Winston diagrams will consist of objects of type *network* (rectangle with double outline), *process* (normal rectangle), *channel* (line segment connecting rectangles), and *external port* (small black-filled rectangles). Process objects will contain objects of type *state* (circle) and *transition* (line segment connecting states). The parenthetical remarks refer to Winston's graphical representation for each of the objects.

The hierarchical structure of networks is captured in Winston by decomposing a diagram into *pages* (one for each network and process object), and maintaining the parent-child relationships between pages. The resulting tree structure is called the diagram's *page hierarchy*. The user can directly instruct Winston which page to edit next simply by pointing to the desired node in the page hierarchy. The page hierarchy for the AB protocol is depicted in Figure 4.

The actual editing of a Winston page is accomplished in a menu-driven fashion, where the menu items are icons of the various Winston objects. Instances of objects are placed on the screen with the aid of a mouse, and may then be resized and reshaped. Channels and transitions are automatically redrawn if the position or shape of either of the objects they interconnect changes.

Text may be associated with any object for documentary purposes. For example, the text appearing within the states of the AB protocol FSPs, and the CCS expressions accompanying each of these FSPs (Figures 2 and 3), serve as an aid to the user in understanding the protocol's behavior. Such text is an example of one numerous user-defined attributes that can be assigned to objects within the environment. Other examples include the action alphabet of a channel, and the initial/non-initial determination of FSP states. User interaction with the environment for assigning attributes to objects is via dialogue boxes, which are common to most Macintosh applications.

3.1 Top-Down Design in Winston

In the current implementation of Winston system specification proceeds in a top-down fashion. The user first specifies the decomposition of the root network into subnetworks by placing network or process objects within the root network's page. Next, channel objects are drawn, each connecting a mutually distinct pair of networks, and their communication alphabets specified by way of dialogue. Here Winston constrains the user to comply with Definition 2, so that an action symbol is shared by exactly two networks on the page. The user may then navigate downward in the page hierarchy by similarly decomposing subnetworks of the root network into subnetworks, or by specifying the internals (transition diagram) of a process object. This editing process normally continues until the entire page hierarchy and the internals of all processes have been specified.

Winston also supports *incomplete* or *partial specifications* of systems. For example, in the AB protocol, the behavior of network *user1 and user2* is never specified, as it is not relevant to the design of the protocol. We need only know that these users exchange *send* and *rec* messages with *AB-prot*.

When editing the transition diagrams of process objects, the set of symbols from which the user may choose to label transitions is dictated by the communication structure in which that process appears. In particular, this set is the union of the communication alphabets of all incident channels, and is presented to the user in a menu-oriented fashion. For example, the communication structure of the AB protocol (Figure 1) indicates that transitions in the FSP *Sender* may be chosen only from the set {*send, sm1, sm0, ea, as0, as1*}.

Future versions of Winston will allow the user the user to specify a system in both a top-down and bottom-up (i.e., starting from the FSPs) fashion.

4 Multi-Level Simulation in Winston

Simulation in Winston is essentially a visualization of the symbolic execution of the processes specified in a Winston diagram. The rules of execution are given by the composition operator, defined formally in Section 2 (Definition 3).

Simulation is multi-level in that the hierarchical structure of Winston diagrams allows the user to selectively view the execution at any level of abstraction, and different levels of the network hierarchy can be simulated simultaneously. Additionally, any *subset* of processes can be chosen for simulation by the user.

The definition of composition given in Section 2 was for two processes only, and is readily extended to an arbitrary number of processes. To do this, consider a network N containing (as its leaves) the set of processes $P_1, \ldots, P_m$. Here N can be any subnetwork of a Winston diagram. A *global state* of N is an m-tuple of states, indicating the current locus of control for each of the m processes in the network. Simulation begins with each process in its initial state.

A "step" of the simulation corresponds to the occurrence of one of two types of possible events (see Definition 3).

1. A single process P_i makes a transition independently of the rest of the the the network. In this case, the action symbol labeling this transition must not belong to the alphabets of any of the processes in N. This event can be interpreted as an offering by the network to engage in a communication with a process in another network.

2. Two processes P_i and P_j make simultaneous transitions. In this case, the action symbols labeling the two transitions must be the same. This event corresponds to a synchronized (handshake) communication between P_i and P_j.

When such events occur, and the involved FSPs are displayed on the screen, Winston highlights the transitions in question along with the symbol being communicated. This allows the user to visually follow the simulation.

The execution rules introduce nondeterminism in Winston simulations, as more than one event may be possible from a given global state. In such cases, the user can ask Winston to randomly select the next event, or tell Winston (through menu selection) which event to execute next. More details on the user interface to simulation are given below.

From certain global states, no further events will be possible, indicating that the system has terminated normally, or that a deadlock situation has arisen. In fact, one of the purposes of simulation in Winston is to inspect for the occurrence of such special global states.

Let us now consider simulation at the network level. The intent here is to allow the user to alternatively view communication at a level higher than that of processes making transitions. The communication structures of networks mesh well with this goal. Consider a network N having as its communication structure the labeled undirected graph C_N. A communication involving processes in two different subnetworks of N, say *SN1* and *SN2*, will be conveyed to the user graphically by highlighting in C_N the channel connecting *SN1* and *SN2*. The symbol exchanged will also be displayed. If the communication involves *SN1* and a process external to N, then the channel connecting *SN1* to an external port object will be highlighted and the exported symbol displayed.

To illustrate multi-level simulation, consider once again the AB protocol. Figures 5 and 6 depict two global states of the protocol. Simultaneous views of the communication structure of *AB-prot* and its four FSPs are given. Figure 5 shoes the global state of the protocol after *user1* has transmitted a message to *Sender*. Figure 6 is derived from Figure 5 through one step of simulation in which *Sender* and *Medium M1* have exchanged an *sm0* message. This event will cause Winston to highlight the (unique) *sm0* transitions in the *Sender* and *M1* FSPs. Concomitantly, the channel connecting these two processes in the protocol's communication structure will be highlighted and the symbol *sm0* prominently displayed. Note that stepping up one more level in the hierarchy to

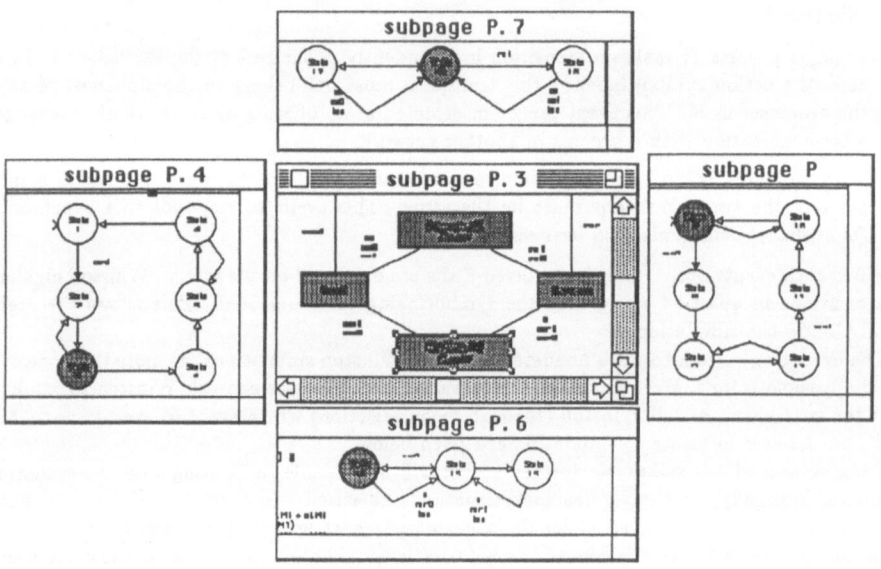

Figure 6: State of the system after *Sender* and *Medium 1* have exchanged the *sm0* message

the root network of the system (Figure 1), would not reveal any additional information, as the communication is internal to the network *AB-prot*.

Regarding the user interface to simulation, the user is first asked to choose which networks are to be included in the simulation. This is accomplished by clicking on the desired network objects in the communication structures. Adding a network has the effect of adding to the simulation all of the processes at the leaves of the network. Subsequently, the user may choose between menu options "step mode" – where one event at a time is simulated under user direction, and "run mode" – where the user can sit back and watch, and let Winston take control of the simulation. In step mode, the user is presented with a menu of possible events from which to choose. In run mode, an event is chosen randomly by Winston.

5 Verification in Winston

Currently, Winston supports verification of concurrent systems in the form of efficient decision procedures for deciding Milner's strong and weak observational equivalence among FSPs. These procedures, presented originally in [KS88], allow the user to test whether an implementation (presumably a concurrent system) meets its specification (often a sequential system). For example, the specification of the AB protocol is given by the simple two-state FSP derived from the expression $SS = send?.rec?.SS$ (see Figure 1). Applying the procedure for weak observational equivalence to this specification and the process resulting from the composition of the four processes in *AB-prot*, produces a positive result. Such composition is performed automatically by Winston when a network (rather than a single process) is chosen by the user for testing.

In [KS88], efficient procedures were developed for deciding properties of loosely connected networks of finite tree processes, such as potential blocking, termination, and lockout. A natural extension of the method to cyclic processes was given. Although the problems in the cyclic case are intrinsically difficult from a computational point of view, the extended method should provide a practical heuristic. Incorporation of this technique into Winston is underway.

6 Availability

Winston is written in C and is currently implemented on the Apple Macintosh and IBM PC. Future plans include embedding Winston in Clara, an interactive environment that supports full CCS [GS88].

References

[des88] *Design 2.0 Manual.* Meta Software Corporation, 150 Cambridge Park Drive, Cambridge, MA 02140, 1988.

[GS88] A. Giacalone and S. A. Smolka. Intregrated environments for formally based design and simulation of concurrent systems. *Special Issue on Integrated Software Engineering Environments, IEEE Transactions on Software Engineering,* 1988.

[KS83] P. C. Kanellakis and S. A. Smolka. CCS expressions, finite state processes, and three problems of equivalence. In *Proceedings of the 2nd ACM Symposium on Principles of*

Distributed Computing, Montreal, Canada, pages 228–240, August 1983. Revised version to appear in Information and Computation.

[KS88] P. C. Kanellakis and S. A. Smolka. On the analysis of cooperation and antagonism in networks of communicating processes. *Algorithmica*, 1988.

[Mil80] Robin Milner. *A Calculus for Communicating Systems*. Volume 92 of *Lecture Notes in Computer Science*, Springer Verlag, 1980.

[Mil84] R. Milner. A complete inference system for a class of regular behaviors. *Journal of Computer and System Science*, 28:439–436, 1984.

[OA88] *Design Open Architecture Development System Manual*. Meta Software Corporation, 150 Cambridge Park Drive, Cambridge, MA 02140, 1988.

[Par85] J. Parrow. *Fairness properties in process algebra with applications in communication protocol verification*. Ph.D. Thesis DoCS 85/03, Department of Computer Science, Department of Computer Science, Uppsala University, Uppsala, Sweden, 1985.

A Specification-Verification Framework for Distributed Applications Software

Donal Roantree and Maurice Clint

Department of Computer Science

The Queen's University of Belfast

University Road

Belfast BT7 1NN

N. Ireland

Abstract

A method for designing and proving systems of distributed deterministic processes is presented. The type of communication considered is synchronous direct communication. Much of the notation and many of the ideas are derived from CSP[Hoare1] and CCS[Milner]. Part of the method requires, for a system of processes, the design of a behaviour expression which is akin to the 'traces' of a system as described in [Hoare2]. Hoare's traces are derived from the traces of each process of a system. However, in the proposed method, the behaviour expression of the whole system is designed first and from this is derived the behaviour expresion of each constituent process.

The proposed method is constructed with the view that considerations of deadlock freedom are of crucial importance in a well-structured design method for distributed systems. Use of the proposed method ensures that the combined specifications of the constituent processes guarantee the deadlock freedom of the whole system.

1   Introduction

A system of distributed processes using synchronous direct communication is, unfortunately, much more difficult to design correctly than a single sequential process. For example, to design numerical algorithms which efficiently utilise a distributed system requires consideration of the number and communication network of processors available for their execution and, given this knowledge, decisions about what the process on each will do.

In addition to the problems that may occur in the realisation of a sequential process, special problems may arise in the realisation of a distributed system. These special problems arise from the need of processes to receive values from other processes. In addition, the processes must 'cooperate'— meaning that the values a process is expecting to receive are indeed the values that the other processes will send. A situation that it is crucial to avoid is *deadlock* in which some processes are waiting to communicate while the others have terminated but none of the offered communications 'match'. In this situation, no more work will be done even though not all of the processes have completed.

When speaking about concurrent systems in general, Roscoe[Roscoe] states that

> ... a proof of deadlock freedom for such a system is an integral part of a total correctness proof, and is often a desirable first step towards the latter.

Since proof considerations should guide the design of systems, the proposed method has been developed with the view that, in the development of a system, freedom from deadlock should be guaranteed from the earliest possible stage.

The ideas and motivation behind the method are introduced in the next section.

2   Motivation

The motivation for the proposed method is based on the following observation embodied in the theorem below. In order to state the theorem concisely, some preliminary definitions are required. Suppose that a communication between two processes is established by their engaging in the same event simultaneously. (The direction of the communication is not represented at this level of abstraction.) The events that a process can engage in define its *alphabet*. In this paper, only systems for which no event belongs to the alphabet of more than two processes are considered i.e. only two processes can engage in any communication.

Definition A process is *simple* if it can have its communications and their order described by a finite sequence of events, after which it terminates.

The choice of the 'next' event for a simple process does not depend on any values received during its participation in previous events.

Definition A system of simple processes is *closed* if no event is contained in the alphabet of just one of the processes.

Usually communications between a process in a system and the system's environment are indicated by an event which is contained solely in the alphabet of that process. To 'close' such a system requires including another simple process in the system which represents the environment.

Definition A *trace* T of a closed system of simple processes $A_1, \ldots, A_n$ with event sequences $E_1, \ldots, E_n$ is a sequence $< \alpha_1, \alpha_2, \ldots, \alpha_m >$ such that $< \alpha_1, \alpha_1, \alpha_2, \alpha_2, \ldots, \alpha_m, \alpha_m >$ is an interleaving of event sequences $E'_1, \ldots, E'_n$ where, for each i, E'_i is an initial subsequence of E_i. T is *complete* if, for each i, $E'_i = E_i$.

Thus any sequence of events recorded from a particular concurrent execution of the processes in the system, up to a certain time, should be a trace of the system. If at this time all of the processes have terminated then this sequence should be a complete trace of the system.

Definition A closed system of simple processes is *blocked* after trace s if s is not a complete trace and there is no other trace of the system which has s as an initial subsequence. A system is *deadlock free* if it is not blocked after any of its traces.

It is clear that if a closed system of simple processes is deadlock free then there must exist a complete trace. The interesting and important point is that the existence of a complete trace is also a sufficient condition for deadlock freedom.

Theorem A closed system of simple processes is deadlock free iff there exists a complete trace of the system.

This theorem can be used to prove that a system of simple processes, each specified by a communication sequence, is deadlock free. However, it can in addition be used as a justification for the following design method. As a first step of the design, produce a complete trace of the system. Then 'extract' from this the communication sequence of each process. Each of these sequences forms the specification of the communication of each process. Then the system, comprising these processes is guaranteed by the theorem not to deadlock.

Of course most processes are not going to be as simple as those described above. In particular, most processes cannot have their communication behaviour described by a communication sequence since it is usual for the choice of 'next' communication of a process to depend on values received during previous communications. Thus the behaviour of processes must be defined by more general expressions than communication sequences in order to allow the specification of values communicated and to accomodate conditional behaviour. Freedom from deadlock for such a system of processes is guaranteed by the existence of a behaviour which is a valid 'merging' of the behaviours of each process.

We will confine our present discussion to deterministic processes. The type of determinism needed is that which requires a process to be ready to communicate with only one other process at any particular time.

3 Behaviour Notation

The behaviour of the system of processes is defined by its *behaviour expression*. A behaviour expression is a sequence of simple terms and structured terms separated by commas. Within these terms, process variables may appear. To simplify the development of the method, it is assumed that these variables are all of integer type.

3.1 Simple Terms

The simple terms are (i) the communication term, (ii) the term 'skip' and (iii) the assertion. The communication term defines which two processes are involved in the communication and may contain process variables whose values may be affected by the communication. The term 'skip' denotes inaction and will usually only be used within structured terms to specify that, under certain conditions, no communication will take place within that term. The assertion is a predicate enclosed in curly brackets. It states certain relationships between process variables which are required by some of the proofs involved when using the method. These assertions also provide good documentation. A communication term takes the form

$$(A \to B \mid R , E)$$

which specifies that process A sends a message to process B. R is a predicate which specifies the value passed in the communication. Variables of the processes A and B may appear in R in addition to the special symbol θ , which represents the value passed. E is a predicate which specifies the side-effect of the communication on the variables of the processes A and B. Within E, dashed occurrences of variables refer to the new values of the variables and the undashed occurrences refer to the old values. If either of the predicates R or E is not required then 'true' can be put in its place. If neither is needed then the communication term takes the form ($A \to B$).

For example, the following behaviour specifies that process A sends two values, y_1 and y_2, to process B, which then sends back their lowest common multiple.

$(A \to B \mid \theta = y_1 , z_1' = \theta)$
$\{ x_1 = y_1 \}$
$(A \to B \mid \theta = y_2 , z_2' = \theta)$
$\{ x_1 = y_1 \wedge x_2 = y_2 \}$
$(B \to A \mid \theta = \mathrm{lcm}(x_1, x_2) , y' = \theta)$
$\{ y = \mathrm{lcm}(y_1, y_2) \}$

Assertions may not contain the symbol θ nor any dashed variables. Such assertions might be valid but they are not necessary.

3.2   Structured Terms

(i) The *conditional term* takes the form

if

    b_1 then S_1

    $|\, b_2$ then S_2

    ...

    $|\, b_n$ then S_n

endif

where $b_1, \ldots, b_n$ are boolean expressions guarding the behaviour expressions $S_1, \ldots, S_n$ respectively. The intended meaning is that the system behaves according to the expression S_i which corresponds to a guard b_i that is true. Since we are restricting ourselves to deterministic processes, we insist that one, and only one, of the guards is true at this point in the behaviour of the system.

For example, the following piece of behaviour specifies that process A first passes a value to process B. Then process B passes the same value on to a process C if it is positive, passes it on to a process D if it is negative and does nothing if it is zero.

($A \rightarrow B$ |true, $x = \theta$)

if

    $x > 0$ then ($B \rightarrow C$ | $\theta = x$, true)

    $|\, x < 0$ then ($B \rightarrow D$ | $\theta = x$, true)

    $|\, x = 0$ then skip

endif

(ii) The *guarded repetition term* takes the form

while b do

    S

endwhile

where b is a boolean expression guarding the behaviour S. The intended meaning is that the system should repeat the behaviour S as long as the guard b is true.

(iii) The *constant repetition term* takes the form

for i **upto** e **do**

    S

endfor

where e is an integer expression and S is a behaviour expression. The intended meaning is that the system should repeat the behaviour S exactly e times, where e is evaluated at the point in the behaviour expression when this term is reached. i is an integer variable which may appear in S and has the value k in the kth iteration of S.

(iv) The *block* takes the form

(**var**

    'declarations'

    S

)

where the 'declarations' part consists of lines of the form

    $x_1, \ldots, x_n$ of $A \mid O$;

A is a process to which the variables $x_1, \ldots, x_n$ are declared to belong. O is a predicate restricting the initial values of $x_1, \ldots, x_n$.

4   Restrictions on Behaviour Expressions

We abbreviate the term 'behaviour expression' to 'b.e.' in the remaining discussion.

All assertions that appear in a b.e. must be validated. The rules for the validity of assertions are given in the next section.

For the sake of the simplicity of the rules to be given later we insist that each variable is declared, in some block, to belong to a particular process. A variable may not be declared twice in the same declaration section. Ordinary scope rules apply so that the scope of this declaration extends to the end of the block but does not include any sub-blocks which contain another declaration of this variable. Thus each occurrence of a variable name can be uniquely associated with a declaration.

For each communication term $(A \rightarrow B \mid R, \ E)$, all variables appearing free in R must belong to either A or B. R may not include any dashed variables. E must be in the form of a conjunction of terms, each of which may include θ but must not include variables of both A and B.

A further restriction on communication predicates is needed. Consider the communication term $(A \rightarrow B \mid \theta^2 = y, \ \text{true})$ which we assume is validly preceded by the assertion $\{ y \neq 0 \}$. The communication term has subtly restricted the possible values of y to positive numbers and this restriction is not guaranteed by the preceding assertion. It must be the case that anything predicated of process variables be guaranteed by the preceding assertion. So we demand that, for each communication term $(A \rightarrow B \mid R, \ E)$ there must be an assertion P which could validly precede it such that $P \Rightarrow \exists \theta . R$. Similarly E should not be stronger P; in addition, it should not restrict the value of θ any more than R does. For this reason we also require that $P \wedge R \Rightarrow \exists l'_1, \ldots, l'_n . E$ where $l'_1, \ldots, l'_n$ are all of the variables which appear dashed in E.

A declaration in a block has the form $x_1, \ldots, x_n$ of $A \mid R$. R may contain the variables $x_1 \ldots, x_n$ and any variable of the process A which is global to the block. The latter must not be restricted further by R than is guaranteed by some valid assertion preceding the block. So we insist that there is such an assertion P satisfying $P \Rightarrow \exists x_1, \ldots, x_n . R$.

As mentioned before, the guards in each branching term must be mutually exclusive and one of them must evaluate to true. Thus we require that for each term of the form

$$\text{if } b_1 \text{ then } S_1 \mid \ldots \mid b_n \text{ then } S_n \text{ endif}$$

there must be an assertion P which could validly precede it such that

$$P \Rightarrow ((b_1 \vee \ldots \vee b_n) \wedge ((1 \leq i, j \leq n \wedge i \neq j) \Rightarrow \neg (b_i \wedge b_j)))$$

In the constant repetition term

$$\text{for } i \text{ upto } e \text{ do } S \text{ endfor}$$

i is considered implicitly to be declared local to this term though it does not belong to any particular process. Thus it can be referred to in any communication or assertion. For the sake of the simplicity of the rules in the next section we insist that neither i nor any variable which appears free in e can appear dashed in S.

5 Rules for the Validity of Assertions

In this section we present the rules defining the validity of assertions in b.e.'s.

In these rules the expression $\| S \|$ is used, where S is a b.e. within which assertions may appear underlined or overlined. The meaning of $\| S \|$ is that if all the underlined assertions in S are valid in the positions in which they occur then so are all the overlined assertions. The rules are usually given in the form

$$\frac{P_1, \ldots, P_n}{\| S \|} \quad \text{where } P_1, \ldots, P_n \text{ are predicates,}$$

This notation has the usual meaning that, if the assumptions $P_1, \ldots, P_n$ are true, then so is $\| S \|$. If $n = 0$ then the rule is given by the expression $\| S \|$ alone.

It is assumed in the following rules that the occurrences of the assertions conform to the syntax of b.e.'s.

The first rule states that $\{true\}$ is a valid assertion anywhere.

VA 1

$$\| \overline{\{true\}} \|$$

Any consequence of a valid assertion can validly follow that assertion.

VA 2

$$\frac{S \Rightarrow T}{\| \{S\}, \overline{\{T\}} \|}$$

The bottom line of this rule states that if $\{S\}$ is a valid assertion at a point in a b.e., then $\{T\}$ *could* validly be placed after $\{S\}$. There is no necessity for the assertion $\{S\}$ actually to be present in the b.e. to prove the validity of $\{T\}$. It is sufficient that $\{S\}$ be valid at that point.

The term **skip** does not alter the values of any variables.

VA 3

$$\| \ \{S\} \ , \ \mathbf{skip} \ , \ \overline{\{S\}} \ \|$$

The most complicated rule is that defining the validity of assertions following communication terms. Consider the communication term

$$(\ A \rightarrow B \ | \ \theta = 72 \ , \ z' = z + k \ )$$

which we assume is validly preceded by the assertion $\{ \, z = 1 \wedge k = 3 \, \}$. The strongest assertion that should follow the communication is $\{ \, z = 4 \wedge k = 3 \, \}$. To derive this we first consider the conjunction of the communication's predicates and its preceding assertion

$$z = 1 \wedge k = 3 \wedge \theta = 72 \wedge z' = z + k \ . \ (\dagger)$$

Now the symbol θ cannot appear in the assertion following the communication. Also the presence of z' indicates that the value of z gets changed and its new value is given by the value of z' in this predicate. (This is the definition of dashed variables, as explained in the section on behaviour notation.) Hence we should deduce from ($\dagger$) an intermediate predicate which involves neither θ nor z, e.g. $k = 3 \wedge z' = 4$. Now, an assertion which would validly follow the communication can be derived from this predicate by substituting z for z', i.e. $k = 3 \wedge z = 4$.

We generalise this in the following rule where T corresponds to the predicate which we called 'intermediate' above and $l_1, \ldots, l_n$ are the variables which appear dashed in the communication predicate E.

VA 4

$$\frac{P \wedge R \wedge E \Rightarrow T, \quad \theta, l_1, \ldots, l_n \notin \mathrm{free}(T), \quad P' = T[l_1, \ldots, l_n / l'_1, \ldots, l'_n]}{\| \ \{P\} \ , \ (\ A \rightarrow B \ | \ R \ , \ E \ ) \ , \ \overline{\{P'\}} \ \|}$$

where $\mathrm{free}(T)$ denotes the set of free variables of the predicate T and $T[u_1, \ldots, u_m / v_1, \ldots, v_m]$ denotes the predicate derived from T by simultaneously replacing each occurrence of variable v_i by the variable u_i. (The v_i are assumed to be distinct.)

The next rule shows how valid assertions can be placed within conditional terms.

VA 5

$$\| \ \{P\} \ , \ \mathbf{if} \ b_1 \ \mathbf{then} \ S_1 \ | \ldots | \ b_i \ \mathbf{then} \ \overline{\{P \wedge b_i\}} \ , \ S_i \ | \ldots | \ b_n \ \mathbf{then} \ S_n \ \mathbf{endif} \ \|$$

If $\{P_i\}$ is a valid assertion following the ith branch of a conditional term, for $i \in 1 \ldots n$, then following the complete term one of the P_i must be true.

VA 6

$$\| \text{ if } b_1 \text{ then } S_1, \ \underline{\{P_1\}} \mid \ldots \mid b_n \text{ then } S_n, \ \underline{\{P_n\}} \text{ endif }, \ \overline{\{S_1 \vee \ldots \vee S_n\}} \|$$

Consider the guarded repetition term **while** b **do** S **endwhile**. The idea behind the rule is the familiar one of the invariant.

Definition Assertion $\{I\}$ is called an *invariant* of the term **while** b **do** S **endwhile** if $\| \underline{\{I \wedge b\}}, \ S, \ \overline{\{I\}} \|$.

If $\{I\}$ is a valid assertion preceding the term **while** b **do** S **endwhile**, of which it is an invariant, then $I \wedge b$ will be true before each occurrence of S and $I \wedge \neg b$ will be true following the last occurrence of S.

VA 7

$$\frac{\| \{I \wedge b\}, \ S, \ \overline{\{I\}} \|}{\| \underline{\{I\}}, \ \text{while } b \text{ do } \{I \wedge b\}, \ S \text{ endwhile}, \ \overline{\{I \wedge \neg b\}} \|}$$

The rule for the constant repetition term **for** i **upto** e **do** S **endfor** is similar except that, rather than having an 'invariant' which is kept true by every occurrence of S, we have a sequence of predicates $I(0), \ldots, I(e)$. $I(k)$ is required to be true after the kth iteration of S. Recall that the meaning of this construct is such that i, which may appear in S, has the value k in the kth iteration of S. Hence we require that

$$1 \leq i \leq e \Rightarrow \| \underline{\{I(i-1)\}}, \ S, \ \overline{\{I(i)\}} \| .$$

$\{I(0)\}$ must be a valid assertion preceding the constant repetition term. $\{I(e)\}$ will be a valid assertion after the constant repetition term. Note, however, that if $e < 1$ then this term should be equivalent to **skip** so we require that $(I(0) \wedge (e < 1)) \Rightarrow I(e)$.

VA 8

$$\frac{(I(0) \wedge (e < 1)) \Rightarrow I(e), \ 1 \leq i \leq e \Rightarrow \| \{I(i-1)\}, \ S, \ \overline{\{I(i)\}} \|}{\| \underline{\{I(0)\}}, \ \text{for } i \text{ upto } e \text{ do } \{I(i-1)\}, \ S \text{ endfor}, \ \overline{\{I(e)\}} \|}$$

Consider the block (**var** $D\ S$) where D is the declaration of variables local to the block. Suppose that O is the conjunction of the predicates in D defining the initial values of the local variables. If no variables which exist outside this block are re-declared in D, then the rule is straightforward.

$$\frac{\|\,\{P \wedge O\}\,,\ S\,,\ \overline{\{Q\}}\,\|\,,\ v\ \text{in}\ D \Rightarrow v \notin \text{free}(Q)}{\|\,\{P\}\,,\ (\,\textbf{var}\ D\ \overline{\{P \wedge O\}}\,,\ S\,)\,,\ \overline{\{Q\}}\,\|}$$

where v in D is true iff variable v is declared in D.

Alternatively, suppose that $v_1, \ldots, v_n$ are the variables which are global to the block and also declared in D. Since it is impossible to refer to the values of these global variables from within the block we should deduce from P a predicate, say P', which does not involve the variables $v_1, \ldots, v_n$. Then $\{P' \wedge O\}$ can validly precede S in the block. To construct an assertion which could follow the block we should look for an assertion, say Q, which validly follows S within the block but does not include local variables or $v_1, \ldots, v_n$. Note also that if $P \Rightarrow P''$ where P'' involves only the variables $v_1, \ldots, v_n$, then P'' will be true following the block. So $\{Q \wedge P''\}$ can validly follow the block.

VA 9

$$\frac{\begin{array}{c} P \Rightarrow P'\,,\ v_1, \ldots, v_n \notin \text{free}(P') \\[4pt] P \Rightarrow P''\,,\ \text{free}(P'') \subseteq \{v_1, \ldots, v_n\} \\[4pt] \|\{P' \wedge O\}\,,\ S\,,\ \overline{\{Q\}}\,\|\,,\ v_1, \ldots, v_n \notin \text{free}(Q) \end{array}}{\|\,\{P\}\,,\ (\,\textbf{var}\ D\ \overline{\{P' \wedge O\}}\,,\ S\,)\,,\ \overline{\{P'' \wedge Q\}}\,\|}$$

If this rule is not strong enough then some of the local variables $v_1, \ldots, v_n$ should be renamed. For example, consider the following behaviour sequence.

$$
\begin{aligned}
(*) \qquad & \{x = a\}, \\[4pt]
& (\,\textbf{var}\ x : P \,|\, x = 0\ : \\[4pt]
& \quad (\,A \rightarrow B \,|\, \text{true}\,,\ a' = a + 1\,) \\[4pt]
&) \\[4pt]
(**) \qquad & \{x + 1 = a\}
\end{aligned}
$$

It cannot be shown from the above rule that if (*) is valid then so is (**). It is necessary to rename x inside the block.

6   Matching Rules

In this section we address the task of 'extracting' the b.e. of a single process from the b.e. of a system of processes.

The following rules are given in the form 'X matches Y' where X and Y are b.e.'s. It is assumed that X is a part of the b.e. for a process A and that Y is a part of the b.e. for a system of processes of which A is a part. P is assumed to be a valid assertion preceding Y.

MR 1 T_1, T_2 matches S_1, S_2 if T_1 matches S_1 and T_2 matches S_2.

MR 2   **skip** matches S if S involves no communications of the process A.

MR 3 $(A{\to}B\,|\,R_A, E_A)$ matches $(A{\to}B\,|\,R, E)$ where $(P{\wedge}R_A){\Rightarrow}R$ and E_A is derived from E by dropping the conjuncts involving variables of the process B. Likewise, R_A must not involve variables of B.

MR 4 $(B{\to}A\,|\,R_A, E_A)$ matches $(B{\to}A\,|\,R, E)$ where $P{\wedge}R \Rightarrow R_A$ and E_A is derived from E by dropping the conjuncts involving variables of the process B. Likewise, R_A must not involve variables of B.

The subtle difference between these last two rules arises from the fact that when A is *sending* a value that value must satisfy the predicate R but when A is *receiving* a value, it may assume less about the value than is guaranteed by R.

MR 5   **if** c_1 **then** T_1 $|...|$ c_m **then** T_m **endif**   *matches*   **if** b_1 **then** S_1 $|...|$ b_n **then** S_n **endif**   *if*

$$(A \wedge (1 \le i \le m) \wedge (1 \le j \le n)) \Rightarrow (\neg(c_i {\wedge} b_j) \vee T_i \text{ matches } S_j)$$

This rule says that every combination of branches T_i and S_j must match unless a preceding assertion implies that their guards are mutually exclusive. The expression T matches a conditional term if the above rule shows that it can be matched by **if true then** T **endif**.

MR 6   for i upto e' do T endfor   *matches*   for i upto e do S endfor   *if* $P \Rightarrow (e = e')$ *and T matches S.*

Here e may be an expression involving variables of many of the processes whereas e' can only involve variables of the process A. Thus, for the process A to repeat the behaviour T for the number of times indicated by the system behaviour, it must be the case that e and e' are equal.

MR 7   while c do T endwhile   *matches*   while b do S endwhile   *if* $P \Rightarrow I$ *where I is an invariant of*   while b do S endwhile   *and $I \Rightarrow (c \Leftrightarrow b)$ and T matches S.*

As with e and e' in the previous rule, the expressions c and b in this rule will usually involve different variables. P is again a valid preceding assertion. Hence we must ensure that c and b have the same truth values before each iteration of the loops.

MR 8   (var D' : T)   *matches*   (var D : S)   *if* T *matches S and D' consists of that part of D which declares variables of the process A.*

Of course, if no new variables of A are declared in D then D' will be empty and T itself will match the block (var D S).

7   Proof Rules

In this section rules are introduced for proving that a sequential process described in a CSP-like notation will satisfy its behaviour specification. However, it is hoped that using the method as it has been described so far, for the purpose of designing the behaviour of each of a number of processes, will, by itself, help significantly in the construction of a reliable distributed algorithm for the solution of a given problem. The method guarantees that the system of processes will not deadlock if implemented correctly; and the clarity and detail of the description of their behaviour should justify having a high degree of confidence in the correctness of the implementation without having to go so far as to prove it using the proof rules which are introduced in this section.

Given a behaviour expression specification of a single process, the last requirement is to produce an implementation of this process and prove that it satisfies this specification. We assume that the implementation language is an imperative language in which several named processes may be

defined. We further assume that the only means a process has of communicating with another is by
executing an output or input statement. We will copy the notation of [Hoare1] and represent the
statement which outputs a value to a process A by $A!e$ and the statement which inputs a value from
a process A and stores the value in a variable x by $A?x$. We assume that the informal semantics
of these output and input statements require that a process, say B, executing them must wait until
the process A executes a corresponding input, $B?y$, or output, $B!f$. We assume that an axiomatic
semantics is given for the sequential statements of the language in the form of Hoare proof rules. In
this section we augment this semantics with proof rules for the input and output statements. For
the purposes of these extra rules, we suppose that every process has an implicitly declared variable
which is a behaviour expression. For a process A, we denote this variable by BE_A. The semantics
of the input and output statements are such that they have the effect of adding a communication
term to BE_A. Then a process A, defined by the code $CODE_A$ satisfies its behaviour expression
specification S if

$$\{ BE_A = \textbf{skip} \}\ CODE_A\ \{ BE_A = S \}$$

The rules given could be applied to an Occam[Occam] program which does not use the Alt
construct or to a CSP program in the notation of [Hoare2] which does not have input guards.

Consider the design of a process A given its behaviour expression S. Suppose that $S = S_1, S_2$ and
$code_1$ has been written such that $\{ BE_A = \textbf{skip} \}\ code_1\ \{(BE_A = S_1) \wedge P\}$. P will be a program
predicate relating the values of program variables and possibly those of variables of the behaviour
expression BE_A with the omission of the b.e. variable BE_A itself. b.e. variables appearing in
program predicates can be considered to be *ghost* variables, (see [Clint]), as can BE_A. Program
variables and b.e. variables must always be distinct. Suppose that the first term in S_2 is the
communication ($A{\rightarrow}B \mid R$, E). Clearly the next communication in the code for process A must
be a statement $B!e$ where e is an integer expression involving program variables. The value of
e, when substituted for θ , must satisfy the predicate R. R will, of course, contain only θ and
variables of the b.e.. Thus we may have to rely on the predicate P to guarantee this satisfaction i.e.
$(P \wedge (e = \theta)) \Rightarrow R$. After this output statement, P may no longer hold since E may contain dashed
variables which, since these may appear in P, will not reflect the changes in these variables. In this
case we must conjoin P,R and E and from this deduce a predicate P'. If the variable l appears
dashed in E then we do not wish it to appear undashed in P' since we are concerned only with its

new value. Then we substitute l for l' in P' to derive a predicate which will be true after the output statement. Thus we have the following proof rule.

$$\frac{(P \wedge (e = \theta)) \Rightarrow R, \ (P \wedge R \wedge E) \Rightarrow P', \ l_1, \ldots, l_n \notin \text{free}(P')}{}$$

$$\{ P \wedge (BE_A = S) \}$$

$$B!e$$

$$\{ P'[l_1, \ldots, l_n / l'_1, \ldots, l'_n] \wedge (BE_A = S, (A \rightarrow B | R, E)) \}$$

where $l_1, \ldots, l_n$ are the variables that appear dashed in E and the variable BE_A does not appear free in P.

This reasoning is very similar to that used to derive the rule for the validity of assertions following communication terms (**VA 4**).

Suppose instead that the first term in S_2 is the communication $(B \rightarrow A \mid R , E)$. Then the next communication in the code for process A must be an input statement $B?x$ for some variable x. Again, we suppose that P is a program predicate at the end of code $code_1$ which possibly relates program variables to b.e. variables. We look for a program predicate that will be true after executing the statement $B?x$. This must reflect the change in x and also in the b.e. variables as given by E. Firstly, we derive from P a predicate P' which does not involve x since its old value is overwritten with the value input. Then from $P' \wedge R \wedge E \wedge (x = \theta)$ we deduce P'' which, as in the previous rule, does not contain any occurrences of any variable l if l' appears in E. Then $P''[l_1, \ldots, l_n / l'_1, \ldots, l'_n]$ is true after execution of the input statement. Thus we have the following proof rule.

$$\frac{P \Rightarrow P', \ x \notin \text{free}(P'), \ (P' \wedge R \wedge E \wedge (x = \theta)) \Rightarrow P'', \ l_1, \ldots, l_n \notin \text{free}(P'')}{}$$

$$\{ P \wedge (BE_A = S) \}$$

$$B?x$$

$$\{ P''[l_1, \ldots, l_n / l'_1, \ldots, l'_n] \wedge (BE_A = S, (B \rightarrow A | R, E)) \}$$

Although these two rules would permit us to prove that an implementation satisfies its b.e. specification they are only sufficient if the b.e. is a sequence of communication terms which do not involve variables. Variables may not be present because all variables must be declared in a block and neither of these rules allows for the occurrence of structured terms in the b.e. . As a first step to remedying the situation, we could allow assertions to appear in BE_A. In particular, if we can deduce that $BE_A = S$ after a piece of code for a process A then it should also be deducible that $BE_A = S, \{P\}$ if P is a valid assertion following S. In order to introduce blocks we would like to be able to assert that $BE_A = $ 'S, (var $z_1, \ldots, z_n$ of $A \mid O$' . Note that BE_A will not here be a properly formed b.e. because there is an unclosed block. We will call a b.e. like this a *partial* b.e. (p.b.e.) . Thus, in this case, BE_A will be a p.b.e. .

We cannot simply define these p.b.e.'s to be equivalent in order to introduce the block. Suppose, for example, that the program predicate $\{ (a = z = 1) \wedge (BE_A = S) \}$ is true after a piece of code, where a is a program variable and z is a variable declared in BE_A. Use of the equivalence to deduce the predicate $\{ (a = z = 1) \wedge (BE_A = $ 'S, (var z of $A \mid z = 0$' $\}$ leads to the contradiction $z = 0 \wedge z = 1$. The problem arises because introduction this block is not accompanied by a change of the value of the newly declared variables which appear in the rest of the program predicate. Later, we present a method for validly introducing blocks into BE_A.

For the purpose of introducing other structured terms we first allow the insertion into the p.b.e. of predicates enclosed in square brackets, which contain only variables of the p.b.e.. The difference between these predicates and the assertions used in b.e.'s up till this point is that the validity of the latter depend on the b.e. alone whereas the new assertions may possibly only be deducible from program predicates. We call these new assertions p-assertions and refer to the previously introduced kind as b-assertions.

For example, consider the following piece of code for the process A, documented with program predicates in the usual way when using the Floyd-Hoare approach to proving correctness.

$\vdots$

$\{ (z = u) \wedge (BE_A = \ldots, (B{\to}A \mid true, u' = 0)) \}$

if $z > 0$

then

$\qquad \{ (z = u) \wedge (z > 0) \wedge (BE_A = $ '$\ldots, (B{\to}A \mid true, u' = 0), [u > 0]$' $) \}$

$\qquad \vdots$

else

$\qquad \vdots$

The b-assertion $\{ u > 0 \}$ could not validly appear above in the b.e. in place of the p-assertion $[u > 0]$ because it is only the relation between the *program* variable z and the b.e. variable u that allows us to deduce that $u > 0$.

Although some of the rules needed to introduce structured terms require consideration of the program context (as we saw with the introduction of the block), most can be introduced without such deliberation. As a result we give some transformation rules for p.b.e.'s in the form $S_1 \gg S_2$ meaning that S_1 can be transformed into S_2. The intention is that if $\{BE_A = S_1\}$ is a valid program predicate at some point in the code of a program, then so is $\{BE_A = S_2\}$ if $S_1 \gg S_2$.

TR 1

$$S \gg S, \{P\} \qquad \text{if } \{P\} \text{ validly follows } S.$$

TR 2

$$S_1 \gg S_2$$

where S_2 is derived from S_1 by removing p-assertions.

TR 3

$$S, [p_1], S_1 \quad \gg \quad S, [p_1], \text{ if } p_1 \text{ then } S_1 \mid \ldots \mid p_n \text{ then } S_n \text{ endif}$$

provided that this conditional term satisfies the restrictions mentioned in section 4 viz. that a preceding assertion must imply that the predicates $p_1, \ldots, p_n$ form a partition of truth.

TR 4

$$S, ([p], S_1)^n , [\neg p] \quad \gg \quad S, \text{ while } p \text{ do } S_1 \text{ endwhile}, [\neg p]$$

where n is an arbitrary number.

TR 5

$$S, S(1), S(2), \ldots, S(e) \quad \gg \quad S, \text{ for } i \text{ upto } e \text{ do } S(i) \text{ endfor}$$

provided no variable free in e appears in $S(i)$.

TR 3 states that if p_1 is known to be always true at a point in the behaviour, then it is allowable to attribute any behaviour to the process at this point when p_1 is not true. Use of this rule allows for the introduction of a conditional term as illustrated by the following example. The predicates preceded by (*) are deduced from the immediately preceding assertion, by applications of this transformation rule.

Suppose that code is to be written to implement the following behaviour fragment for a process A.

$$\text{if } u = 0 \text{ then } (A \to C) \mid u \neq 0 \text{ then } (A \to D) \text{ endif}$$

The section of code might be as follows.

$\{ z = u \wedge BE_A = S \}$

if $z = 0$

then

$\qquad \{ BE_A = S, [u = 0] \}$

$\qquad C!1$

(†) $\quad \{ BE_A = S, [u = 0], (A \to C) \}$

(*) $\quad \{ BE_A = S, \text{if } u = 0 \text{ then } (A \to C) \mid u \neq 0 \text{ then } (A \to D) \text{ endif} \}$

else

$\qquad \{ BE_A = S, [u \neq 0] \}$

$\qquad D!1$

$\qquad \{ BE_A = S, [u \neq 0], (A \to D) \}$

(*) $\quad \{ BE_A = S, \text{if } u = 0 \text{ then } (A \to C) \mid u \neq 0 \text{ then } (A \to D) \text{ endif} \}$

endif

$\{ BE_A = S, \text{if } u = 0 \text{ then } (A \to C) \mid u \neq 0 \text{ then } (A \to D) \text{ endif} \}$

As explained earlier, we cannot introduce structured terms by defining an equivalence relation on p.b.e.'s. Consequently, we suppose that there is an invisible statement—transform—in the implementation language. We allow the prover to assume that this statement is executed automatically whenever he wishes to assert that BE_A has been transformed, for example, from the assertion (†) above to the assertion (*) following it. We require that the semantics of transform guarantee, among others things, the truth of the following rules.

$$\{ P \wedge (BE_A = S) \} \text{ transform } \{ P \wedge (BE_A = S') \} \qquad \text{if } S \gg S'$$

$$\{ P \wedge BE_A = S \} \text{ transform } \{ P \wedge BE_A = S, [p_1] \}$$

if $P \Rightarrow p_1$ and p_1 contains only b.e. variables.

The only other rules that transform must guarantee are the two rules which allow the introduction of a block into BE_A and the ability to close it. These rules are very similar to the rule **VA 9** for the validity of assertions in blocks and following blocks.

$$P \Rightarrow P' \ , \ x_1, \ldots, x_n \notin \text{free}(P')$$

$$P' \wedge O \Rightarrow P''$$

$$\{ P \wedge BE_A = S \} \ \texttt{transform} \ \{ P'' \wedge BE_A \ = \ S, (\texttt{var} \ x_1, \ldots, x_n \ \text{of} \ A \,|\, O \ \}$$

$$P \Rightarrow P' \ , \ x_1, \ldots, x_n \notin \text{free}(P')$$

$$p_1 \Rightarrow p_1' \ , \ \text{free}(p_1') \subseteq \{x_1, \ldots, x_n\}$$

$$\{ P \wedge BE_A \ = \ S, [p_1], (\texttt{var} \ x_1, \ldots, x_n \ \text{of} \ A \,|\, O \,;\, S_1 \ \}$$

$$\texttt{transform}$$

$$\{ P' \wedge p_1' \wedge BE_A \ = \ S, [p_1], (\texttt{var} \ x_1, \ldots, x_n \ \text{of} \ A \,|\, O \,;\, S_1 \ ) \}$$

The semantics of the 'invisible' statement transform is given in the unusual manner—by the *four* rules above. It is also unusual in that it is not intended to satisfy the law of the *distributivity of conjunction* (for the statement St)

$$\{ P \} \, St \, \{ Q_1 \} \ \wedge \ \{ P \} \, St \, \{ Q_2 \} \quad \Leftrightarrow \quad \{ P \} \, St \, \{ Q_1 \wedge Q_2 \}$$

which is a fundamental law of axiomatic semantics. Normally a statement which does not satisfy this law is unimplementable but, in the case of transform, the rules do not require its effect to involve the changing of the value of any program variable; only BE_A and its variables which are all *ghost* variables. Thus implementability does not pose a problem and the four rules can be used as they were implicitly in the last example in which the first rule was used.

8 Example

The example we give is an algorithm which implements the Newton-Raphson iterative method for finding a root of a given function f. We suppose that process A is required to compute the root and that process E is the process requesting the root. E is viewed as the *environment* process. E might represent the other processes in a large system of processes or it might represent a user who communicates imformation to A via a keyboard and who accepts output from A via a screen.

We suppose that E decides on some initial estimate, a_0, for a root, which it sends to A. The Newton-Raphson formula then defines an infinite sequence of estimates $a_0, a_1, a_2, \ldots$ where

$$a_{n+1} = a_n - \frac{f(a_n)}{f'(a_n)}$$

This is guaranteed to converge (under certain conditions which we assume to be satisfied) to a root of the function f. Thus, given a tolerance ε, there is a natural number n such that $|f(a_n)| < \varepsilon$.

The top-level specification of this system of two processes A and E is as follows.

(var

       a of $E \mid a = a_0$;

       est_1 of A ;

    $(E{\rightarrow}A \mid \theta = a , \, est'_1 = \theta)$

    $(A{\rightarrow}E \mid \exists n \cdot \theta = a_n \wedge |f(a_n)| < \varepsilon)$

)

We decide to refine this by using two other processes B and C, whose tasks are to evaluate the functions f and f' respectively. Our informal strategy is to have process A send the latest estimate e to both processes B and C. The process B then passes back the value $f(e)$. B also sends a message to C telling it whether or not an acceptable estimate has been reached, depending on whether $|f(e)| < \varepsilon$ or not. If $|f(e)| \geq \varepsilon$ then C sends the value $f'(e)$ back to A which calculates the new estimate using $e, f(e)$ and $f'(e)$ and then repeats its behaviour.

In the formal behaviour expression of the system given below, it may seem initially that there are many more variables than are needed. The reason for the large number of variables is that each process has its own variables and, in communication terms, only variables of the two processes concerned in the communication can appear. Use of this number of variables does, however, greatly simplify the extraction of the b.e. of each process. In the system b.e. below, all variables are of type real. The processes A, B and C have variables $stop_1$, $stop_2$ and $stop_3$ respectively. If any of these variables has the value zero then this indicates that the corresponding process should stop. A, B and C also have variables $est1$, $est2$ and $est3$, respectively, in which each keeps a copy of the current estimate. The invariant of the loop of the b.e. must therefore ensure that

$stop_1 = 0 \Leftrightarrow stop_2 = 0 \Leftrightarrow stop_3 = 0$. We should initialise each of the $stop_i$ to an arbitrary non-zero value.

The refined system behaviour expression is as follows. The numbers appearing at the start of lines denote the VA rule used to show the validity of the assertion on that line.

(var

        a of $E \mid a = a_0$;

        $est1, \ fest, stop1$ of $A \mid stop1 = 1 \wedge i = 0$;

        $est2, \ stop2$ of $B \mid stop2 = 1$;

        $est3, \ stop3$ of $C \mid stop3 = 1$;

9      $\{ \, (stop1 = stop2 = stop3 = 1) \wedge (a = a_0) \wedge (i = 0) \, \}$

   $(\, E \rightarrow A \mid \theta = a \, , \ est1' = \theta \,)$,

4      $\{ \, (stop1 = stop2 = stop3 = 1) \wedge (est1 = a_0) \wedge (i = 0) \, \}$

2      $\{ \, I \, : \, (est1 = a_i) \wedge (\, stop1 = 0 \Leftrightarrow stop2 = 0 \Leftrightarrow stop3 = 0 \,) \wedge (\, stop1 = 0 \Rightarrow |f(est1)| < \epsilon \, \}$

   while $\neg(stop1 = 0)$ do

7        $\{ \, I \wedge \neg(stop1 = 0) \, \}$

     $(\, A \rightarrow B \mid \theta = est1 \, , \ est2' = \theta \,)$,

4        $\{ \, I \wedge est1 = est2 \, \}$

     $(\, A \rightarrow C \mid \theta = est1 \, , \ est3' = \theta \,)$,

4        $\{ \, I \wedge (est1 = est2 = est3) \, \}$

     $(\, B \rightarrow A \mid \theta = f(est2) \, , \ (fest' = \theta) \wedge (|\theta| < \epsilon \Leftrightarrow stop1' = 0) \wedge (|\theta| < \epsilon \Leftrightarrow stop2' = 0) \,)$,

4        $\{ \, (est1 = est3 = a_i) \wedge (fest = f(est1)) \wedge (|fest| < \epsilon \Leftrightarrow stop1 = 0 \Leftrightarrow stop2 = 0) \, \}$

     $(\, B \rightarrow C \mid \theta = stop2 \, , \ stop3' = \theta \,)$,

4        $\{ \, I \wedge (est1 = est3) \wedge (fest = f(est1)) \, \}$

     if $stop1 = 0$ then

5,2 $\{ \, I \, \}$

       skip

3          $\{ \, I \, \}$

    $\mid stop1 \neq 0$ then

5,2 $\quad\quad\quad \{ I \wedge (est1 = est3) \wedge (fest = f(est1)) \}$

$\quad\quad\quad (C \rightarrow A \mid \theta = f'(est3) , (i' = i + 1) \wedge (est1' = est1 - fest/\theta))$

4 $\quad\quad\quad\quad \{ I \}$

$\quad\quad$ **endif**

6 $\quad\quad\quad \{ I \}$

$\quad$ **endwhile**

7 $\quad\quad \{ I \wedge stop1 = 0 \}$

2 $\quad\quad \{ (est1 = a_i) \wedge (|f(est1)| < \varepsilon) \}$

$\quad\quad (A \rightarrow E \mid \exists n \cdot (\theta = a_n \wedge |f(a_n)| < \varepsilon) , true)$

$\quad)$

It can be easily checked that this b.e. satisfies the restrictions given in section 4. I is shown, from the text, to be an invariant of the loop by its validity immediately preceding the loop and at the end of the loop at line (*).

The next stage is to extract the b.e. of each of the four processes. For any of these processes, this task involves, in the main, simply taking out the terms that involve that process and taking out those conjuncts in the predicates which involve its variables. A less mechanical part of the procedure requires the extractor to find guards for the repetition and conditional terms, which include only variables of that process, but which have the same truth value as the corresponding guards in the system b.e. . It is at this extraction stage that many problems with the design can be detected. For example, suppose that the communication from process B to process C was not present in the system b.e. . Then, when the b.e. for process C is being extracted, it will be discovered that there is no predicate, which includes only variables of C, and which provably has the same truth value as the guard $stop1 = 0$ in the conditional term. (In fact this error could (with just a little foresight) have been easily spotted when designing the system b.e.) Such defects usually indicate that a process has not received enough information for it to be able to make the choices of behaviour that are required of it by the system behaviour. These errors can usually also be corrected easily since the design/implementation is at a very early stage. In this case, the extra communication can be added in to the system b.e. and only slight changes would be required in the assertions following it. It would not be necessary to re-check the validity of the whole system b.e. .

The extracted b.e. for the four processes are as follows.

Process E :

(var a of E ;

    ($E \rightarrow A \mid \theta = a$, $true$),

    ($A \rightarrow E \mid \exists n \; \cdots \; (\theta = a_n \wedge |f(a_n)| < \epsilon)$, $true$)

)

Process C :

(var $est3$, $stop3$ of $C \mid stop3 = 1$)

    while $\neg(stop3 = 0)$ do

       ($A \rightarrow C \mid true$, $est3' = \theta$),

       ($B \rightarrow C \mid true$, $stop3' = \theta$),

       if $stop3 = 0$ then skip

       | $stop3 \neq 0$ then

          ($C \rightarrow A \mid \theta = f'(est3)$, $true$)

       endif

    endwhile

)

The loop guard $\neg(stop3 = 0)$ is valid, by rule **MR 7**, because I, the invariant of the loop in the system b.e., implies that $stop3 = 0 \Leftrightarrow stop1 = 0$. Similarly, the guards in the conditional term are valid because the assertion before the conditional term in the system b.e. implies that they are equal to the guards in the system b.e.'s conditional term. Thus they satisfy the conditions for rule **MR 5**. The other parts of this b.e. match the system b.e. because they are merely syntactic extractions.

Process B

(**var** $est2$, $stop2$ **of** B | $stop2 = 1$;

     while $\neg(stop2 = 0)$ **do**

         ($A{\rightarrow}B$ | $true$, $est2' = \theta$),

         ($B{\rightarrow}A$ | $\theta = f(est2)$, $(|\theta| < \varepsilon \Leftrightarrow stop2' = 0)$),

         skip

     endwhile

)

As for the last process, the loop guard matches the loop guard in the system b.e. . Using MR 2, the term **skip** matches the conditional term in the system b.e. because the process B does not appear in it.

Process A

(**var** i, $est1, fest$, $stop1$ **of** A | $stop1 = 1 \wedge i = 0$;

     ($E{\rightarrow}A$ | $\theta = a_0$, $est1' = \theta$),

     while $\neg(stop1 = 0)$ **do**

         ($A{\rightarrow}B$ | $\theta = est1$, $true$),

         ($A{\rightarrow}C$ | $\theta = est1$, $true$),

         ($B{\rightarrow}A$ | $\theta = f(est1)$, $(fest' = \theta) \wedge (|\theta| < \varepsilon \Leftrightarrow stop1' = 0)$),

         if $stop1 = 0$ **then**

             skip

         | $stop1 \neq 0$ **then**

             ($C{\rightarrow}A$ | $\theta = f'(est1)$, $(i' = i + 1) \wedge (est1' = est1 - fest/\theta)$)

         endif

     endwhile ,

     ($A{\rightarrow}E$ | $\exists n \cdot (\theta = a_n \wedge |f(a_n)| < \varepsilon)$, $true$)

)

It is hoped that anyone familiar with the notation of b.e.'s could implement with confidence the b.e. specifications of the four processes omitting the proof that the implementations satisfy the specifications. Of course, proving this satisfaction establishes absolute confidence in the implementation.

We give now an implementation together with a proof of the process C. The keywords of the implementation language and the program variables will appear in boldface. The following abbreviations will be used to shorten the predicates.

D   for $est3$, $stop3$ of $C \mid stop3 = 1$;

g   for $\neg(stop3 = 0)$

T   for $[g]$, $(A{\to}C \mid true$, $\theta = est3')$, $(B{\to}C \mid true$, $\theta = stop3')$

F   for   **if** $stop3 = 0$ **then skip**

       $\mid stop3 \neq 0$ **then**

            $(C{\to}A \mid \theta = f'(est3)$, $true$)

       endif

S   for   **T** , **F**

Process C ;

Var i, pest3, pstop3, ans : **Real** ;

Function fdash(z : **Real**) : **Real** ;

$$\vdots$$

Begin

     $\{\ BE_A = \text{'skip'}\ \}$

   pstop3 := 1 ; **i** := 0 ;

     $\{\ (\textbf{pstop3} = 1) \wedge (\textbf{i} = 0) \wedge (BE_A = \text{'skip'})\ \}$

   Transform

     $\{\ (\textbf{pstop3} = stop3) \wedge (\textbf{i} = 0) \wedge (BE_A = \text{'(var \textbf{D}')})\ \}$

     $\{\ \textbf{I} : (\textbf{pstop3} = stop3) \wedge$

          $(BE_A = \text{'(var \textbf{D} ([g], \textbf{S})}^{\textbf{i}'}\ )\ \}$

   While pstop3 $\neq 0$ **Do**

     Begin

        $\{\ \textbf{I} \wedge (\textbf{pstop3} \neq 0)\ \}$

Transform

$$\{\ BE_A = \text{'}(\ \text{var}\ D\ ([g], S)^i, [g]'\ \}$$

A?pest3 ;

$$\{\ (pest3 = est3) \land BE_A = \text{'}(\ \text{var}\ D\ ([g], S)^i, [g], (A{\to}C \mid true,\ \theta = est3')'\ \}$$

B?pstop3 ;

$$\{\ (pest3 = est3) \land (pstop3 = stop3) \land BE_A = \text{'}(\ \text{var}\ D\ ([g], S)^i, T'\ \}$$

i := i + 1 ;

$$\{\ (pest3 = est3) \land (pstop3 = stop3) \land BE_A = \text{'}(\ \text{var}\ D\ ([g], S)^{i-1}, T'\ \}$$

If pstop3 = 0 **Then**

$$\{\ (pstop3 = stop3 = 0) \land BE_A = \text{'}(\ \text{var}\ D\ ([g], S)^{i-1}, T'\ \}$$

Ttransform

$$\{\ (pstop3 = stop3 = 0) \land BE_A = \text{'}(\ \text{var}\ D\ ([g], S)^{i-1}, T, [\neg g]'\ \}$$

Transform

$$\{\ (pstop3 = stop3 = 0) \land BE_A = \text{'}(\ \text{var}\ D\ ([g], S)^{i-1}, T, F'\ \}$$

$$\{\ I\ \}$$

Else

$$\{\ (pstop3 = stop3) \land (pest3 = est3) \land (pstop3 \neq 0) \land BE_A = \text{'}(\ \text{var}\ D\ ([g], S)^{i-1}, T'\ \}$$

Transform

$$\{\ (pstop3 = stop3) \land (pest3 = est3) \land BE_A = \text{'}(\ \text{var}\ D\ ([g], S)^{i-1}, T, [g]'\ \}$$

ans := fdash(pest3) ;

$$\{\ (pstop3 = stop3) \land (ans = f'(est3)) \land BE_A = \text{'}(\ \text{var}\ D\ ([g], S)^{i-1}, T, [g]'\ \}$$

A!ans

$$\{\ (pstop3 = stop3) \land BE_A = \text{'}(\ \text{var}\ D\ ([g], S)^{i-1}, T, [g], (C{\to}A \mid \theta = f'(est3),\ true)'\ \}$$

Transform

$$\{\ (pstop3 = stop3) \land (ans = f'(est3)) \land BE_A = \text{'}(\ \text{var}\ D\ ([g], S)^{i-1}, T, F'\ \}$$

$$\{\ I\ \}$$

Endif

$$\{\ I\ \}$$

Endwhile

$\{\, I \wedge (pstop3 = 0)\, \}$

Transform

$\{\, BE_A = \text{'(var D ([g], S)}^{\mathbf{i}}, [\neg g]\text{'}\, \}$

Transform

$\{\, BE_A = \text{'(var D while g do S'}\, \}$

Transform

$\{\, BE_A = \text{'(var D while g do S)'}\, \}$

The implementation was constructed along with its proof. The code can now be taken out. The statement **transform** should not appear in the code since it is it is included only for the sake of the proof. In addition, the variable i above can be considered a ghost variable because it was only used to help the proof and can therefore be omitted. Thus the code for process C might be as follows.

Process C

var pest3 , pstop3 , ans : Real ;

Function   fdash(x : Real) : Real ;

$\vdots$

Begin

   pstop3 := 1 ;

   While (pstop3 $\neq$ 0) **Do**

      Begin

         A?pest3 ;

         B?pstop3 ;

         If (pstop3 $\neq$ 0) **Then**

            Begin

               ans := fdash(pest3) ;

               A!ans

            End

End

End

Correct implementations of the other processes can be constructed in a similar way.

9 General Comments

- The proposed method helps to relieve the designer of the burden of trying to maintain intellectual control over a number of concurrent processes by allowing him to think of the communications as being strictly ordered by his system b.e.. In fact the communications may occur, in a particular execution, in a different order to that suggested by the system b.e. but the implementation will still be correct. A drawback of this sequential conception of a concurrent program may have the unfortunate side-effect of obscuring other aspects of the concurrent execution of the processes involved. For example, it may discourage the designer from considering whether the processes can be made to depend less on each other in order not have to wait as long for each other.

- In addition to guaranteeing freedom from deadlock, the proposed method provides a useful way of proving the partial correctness of a system of processes.

- The notation and rules can be easily extended to allow certain well-defined forms of non-determinism to be exhibited by the processes of a system.

- It is possible to use the proposed method to design a deadlock-free system without attempting to prove any other aspects of its correctness. This can be achieved by ignoring as far as possible the actual values passed in communications and concentrating on the occurrences of the communications themselves.

- We have not mentioned the possibility of processes looping indefinitely. To prove the absence of such behaviour, it will often be enough to show from its b.e. or its code that each individual process cannot loop indefinitely. Sometimes however termination of respective constructs of a process depends on another process. In such cases, it is enough to show that one process, which is guaranteed to communicate during each iteration of a loop in the system b.e., will terminate.

In the proof of part of the implementation of the Newton-Raphson algorithm given earlier, the extensiveness of the assertions was mainly due to the occurrence of a conjunct of the form $(BE_A = \ldots)$ in each. However, suppose that our b.e. notation forms a subset of a language which is both a specification and programming language. Indeed, it is easy to see how the b.e.'s would be implemented—with the exception of the communications because of the possible occurrence of quantifiers in the communication predicates. Suppose that the input, output and the usual imperative language statements are also part of this language. Then, in a way, somewhat similar to that proposed in [Morris], we can consider the communication to be a *prescription* and give a refinement rule to allow a communication term, expressed in the b.e. notation, to be refined to an input or output statement with possibly a piece of code involving the common imperative language statements. For example, the communication term for process A

$$(A {\rightarrow} B \mid \theta = x - y, \ x' = x - 1 \wedge (y' = 0 \Leftrightarrow \theta > 10))$$

could be refined to

```
A!x − y ;

x := x − 1 ;

if  x − y > 10  then  y := 0

                 else  y := 1

endif
```

after it is shown that the code $x := x - 1$; if ... endif satisfies the prescription $x' = x - 1 \wedge (y' = 0 \Leftrightarrow \theta > 10)$.

These refinements may always be accomplished within the immediate context of the communication in the b.e.. That is, in order to refine any communication, it is only necessary to know the preceding valid assertion. Thus this method of *refining to code* is much shorter, cleaner and therefore less error prone than the method used in the example earlier. However, it assumes the existence of the language of which the b.e. notation is a part and, as a result, the refinement rules may not be altered easily to apply to another language.

183

Acknowledgements

We wish to thank Stephen Gilmore for the many helpful suggestions and comments he made during the development of this work.

References

[Hoare1] C.A.R. Hoare, *Communicating Sequential Processes* (Prentice-Hall International U.K. Ltd. 1985)

[Hoare2] C.A.R. Hoare, *Communicating Sequential Processes* (Comm. ACM 21 (8), 666-677 (1978))

[Milner] R. Milner, *A Calculus of Communicating Systems* (Springer Verlag, LNCS 92, 1980)

[Roscoe] A.W. Roscoe and Naiem Dathi, *The Pursuit of Deadlock Freedom* (Oxford University Computing Laboratory, Technical Monograph PRG-57)

[Occam] Inmos Ltd., *the occam programming manual* (Prentice-Hall International, 1984)

[Clint] M. Clint, *Program Proving: Coroutines* (Acta Informatica 2, 50-63 (1973))

[Morris] Joseph M. Morris *A Theoretical Basis for Stepwise Refinement and the Predicate Calculus* (Science of Computer Programming 9 (1987) 287-306)

Dynamic Communication Links

C. M. Holt

Computing Laboratory
University of Newcastle
Newcastle upon Tyne NE1 7RU

Abstract

Communication links are viewed as mathematical variables with additional logical operations relating the success of various references to the link, one to another. They may be described as having a tree structure, with arcs associated with success or failure, and nodes associated with the logical operations. Dynamic links are those which are created and manipulated explicitly as data objects, before being evaluated. Arcs as values may be transmitted via other links; thus an intermediate process can introduce two other processes to one another by giving them arcs of a link. The other processes may then "Eval" the arcs, in the Lisp sense, and communicate with one another directly. The behaviour of links is considered when time operations are available, both for defining sequences of events and for indicating relations that hold over intervals.

CR Categories

D.3.3 Language Constructs: Concurrent programming structures
F.3.2 Semantics of Programming Languages

Keywords

Theory of concurrency, time-free communication, dynamic links, formal models of programming, nature of mathematical variables.

I. Introduction

In common parlance dynamic communication links are message paths that may be created, changed, and destroyed. They are needed to communicate with dynamic processes, and to short-circuit switchboards once two processes have been introduced. There seems at first glance to be an unavoidable time component in their analysis, since something cannot be "created" or "changed" outside of a temporal environment. However, the word dynamic can also be understood as antinomic to static: just as static scoping is determined by a program text, static links are those whose connection patterns are evident from the text. Dynamic links may be considered those for which the text is inadequate in determining the processes they are used to connect; they may depend upon the environment. In point of fact, in the presence of temporal sequences, dynamic links as defined here do indeed imply the standard usage - this was the underlying motivation - but such a restriction is not necessary.

The study of communication between concurrent processes has often proceeded in three stages:
  i. the semantics of a single time sequence of operations is defined;
  ii. sequences are called processes, and combined using a parallel operation; and
  iii. additional, communications operations are defined to link processes together.
This approach is execution-driven; the overriding importance of time sequences has its origins in the history of machine architecture. It has permeated a variety of theoretical treatments, e.g. [1,2,6,11,14].

The rise in popularity of functional and logical programming languages has led to a second approach, in which the introduction of time sequences is omitted:
  i. the semantics of functional or logical expressions are defined;
  ii. a means is provided for indicating when these can be evaluated in parallel; and
  iii. communication is defined using variables available in "parallel" expressions.
Implementation considerations have been deferred to the second stage, and execution has been renamed evaluation to emphasize its time-independent nature. Time sequences and changes in link values may be simulated using lazy evaluation on successive components of sequences. This kind of approach is favoured in the parallel Prologs, e.g. [4,17].

There has been a shift in design methodology corresponding to this change of approach. Rather than beginning with a knowledge of what can be implemented, followed by finding a means to express it (a broadening, incremental style), it is becoming increasingly acceptable to define a domain of discourse intended to span the application area, and then implement only a portion of this (a narrowing, restrictive style). The design shift can be carried further by generalising the nature of mathematical variables, and so allowing communication to be defined independently of evaluation [8]. The resulting approach differs from the above by ignoring implementations altogether:

   i. the semantics of values and individual functional/relational applications are defined;

   ii. communication and mathematical variables are defined;

   iii. interrelationships among components of structures are studied.

This is not to dismiss implementations, but rather to defer such considerations. Only once the domain has been defined does it become desirable to consider evaluation, noting that many structures cannot be resolved efficiently; and only then should time be introduced. (One benefit of this approach is that the initial, broad algebra can be used as the basis for a specification language; program development consists of the refinement of statements in this language into a subset of the same language that has been implemented.)

Section II of this paper is concerned with the above programme. Communication variables are a generalisation of mathematical variables, obtained by allowing the success of a variable to depend upon the success of given uses of that variable. A static link is a variable associated with a given name (allowing for the usual aliasing via formal parameters); the name is a constant in the sense that it always refers to the same link. There are two ways in which this may be generalised: a link may be associated with more than one name, or a name may be associated with more than one link. The former may be described in terms of equalities and disjunctions in a suitable language (e.g. an equational logic based on the approaches of [9,13]). For example, if the value of a link x is to be identified with either that of link y or link z, one might say

   $(x=y)$ or $(x=z)$.

The second generalisation is less trivial. If a name may refer to any of a number of links, it is essentially a variable of variables, and care is needed to ensure that evaluation is at the intended level. It is necessary to be able to refer to links as values, in order to pass them from one position to another; but it must also be possible to use them as links. It is argued that this distinction between a variable as a data object in itself, and a variable as standing for another value is similar to that controlled by the

Lisp operations of Quote and Eval. When a variable is to be viewed as a data object, one is at a meta-level. The use of an evaluation function causes its argument to be evaluated, taking a meta-object and reducing its level by one. The construction of a variable requires functions that manipulate the structure of that variable; this is discussed in Section III. To illustrate more ordinary examples of dynamic links, a rather simple event/interval calculus is then sketched out, based on common temporal logics [15,16], and interactions among links in events and intervals are considered.

II.   Communication Variables

An arbitrary algebra defined in the usual way (as a triple of Sorts, Operations, and Equations) can be understood as a domain Dom of values, with functions Fun relating values to one another. Some values are associated with names, while others may be denoted only by applying functions to named values (e.g. negative integers). For this discussion, questions of denotation are not of interest; it shall be assumed that there is a name d_i for every value of Dom, and a name f_i for every function of Fun. There is an obvious mapping from functions into relations, such that if f_i is an n-adic function, r_i is the relation whose first n arguments are the arguments of f_i, and whose final (unique) argument is the result of f_i when applied to those arguments. This mapping is here called R; i.e. $R(f_i) = r_i$. It should be noted that since the values d_i may be viewed as anadic functions, $R(d_i)$ yields a monadic "relation" that accepts the value d_i as its argument. There is an inverse mapping $F=R^{-1}$, that maps relations onto functions. Of course, if F is applied to a relation that has more than one possible final argument given its first n, the result may be viewed as nondeterministic or undefined.

A relation defines an environment within which a limited number of patterns may be accepted. It is effectively a small language acceptor on a semantic rather than a syntactic level. For example, R(succ) is a relation that accepts only pairs of integers of the form (n,n+1). An application of a relation to a pattern that it can accept is said to succeed; an application of a relation to any other pattern is said to fail (this is quite like the Snobol approach to conditionals [5]). (Note: The concern here is only with applications in which success or failure can be determined. Undecided and undecidable applications are ignored.) The application of a function to arguments succeeds if and only if its equivalent relational application can succeed; this depends both upon its arguments and the context within which its result occurs. For example, the success of the expression

(3*4)+true

may be checked as follows:

    a. Primitive values can always succeed on their own; so R(3) succeeds given the value 3, R(4) succeeds given 4, and R(true) succeeds given true.

    b. Looking at the product term, we form R(*) and apply it to its arguments. The only possible third argument is 12; i.e. R(*)(3,4,12) can succeed.

    c. The result of the product is the first argument to +, so it is necessary to determine whether or not there is a value v such that R(+)(12,true,v) succeeds. There is not, so the + application fails.

    d. There is no other possible third argument for the * application; it fails as well. Since there are no other possible arguments for the primitive values, they fail.

Thus, not only does the overall application fail, but all of the components fail also.

There are two ways in which applications may be related to one another. The first is by defining general operations over the domain of applications. If these operations take no account of the actual relations and values of the applications, but succeed or fail depending upon the success or failure of their arguments, they are called logical. Conjunction (&) succeeds only if all its arguments succeed; disjunction (|) succeeds if any of its arguments succeed; and negation ($\neg$) succeeds if its argument fails, and fails if its argument succeeds. For example,

$$R(+)(3,4,7) \quad \& \quad \neg(R(*)(3,4,7))$$

succeeds: the first argument succeeds since 3+4=7, and the second succeeds since 3*4≠7 (which means that the inner application fails and so its negation succeeds). It is possible to define extensions to logical operations so that they are functions returning values, rather than pure relations [7]; this is one (perhaps idiosyncratic) interpretation of type theory [3,10,12].

The second way in which applications may be related is to provide a means for insisting that values used in various positions (i.e. as different arguments or in different applications) be identical. This has been done above in the interpretation of results of functions; a nested application f(g(a)) is broken into applications of the relations R(f) and R(g). In evaluating the pair, an attempt is made to find some value x such that R(g)(a,x) succeeds; and that same x is used as the first argument of R(f). Mathematical variables fill this role; there is an implicit existential quantification over the domain of values, such that each use of the name of the quantifier refers to the same value. The statement f(g(a))=y is equivalent to

$$\exists x:Dom . R(f)(x,y) \quad \& \quad R(g)(a,x).$$

Of course, if the system of applications is inconsistent then no values can be found for the variables and the system fails. One may imagine each use of a variable to define a possible set of solutions for that variable (perhaps dependent upon others), with the solution of a system of applications being the unification of the constraints of the applications.

The presence or absence of a variable in a given application is ordinarily important only insofar as that application restricts the possible set of values of the variable. For example, if an application A forces a variable x to be one of a given set S of possible values, then the introduction of another, weaker application, conjoined with A, that forces x to be in a superset of S has no effect on the solution of the system of relations (assuming no other variables are then more strongly constrained). Communication does not preserve this; the presence or absence of a variable in an application can affect the success or failure of the variable. In this sense, communication variables bind applications more closely together than mathematical variables; they "transfer" success and failure as well as values. For example, consider the system of equations:

$$x + y = 1 \ \& \ (x - y = 1 \ | \ x = 0).$$

Viewing x and y as mathematical variables, there are two possible solutions; (x,y) can be (1,0) or (0,1). If x is a communication variable, defined such that it must occur on both sides of the conjunction, then again both solutions are possible. However, if y is such a communication variable, the second solution is not allowed, because y is not successful in the second term. The expression x-y=1 fails for (0,1); and the expression x=0 does not contain a successful use of y. It is this notion of requiring a successful use of a variable that distinguishes communication variables from more traditional ones.

There are a number of ways in which usage constraints imposed on variables can be formalized. One may associate an alphabet of variables with each conjunction and disjunction, and indicate whether any given variable must or may be used successfully in each argument; one may subscript each use of a variable, and define its overall success in terms of logical operations applied to instances of those subscripted uses; or one may define scopes or environments as subsystems of applications, requiring that a variable be used exactly once within such a scope, and defining logical operations on the scopes of a variable to indicate its overall success. The first of these is overly complicated, trying to fit one structure onto another; the other two can be viewed as special cases of objects here called communication trees.

A variable is defined as a tree, with a single root arc (in fact, a directed acyclic graph with a single root would be more appropriate; but a tree is simpler and is usually an adequate first approximation). Each node of the tree is associated with a logical operation; each arc of the tree is associated with either success or failure for a given value. The success of each arc leaving a node is determined by applying the node's operation to the arcs entering the node. The leaves of the tree are positions, that may be individual uses of a variable, subsequences of arguments in an application, an entire application, or a logical expression containing applications. The granularity of a leaf does not matter (much as a definition of a function/procedure may have an arbitrarily large body); but some notion of contiguity or connectedness is desirable.

As an example, consider a system of three conjoined applications S, T, and U connected by the variables x, y, and z. The communication variable x is to succeed in each application, so it has the graph

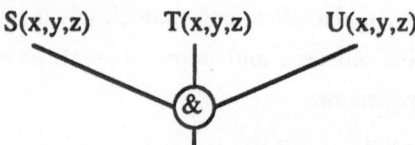

The communication variable y is to succeed in S and either T or U:

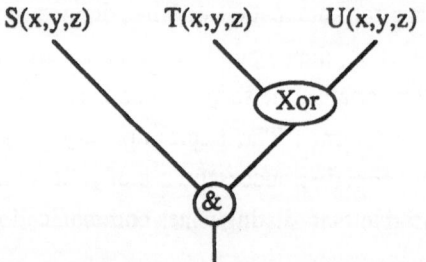

Finally, z is a mathematical variable upon which no additional communications constraints are imposed. Wherever z is present, its graph matches that of the logical operations used to combine applications; so no graph is necessary to describe its link pattern. (Note: if this were to form the basis of a language, a non-planar overlaying style of description would be useful, in which a user could look at the various graphs at will.)

III. Link Manipulation

Given that communication links are viewed as logical graphs, the question arises as to how to create and manipulate them. (This follows from the argument that it should always be possible to program a task that can be done at a terminal; and if a person can construct programs with communicating processes, then so should a machine be able to.) In fact, many cases involving dynamic links can be mapped to the following scenario:

> Given processes A, B, and C, with link ab between A and B, and link bc between B and C. Provide a means by which A and C can communicate without requiring any intervention on the part of B (after an initialization stage).

This necessitates the creation of a link as a passive data structure whose leaves can be transmitted to the points at which they are to be activated. A variable begins (or ends) with the arc leaving its root node, that indicates its ultimate success or failure in finding an acceptable value (and pattern of uses). The function Root is defined to generate such an arc, returning the single leaf at its end. This is a single-use variable, that if activated can take any value. The tree may be extended by transforming the leaf to a node, introducing additional arcs and returning the new leaves so generated. The function Arcs is introduced to do this; it takes as arguments a leaf, a logical operation, and number of branches that are desired, and returns the specified number of leaves, attached by arcs that are related by the logical operation in the node. For example, the expression

Arcs(Root,&,3)

returns the three leaves (here labelled L1, L2, and L3) of the following structure:

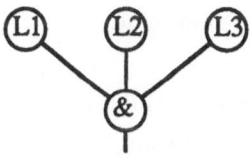

The addition of an activation (evaluation) function Eval is enough to permit the solution of the given scenario as follows:

in A: ... CLink=Eval(ab) ...
in B: ... (ab,bc)=Arcs(Root,&,2) ...
in C: ... ALink=Eval(bc) ...

That is, the link ab has as its value the first arc generated by the Arcs function in B, and the link bc has as its value the second such arc. Within A the variable CLink is identified with the value of the new variable, and within C the variable ALink is so identified. The effect here is exactly the same as would result from:

in A: ... CLink=ab ...
in B: ... ab=bc ...
in C: ... ALink=bc ...

but in the latter case the links are all live; no process can manipulate them further.

Suppose now that C contains two parts C1 and C2, and that exactly one of those parts is to be linked with A. Given the use of Arcs and Eval as above, C can be modified to contain:

in C: ... (A1Link,A2Link)=Arcs(bc,Xor,2) & C1(A1Link) & C2(A2Link) ...

This latter case cannot be described using static links; it requires some form of dynamic linking because the link graph structure does not reflect the application structure.

More interesting examples arise when links are used in temporal contexts. An event sequence operation is defined in the usual way, and associated with ";"; the expression

a;b

indicates that the event b occurs after the event a. That is, there is no chance for b to succeed until some time after a has succeeded. The sequence as a whole succeeds at each time step, as there is an implicit alternative with an empty successful application Ø (i.e. skip); the expression (a;b) can be defined recursively as

a followed_by (b else (Ø;b)),

where followed_by is a temporal relation whose first argument is immediately followed by the second over the minimum discreteness of time, or (taking the limit) to the point at which time is continuous. Events or expressions may be labelled by either single times or intervals. For example,

t1:a ; t2:b ; c ; t3:d

associates t1, t2, and t3 with the times of the events a, b, and d respectively; the time of event c is not named. On the other hand,

(t1..t2):a ; t3:b

indicates that a succeeds over the interval t1 to t2, and only after that can b occur at t3. Temporal inconsistencies result in errors (e.g. (t1:a ; t1-1:b)). The truth of an interval expression is necessary for any event at the start of an interval to succeed. Thus,

t1: a;

((t2..t7): b

& t2: c)

begins as sooon as event a succeeds; that time is associated with t1. Then, when both b and c succeed, t2 occurs. Because b is labelled with an interval, it persists until t7 (perhaps an event in another time sequence).

One problem with links and variables is to distinguish between cases in which a link's value is to persist over time, versus those in which it changes. If a link is used in an event application, then its value is transitory. For example,

t1: x=3;

t2: x=4;

t3: x=5

indicates that the link x has three distinct values, 3, 4, and 5, at the various times t1, t2, and t3. An interval indicates that the value of the link does not change; e.g.

(t1..t3): x=3

specifies that the value of the link x is 3 throughout the entire interval.

A more difficult problem is to specify that two links always have the same value as each other within an interval, but that that value may change; for example, the equation x=y should always hold within (t1..t2), but the actual value of x and y may be (3;4;5) (i.e. x=y=3;x=y=4;x=y=5). It is necessary to break the interval (t1..t2) into sub-intervals, one for each different value of the links. As the number of different values cannot be known in advance, a recursive structure is required. Where (loop::x) associates the name loop with the expression x over the scope of x, and the function "in" returns a value in the given interval, then the expression

t1: loop:: (in(t1..t2): x=y; loop

　　　　　| in(t2...): Ø)

fulfills the desired goal. One might of course devise a shorthand notation for this.

Let us return to the problem of a switchboard process, whose purpose is to connect callers to one another via dynamic links. Each user is plugged into the switchboard via a link, associated with an ID. If user1 wishes to call user2, then it gives the switchboard the ID of user2, and receives a link, which can then be used for the call:

user1:: ... Switchboard=user2_ID; Switchboard=wire; Eval(wire)=message ...

where message is the value to be sent. The switchboard receives the ID over one of its lines, creates a link, and sends one end of the link to the caller and the other to the user being called:

Switchboard:: ... user[i]=ID; (user[i],user[ID])=Arcs(Root,&,2) ...

The called user checks the switchboard occasionally to determine whether or not there are any incoming calls; at one such check, the call is received:

user2:: ... Switchboard=incoming_call; Eval(incoming_call)=messagevar ...

where messagevar is a variable. It is necessary to ensure that a user listening for a call does not connect with the Switchboard waiting for a user to make a call. This is trivially done in any of a number of ways, be it by having separate call and listen links, introducing priorities among the alternative communications, or perhaps most easily by requiring calls and listens to have fields with distinct values (e.g. a caller would set Switchboard=(call,ID) where call is a value). This last corresponds to the use of typing constructors, and depends upon communication being resolved by unification. With such a mechanism, there is no need to wait for the new line; a call could take the form

> Switchboard = (call, ID, wire)

where ID is sent to the switchboard and wire is received from it in the same communication.

IV. Conclusion

One approach to the definition of communication between processes is to defer any notion of evaluation, and view links as variables with constraints imposed such that success of usage affects the overall success of the variable. This leads to the view of a variable in terms of a tree of logical operations applied to references to that variable, with the root arc indicating overall success or failure. This image of a variable as a structure is exploited in the manipulation of dynamic links, by defining functions that can create and extend such graphs, and then cause sub-arcs to "come to life". Arcs that have not been awakened can be communicated as values, and so used to produce direct communication links between processes that would otherwise have to go through intermediaries.

The use of a meta-level, in which variables, applications, and the like are treated as data objects, is quite common in Lisp, but seldom met elsewhere. It is gaining ground in the development of verification techniques, since a proof of correctness may be viewed as a meta-function from a program to its specification; and it is hoped that it will also become popular as a means for describing the rearrangement of communication links in machines that allow for a number of interconnection network patterns.

References

[1]   H. Alexander, Formally-Based Tools and Techniques for Human-Computer Dialogues (Ellis Horwood Ltd., Chichester England, 1987).

[2]   J. A. Bergstra and J. W. Klop, Algebra of Communicating Processes with Abstraction, Report CS-R8403 (Dept. of Comp. Sci., Centre for Math. and Comp. Sci., Amsterdam, Jan. 1984).

[3] R. L. Constable et al., Implementing Mathematics with the Nuprl Proof Development System (Prentice-Hall, Englewood Cliffs NJ, 1986).

[4] S. Gregory, Parallel Logic Programming in Parlog (Addison-Wesley, Wokingham England, 1987).

[5] R. E. Griswold, String and List Processing in Snobol 4 (Prentice-Hall, Englewood Cliffs NJ, 1975).

[6] C. A. R. Hoare, Communicating Sequential Processes (Prentice-Hall, Englewood Cliffs NJ, 1985).

[7] C. M. Holt, An Associative Constructive Logic, Tech. Report 250 (Comp. Lab., U. of Newcastle upon Tyne, Jan. 1988).

[8] C. M. Holt, Concurrent Constructive Logic, Tech. Report 255 (Comp. Lab., U. of Newcastle upon Tyne, Apr. 1988).

[9] B. Jayaraman, Semantics of EqL (IEEE Trans. Soft. Eng. 14 4, April 1988) 472-480.

[10] P. Martin-Lof, Constructive Mathematics and Computer Programming, in: Meth. and Phil. of Sci. VI, Proc. of 6th Int. Cong. Hanover (North-Holland, Amsterdam, 1979).

[11] R. Milner, A Calculus of Communicating Systems, Lecture Notes in Comp. Sci. 92 (Springer-Verlag, Berlin, 1980).

[12] B. Nordstrom and J. Smith, Propositions and Specifications of Programs in Martin-Lof's Type Theory (BIT 24, 1984) 288-301.

[13] M. J. O'Donnell, Equational Logic as a Programming Language (MIT Press, Cambridge MA, 1985).

[14] D. Peleg, Concurrent Dynamic Logic (JACM 34 2, April 1987) 451-479.

[15] A. Pnueli, The Temporal Logic of Programs, in: Proc. 18th IEEE Symp. on Found. of Comp. Sci. (IEEE, New York, 1977) 46-57.

[16] F. Sadri, Three Recent Approaches to Temporal Reasoning, in: A. Galton, Ed., Temporal Logics and Their Applications (Academic Press, Orlando FL, 1987) 121-168.

[17] E. Y. Shapiro, Systems Programming in Concurrent Prolog, in: Proc. 11th Symp. on Principles of Prog. Lang. (ACM, New York, 1984) 93-105.

Formal Environment and Tools Description

for the Analysis of Real time Concurrent Systems

Vangalur S. Alagar

Geetha Ramanathan

Department of Computer Science

Concordia University

1455 de Maisonneuve Blvd West

Montreal, Quebec H3G 1M8.

Canada

Abstract

A formal functional model for describing and reasoning about the behavior of real-time concurrent systems is described. A summary of tools and their operations are given. The usefulness of the model is illustrated through the formal specification of the design and proof of correctness of the design of a robotic navigation controller installed in a rectangular shaped common workspace.

1. Introduction

This paper presents a formal functional model for describing and reasoning about the behavior of real-time concurrent systems. The model is event-based and its basic elements are events, functions, sets and sequences. Generalizing the model due to Caspi and Halbwachs [3], we enrich it with tools for expressing and proving properties of problems in a spectrum of areas such as hardware design, distributed database and concurrent robotic assemblies.

The classical approach to proving the correctness of sequential programs is to model them as finite state transformers and embed the set of input-output assertions in predicate logic wherein correctness or contradictions can be established. In the context of parallel programming, timing constraints induced by synchronization and communication do arise. Several approaches for parallel programming systems [4,6,7] extend the classical proof method by inventing tools to model the behavior of parallel programs that are independent of actual execution times. In such models, two systems exhibiting the same output behavior on identical input streams will be declared identical, although the durations of intermittent or interleaving computations may differ. Hence this modeling approach may not be acceptable for real time concurrent systems, hardware systems and distributed systems.

We must also acknowledge that a formal model supporting an abstract specification should remain independent of specific hardware performances and the load of host computers unless these can be specified in the formal model. Only when this requirement, although stringent, is met, correctness proofs given within the formal model remain valid. Hence, for a correct study of the functional behavior of events in a real time or distributed system, the proposed formalism should be capable of adequately expressing events, event orderings, event executions and event histories in terms of the tools defined within the formal system.

In [3], Caspi and Halbwachs introduced a formalism in which time sequences (which indicate instances of event occurrences) are monotonic increasing functions from integers to time (real numbers). We believe that only a limited set of primitive actions at the hardware level can be instantaneous. As an example, the event send (receive) can be considered instantaneous at the hardware specification level but the actual transmission time is a function of message length. In high level specification of real time or distributed systems, the events send, receive or assign are not instantaneous events. Thus it seems natural to associate events with intervals so that the end points and the length of the interval denote the starting time, the completion time and the duration of the event. This extended formalism, as shown in [1,2], seems sufficient to define and describe the behavior of events in low level hardware systems and is necessary for a high level modelling and description of the behavior of real-time concurrent systems, distributed systems and network architecture.

A brief summary and organization of this paper are the following: Section 2 provides a summary of the formal model and a list of the formal tools. Section 3 lists five problems and illustrates the use of formalism on one of these problems; see [1,2] for other problems. The paper concludes in section 4 with few remarks on related approaches and the future direction of this research.

2. Description of the Formal Model

A complete description of the formal model appears in [1]; in this section we omit formal proofs and provide only a brief outline. The formal model is intended to support distributed and real time behavior in the entire spectrum of systems extending between hardware simulation and industrial process control such as robotics. The primitive objects that we wish to characterize are formalized as *events*. An event is characterized by a sequence of intervals denoting the occurrences of that event. In [7], Lamport considers different occurrences as different events; but in our model these occurrences denoted by the intervals refer to a single event. For example, assigning a value to a variable, sending and receiving messages are events in the system. The activities or the occurrence of events are subject to time constraints due to two important factors: there is an interaction with a physical process, such as sensor, that must be read periodically; or, time is imposed by mutual exclusion, synchronization and ordering of the events. Thus 'what is the effect of an event and when the event causing the effect occurs' are important to be considered together.

The occurrences, executions and the effects of events can be continuous and need not be instantaneous. Although the entire process itself may be nonterminating, the individual events making up the process may be ideally viewed as happening continuously over piecewise continuous intervals. Hence we identify time Π with $\mathbb{R}$, the real line. To deal effectively with extreme cases, we let $\overline{\Pi} = \mathbb{R} \cup \{\infty\} \cup \{-\infty\}$ and $\overline{N} = \mathbb{N} \cup \{0\} \cup \{+\infty\}$.

2.1 Basic concepts of event and time

The basic assumptions on time and events are:

[A1] An event can occur any number of times within a system; however within a finite period of time, there can be only a finite number of occurrences.

[A2] An event may occur continuously between its start time and completion time. So, each occurrence associates an interval with it and the event history is given by the associated sequence of intervals.

The following definition is based on these assumptions:

Let INC denote the set of all non-decreasing functions from $\overline{N}$ to $\overline{\Pi}$. Define the higher order functions TIME_1 and TIME_2,

$$\text{TIME}_j : E \rightarrow (\overline{N} \rightarrow \overline{\Pi}), \ j = 1,2$$

where E is the set of all events and $\text{TIME}_2(e) \geq \text{TIME}_1(e)$. Thus the set of all events is embedded into the set

$$L = \{(f_1, f_2) | \ f_1, f_2 \in \text{INC}, \ f_2 \geq f_1\}.$$

The function comparison is done pointwise. Notice that $(\text{INC}, \leq)$ is a poset and every interval of this poset is a member of L. Although every event in E is mapped onto an interval in L (and hence a sequence of intervals on the real line), corresponding to an arbitrary interval in L there may not be an event in the system. By admitting empty (vacuous) events that do not have any effect in the system and letting them correspond to such intervals, we can overcome this and thus identify E with L.

Example 1

Let $a_i(c_i)$ and $b_i(d_i)$ denote the start and finish times of the i-th occurrence of the event $e(f)$. Then,

$$\text{TIME}_1(e)(i) = a_i \ ,$$
$$\text{TIME}_2(e)(i) = b_i \ ,$$
$$\text{TIME}_1(f)(i) = c_i \ ,$$
$$\text{TIME}_2(f)(i) = d_i \ ,$$

$a_i \leq b_i$, $c_i \leq d_i$. Figure 1 shows a sample of event occurrences.

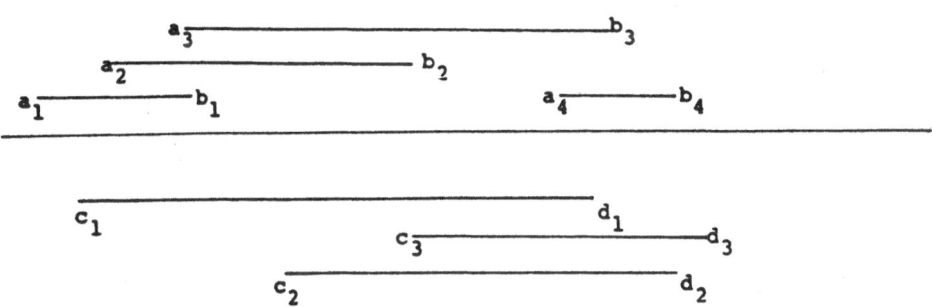

Figure 1

The following properties of $TIME_j$, $j = 1, 2$ are assumed:

[t_1] $TIME_j(e)$, $j = 1, 2$ are monotonic non-decreasing functions.

[t_2] $TIME_2(e)(n) = + \infty$, $n \neq \infty$ means that the n-th occurrence of e is not yet complete.

[t_3] $TIME_j(e)(0) = - \infty$

[t_4] $TIME_j(e)(+ \infty) = + \infty$

Due to [t_3] and [t_4] it is sufficient to consider $TIME_1(e)(n)$ and $TIME_2(e)(n)$, $1 \leq n < \infty$.

The history of a variable in a system is captured by the sequence of values and the times (with duration) of assignment of such values. If $v \in V$ is a variable, then $ASSIGN(v)$ is an event in E and $TIME_1(ASSIGN(v))(k)$ and $TIME_2(ASSIGN(v))(k)$ denote the start and completion times of k-th assignment to v, a value from its domain. Thus we have the functions,

ASSIGN : V → E

VALUE : V → (N → DOM),

where DOM is the set of values necessary for describing the system.

Although the time sequences associated with $TIME_1(e)$ and $TIME_2(e)$ are monotonic non-decreasing, $TIME_1(e)(n)$ and $TIME_2(e)(n)$ need not correspond to the start time and finish time of the n-th occurrence of e. That is, the durations of different occurrences of an event need not be the same; however for every n, there is a k ∈ $\overline{N}$ such that $[TIME_1(e)(n), TIME_2(e)(k)]$ is the interval corresponding to the n-th occurrence of e. Thus, there exists a bijective function $P_e : \overline{N} \to \overline{N}$ with $P_e(0) = 0$, $P_e(\infty) = \infty$ such that the composite function $CTIME_2(e) = TIME_2(e) \circ P_e$ is *not* in general monotonic, but ∀n ∈ N, $[TIME_1(e)(n), CTIME_2(e)(n)]$ define the intervals corresponding to the occurrences of e. Since P_e is bijective, it is easy to see that ∀n, the intervals $\{[CTIME_1(e)(n), TIME_2(e)(n)]\}$, where $CTIME_1(e) = TIME_1(e) \circ P_e^{-1}$, define the same set of intervals as $\{[TIME_1(e)(n), CTIME_2(e)(n)]\}$.

An event e is called single occurrent if $TIME_1(e)(n) \geq TIME_2(e)(n-1)$, n ≥ 1. *Example 2*

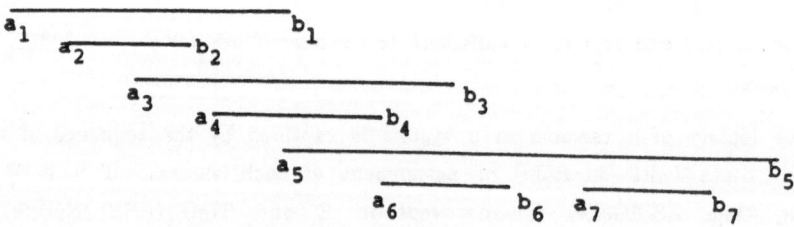

Figure 2

The occurrences of e are the intervals $[a_n, b_n]$, $n \geq 1$; Figure 2 shows the first seven occurrences. The sequences and the bijective map are:

$$\text{TIME}_1(e) \leftrightarrow \{a_1, a_2, a_3, a_4, a_5, a_6, a_7 \ldots\}$$

$$\text{TIME}_2(e) \leftrightarrow \{b_2, b_1, b_4, b_3, b_6, b_7, b_5, \ldots\}$$

$P_e : \mathbf{N} \rightarrow \mathbf{N}$,

$\quad P_e(1) = 2, \ P_e(2) = 1, \ P_e(3) = 4$

$\quad P_e(4) = 3, \ P_e(5) = 7, \ P_e(6) = 5,$

$\quad P_e(7) = 6, \ \ldots$.

The next two examples show the usefulness and the expressive power of the formalism in the specification of timing constraints.

Example 3

In robotics certain global sensor variables must be bound to particular sensors. The value of the sensor variable at any instant is the current measurement of the sensor. Such values may have to be monitored continuously or once measured may remain constant over a period of time. Thus, if we want to state that sensor variables are piecewise constants (step functions) with periodicity δ, we can express as:

$$\forall n \in \mathbf{N}, \ \text{TIME}_1(\text{ASSIGN}(s))(n + 1)$$
$$= \text{CTIME}_2(\text{ASSIGN}(s))(n) + \delta.$$

Hence, $\forall t \in [\text{CTIME}_2(\text{ASSIGN}(s))(n), \ \text{TIME}_1(\text{ASSIGN}(s))(n + 1)]$
the value of s, namely $\text{VALUE}(s)(n)$ is a constant.

Example 4

Effector variables denote the parameters of a robot's end effectors and a change in the value of this variable causes the robot to move. Since the end effector should move continuously, the function $y = f(x)$ describing the motion should be computed in real time; that is, $y_n = f(x_n)$ where y_n, x_n are n-th values computed for y and x respectively. However, a response time δ between the receipt of the value x and the sending of the corresponding value y (to the end effector) is frequently assumed. Similar to the remarks in [1], we give three different interpretations:

a) The n-th value of y must be computed from the n-th value of x:

$\forall n \in \mathbf{N} \quad VALUE(y)(n) = f(VALUE(x)(n))$ and $CTIME_2(ASSIGN(x))(n) \leq TIME_1(ASSIGN(y))(n) \leq CTIME_2(ASSIGN(x))(n) + \delta$

b) The value of y may be obtained at a lower frequency than the value of x, but a value of y cannot be computed from a value of x issued at time exceeding δ:

$\forall n \in \mathbf{N}, \exists m \in \mathbf{N},$
$VALUE(y)(n) = f(VALUE(x)(m)),$
$CTIME_2(ASSIGN(x))(m) \leq TIME_1(ASSIGN(y))(n) \leq CTIME_2(ASSIGN(x))(m) + \delta.$

c) The value of y must be computed more frequently than x.

$\forall m \in \mathbf{N}, \exists n \in \mathbf{N},$
$VALUE(y)(n) = f(VALUE(x)(m)),$
$CTIME_2(ASSIGN(x))(m) \leq TIME_2(ASSIGN(y))(n) \leq CTIME_2(ASSIGN(x))(m) + \delta.$

2.2 An Algebraic Structure for E

An algebraic system is a set together with certain relations and operations defined on the set. An ordered set as well as the set of complex numbers are simple examples of algebraic systems in mathematics. In this section we shall first make E an algebraic system and then characterize its properties.

Two events $e, f \in E$ are equal, written $e = f$, if $TIME_1(e) = TIME_1(f)$ and $CTIME_2(e) = CTIME_2(f)$. Note that two equal events may have different effects in the system and hence are not identical. For example, if for two variables x, y in the system, $TIME_1(ASSIGN(x))(n) = TIME_1(ASSIGN(y))(n)$ and $CTIME_2(ASSIGN(x))(n) = CTIME_2(ASSIGN(y))(n)$ then the two events of assigning values to x and y are equal but not identical events. In short, equal events simultaneously occur all the time.

For two events e and f in E, denote $e \leq f$, if $\forall n \in N$, the start and completion times of the n-th occurrence of e do not happen earlier than the start and completion times of the n-th occurrence of f. Notice that the start time of e may happen before the completion time of f:

$$e \leq f \text{ if } [TIME_1(e) \geq TIME_1(f)] \wedge [CTIME_2(e) \geq CTIME_2(f)]$$

Based on this definition and our discussion on time functions, we state the next theorem.

Theorem 1
i) $(E, \leq)$ is a poset
ii) Let $INC = \{f | f : \overline{N} \to \overline{\Pi}, f \text{ is monotonic non-decreasing}\}$
     and $L = \{(f_1, f_2, P) \mid f_1, f_2 \in INC, P : \overline{N} \to \overline{N}, P \text{ is bijective},$
     $f_2 \geq f_1, f_2 \circ P \geq f_1\}.$

The poset $(E, \leq)$ is isomorphic to the algebraic system L.

Two events e and f in E are not related by $\leq$ if and only if there is an interval in the sequence characterizing e (f) which covers the corresponding interval characterizing f (e). That is, $\exists$ n such that

either
$$TIME_1(e)(n) < TIME_1(f)(n) < CTIME_2(f)(n) < CTIME_2(e)(n)$$
or
$$TIME_1(f)(n) < TIME_1(e)(n) < CTIME_2(e)(n) < CTIME_2(f)(n).$$

In order to compare every pair of elements in $(E, \leq)$, we define glb, the greatest lower bound and lub, the least upper bounds for events in E. For two events $e, f \in E$, define $g \in E$, $g = glb(e, f)$:

$$TIME_1(g) = max\{TIME_1(e), TIME_1(f)\}$$
$$CTIME_2(g) = max\{CTIME_2(e), CTIME_2(f)\}.$$

Both $TIME_2(g)$ and P_g are obtained from $CTIME_2(g)$:

$TIME_2(g)$ is the sorted sequence of $CTIME_2(g)$ and P_g is the bijection forcing the sorted sequence.

Similarly, we define $h = lub(e, f)$:

$$TIME_1(h) = min\{TIME_1(e), TIME_1(f)\}$$
$$CTIME_2(h) = min\{CTIME_2(e), CTIME_2(f)\}.$$

The functions $TIME_2(h)$ and P_h are obtained from the sorted sequence of $CTIME_2(h)(n)$, $n \geq 1$.

It is easy to see that $(E, \leq, glb, lub)$ is a lattice having a greatest element. The lattice is not complete because it does not have a least element.

Since event occurrences need not overlap, we now study the structure of E under a strict precedence relation '$\ll$' (follows). We say that an event e 'follows' an event f and write $e \ll f$, if $\forall n \in N$, the start time of the n-th occurrence of e is greater than or equal to the completion time of the n-th occurrence of f:

$$e \ll f \text{ if } CTIME_2(f) \leq TIME_1(e)$$

The next theorem gives two properties of $\ll$.

<u>Theorem 2</u>

(i) $e \ll f \rightarrow e \leq f$

(ii) $(E, \ll)$ is an irreflexive poset.

In general, any two events e, f in E need not have a glb in $(E, \ll)$; however, in later section, we give a criteria to be satisfied by e, f for $glb(e,f)$ (under the relation $\ll$) to be a member of E.

2.3   <u>Operators, Relations, Counters and Properties</u>

In this section we first state the generalization of some of the operators and functions introduced in [1] and then prove new theorems characterizing events in our generalized formalism.

2.3.1   Delay Relation and Shift Function

For events $e \in E$ such that $P_e = I$ and for $\delta \geq 0$, we define the delay relation $\Delta_\delta(e)$ as follows :

$(e,f) \in \Delta_\delta(e)$ if $P_f = I$ and $TIME_1(f) = TIME_2(e) + \delta$.
It is clear that $f \ll e$ if $(e,f) \in \Delta_\delta(e)$.

The shift function S_δ, $\delta \geq 0$ is defined on E by

$$S_\delta(e) = f \text{ if } TIME_1(f) = TIME_1(e) + \delta \text{ and } \qquad TIME_2(f) = TIME_2(e) + \delta.$$

It is clear that S_δ shifts every occurrence of e by a constant amount δ and hence $S_\delta(e) \leq e$. If δ exceeds the maximum duration of all occurrences of e, $S_\delta(e) \ll e$.

2.3.2 Subevents and Subsequences Relations

An event f is called a subevent of e, $f \subseteq e$, if there exists an increasing function $r \in INC$ such that

$$TIME_1(f) = TIME_1(e) \circ r$$
and
$$CTIME_2(f) = CTIME_2(e) \circ r.$$

Example 5

In Fig. 3, the occurrences of an event e and a subevent f are shown.

$$TIME_1(e) \leftrightarrow (a_1, a_2, a_3, a_4, a_5, a_6)$$

$$TIME_2(e) \leftrightarrow (b_1, b_3, b_2, b_5, b_4, b_6)$$

$$P_e = \begin{pmatrix} 1 & 2 & 3 & 4 & 5 & 6 \\ 1 & 3 & 2 & 5 & 4 & 6 \end{pmatrix}$$

$$TIME_1(f) = (a_2, a_4, a_5)$$

$$TIME_2(f) = (b_2, b_5, b_4)$$

$$r : r(1) = 2, \ r(2) = 4, \ r(3) = 5$$

It is easy to verify that

$$\text{TIME}_1(f) - \text{TIME}_1(e) \circ r$$

$$\text{CTIME}_2(f) - \text{CTIME}_2(e) \circ r.$$

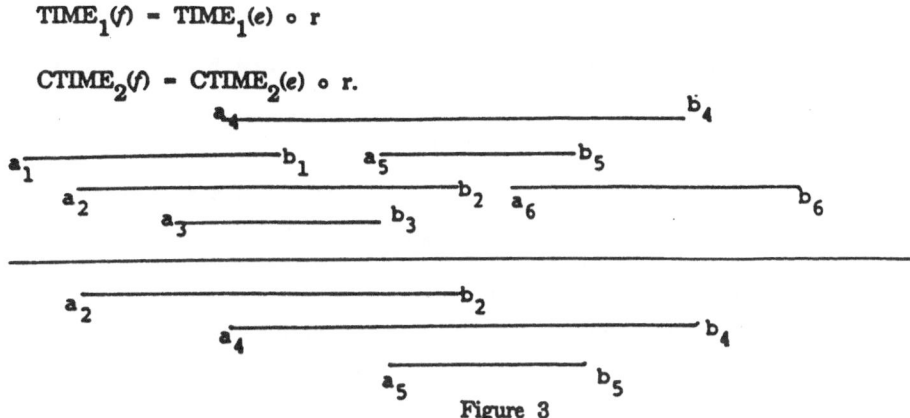

Figure 3

Theorem 3

$f \subseteq e \Rightarrow f \leq e$, iff $P_e - \text{I}$.

2.3.3 Counters

Loose and strict counters are inverse functions to the time functions. Using the Galois representation results [10], we know that $\text{TIME}_1(e)$ and $\text{TIME}_2(e)$ are infinitely supremum and infinitely infimum distributive functions. That is, for any subset E_1 of E,

$$\sup\{\text{TIME}_1(e) \mid e \in E_1\} - \text{TIME}_1\{\sup(e) \mid e \in E_1\}$$

and

$$\inf\{\text{TIME}_2(e) \mid e \in E_1\} - \text{TIME}_2\{\inf(e) \mid e \in E_1\}$$

It is proved in [10] that such functions admit inverses; that is, there are two functions $\text{TIME}_1(e)^-$, $^-\text{TIME}_1(e)$ which represent *loose* and *strict* counters:

$TIME_1(e)^- \equiv LCOUNT_1(e) : \overline{\Pi} \to \overline{N}$ where

$LCOUNT_1(e)(t)$ is the number of initiations of e upto and including the time t.

$^-TIME_1(e) \equiv COUNT_1(e) : \overline{\Pi} \to \overline{N}$, where

$COUNT_1(e)(t)$ is the number of initiations of e strictly before t.

In the same fashion, we can define two inverse functions for $TIME_2(e)$:

$TIME_2(e)^- \equiv LCOUNT_2(e) : \overline{\Pi} \to \overline{N}$, where

$LCOUNT_2(e)(t)$ is the number of completed occurrences of e upto and including t.

$^-TIME_2(e) \equiv COUNT_2(e) : \overline{\Pi} \to \overline{N}$, where

$COUNT_2(e)(t)$ is the number of completed occurrences of e strictly before t.

The next theorem summarizes the precedence relations on counters induced by the precedence relations on the events E.

<u>Theorem</u> 4

(1) For $e \in E$,

$COUNT_1(e) \leq LCOUNT_1(e)$ and $COUNT_2(e) \leq LCOUNT_2(e)$

(2) If $e, f \in E$ and $f \leq e$ then the precedence among the counters is shown in Fig. 4, where $x \ldots \to y$ means $x \leq y$.

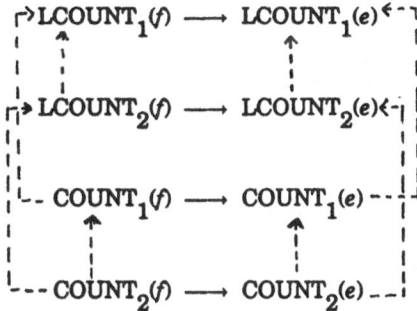

Figure 4

In general, $[COUNT_1(e) \leq COUNT_1(f)]$ and $[COUNT_2(e) \leq COUNT_2(f)]$ does not imply $e \leq f$; however, if $P_e = I$ and $P_f = I$ then the result is true.

There are several precedence relations on the counters induced by the strict precedence operator $\ll$ on E. For example, $e \ll f$ implies:

$$COUNT_2(e) \leq COUNT_1(e) \leq COUNT_2(f) \leq COUNT_1(f)$$
and
$$LCOUNT_2(e) \leq LCOUNT_1(e) \leq LCOUNT_2(f) \leq LCOUNT_1(f).$$

However, $[COUNT_1(e) \leq COUNT_2(f)] \rightarrow e \ll f$ if $P_f = I$.

Theorem 5

For $e \in E$, $\delta > 0$, $f = S_\delta e$

$$COUNT_1(f) = COUNT_1(e) \circ (I - \delta)$$

$$COUNT_2(f) = COUNT_2(e) \circ (I - \delta)$$

$$\text{LCOUNT}_1(f) - \text{LCOUNT}_1(e) \circ (I - \delta)$$

$$\text{LCOUNT}_2(f) - \text{LCOUNT}_2(e) \circ (I - \delta),$$

where I is the identity function. If $(e,f) \in \Delta_\delta(e)$, then

$$\text{COUNT}_1(f) = \text{COUNT}_2(e) \circ (I - \delta)$$

$$\text{LCOUNT}_1(f) - \text{LCOUNT}_2(e) \circ (I - \delta).$$

2.3.4 Sum of events

If an event g is initiated whenever e is initiated or f is initiated and the initiated occurrence of g terminates when the corresponding occurrence of e or f terminates, then g is called the sum of events e and f. Hence it follows that the TIME_j sequence of g is the merge sequence of the TIME_j sequences of e and f; a more formal definition follows Example 6. Based on the merge sequence, a definition of the sum in terms of the counters can also be given: g is the sum $e + f$ if

$$\text{LCOUNT}_1(g) - \text{LCOUNT}_1(e) + \text{LCOUNT}_1(f)$$

$$\text{LCOUNT}_2(g) - \text{LCOUNT}_2(e) + \text{LCOUNT}_2(f)$$
or equivalently

$$\text{COUNT}_1(g) = \text{COUNT}_1(e) + \text{COUNT}_1(f)$$

$$\text{COUNT}_2(g) = \text{COUNT}_2(e) + \text{COUNT}_2(f).$$

The next example illustrates the computation of $e + f$ from e and f.

Example 6

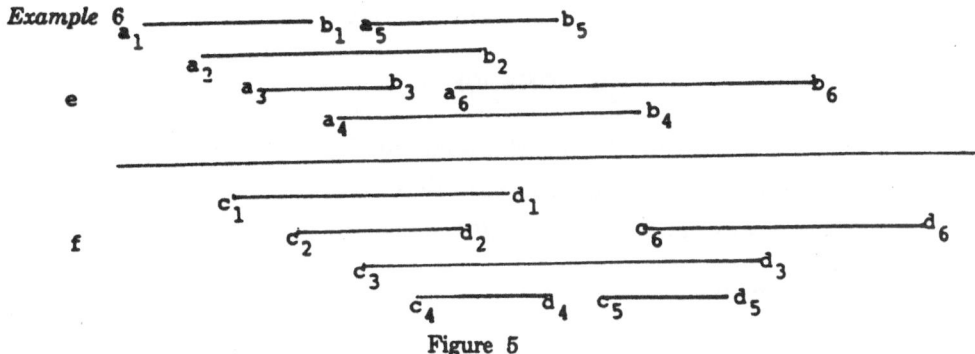

Figure 5

From the occurrences of events e and f shown in Fig. 5, we have

$\text{TIME}_1(e) \leftrightarrow \{a_1, a_2, a_3, a_4, a_5, a_6\}$

$\text{TIME}_1(f) \leftrightarrow \{c_1, c_2, c_3, c_4, c_5, c_6\}$

$\text{TIME}_2(e) \leftrightarrow \{b_1, b_3, b_2, b_5, b_4, b_6\}$

$\text{TIME}_2(f) \leftrightarrow \{d_2, d_1, d_4, d_5, d_3, d_6\},$

$P_e = \begin{pmatrix} 1 & 2 & 3 & 4 & 5 & 6 \\ 1 & 3 & 2 & 5 & 4 & 6 \end{pmatrix}$

$P_f = \begin{pmatrix} 1 & 2 & 3 & 4 & 5 & 6 \\ 2 & 1 & 5 & 3 & 4 & 6 \end{pmatrix}$

Let $g = e + f$.

$\text{TIME}_1(g) = \{a_1, a_2, c_1, a_3, c_2, a_4, c_3, a_5, c_4, a_6, c_5, c_6\}$

$\text{TIME}_2(g) = \{b_1, b_3, d_2, b_2, d_1, d_4, b_5, b_4, d_5, d_3, b_6, d_6\}$

$P_g = \begin{pmatrix} 1 & 2 & 3 & 4 & 5 & 6 & 7 & 8 & 9 & 10 & 11 & 12 \\ 1 & 4 & 5 & 2 & 3 & 8 & 10 & 7 & 6 & 11 & 9 & 12 \end{pmatrix}$

In general, we compute $g = e + f$ as described below:

1) $\text{TIME}_1(g) = \text{merge}\{\text{TIME}_1(e), \text{TIME}_1(f)\}$

Among the occurrences having the same starting time, the order is defined depending on their corresponding completion time.

$\text{TIME}_2(g) = \text{merge}\{\text{TIME}_2(e), \text{TIME}_2(f)\}$

2) Define the functions

$$M_1 : \mathbf{N} \to \{1, 2\} \times \mathbf{N}$$

$$M_2 : \{1, 2\} \times \mathbf{N} \to \mathbf{N}$$

such that

$M_1(n) = (1, i)$, if the n-th element in $\text{TIME}_1(g)$ sequence is the i-th element from $\text{TIME}_1(e)$ sequence.

$\quad = (2, j)$, if the n-th element in $\text{TIME}_1(g)$ sequence is the j-th element from $\text{TIME}_1(f)$ sequence.

$M_2(1, i) = n$, if the n-th element in $\text{TIME}_2(g)$ sequence is the i-th element from $\text{TIME}_2(e)$; $M_2(2, j) = n$, if the n-th element in $\text{TIME}_2(g)$ sequence is the j-th element of the sequence $\text{TIME}_2(f)$.

The bijection P_g is defined by,

$P_g(n) = M_2(1, P_e(i))$, if $M_1(n) = (1, i)$

$\quad = M_2(2, P_f(j))$, if $M_1(n) = (2, j)$.

It is clear from the definition and Example 6, that $e \subseteq e + f$ and $f \subseteq e + f$.

Further for $e, f \in E$, if $P_{e+f} = I$ then $glb(e,f)$ and $lub(e,f)$ can be defined in the poset $(E,<<)$ as follows :

$$g = glb(e,f) \text{ where}$$
$$\text{TIME}_1(g) = \max \{\text{TIME}_2(e), \text{TIME}_2(f)\},$$
$$\text{TIME}_2(g) = \text{TIME}_1(g),$$

and $P_g = I$.

$$h = lub(e,f), \text{ where}$$
$$\text{TIME}_1(h) = \min \{\text{TIME}_1(e), \text{TIME}_1(f)\},$$
$$\text{TIME}_2(h) = \text{TIME}_1(h),$$

and $P_h = I$.

Then, $g, h \in (E, <<)$.

A complete characterization of the sum structure is given by five theorems stated in the next section. Now, we shall show that the extended formalism admits other relations and functions similar to those discussed in [1].

For every $j \in N$, we define an integer event j as

$$\text{TIME}_1(j)(n) = \begin{cases} -\infty, & n \leq j \\ +\infty & n > j \end{cases}$$

$$\text{TIME}_2(j)(n) = \text{TIME}_1(j)(n)$$

and

$P_j = I$, the identity function.

In particular the integer event 1 is characterized by $(\text{TIME}_1(1), \text{TIME}_1(1), I)$, where

$$\text{TIME}_1(1) \leftrightarrow (-\infty, +\infty, +\infty, ...)$$

$$\text{TIME}_2(1) \leftrightarrow (-\infty, +\infty, +\infty, ...)$$

Thus for any event $e \in E$ with

$$TIME_1(e) \leftrightarrow (a_1, a_2, a_3, ...)$$

$$TIME_2(e) \leftrightarrow (b_1, b_2, b_3, ...)$$

we have $f = e + 1$ defined by the functions

$$TIME_1(f) = (-\infty, a_1, a_2, ...)$$

$$TIME_2(f) = (-\infty, b_1, b_2, ...)$$

$$P_f(n) = \begin{cases} 1, & n = 1 \\ P_e(n-1) + 1, & n > 1 \end{cases}$$

It is easy to show that
$$e \leq e + 1 \text{ iff } P_e = I,$$
and $\quad e \ll e + 1$ iff e is single-occurrent.

Several useful functions arise from functional compositions of counter functions with those already defined. Three of them are described next.

[F1]   A variable in the system has a sequence of values. The value of x at any instant should be dependent on the event ASSIGN(x), especially its completion time.

Thus there are two functions

CURRENT(x), LCURRENT(x) : $\overline{\Pi} \rightarrow$ DOM(x)
defined by
$$CURRENT(x) = VALUE(x) \circ COUNT_2(ASSIGN(x))$$

$$LCURRENT(x) = VALUE(x) \circ LCOUNT_2(ASSIGN(x)).$$

Hence if e - ASSIGN(x) and $COUNT_2(e)(t) \neq LCOUNT_2(e)(t)$, then LCURRENT($x$)($t$) is a more recent value of x than CURRENT(x)(t). Moreover if y - $f(x)$, then CURRENT(y) and LCURRENT(y) will have three interpretations depending on the renewal period as explained in Example 4.

[F2] For every event $e \in E$, we define $LAST_1(e)$, $LAST_2(e)$ and $NEXT_1(e)$, $NEXT_2(e)$ giving the last and next occurrences of e:

$$LAST_1, \ LAST_2, \ NEXT_1, \ NEXT_2 \ : \ E \rightarrow (\overline{\Pi} \rightarrow \overline{\Pi})$$

$LAST_1(e)(t)$ - t_1, if the last initiated time of the event e strictly before t was t_1.

$LAST_2(e)(t)$ - t_2, if the last completed time of the event e strictly before t was t_2.

$NEXT_1(e)(t)$ - t_3, if t_3 is the first initiation time of e strictly after time t.

$NEXT_2(e)(t)$ - t_4, if t_4 is the first completion time of e strictly after time t.
It is easy to see that

$$LAST_1(e) \ - \ TIME_1(e) \ \circ \ COUNT_1(e)$$

$$LAST_2(e) \ = \ TIME_2(e) \ \circ \ COUNT_2(e)$$

$$NEXT_1(e) \ - \ TIME_1(e) \ \circ \ LCOUNT_1(e + 1)$$

$$NEXT_2(e) \ - \ TIME_2(e) \ \circ \ LCOUNT_2(e + 1).$$

Some applications require the starting time of the last completed occurrence of e strictly before t or/and the starting time of the first completed occurrence of e strictly after t. These are given by

$$SLAST_2(e) - TIME_1(e) \circ P_e^{-1} \circ COUNT_2(e)$$

and

$$SNEXT_2(e) - TIME_1(e) \circ P_e^{-1} \circ LCOUNT_2(e + 1).$$

By relaxing strictly before (or strictly after) to include the time of observation, we get the following six functions:

$$LLAST_1(e) - TIME_1(e) \circ LCOUNT_1(e)$$

$$LLAST_2(e) - TIME_2(e) \circ LCOUNT_2(e)$$

$$SLLAST_2(e) - TIME_1(e) \circ P_e^{-1} \circ LCOUNT_2(e)$$

$$LNEXT_1(e) - TIME_1(e) \circ COUNT_1(e + 1)$$

$$LNEXT_2(e) - TIME_2(e) \circ COUNT_2(e + 1)$$

$$SLNEXT_2(e) - TIME_1(e) \circ P_e^{-1} \circ COUNT_2(e + 1)$$

These functions show the importance of integer event and event sum in their definition.

[F3]   A condition C is a function from $\overline{\Pi}$ to {true, false} with $C(-\infty)$ - true. If x is a variable with $DOM(x)$ = {true, false}, then CURRENT(x) and LCURRENT(x) are conditions. For $e, f \in E$,
$(LAST_1(e) - LAST_2(f))$ and $(COUNT_2(e) - COUNT_2(f))$ are conditions.

An event f which is initiated whenever C is true at the completion of e, where P_e - I, is denoted by $e \mid C$. Thus a formal definition is the following:

Let

$X = \{f \mid f \in E, \exists\, r : N \to N$ such that

$\quad$ $[r \circ COUNT_2(e) = COUNT_1(f)] \wedge [C \circ LAST_1(f) = true] \wedge$

$\quad$ $[[LAST_2(e) \neq LAST_1(f)] \to [C \circ LAST_2(e) = false]]\}$.

Now, every member of X is an event $e \mid C$; that is, X denotes a class of events which begin whenever e finishes and C is true.

$\quad$ It is clear that if $f = e \mid C$ and $g = e \mid \sim C$ then

1. $\quad COUNT_2(e) = COUNT_1(f) + COUNT_1(g)$
2. $\quad C \circ LAST_1(f) = \sim C \circ LAST_1(g) = true$.

$\quad$ Let $e \mid_s C$ denote an event f which starts whenever e starts and the condition C is true. We define

$Y = \quad \{f \mid f \in E, \exists\, r : N \to N$ such that

$\quad$ $[r \circ COUNT_1(e) = COUNT_1(f)] \wedge [C \circ LAST_1(f) = true] \wedge$

$\quad$ $[[LAST_1(e) \neq LAST_1(f)] \to [C \circ LAST_1(e) = false]]\}$.

Any member of Y is an event $e \mid_s C$. If $f = e \mid_s C$ and $g = e \mid_s \sim C$ then

$\quad COUNT_1(e) = COUNT_1(f) + COUNT_1(g)$.

$\quad$ Continuous monitoring of significant events give rise to exception handlings in real time systems. Thus, if S_1 must be activated when a condition C becomes true, S_1 must remain active when C remains true and S_1 must be aborted and S_2 simultaneously activated when C becomes false, the specification can be given as follows :

$\quad$ Let *clock* be the event of ticking of the system clock. Let s_1 and s_2 be the events denoting the invocation of S_1 and S_2 respectively.

$$s_1 = clock \mid C \wedge [COUNT_2(s_1) = COUNT_1(s_1)]$$

$$s_2 = clock \mid {\sim}C \wedge [COUNT_2(s_2) = COUNT_1(s_2)]$$

$$[[COUNT_1(s_1) \circ LAST_1(s_2) \neq COUNT_2(s_1) \circ LAST_1(s_2)]$$

$$\rightarrow [TIME_2(s_1) \circ COUNT_1(s_1) \circ LAST_1(s_2) = LAST_1(s_2)]]$$

During an occurrence of the event s_1, if C becomes false then we define the completion time of that occurrence to indicate that s_1 is aborted.

2.3.5   The Structure of Sum of Events Induced by the Precedence Relations

For our discussion in this section we let $e_i = TIME_1(e)(i)$, $e_i' = CTIME_2(e)(i)$, $i \geq 1$, $e \in E$. Moreover we identify an event e with these two time sequences and let $dur(e_i)$ denote the duration of i-th occurrence of e; that is, $dur(e_i) = CTIME_2(e)(i) - TIME_1(e)(i)$, $i \geq 1$.

For instantaneous events, the sum is order preserving; but this is not so in our model. The following theorems give a set of sufficient conditions for preserving order in the sum of events.

Theorem 6

For any three events $e, f, g \in E$, $e \ll f \Rightarrow (e + g) \ll (f + g)$ if for every i, $f_i' \leq g_i$, $g_i' \leq e_i$, and $P_f = P_g = I$, $P_{f + g} = I$.

The event occurrences shown in Fig. 6 suggest that $e \leq f$ does not imply $e + g \leq f + g$; the next theorem provides a set of sufficient conditions for preserving the sum of events under the partial order $\leq$.

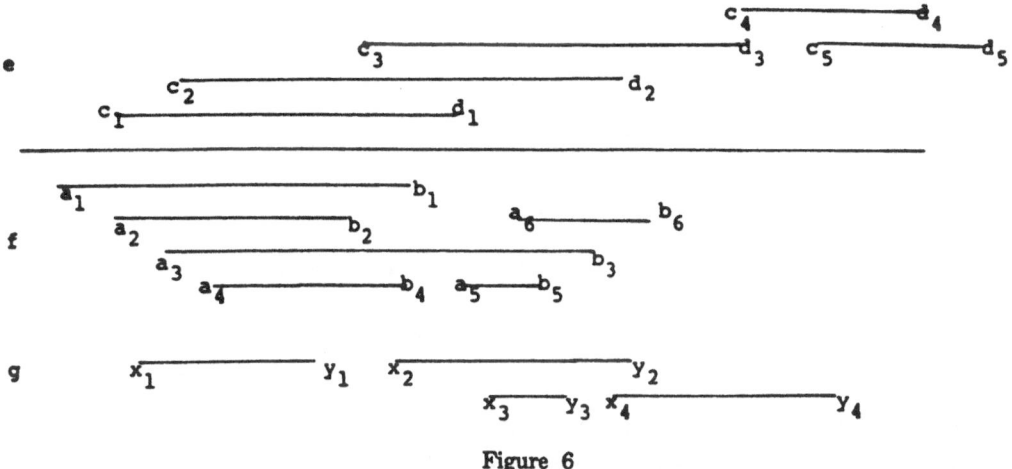

Figure 6

Theorem 7

Let $MAX = \max\{dur(f_i)\}$ and $MIN = \min\{dur(e_i)\}$.
For $e, f, g \in E$, $e \leq f \Rightarrow e + g \leq f + g$ if $MAX \leq dur(g_i) \leq MIN$, $\forall i$ and $P_g = I$.

Theorem 8

Let $e \ll f$, $g \ll h$. If $P_{f+h} = I$, then $e + g \ll f + h$.

Theorem 9

$\forall e, f, g \in E$, $e \ll g$ and $f \ll g \Rightarrow e + f \ll g$

if $TIME_2(g)(n) \leq TIME_1(e)(k)$, and $TIME_2(g)(n) \leq TIME_1(f)(k)$, $n, k \geq 1$.

The next and final theorem of this section gives the strict precedence additivity of the last theorem under subadditive property of strict counters.

<u>Theorem</u> 10

For events e, f, $g \in$ E, let

$$\text{COUNT}_1(e) + \text{COUNT}_1(f) \leq \text{COUNT}_2(g).$$

Then $e \ll g$ and $f \ll g \Rightarrow e + f \ll g$ if $P_g = I$.

(The theorem asserts that for each occurrence of g there is utmost one occurrence of e or one occurrence of f but not both.)

3. Specification and Proof of Correctness

The tools of the formalism introduced in Section 2 have been applied in formulating the specification and deriving the proof of correctness of five problems :

1) a distributed asynchronous bus arbiter design

2) a multiple copy update problem and a synchronous broadcast based solution

3) automatic assembly of parts with two arms of a robot

4) the distributed control and majority consensus approach to concurrency control for multiple copy databases, and

5) the design of an asynchronous navigation controller for the coordinated motion of multiple robots in a common workspace.

See [1] for a discussion of the first three problems and [2] for a formal description and proof of the fourth problem. Below we give an informal statement of problem 5 and follow it up with a formal treatment.

<u>Design of a navigation controller</u>

Many robotic applications exhibit a significant degree of concurrency and sensor-based robots require real-time concurrency control. For example consider the coordinated motion of several robots in a common workspace. Some of the significant actions characterizing this coordinated workspace are communication between robots, communication between robots and external devices, synchronization with external events, waiting and monitoring for some event occurrences and dealing with concurrent activities. For simplicity, we assume that cartesian robots move in a large rectangular-shaped workspace and the only requirement is that collisions must be avoided. Let us also assume that the workspace has sufficient width so that three robots, each in a lane of its own, can move simultaneously. Under this assumption we give the design specification and the proof of correctness of the design of a distributed navigation controller that is to be installed at the intersections.

This problem can be further simplified and abstracted to the design of a traffic controller that is to be installed at the intersection of two two-way highways (east-west and north-west) so that cars coming at the intersection can pass the intersection in a finite amount of time with no collision. Notice that both in the original problem (with robots) and in the abstracted problem (with cars), the speed and the space (width of lane) required are not required in the specification of the controller.

We assume the following : 1) at a predetermined proximity to the intersection there are three lanes on each highway; 2) a car in the right lane must turn right; a car in the middle lane goes straight and a car in the left lane must turn left; 3) the traffic controller at the intersection should allow the car that arrives at the intersection first to cross the intersection first (without collision); in case of a tie, traffic on east-west highway has priority over the traffic on north-south highway; 4) to enhance the flow of traffic, cars that can cross the intersection in parallel should be allowed to do so; and 5) cars, when allowed to cross the intersection, do so in a finite amount time (they do not break down at the intersection).

Events

east :   event that the sensor at the intersection goes high indicating the arrival of a car to go east.

west :   event that the sensor at the intersection goes high indicating the arrival of a car to go west.

north :   event that the sensor at the intersection goes high indicating the arrival of a car to go north.

south :   event that the sensor at the intersection goes high indicating the arrival of a car to go south.

en :   event that the sensor at the intersection goes high indicating the arrival of a car from west to go north (to make a left turn at the intersection)

es :   event that the sensor at the intersection goes high indicating the arrival of a car from west to go south (to make a right turn at the intersection)

wn :   event that the sensor at the intersection goes high indicating the arrival of a car from east to go north (to make a right turn at the intersection)

ws :   event that the sensor at the intersection goes high indicating the arrival of a car from east to go south (to make a left turn at the intersection)

ne :   event that the sensor at the intersection goes high indicating the arrival of a car from south to go east (to make a right turn at the intersection)

nw :   event that the sensor at the intersection goes high indicating the arrival of a car from south to go west (to make a left turn at the intersection)

se :   event that the sensor at the intersection goes high indicating the arrival of a car from north to go east (to make a left turn at the intersection) *sw* :   event that the sensor at the intersection goes high indicating the arrival of a car from north to go west (to make a right turn at the intersection)

e-high :   event that the sensor at the intersection indicating the arrival of a car to go east, remains high.

w-high :   event that the sensor at the intersection indicating the arrival of a car to go west, remains high.

n-high :   event that the sensor at the intersection indicating the arrival of a car to go north, remains high.

s-high :   event that the sensor at the intersection indicating the arrival of a car to go south, remains high.

en-high : event that the sensor at the intersection indicating the arrival of a car from west to go north, remains high.

es-high : event that the sensor at the intersection indicating the arrival of a car from west to go south, remains high.

wn-high : event that the sensor at the intersection indicating the arrival of a car from east to go north, remains high.

ws-high : event that the sensor at the intersection indicating the arrival of a car from east to go south, remains high.

ne-high : event that the sensor at the intersection indicating the arrival of a car from south to go east, remains high.

nw-high : event that the sensor at the intersection indicating the arrival of a car from south to go west, remains high.

se-high : event that the sensor at the intersection indicating the arrival of a car from north to go east, remains high.

sw-high : event that the sensor at the intersection indicating the arrival of a car from north to go west, remains high.

e-go :   event that the car at the intersection bound to go east is allowed to cross the intersection.

w-go : event that the car at the intersection bound to go west is allowed to cross the intersection.

n-go : event that the car at the intersection bound to go north is allowed to cross the intersection.

s-go : event that the car at the intersection bound to go south is allowed to cross the intersection.

en-go : event that the car at the intersection from west bound to go north is allowed to cross the intersection.

es-go : event that the car at the intersection from west bound to go south is allowed to cross the intersection.

wn-go : event that the car at the intersection from east bound to go north is allowed to cross the intersection.

ws-go : event that the car at the intersection from east bound to go south is allowed to cross the intersection.

ne-go : event that the car at the intersection from south bound to go east is allowed to cross the intersection.

nw-go : event that the car at the intersection from south bound to go west is allowed to cross the intersection.

se-go : event that the car at the intersection from north bound to go east is allowed to cross the intersection.

sw-go : event that the car at the intersection from north bound to go west is allowed to cross the intersection.

e-cross : event that the car at the intersection bound to go east crosses the intersection.

w-cross : event that the car at the intersection bound to go west crosses the intersection.

n-cross : event that the car at the intersection bound to go north crosses the intersection.

s-cross : event that the car at the intersection bound to go south crosses the intersection.

In a similar fashion, we define events *ne-cross*, *nw-cross*, *se-cross*, *sw-cross*, *en-cross*, *es-cross*, *wn-cross* and *ws-cross*.

> *e-low* : event that the sensor which senses the arrival of the east-bound car going low.

Similarly, we define events *w-low*, *n-low*, *s-low* ...

Formal Specification of the problem

Any design of the traffic controller should satisfy the following two conditions.

1) At any time t, there should not be any collision between the cars at the intersection.

2) Cars that go to the intersection should be allowed to cross the intersection within a finite amount of time. That is, the system should be starvation-free. This ensures that there is no deadlock in the system.

Next, we define the conditions using the above events that in turn lead to the predicate for a collision free motion of the cars.

$$\text{INTER-E} = \text{COUNT}_1(\text{e-go}) > \text{COUNT}_2(\text{e-cross})$$
$$\text{INTER-W} = \text{COUNT}_1(\text{w-go}) > \text{COUNT}_2(\text{w-cross})$$
$$\text{INTER-N} = \text{COUNT}_1(\text{n-go}) > \text{COUNT}_2(\text{n-cross})$$
$$\text{INTER-S} = \text{COUNT}_1(\text{s-go}) > \text{COUNT}_2(\text{s-cross})$$
$$\text{INTER-EN} = \text{COUNT}_1(\text{en-go}) > \text{COUNT}_2(\text{en-cross})$$
$$\text{INTER-ES} = \text{COUNT}_1(\text{es-go}) > \text{COUNT}_2(\text{es-cross})$$
$$\text{INTER-WN} = \text{COUNT}_1(\text{wn-go}) > \text{COUNT}_2(\text{wn-cross})$$
$$\text{INTER-WS} = \text{COUNT}_1(\text{ws-go}) > \text{COUNT}_2(\text{ws-cross})$$
$$\text{INTER-NE} = \text{COUNT}_1(\text{ne-go}) > \text{COUNT}_2(\text{ne-cross})$$
$$\text{INTER-NW} = \text{COUNT}_1(\text{nw-go}) > \text{COUNT}_2(\text{nw-cross})$$
$$\text{INTER-SE} = \text{COUNT}_1(\text{se-go}) > \text{COUNT}_2(\text{se-cross})$$
$$\text{INTER-SW} = \text{COUNT}_1(\text{sw-go}) > \text{COUNT}_2(\text{sw-cross})$$

The meaning of these conditions should be clear; for example, the condition INTER-E is true if there is an east-bound car that is allowed to cross the intersection but has not yet crossed the intersection.

E = INTER-S ∨ INTER-N ∨ INTER-WS ∨ INTER-NW

EL = INTER-S ∨ INTER-WS ∨ INTER-NW ∨ INTER-W ∨ INTER-SE

W = INTER-S ∨ INTER-N ∨ INTER-EN ∨ INTER-SE

WL = INTER-N ∨ INTER-EN ∨ INTER-E ∨ INTER-NW ∨ INTER-SE

S = INTER-E ∨ INTER-W ∨ INTER-NW ∨ INTER-EN

SL = INTER-W ∨ INTER-NW ∨ INTER-N ∨ INTER-EN ∨ INTER-WS

N = INTER-E ∨ INTER-W ∨ INTER-SE ∨ INTER-WS

NL = INTER-E ∨ INTER-SE ∨ INTER-S ∨ INTER-EN ∨ INTER-WS

NO-COLLISION = [INTER-E → ~E] ∧ [INTER-EN → ~EL] ∧ [INTER-W → ~W] ∧ [INTER-WS → ~WL] ∧ [INTER-S → ~S] ∧ [INTER-SE → ~SL] ∧ [INTER-N → ~N] ∧ [INTER-NW → ~NL].

Liveness condition (Starvation-free system) can be stated as:

The duration of each occurrence of the events *e-high*, *w-high* ... is finite. That is,

$dur_i(e\text{-}high)$ = finite, ∀ i.

$dur_i(w\text{-}high)$ = finite, ∀ i and so on.

Specification of the design of the traffic controller

Some of the relations governing the events are the following:

e-high ≪ *east*

w-high ≪ *west*

n-high ≪ *north*

s-high ≪ *south*

en-high ≪ *en*

es-high ≪ *es*

wn-high ≪ *wn*

ws-high ≪ *ws*

ne-high ≪ *ne*

nw-high ≪ *nw*

se-high ≪ *se*

sw-high ≪ *sw*

If an east-bound car arrives at the intersection, it should be allowed to cross the intersection, provided the intersection is clear. The following conditions are defined for specifying this requirement:

Let,

E-BEFORE-N = $[COUNT_2(n\text{-}go)$ = $COUNT_1(n\text{-}high)]$ ∨

$\qquad$ $[[COUNT_2(n\text{-}go) < COUNT_1(n\text{-}high)]$ ∧

$\qquad$ $[TIME_1(n\text{-}high)$ ∘ $(COUNT_2(n\text{-}high)$ + 1) ≥ $TIME_1(e\text{-}high)$ ∘

$(COUNT_2(e\text{-}high)$ + 1)]]

E-BEFORE-S = $[COUNT_2(s\text{-}go)$ = $COUNT_1(s\text{-}high)]$ ∨

$\qquad$ $[[COUNT_2(s\text{-}go) < COUNT_1(s\text{-}high)]$ ∧

$\qquad$ $[TIME_1(s\text{-}high)$ ∘ $(COUNT_2(s\text{-}high)$ + 1) ≥ $TIME_1(e\text{-}high)$ ∘

$(COUNT_2(e\text{-}high)$ + 1)]]

E-BEFORE-WS = $[COUNT_2(ws\text{-}go)$ = $COUNT_1(ws\text{-}high)]$ ∨

$\qquad$ $[[COUNT_2(ws\text{-}go) < COUNT_1(ws\text{-}high)]$ ∧

$\qquad$ $[TIME_1(ws\text{-}high)$ ∘ $(COUNT_2(ws\text{-}high)$ + 1) ≥ $TIME_1(e\text{-}high)$ ∘

$(COUNT_2(e\text{-}high)$ + 1)]]

E-BEFORE-NW = $[COUNT_2(nw\text{-}go)$ = $COUNT_1(nw\text{-}high)]$ ∨

$\qquad$ $[[COUNT_2(nw\text{-}go) < COUNT_1(nw\text{-}high)]$ ∧

$\qquad$ $[TIME_1(nw\text{-}high)$ ∘ $(COUNT_2(nw\text{-}high)$ + 1) ≥ $TIME_1(e\text{-}high)$

∘ $(COUNT_2(e\text{-}high)$ + 1)]]

E-BEFORE-SE = $[COUNT_2(se\text{-}go)$ = $COUNT_1(se\text{-}high)]$ ∨

$$[[COUNT_2(se\text{-}go) < COUNT_1(se\text{-}high)] \wedge$$
$$[TIME_1(se\text{-}high) \circ (COUNT_2(se\text{-}high) + 1) \geq TIME_1(e\text{-}high) \circ$$
$$(COUNT_2(e\text{-}high) + 1)]]$$

In a similar fashion, we define functions W-BEFORE-N, W-BEFORE-S, W-BEFORE-EN, W-BEFORE-NW, W-BEFORE-SE, N-BEFORE-NE, N-BEFORE-EN, N-BEFORE-WS, S-BEFORE-NW, S-BEFORE-EN, N-BEFORE-WS, EN-BEFORE-WS, SE-BEFORE-NW, SE-BEFORE-EN, SE-BEFORE-WS, NW-BEFORE-EN, NW-BEFORE-WS.

Due to symmetry we have,

$$N\text{-}BEFORE\text{-}E = [COUNT_2(e\text{-}go) = COUNT_1(e\text{-}high)] \vee$$
$$[[COUNT_2(e\text{-}go) < COUNT_1(e\text{-}high)] \wedge$$
$$[TIME_1(e\text{-}high) \circ (COUNT_2(e\text{-}high) + 1) > TIME_1(n\text{-}high) \circ$$
$$(COUNT_2(n\text{-}high) + 1)]]$$

Similarly, we define functions N-BEFORE-W, S-BEFORE-E, S-BEFORE-W, EN-BEFORE-S, EN-BEFORE-N, EN-BEFORE-W, WS-BEFORE-S, WS-BEFORE-N, WS-BEFORE-E, SE-BEFORE-E, SE-BEFORE-W, SE-BEFORE-N, NW-BEFORE-E, NW-BEFORE-W, NW-BEFORE-S, EN-BEFORE-NW, EN-BEFORE-SE, WS-BEFORE-NW, WS-BEFORE-SE, WS-BEFORE-EN, NW-BEFORE-NE.

All events are not independent; for example we have:

$$e\text{-}go = (e\text{-}high + n\text{-}high + s\text{-}high + ws\text{-}high + nw\text{-}high + 1) \mid$$
$$[COUNT_2(e\text{-}go) < COUNT_1(e\text{-}high)] \wedge$$
$$E\text{-}BEFORE\text{-}N \wedge E\text{-}BEFORE\text{-}S \wedge E\text{-}BEFORE\text{-}WS \wedge E\text{-}BEFORE\text{-}NW.$$

$$w\text{-}go = (w\text{-}high + n\text{-}high + s\text{-}high + en\text{-}high + se\text{-}high + 1) \mid$$
$$[COUNT_2(w\text{-}go) < COUNT_1(w\text{-}high)] \wedge$$
$$W\text{-}BEFORE\text{-}N \wedge W\text{-}BEFORE\text{-}S \wedge W\text{-}BEFORE\text{-}EN \wedge W\text{-}BEFORE\text{-}SE.$$

en-go = (*en-high* + *s-high* + *ws-high* + *nw-high* + *se-high* + *w-high* + 1) |
$[COUNT_2(en\text{-}go) < COUNT_1(en\text{-}high)] \wedge$
EN-BEFORE-S ∧ EN-BEFORE-WS ∧ EN-BEFORE-NW
∧ EN-BEFORE-W ∧ EN-BEFORE-SE.

ws-go = (*ws-high* + *n-high* + *en-high* + *nw-high* + *se-high* + *e-high* + 1) |
$[COUNT_2(ws\text{-}go) < COUNT_1(ws\text{-}high)] \wedge$
WS-BEFORE-N ∧ WS-BEFORE-EN ∧ WS-BEFORE-NW
∧ WS-BEFORE-E ∧ WS-BEFORE-SE.

In a similar fashion, we give the specification of the events *s-go*, *se-go*, *n-go* and *nw-go*.

ne-go = (*ne-high* + 1) | $[COUNT_1(ne\text{-}high) > COUNT_2(ne\text{-}go)]$
sw-go = (*sw-high* + 1) | $[COUNT_1(sw\text{-}high) > COUNT_2(sw\text{-}go)]$
es-go = (*es-high* + 1) | $[COUNT_1(es\text{-}high) > COUNT_2(es\text{-}go)]$
wn-go = (*wn-high* + 1) | $[COUNT_1(wn\text{-}high) > COUNT_2(wn\text{-}go)]$

Some of the partial orderings and the completion time sequences of these new events are:

e-cross << *e-go*

w-cross << *w-go*

n-cross << *n-go*

s-cross << *s-go*

en-cross << *en-go*

es-cross << *es-go*

wn-cross << *wn-go*

ws-cross << *ws-go*

ne-cross << *ne-go*

nw-cross << *nw-go*

se-cross << *se-go*

sw-cross $\ll$ *sw-go*

e-low $\ll$ *e-cross*

w-low $\ll$ *w-cross*

n-low $\ll$ *n-cross*

s-low $\ll$ *s-cross*

$\mathrm{TIME}_2(e\text{-}high) = \mathrm{TIME}_2(e\text{-}low)$

$\mathrm{TIME}_2(w\text{-}high) = \mathrm{TIME}_2(w\text{-}low)$

$\mathrm{TIME}_2(n\text{-}high) = \mathrm{TIME}_2(n\text{-}low)$

$\mathrm{TIME}_2(s\text{-}high) = \mathrm{TIME}_2(s\text{-}low)$

In a similar fashion, we define events *en-low*, *ws-low* ... and the completion time sequence of the events *en-high*, *ws-high* and so on.

Proof of correctness of the traffic controller design

We prove that the design of the traffic controller meets the two criteria, namely, 1) there is no collision and 2) the system is starvation-free.

Proof of 1) :

From the definition of the conditions E, EL, W, WL, N, NL, S and SL, it is clear that the condition for collision free system is equivalent to the following condition:

[[INTER-E $\rightarrow$ ~E] $\wedge$ [INTER-EN $\rightarrow$ ~EL]] $\vee$

[[INTER-E $\rightarrow$ ~E] $\wedge$ [INTER-W $\rightarrow$ ~W]] $\vee$

[[INTER-E $\rightarrow$ ~E] $\wedge$ [INTER-SE $\rightarrow$ ~SL]] $\vee$

[[INTER-EN $\rightarrow$ ~EL] $\wedge$ [INTER-N $\rightarrow$ ~N]] $\vee$

[[INTER-W $\rightarrow$ ~W] $\wedge$ [INTER-WS $\rightarrow$ ~WL]] $\vee$

[[INTER-W $\rightarrow$ ~W] $\wedge$ [INTER-NW $\rightarrow$ ~NL]] $\vee$

[[INTER-WS $\rightarrow$ ~WL] $\wedge$ [INTER-S $\rightarrow$ ~S]] $\vee$

[[INTER-N $\rightarrow$ ~N] $\wedge$ [INTER-NW $\rightarrow$ ~NL]] $\vee$

[[INTER-S $\rightarrow$ ~S] $\wedge$ [INTER-SE $\rightarrow$ ~SL]].

We prove that [[INTER-E $\rightarrow$ ~E] $\wedge$ [INTER-EN $\rightarrow$ ~EL]]; a similar proof can be given for other clauses. Proving INTER-E → ~E is equivalent to proving the events

i) *e-go* + *s-go*, ii) *e-go* + *n-go*, iii) *e-go* + *ws-go* and iv) *e-go* + *nw-go* are single-occurrent events.

From the definition of conditions E-BEFORE-S and S-BEFORE-E, they cannot be true simultaneously. At any time t, at most one of them can be true. For the event *e-go* to start occurring, E-BEFORE-S should be true at that time. For the event *s-go* to start occur, S-BEFORE-E should be true at that time. Suppose E-BEFORE-S is true at time t and the event *e-go* starts at that time, then S-BEFORE-E is false at least until time t′ at which the corresponding occurrence of the event *e-high* finishes. Hence *s-go* cannot start up until that time. Similarly, if an occurrence of *s-go* starts, until the corresponding occurrence of *s-high* finishes, an occurrence of *e-go* cannot start. This implies, *e-go* + *s-go* is single-occurrent. Similarly we can prove (ii), (iii) and (iv). Similar argument can be carried out to prove the other condition [INTER-EN → ~EL].

Proof of 2) :

We prove that $dur_i(ws\text{-}high)$ is finite $\forall\ i \in \mathbf{N}$. When an occurrence of the event *ws-high* starts happening (a west-bound car that is about to make a left turn at the intersection is sensed), there might be a maximum of six cars that have to cross the intersection before this car is allowed to cross the intersection. According to our assumption, cars do not break down at the intersection and they take only finite amount of time to cross the intersection. This implies, $dur_i(ws\text{-}high)$ is finite $\forall\ i$.

4.   Concluding Remarks

In this paper a brief summary of the formalism investigated in [1] for the specification of real-time and concurrent systems is given. This study concentrated on the aspects: 1) relaxing the constraint [3] that events are instantaneous and proving the algebraic structure of events in the system; 2) providing tools for problem specification and proving that algorithm specified within the formal model is correct (in the sense that it meets the specification of the problem) and 3) showing the applicability of the formalism at several levels of specifications.

Our interest in this study originated from an attempt to study and formally reason about the behaviour of real-time systems such as robotics. We noticed that the traditional approach of modeling external stimuli versus system response as finite-state machines takes into account only a discrete notion of time, and as pointed out in [3] is likely to give rise to unnecessary modeling problems. Moreover the assumption that events are instantaneous, although seems quite adequate for the specification of low level hardware systems, is likely to provide incorrect high level specifications. Hence our formalism, seems to provide a natural approach for capturing and reasoning about the behavior of real-time concurrent systems.

Finally, an extended comparison with other alternate approaches such as algebra of concurrency [8], calculus of communication systems [7] and temporal logic [5] is not attempted here. In some sense, such a comparison is premature and this belief is supported by the remark [5] 'we find it difficult to achieve certainty that our specifications correctly reflect our abstract requirements' after comparing four specification methods on alternating bit protocol. It is plausible that any extension of state machine model for specification and validation of real-time constraints can only enormously increase the complexity of proof; on the other hand the undecidability of proofs in many formal models is also well known. In this context notice also that Koymans and de Roever [5] have given only temporal logic specifications and no correctness proofs for three STL/SERC problems. The primary innovation here is to demonstrate the use of the formal model in encapsulating robotic navigation controller and in providing the correctness proof of its design. Examples in section 2 suggest that formal denotational style semantics can be defined for real-time and distributed constructs embedded in a functional language. A more detailed report on this study will appear elsewhere.

References

1. V.S. Alagar and G. Ramanathan, "A functional model for specification and analysis of distributed real-time systems : Formalism and Applications", submitted for publication, March 1988.

2. V.S. Alagar and G. Ramanathan, "Formal specification and proof of correctness of the majority consensus solution to the database update synchronization problem", in preparation.

3. P. Caspi and N. Halbwachs, "Functional model for describing and reasoning about time behavior of computing systems", Acta Informatica, vol. 22, pp. 595-627, 1986.

4. G. Kahn, "The semantics of a simple language for parallel processing", Proc. IFIP Congress, 1974.

5. R. Koymans and W.P. de Roever, "Examples of real-time temporal logic specification", Workshop on the analysis of Concurrent systems, Cambridge, pp. 231-251, 1983.

6. L. Lamport, "Time, clocks and the ordering of events in a distributed system", CACM vol. 21, pp. 558-565, July 1978.

7. R. Milner, "A calculus of communicating systems", LNCS, vol. 92, 1980.

8. R. Milner, "Using algebra for concurrency : some approaches", Workshop on the analysis of concurrent systems, pp. 7-25, 1983.

9. L.E. Sanchis, "Data types as lattices : Retractions, closures and projection", RAIRO Theor. Comput. Sci., vol. 11, pp. 329-344, 1977.

An Equivalence Decision Problem in Systolic Array Verification*

Parosh Abdulla[†]   Stefan Arnborg[‡]

June 17, 1988

Abstract

Verification of linear algebra and signal processing systolic circuit families is reduced to a zero equivalence problem for expressions containing indefinite summation. This problem is solved using a linear case of algebraic decomposition and reducing it to the zero equivalence problem for a finite set of multivariate polynomials.

1   Introduction

Regular circuit structures have drawn considerable attention recently as a means of implementing parallel algorithms in areas like linear algebra, signal processing, pattern matching, etc. There has been a big gap between the practical justifications of such structures used and formal verification. Although many examples of systolic circuits are easy to both synthesize and verify using simple algebraic manipulation [10,11], some are not. From a mathematical point of view, a correctness assertion for a systolic circuit is a statement about the solutions to a family of triangular systems of equation, one for each admissible size of the circuit.

In practice, all circuits claimed to be systolic have their computation elements on a lattice of one, two or (in few cases) more dimensions, and the region where cells are placed is a finite polytope in this space. If we can find an explicit expression for the signal appearing on every wire at every time instant, then it is an easy matter to systematically verify, for each cell, that the output computed from the input is correct, and this constitutes a verification of the structure. This was explicitly stated by Bryant[3] for discrete logic. In a regular circuit there are only finitely many cases to consider, since we have *e.g.*, for a three-dimensional structure a triply indexed cell in the interior, a doubly indexed cell on each of the faces, a singly indexed cell on each edge and a single cell on each corner. Thus, an infinite family of circuits can be verified by a finite number of cases. The tractability of this method depends on our ability to actually describe the internal signals concisely using such second-order tools as summation and product operators, and manipulate them adequately. This leads

---

*Research supported by the Swedish Board for Technical Development

[†]Department of Computer Systems, Uppsala University, Box 520, S-751 20 Uppsala, Sweden

[‡]Department of Numerical Analysis and Computing Science, The Royal Institute of Technology, S-100 44 Stockholm, Sweden and the Swedish Institute for Computer Science.

sometimes to rather messy expressions, and might require a more expressive language than the language used to specify the circuit.

An alternative method is to use mathematical induction over the size parameters of the structure. Such a technique was explored by Tidén[13], and in some cases it is only necessary to use the specification as inductive hypothesis. We have verified a number of 'tricky' circuits by induction in the HOL system[7], and found that even this technique is difficult to apply in practice for many cases.

For these reasons we have tried to define a completely mechanical procedure for verifying 'most' systolic structures. Needless to say, it is easy to define artificial systolic circuits whose (likewise artificial) correctness depend on deep mathematical conjectures, so we have restricted the family to one that seems to cover most practical cases occurring in the literature. Particular features of the trickier ones are computing modulo a number (to save circuit space) and the use of local storage elements in the cells. Our method finds a sufficient and finite set of cases, based on examination of the circuit structure, so that correctness on these cases implies correctness on all (specified) cases. We can thus see our method as a case of the 'proving by example' paradigm. Other examples of this paradigm are: To decide that a univariate polynomial, with rational coefficients and of degree n, is identically zero, evaluate it on $n + 1$ different points (and check that the values are all zero). To see that a propositional expression is identically true, evaluate it for all value combinations of its variables (and check that all these values are true) . To see that two points in a geometrical construction coincide, put the starting points in general positions and compute the distance between the two points to a given precision, based on the number of steps in the geometrical construction. If the distance is less than the precision, conclude that it is identically zero [8]. In some cases the paradigm yields very efficient algorithms, in some, like the propositional expression example, it does not.

Basically, we define a language in which to describe both the circuit and its specification as systems of equations over a ring. Equations arise first from the computations of individual cells, as a relation between input and output of one cell, secondly from the interconnection structure, as equality between an output of one cell and an input of another. Correctness is a correspondence between the solutions of the two systems, namely that solutions are equal on the intersection between a lattice and a finite polytope in the two systems. When taking account of the permitted size parameters, the problem gets a higher dimensionality since we must take account of the parameter space also. We restrict the design in such a way that solutions to systems can be expressed as expressions in a second order language involving summation signs. From the syntax of the two expressions we can construct a decomposition of the parameter space into polytopes such that the expressions are equal on lattices intersected with these polytopes. To decide if the specification is satisfied we need only check the expressions for equality on a set generating the lattice of the specification within each polytope.

2   Definitions and method

We work in two structures, the field of rationals $\mathbf{Q}$ and a commutative ring $R = (A, +, \times, 0, 1)$. $\mathbf{Z}$ denotes the set of integers. We have two kinds of expressions, index expressions and ring expressions, denoting linear polynomials over $\mathbf{Q}$ and families of polynomials over R, respectively. In many applications we can identify R and $\mathbf{Q}$, but it gives a clearer picture of the power and limit of the method not to do so. *Index expres-*

sions are linear forms in the *index variables* $j_1, j_2, \ldots$, with rational coefficients. *Ring expressions* are built from constants in A, *indexed variables* $a_1^0, a_2^0, \ldots, a_1^1(.), \ldots, a_1^2(.,.), \ldots$, where the dots in argument places are replaced by index expressions. In our small examples we will use a, b, c for indexed variables and i, j, p, q (possibly with subscripts) for index variables. Ring expressions can also be recursively built as $r_1 + r_2$, $r_1 \times r_2$ and $\sum_{i=l}^u r_1$, where r_1 and r_2 are ring expressions, i is an index variable, l and u are index expressions not containing i. The arithmetic operators $+$, $\times$ and $\sum$ stand for ring operations and iterated ring operations in the usual way. We assume that no two summation indices use the same index variable, and that no index variable occurs outside its scope (the ring expression of the corresponding summation). The index variables that are not bound are the parameters of the ring expression. When substituting integers for the parameters we may get a multivariate polynomial over R. The conditions for getting such a polynomial are that every upper and lower bound evaluation generated yields an integer, and such that no lower bound is greater than the corresponding upper bound. The *zero set* for a ring expression with k free variables is the set in $\mathbf{Z}^k$ on which the expression evaluates to a polynomial which is identically zero.

By a *translate of a k-dimensional integer lattice of rank r* we mean a set of k-vectors of integers $\{b_0 + \sum_{i=1}^r c_i b_i | c_i \in \mathbf{Z}\}$, where the c_i are arbitrary integers and $\{b_i\}_{i=1}^r$ is a set of linearly independent k-vectors of integers, the *basis* of the lattice. We will abbreviate this term to *lattice*, although this term is usually reserved for the case $b_0 = (0, \ldots, 0)$. A *polytope* is the intersection of a number of planes and half-spaces. A *constrained lattice* is the intersection of a polytope and a lattice. A *verification problem instance* is a pair (S, I), where S is a ring expression and a constrained lattice, $S = (r, l)$, and I is a set of pairs of a ring expression and a constrained lattice, $I = \{(r_i, l_i)\}_{i \in M}$. The task is to decide if for every point p in l there is some l_i such that $p \in l_i$ and $r_i - r$ evaluates to the zero polynomial on p.

By the *projection* of a set S of index expressions over index variables I *on a set* $F \subset I$ we mean the set of index expressions over index variables F obtained by eliminating, in all possible ways, variables in $I - F$ among elements of S. Elimination is done by forming linear combinations in such a way that coefficients of unwanted variables cancel. Only one index expression from each similarity class need be retained. By a *decomposition of* $\mathbf{Z}^k$, *sign-invariant with respect to* S we mean a decomposition of $\mathbf{Q}^k$ into regions over which each $s \in S$ has constant sign, positive, negative or zero. Projections and decompositions are the central tools in a decision method for elementary algebra and geometry[2]. We are, for reasons that will become apparent, satisfied with the restriction to linear polynomials, and we decompose $\mathbf{Z}^k$ instead of $\mathbf{R}^k$. The *index expression set* for a ring expression r is a set built from its index expressions as follows: Introduce a new index variable for each index position of each indexed variable occurring in r. If index expression v occurs in an indexed variable at a position for which variable s was allocated, add $v - s$ to the index expression set. If r contains a summation $\sum_{i=l}^u$, add $l - i - 1$ and $u - i$ to the index expression set. These are all index expressions in the index expression set for r.

We are not quite ready to prove anything about the zero set of general ring expressions. But an important class of such expressions are the *multilinear ring expressions*, characterized by a structure such that (i) if the subexpression $r_1 \times r_2$ occurs as a subexpression, then no indexed variable occurs in both r_1 and r_2; and (ii) if $r_1 + r_2$ occurs as a subexpression, then every summation index visible at this point occurs both in r_1 and r_2, and similarly for a subexpression $\sum_i r$, the visible summation indices (i among others) must occur in r. Condition (i) ensures that the expression always evaluates to a multilinear form, and (ii) ensures that the coefficients of terms are not effectively polynomials in the circuit parameters (as in $\sum_{i=1}^{m} \sum_{j=1}^{i} a(p) = m(m+1)a(p)/2$).

Main Theorem Let r be a ring expression with free variables F and with index expression set E. Then the zero set of r in a region c of a decomposition of $\mathbf{Q}^k$ w r t the projection of E on F is the intersection of c and a lattice.

Before giving the proof outline we explain the main ideas with two examples. The expression $a(0) + a(1) - a(n) - a(1 - n)$ has index expression set $\{0 - v_a, 1 - v_a, n - v_a, 1 - n - v_a\}$. The projection on $\{n\}$ is $\{1, n, n - 1, 2n - 1\}$. The expression cancels 'accidentally' for $n = 0$ and $n = 1$, regions of the decomposition of the projection. In general, accidental cancellation can only occur on a full polytope of the decomposition of the projection of the index expression set.

Consider now the subexpression

$$\sum_{i=0}^{l_4(p_1,p_2)} \sum_{j=0}^{l_3(i,p_1,p_2)} a(l_1(i, j, p_1, p_2), l_2(i, j, p_1, p_2))$$

of a ring expression. For given integers p, q, the term contains the variable $a(p, q)$ if and only if there is an integer solution for (i, j) to the following system:

$$
\begin{aligned}
p - l_1(i, j, p_1, p_2) &= 0 \\
q - l_2(i, j, p_1, p_2) &= 0 \\
j - l_3(i, p_1, p_2) &\leq 0 \\
i - l_4(p_1, p_2) &\leq 0 \\
i &\geq 0 \\
j &\geq 0
\end{aligned}
$$

The left-hand sides of these equations obviously form the index expression set of the subexpression. The first two equations and the integrality constraints on i and j imply that the set is contained in a lattice and the remaining inequalities constrain it to a polytope in (p, q, p_1, p_2)-space. The projection to (p_1, p_2)-space of the index expression set yields a sign-invariant decomposition whose regions have the property that the boundaries of the polytope vary linearly with p_1 and p_2. Also, the region in parameter space on which lower bounds are not greater than upper bounds are delimited by zero sets of polynomials in the projection of the index expression set. Therefore, the expression evaluates properly on a lattice constrained by a number of entire regions of the decomposition, and this is easily seen to be true in general.

We now outline the proof of the main theorem. For a general multilinear ring expression, we can propagate multiplication signs downwards and plus signs upwards, leaving summation signs in the middle, with the following rewrites:

$$
\begin{aligned}
r_1 \times \sum r_2 &\rightarrow \sum (r_1 \times r_2) \\
\left(\sum r_1\right) \times r_2 &\rightarrow \sum (r_1 \times r_2) \\
\left(\sum r_1\right) \times \left(\sum r_2\right) &\rightarrow \sum \sum (r_1 \times' r_2) \\
(r_1 + r_2) \times r_3 &\rightarrow r_1 \times r_3 + r_2 \times r_3 \\
r_1 \times (r_2 + r_3) &\rightarrow r_1 \times r_2 + r_1 \times r_3 \\
\sum (r_1 + r_2) &\rightarrow \sum r_1 + \sum r_2
\end{aligned}
$$

These rewrites leave the index expression set invariant. Let the index expression set be S and the decomposition of its projection C. Using associativity and commutativity of $+$ and $\times$ we can collect terms in the usual way so that we obtain a sum of monomials, each optionally the operand of a number of summation operators. Each monomial is the product of an integer coefficient and a number of indexed variables (each equipped with the appropriate number of index expressions). By multilinearity restriction (ii) the coefficient is a number, independent of the free variables. Let the *signature* of a monomial be the set of indexed variables (disregarding their actual index expressions) in it (the multilinearity restriction (i) forbids multiple occurences). Monomials with different signatures cannot cancel each other, so we can assume *wlog* that all terms have the same signature. The summation signs enclosing a monomial will, for each assignment of values to free variables, instantiate the monomial over a constrained lattice in the index space of the monomial (in the case of no enclosing summation sign the constrained lattice degenerates to a point). In order that the different constrained lattices cancel each other, their corners must match up. But the corners are either identically coinciding or coincide over regions of C, since each way to match corners corresponds to an equality between index expressions in S.

We can now solve the verification problem. Project the index expression set of the ring expression on its free variables and construct a sign-invariant decomposition $\{c_i\}_{i \in I}$. Find the intersections of the polytopes given with the verification problem with the c_i and find the intersections of the lattice with each such polytope. Each such intersection generates a lattice, possibly of lower rank than the original one. For each such restricted lattice, find a set generating the full lattice , evaluate the ring expression on this set and check that the resulting polynomials are identically zero. For an arbitrary ring this is an expensive operation since it has tautology checking of propositional formulae as a special case (when R is Z_2, the integers modulo 2). Over the ring of integers or rational numbers (the most common domain for signal processing applications), a probabilistic method based on evaluating a homomorphism into Z_p for a large prime p is very efficient[6,12]. Intersecting a polytope with an integer lattice is integer programming, *i.e.*, NP-complete. But the art of systolic array design has not yet reached a level where this poses significant computational problems. The size of the computation problem is fortunately determined by the structure, not the size, of the circuit.

3   Obtaining ring expressions from circuit diagrams

The computation cells are assumed to be placed on a finite, constrained lattice P in the l-dimensional space Z^l. For a linear array, $l = 1$, for a plane array $l = 2$, etc.. P is bounded by a set of hyperplanes defined by a number of index expressions in the k size parameters of the circuit and the d space variables x_i, $1 \leq i \leq d$. Each cell has a number of inputs, outputs and local storage variables. Since we require connections to be between one input and one output, we must apparently have equal number of inputs and outputs in each cell. We assume that each cell computes a number of values at every non-negative time step. So the computations are indexed by elements taken from $P \times Z^+$. Let the inputs of a cell be $i_1, \ldots, i_m$, the outputs $o_1, \ldots, o_m$, and the local storage variables $l_{m+1}, \ldots, l_n$.

We assume that the wires between cells are delay-less and that the delay in a computation step is located within the cells (this is not a significant restriction, since there must be a delay either in the cell or in the connection). The interconnections are described with vectors d_j in the lattice of P:

$$i_j(m, t) = o_j(m - d_j, t), \quad m, m - d_j \in P$$

When $m \in P$ and $m - d_j \notin P$, $i_j(m, t)$ is an input to the circuit. It can be specified to be a ring expression or an index expression on a constrained lattice which is a subset of a boundary of P. Such an expression is indexed by the space and time variables of the circuit.

The outputs o_j and local storage elements l_j in each computation step t depend on the input and local storage variables of the cell at some previous time $t - c_j$. If $t - c_j < 0$, these quantities are initial values for the cell. The computations must have one of the following forms (*i.e.*, one for each $j \in [1, \ldots, n]$):

$$
\begin{aligned}
o_j(m, t) &= i_j(m, t - c_j) \\
o_j(m, t) &= Q_j[i_{j_{j_1}}(m, t - c_j), \ldots, i_{j_{j n_j}}(m, t - c_j), \\
&\qquad l_{k_{j_1}}(m, t - c_j), \ldots, l_{k_{j m_j}}(m, t - c_j)] \\
o_j(m, t) &= i_j(t - c) + Q_j[i_{j_{j_1}}(m, t - c_j), \ldots, i_{j_{j n_j}}(m, t - c_j), \\
&\qquad l_{k_{j_1}}(m, t - c_j), \ldots, l_{k_{j m_j}}(m, t - c_j)] \\
o_j(m, t) &= \text{if } i_{i_j}(m, t - c_j) = 0 \text{ then } f_j(m, t - c_j) \text{ else } g_j(m, t - c_j) \\
l_j(m, t) &= l_j(m, t - c_j) \\
l_j(m, t) &= Q_j[i_{j_{j_1}}(m, t - c_j), \ldots, i_{j_{j n_j}}(m, t - c_j), \\
&\qquad l_{k_{j_1}}(m, t - c_j), \ldots, l_{k_{j m_j}}(m, t - c_j)] \\
l_j(m, t) &= i_j(t - c) + Q_j[i_{j_{j_1}}(m, t - c_j), \ldots, i_{j_{j n_j}}(m, t - c_j), \\
&\qquad l_{k_{j_1}}(m, t - c_j), \ldots, l_{k_{j m_j}}(m, t - c_j)] \\
l_j(m, t) &= \text{if } i_{i_j}(m, t - c_j) = 0 \text{ then } f_j(m, t - c_j) \text{ else } g_j(m, t - c_j),
\end{aligned}
$$

where the Q_j are multilinear polynomials over R and where f_j and g_j have the same forms as some other right-hand sides of the table. The different computations must also have an acyclic dependency structure, *i.e.*, the transitive relation generated by (j, j_{js}), (j, i_j) and (j, k_{js}) must be acyclic. The computation flow must be data-independent, *i.e.*, if an output i_j or storage variable l_j is determined by an if expression, then i_{i_j} must be an index expression propagated through the circuit according to the first rule.

With these requirements the equations will unfold to sums over a linear progression in $P \times Z^+$ (which follows from a fairly long but straightforward argument). Unfortunately, the limits of the progression depend on which face of the polytope is first encountered, and thus it varies over the space of computations. But the expressions have the same form over regions in $k + l + 1$- space of the projection of the following set on circuit parameters, space variables and time: add all linear expressions in space variables describing the boundaries of P, the single polynomial t (which delimits the start of computation), $x_i - v_j d_j^{(i)}$ for $1 \le i \le l$, $1 \le j \le n$ and $t - v_j c_j$ for $1 \le j \le n$. This set also contains the index expression sets of the corresponding ring expressions. Therefore, we can by a syntactic examination of the equations derived from the circuit check that only multilinear ring expressions are generated and obtain an index expression set with the property that it contains all index expression sets of the different multilinear ring expressions that describe the behavior of the circuit. The projection of this set on the space of circuit parameters and time has a decomposition such that circuit correctness follows from correctness in one point of every region which has a non-empty intersection with the specification, if these points also generate the lattice specified for the circuit.

4 An example

Apparently, verification of a systolic circuit family can be performed as follows: Check that the circuit satisfies the requirements, determine the set of index expressions in the ring expressions for the specification and implementation, and project these onto the parameter and time set. Perform a decomposition and determine a full set of sample points such that correctness on these points implies that the circuit is correct for the specified parameter set. The required algebraic capabilities, including Gonnets zero equivalence test[6], are all available in the better Computer Algebra systems.

For the systolic convolution circuit described by Ullman[14] we derive the following equation system by examination of the circuit description:

The polytope P is just the integers from 0 to L, the only circuit parameter. Each cell has four input/output pairs and one local storage variable, and it is indexed by the only space variable x. Data streams 1 to 4 correspond to a, *Traveling-b*, c and *leading-edge* in [14], and the storage variable l corresponds to b. The interconnection structure is:

$$\begin{aligned}
i_1(x,t) &= o_1(x-1,t) \\
i_2(x,t) &= o_2(x-1,t) \\
i_3(x,t) &= o_3(x+1,t) \\
i_4(x,t) &= o_4(x-1,t)
\end{aligned}$$

The computation rules are:

$$\begin{aligned}
o_1(x,t) &= i_1(x,t-2) \\
o_2(x,t) &= i_2(x,t-1) \\
o_3(x,t) &= i_3(x,t-2) + i_1(x,t-2)l(x,t-2) \\
o_4(x,t) &= i_4(x,t-2) \\
l(x,t) &= \text{if } i_4(x,t-2) = 0 \text{ then } l(x,t-2) \text{ else } i_2(x,t-2)
\end{aligned}$$

The initialisation and data input conventions are, where a and b are the vectors convolved to c, $0 \le x \le L$ and $0 \le t \le T$:

$$\begin{aligned}
i_3(L,t) &= 0 \\
o_j(x,0) &= 0 \quad ,1 \le j \le 4 \\
i_1(0,4k+4) &= a_k \quad ,0 \le k < n \\
i_1(0,4k+4) &= 0 \quad ,n \le k < 2n \\
i_2(0,k) &= b_k \quad ,0 \le k < n
\end{aligned}$$

$$
\begin{aligned}
i_2(0,k) &= 0 \quad , n \le k < 2n \\
i_4(0,0) &= 1 \\
i_4(0,k) &= 0 \quad , k > 0 \\
l(x,0) &= 0 \\
c_k &= o_3(0, 4k+6) \quad , 0 \le k < 2n
\end{aligned}
$$

The specification of the convolution is

$$
f_k = \sum_{i=0}^{n-1} a_i b_{k-i}
$$

The above definitions correspond to the MAPLE code of the Appendix, which has free variables L, T and N. Actually, we did not get it quite right the first try, but from a small instantiation of the circuit it was easy to debug the equations. But is it now correct for all values of the parameters, and which are the relations between T, L, and N? We have injected values from the left in such a way that breakpoints occur at times 0, 4, $n-1$ and $4n$. These breakpoints propagate to the right end with speeds 1 and $1/2$, and then return with speed $1/2$. They will hit the leftmost cell at times which are zeros in t of the set: $\{t - 3L, t - 4L, t - 4L - 4, t - n + 1 - 3L, t - n + 1 - 4L, t - 4n + 3L, t - 4n - 4L\}$. But we are interested in the times from 6 (when the first result arrives) to $8n + 2$ (when the last arrives), and the cases where $L > n$ (otherwise the computation will be chopped). In order to verify these cases, we only need to verify one point in every non-empty region defined by a sign assignment to the polynomials above in the specified region. Such a set of (L, N, t)-points (extracted by computer) is: $\{(2,1,6), (3,2,10), (4,3,18), (6,2,18), (8,7,38), (10,9,6)\}$. Thus, for deciding that the convolver satisfies its specification, we only have to check it on this set .

As can be seen already in this simple example, certain polynomials in the projection do not correspond to features of the real computation, but are parasitic. A more detailed (but still easily mechanizable) analysis of the computation will prove correctness of systolic circuits with fewer examples. We are working on a mechanization of a refined sample point extraction in the MAPLE 4.0 computer algebra system. The system was chosen because it has all functions used in this application readily packaged.

5 Conclusion

We have described a method which eliminates the need for induction proofs in verification of a large class of regular circuit families. The class includes all linear algebra and signal processing circuits known to us. The method is based on a technique for deciding equivalence of formulae containing second order constructs like indefinite summation and product. This is, as far as we know, a first step in directions suggested in recent work on automated deduction of equality for first-order expressions [6,8]. One could believe that the method is easily extendable to non-linear index expressions, since the theory of algebraic decomposition treats general multivariate polynomials. This is however not the case since we also have the problem of intersecting a region with a lattice, and this problem is undecidable for general semialgebraic regions by the undecidability of Hilberts tenth problem [4]. This indicates that the equivalence problem for general ring expressions is possibly also undecidable.

From the present application point of view, a number of extensions are both interesting and promising. One is to remove the multilinearity and acyclicity constraints, which would bring certain systolic structures for string-matching[9] and most grid-based electronic structures like arithmetic modules within reach. Another is to treat other regular structures, like shuffle networks, hypercubes and binary trees [10,1]. While the first extension can probably be based on the present theory, the second seems to require some new concepts on a higher level. From a general point of view, it would of course be interesting to have similar methods for zero equivalence testing in other computational structures than rings.

References

[1] Arnborg, S. :Fast circuits for string-matching. Manuscript.

[2] Arnon, D.S., Collins, G.E. and McCallum,S. : Cylindrical algebraic decomposition, *SIAM J. Comp.*, **13**(1984), 865–877, 878–889.

[3] Bryant, R. E.: *Can a simulator verify a circuit?* In Formal Aspects of VLSI, North-Holland 1986.

[4] Davis, M. : Hilberts tenth problem is unsolvable, *Amer. Math. Monthly* 80(1973), 233–269.

[5] Char, B.W., Geddes, K.O., Gonnet, G.H., Watt, S.M. : Maple users guide, WAT-COM Publications, Waterloo 1985.

[6] Gonnet, G. H. : Determining equivalence of expressions in random polynomial time ACM STOC 1984, 334-341.

[7] Gordon,M. : *HOL A proof generating system for higher order logic.* In VLSI specification, verification and synthesis, Kluwer 1987.

[8] Hong, J. W. : Proving by example and gap theorems *IEEE FoCS 1986.*

[9] Lipton, R.J. and Lopresti, D. :*A systolic array for rapid string comparison.* 1985 Chapel Hill Conference on VLSI, 363-376.

[10] Lisper, B. : *Synthesizing synchronous systems by static scheduling in space-time* Ph D Thesis,The Royal Institute of Technology, TRITA-NA-8701.

[11] Moldovan, D. I. : On the analysis and synthesis of systolic arrays *IEEE trans. Comp.* Vol C-31 (Oct 1982) 1121-1126.

[12] Schwartz,J. : Fast probabilistic algorithms for verification of polynomial identities. *JACM 27(1980) 701-717.*

[13] Tidén, E. :*Verification of Systolic Arrays — A Case-study* 1984 Nordic VLSI Symposium, 74-81.

[14] Ullman,J.D.:*Computational aspects of VLSI*, Computer Science Press, 1984

A MAPLE definition of convolution circuit

```
for t from 0 to T do
i_3[L,t]:=0;
i_4[0,t]:=0;
od;
for x from 0 to L do
o_1[x,0]:=0;o_2[x,0]:=0;o_3[x,0]:=0;o_4[x,0]:=0;
l[x,0]:=0;
od;
for k from 0 to N-1 do
i_1[0,4*(k+1)]:=a[k];
i_1[0,4*k+1]:=0;
i_1[0,4*k+2]:=0;
i_1[0,4*k+3]:=0;
i_2[0,k]:=b[k];
c[k]:=o_3[0,4*k+6];
od;
for k from N to 2*N-1 do
i_1[0,4*(k+1)]:=0;
i_1[0,4*k+1]:=0;
i_1[0,4*k+2]:=0;
i_1[0,4*k+3]:=0;
i_2[0,k]:=0;
c[k]:=o_3[0,4*k+6];
od;

i_4[0,0]:=1;
for x from 0 to L do
for t from 0 to T do
if x-1>=0 then i_1[x,t]:=o_1[x-1,t];fi;
if x-1>=0 then i_2[x,t]:=o_2[x-1,t];fi;
if x+1<=L then i_3[x,t]:=o_3[x+1,t];fi;
if x-1>=0 then i_4[x,t]:=o_4[x-1,t]; fi;

if t-2>=0 then o_1[x,t]:=i_1[x,t-2];fi;
if t-1>=0 then o_2[x,t]:=i_2[x,t-1];fi;
if t-2>=0 then o_3[x,t]:=i_3[x,t-2]+i_1[x,t-2]*l[x,t-2];fi;
if t-2>=0 then o_4[x,t]:=i_4[x,t-2];fi;
if t-2>=0 then if i_4[x,t-2]=0 then l[x,t]:=l[x,t-2] else
                                l[x,t]:=i_2[x,t-2];fi;
fi;
od;
od;
for k from 0 to 2*N-1 do print (c[k]);od;
```

Should Concurrency be Specified?

Rosalind L. Ibrahim
John A. Ogden
Shirley A. Williams

University of Reading
Department of Computer Science
Whiteknights Park
PO Box 220
Reading
RG6 2AX

ABSTRACT

Whereas software engineering methodologies have evolved to the point where specification and design activities have been deliberately separated in order to reduce the complexities of large scale software development, hardware engineers are providing us with new architectures eminently suited to concurrent systems. This reality lures the software engineer into exploiting the new potential, but perhaps too soon in the development process.

Should concurrency be specified? In this paper, we provide some thoughts, interpretations, and examples that lead us to conclude that concurrency should rarely, if ever, be specified.

Should Concurrency be Specified?

Rosalind L. Ibrahim

John A. Ogden

Shirley A. Williams

1. Introduction

It can be argued that details of concurrency should not be included in a specification. A specification by its nature is a description of **what** something is supposed to do as opposed to **how** it does it, whereas concurrency **can** be considered an implementation strategy that should not be imposed at the specification stage. This is, however, too much of a simplification. The real questions that must be addressed are:

(i)   what is meant by a specification,

(ii) why use concurrency?

These topics are discussed below. Then follows an overview of parallel architectures and how they affect design decisions. Lastly an example is provided showing how a specification, if it is sufficiently abstract, can allow the designer/implementor the flexibility required to provide a solution that makes efficient use of underlying computing resources. This will be possible only as long as unnecessary ordering constraints are excluded from the specification.

2. Specification and Concurrency

2.1 The Nature of Specification

There is no standard definition of the term specification, but there is general agreement that a software specification expresses the functional and non-functional requirements of a piece of software. It states what a software product is intended to achieve (rather than how it should achieve it), and it states the attributes or features the software must possess.

A specification is used for validation and verification purposes: it is validated to assure that it represents the intended use of the software and that the software as specified will meet its objectives; then it forms a basis for developing a design that can be verified against the validated specification. For any useful piece of software, it is also inherent in the nature of a specification that it will undergo change.

Within this generally accepted context of the nature of specification, many different methods, notations, and techniques exist and are applied to specification levels in the software development process ranging from requirements specification to program (module) specification.

In all cases however the specifier faces some common fundamental problems.

2.2 The Problems of Specification

Three general problems confront the specifier:

(i)   Determining what is actually needed to solve the user's problem

(ii) Distinguishing what is needed from how it can be provided

(iii) Representing the specification so that it is amenable to validation, verification, and change.

Problem analysis is concerned with detecting the true nature of the problem to be solved, extracting it from both symptoms and solutions which may cloud the underlying issue at hand. A user insisting that he/she needs a highly parallel system, should be challenged with the question 'why'. The underlying problem must be specified, not a possible solution which would then (if specified) become a required solution which may not solve the real problem.

Distinguishing the logical computing model of software requirements from a physical design can be difficult unless specification/design distinctions are drawn consciously during specification activities. The emphasis during specification must be on representing the intended effect of the software. How this effect is to be achieved is of no concern. When an application domain under consideration deals with a problem that specifies high performance as a non-functional requirement, more often than not the distinction between design and specification becomes more blurred. It is essential however to avoid introducing any unnecessary constraints that relate to a solution rather than the problem statement.

If a specification is a "natural" representation of a problem, it follows that the specification will probably be easier to validate and modify since validation and modification are activities carried out by those who are familiar with the problem domain. If a specification is abstract and implementation-independent, it will probably lead to a better design since the designer is not falsely constrained and can deal with implementation resources and their management in creative ways possibly not envisaged by a specifier. The difficulty lies in deriving a natural, abstract, implementation-independent specification that is still precise and complete.

We disagree with Swartout and Balzer in their argument that implementation decisions should cause specification modifications. Certainly implementation activities may uncover inadequacies in a specification, but the modifications we envisage here originate with a user, not a developer. They arise from user experience of a system (either a prototype or full system) rather than from implementation concerns.

2.3 Uses of Concurrency

There are two reasons† why concurrency may be used: either for increased performance or because concurrency allows a natural representation of a problem.

Often an implementation on a parallel processor is chosen because this will allow faster execution than on a sequential processor. For instance if a spreadsheet package was required to be 10 times faster than an existing package an implementation could be based on transputers. It is important that the two things are not blurred, i.e. should the specification detail that:

(i)   The implementation should be at least 10 times as fast as package x in specific circumstances.

or

(ii)   The implementation should be based on transputers.

If such details are included as part of a specification they are seen as constraints that can be classified as non-functional requirements.

Many problems are inherently parallel. For instance a banking system that updates individual accounts can essentially update all accounts simultaneously. There is no need to state that first *account 1* is updated then *account 2* etc. Care must be taken not to introduce any unnecessary ordering constraints, for example that actions must be performed either sequentially or simultaneously. A useful specification should indicate whether operations are order-independent or not. Many methods of requirements

---

† Some would argue that there is a third reason - reliability. By replicating processing an opportunity is provided to detect and eliminate errors. This approach was taken, for example, in the Shuttle space programme, and is manifested also in the fly-by-wire software in the A320 Airbus. Computational effort is "wasted", by carrying out a given calculation a number of times in parallel using different hardware and different algorithms. Examination of the (possibly varying) results offers a way of carrying out a calculation to a very high level of confidence in the presence of uncertainties as to the reliability of the individual processors or algorithms. Though this strategy is valid as a route to reliability by introducing redundancy, we choose to regard it as being not strictly an example of computational parallelism. It certainly does not arise from any observation of parallelism inherent in the problem, nor does it offer enhanced running time.

specification provide this flexibility (for example structured analysis [Gane], Petri nets [Reisig], SYSREM [Alford], GYPSY [Ambler], PAISLey[Zave], CORE [Mulley], SADT [Ross], and LOTOS [ISO]) as well as do program specification methods (see Peters for example). The designer is **not** constrained to implement order-independent operations concurrently.

2.4 Should Concurrency be Specified?

Considering the nature and problems of specification, and the reasons for using concurrency, one can legitimately ask, should concurrency by specified? The answer can be found by careful analysis:

- is concurrency required?

or

- is concurrency a way of providing what is required?

We believe there is one problem to be solved, one underlying logical model of what a system is required to do. Thus specification does not involve choice. It states what must be done. If during a specification process there are any questions about whether or not to include concurrency, then this is no longer specification. This concurrency should NOT be specified. If decisions are made during specification regarding for example the identification of functions that might be grouped together for concurrent execution, then this is no longer specification. Again this concurrency should NOT be specified.

A specifier might ask a user "Can these two functions be carried out at the same time?" and the user might well reply "It doesn't matter as long as I get my results on time" or " They can be done whenever the input they need is available." Or, a specifier might ask "How many concurrent users do you expect?" and the response may be some number but the real requirement lies in providing a system that can operate under predetermined performance requirements (constraints). Again, this concurrency should

NOT be specified.

It requires considerable analysis to determine whether certain processes **must** be carried out concurrently, or **may** be carried out concurrently. The overall effect as perceived by the user must be achieved by the system. This is what must be specified and no unnecessary ordering should be part of a specification. Functionality and performance are specification concerns, but once particular computing resources are under consideration, specification has given way to design.

3. Design Considerations

3.1 Parallel Architectures

Several taxonomies of parallel architectures have been proposed [Flynn, Shore, Handler and Hockney]. Flynn's category were SISD (single instruction stream single data stream), SIMD (single instruction stream multiple data stream), MISD (multiple instruction stream single instruction stream) and MIMD (multiple instruction stream multiple data stream). These categories can not be judged to be complete by present day standards, as there is no obvious place to put pipelined computers. However, even with these simplistic categories, it can be seen that a program to be run on an SIMD architecture will require the same instruction to be executed on a number of pieces of data to exploit parallelism, while a target MIMD architecture will be less constrained.

Some problems can be structured in different ways to take advantage of a number of different architectures, whereas other classes of problems may be better suited to using arrays of identical processors that perform the same operation simultaneously (i.e. SIMD), implementation on a pipeline processor, or processing on a loosely coupled distributed system (i.e. MIMD).

The granularity of parallelism feasible on a target hardware will also have to be taken into account. An algorithm that requires frequent communication between processes, will not map well onto an architecture with a heavy overhead on inter-processor communication, unless all the communicating processes can be mapped on to one processor. The number of processing elements available can also affect the design decisions that need to be taken. An algorithm that required 64 processes to execute at one time would, in general be better suited to an architecture with 64 processors than one with 63. Indeed the number of processors available may not be static, in which case such considerations are pointless.

3.2 Languages for Parallel Processing

There are many more programming languages for parallel processing than there are parallel architectures. Some such as Occam [Inmos] are meant for a particular architecture, while others such as Ada [US] are meant for a variety of architectures. For some architectures, such as the Cray 1 [Hockney], a sequential language is used, although in a constrained way so that a vectorising compiler can find the *implicit* parallelism.

There is no accepted taxonomy for languages for parallel processing. It is possible however to identify if a language is specific to a particular architecture or category of architectures. It may also be possible to identify which programming model a language belongs to [Williams]. These models include shared memory (e.g. Modula [Wirth]), message passing (e.g. Occam) and dataflow (e.g. Lucid [Ashcroft]). Some languages exhibit qualities of more than one model (e.g. Ada uses both shared memory and message passing). Additionally languages for parallel processing may exhibit the qualities associated with all languages (e.g. susceptibility to formal methods, strength of data typing etc.).

Thus when the programming language is chosen it will be necessary to determine if it is well suited to the problem and the target architecture. If the target architecture is not yet chosen, the language should not unduly limit the choice. For many projects the approach is to choose the software and hardware architectures concurrently! (This need not be specified however).

4. An Example

Consider the following combinatorial problem:

Given a sequence of values α_i:

$$\alpha_1 \alpha_2 \cdots \alpha_n$$

Find the maximum value of the sums of the subsequences, i.e. the maximum value of S,

where: $S = \sum_{i=j}^{k} \alpha_i$, $\qquad 1 \le j \le k \le n$

Thus for both the following sequences:

$$1\ 2\ 1\ 3\ 4 \quad \text{and} \quad 2\text{-}4\ 5\ 6\text{-}1$$

The answer is 11.

4.1 Specification

This sample has been specified using a variety of techniques in (Ibrahim). Here we extract parts of two sample specifications, one based on Structured Analysis, and the second based on an object-oriented approach.

Structured Analysis has been found to be an effective method for analyzing and defining software requirements. Specifications produced may be elaborated or formalised by means of other methods used during requirements specification or they may suffice at the requirements level and form the basis of a range of design methods (e.g. Structured Design [Stevens et. al.], Object-oriented Design [Booch, Meyer], DARTS [Gomaa], etc).

A functional specification using Structured Analysis typically contains a data flow diagram, data dictionary, and process descriptions. An abbreviated specification for the sample problem is provided in Figures 1 - 3. Three transforms have been denoted on the data flow diagram (Figure 1) and some performance constraints could be added. There is no unnecessary ordering imposed by the specification, but there are various possible implementations that could potentially exploit parallel architectures. The findmaxsubseq

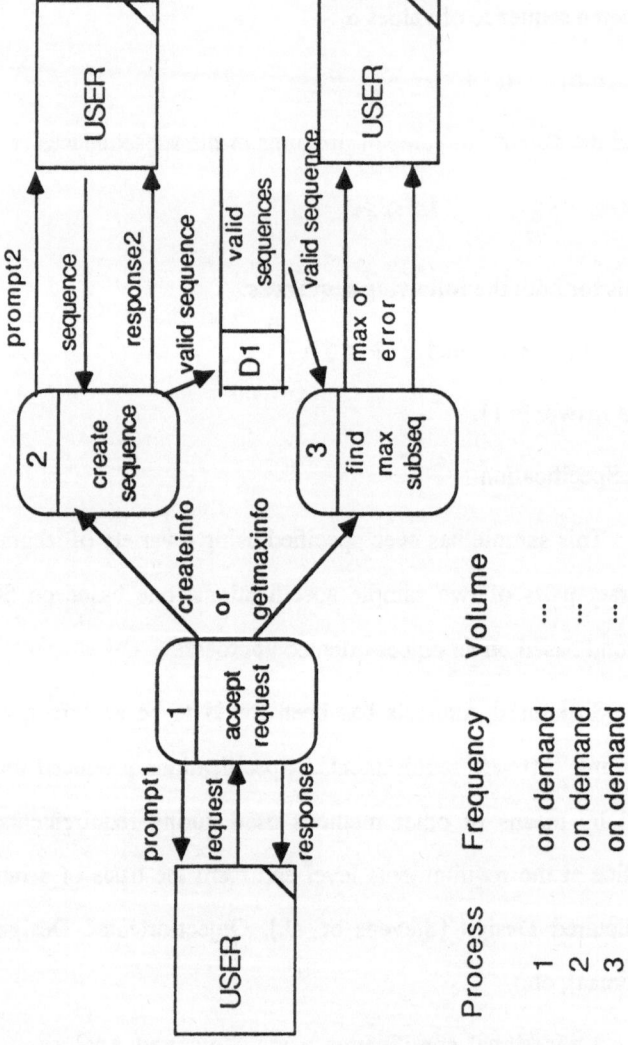

Figure 1. Structured Analysis: Data Flow Diagram

Figure 2. Structured Analysis: Data Dictionary

Name: request
Definition: user request for sequence
  processing
Type: data flow
Format: command +
    [create_info | get_max_info]

Name: response
Type: data flow
Format: text
Value: ["command accepted"|
  "invalid command"]

Name: command
Type: element
Format: string
Value: ["create" | "getmax"| "quit"]

Name: create_info
Type: data flow
Format:
    sequence_name +
    maxlength

Name: sequence_name
Definition: name of a sequence
Type: element
Format: string

Name: maxlength
Definition: maximum length of a
  sequence
Type: element
Format: integer
Value: 1..500

Name: sequence
Alias: numbers, sequence of numbers
Definition: the values to be included
  in the sequence
Type: data flow
Format: numeric
Value: valid integer or real

Name: response2
Type: data flow
Format: string
Value: ["value not numeric, enter again" |
    "maxlength numbers already entered,
      entry ignored" |
    "new sequence (name) has been created"]

Name: valid sequence
Type: data flow
Format:
    sequence_name +
    0 {number} maxlength

Name: get_max_info
Definition: information provided to find max
Alias: sequence name
Type: data flow
Format: sequence_name

Name: max
Definition: the sum of the subsequence
    that has the maximum value
Type: data flow
Format: numeric
Value: valid integer or real

Name: error
Definition: error detected when requesting
    maxsubseq
Type: data flow
Format: string
Value: ["sequence (name) not found" |
    "sequence (name) empty, undefined result"|
    "overflow"]

Name: valid-sequences
Definition: sequences created by user
Type: data store
Format: 0 {valid_sequence} *

Name: USER
Definition: person wanting to process
    sequences
Type: source/sink
Security: ...
Maximum number: ...

Figure 3. Structured Analysis: Process Descriptions

Name: accept request
Definition (purpose): accept and validate a user request, and
    initiate further processing
Type: Process 1
Inputs: request
Process:
  display prompt1
  read request
  while user doesn't quit do
    if valid request then
      display "command accepted"
      and initiate appropriate request process
    else (not valid request) so
      display "invalid command"
    display prompt1
    read request
Outputs:
  response
  create_info
  getmaxinfo
  prompt1

Name: create sequence
Definition (purpose): process user request to create a sequence
Type: Process 2
Inputs:
  create_info
  sequence
Process:
  create sequence with maximum length maxlength and name sequence_name
  display "new sequence (name) has been created"
  display prompt2
  while user doesn't quit do
    accept sequence value
    if not numeric display "value not numeric ,enter again"
    else (numeric) so
      if maxlength exceeded then
        display "(maxlength) numbers already entered, entry ignored"
      else (numeric and maxlength not exceeded) so
        place value in the sequence
    display prompt2
  store valid_sequence
Outputs:
  valid_sequence
  prompt2
  response2

Name: find_max_subsequence
Definition (Purpose): process user request to find the sum of the
  subsequence that has the maximum value
Inputs: get_max_info
    valid sequence
Process:
  get sequence sequence_name
  if not found then
    display "sequence (name) not found"
  else (found) so
    if empty then
      display "sequence (name) empty, undefined result"
    else (found and not empty) so
      find the sum of the subsequence of sequence (sequence_name)
        that is the maximum
      if any sum toobig then
        display "overflow"
      else (sum not toobig) so
        display max
Outputs:
  max
  error

process has been specified (in Figure 3) again avoiding consideration of any particular target hardware.

The next example specification is based on an object oriented approach. A possible view of the objects and their operations is portrayed in Figure 4. The operations of the sequence object are then specified using an interface (input/output) approach in Figure 5. The designer and implementor are not bound by any unnecessary ordering constraints.

4.2 Algorithm Design for Sequential Processing

This problem has been discussed by Bentley. He gives four sequential algorithms of which three are of interest in the present context.

The simplest algorithm is $O(n^2)$, as follows (indentation implies structure):

```
{ initialise max }
max ← 0
{ this loop fixes left boundary of subsequence }
for j ← 1 to n do

    { this loop fixes right boundary of subsequence }
    for k ← j to n do

    sum ← 0

    { calculate sum of subsequence }
    for i ← j to k do
        sum ← sum + a_i

    { compare a new sum with max }
    if sum > max
    then max ← sum

S ← max
```

The disadvantages of the above algorithm are immediately apparent to a sequential programmer: sums are recalculated many times. Bentley discusses an $O(n^2)$ algorithm which we shall not reproduce, and an $O(n \log n)$ which is rather subtle. It is based on the divide-and-conquer principle, and argues that if the array is divided into two sub-arrays A and B of equal length then the maximal subsequence is one of the following:

Figure 4. The Objects and their Operations

User_request
    get_request
    check_request

User_response
    display_response

Sequence
    create
    insert
    max

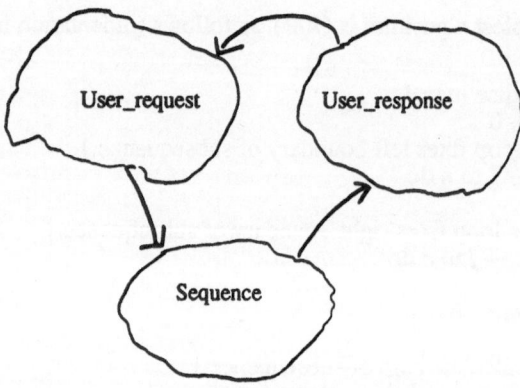

Figure 5. Sequence Operations: An Interface Specification

1. create: maxlength $\rightarrow$ sequence

    input: m: integer

    output: s: {}

2. insert: sequence $\times$ num $\rightarrow$ sequence $\cup$ error

normal:

    input: s :sequence $\{n_1, n_2, \cdots, n_k\} \wedge k <$ maxlength

      n : num

    output: $s: \{n_1, n_2, \cdots, n_k, n\}$

error:

    input: $\{n_1, n_2, \cdots, n_k\} \wedge k =$ maxlength

    output: error ("maximum length exceeded")

3. max: sequence $\rightarrow$ num $\cup$ error

normal:

    input: not empty (s) $\{n_1, n_2 ..., n_n\}$ $n \geq 1$

    output:

$$\forall j, k \text{ s.t. } 1 \leq j \leq k \leq n \quad \text{let } S_{j,k} = \sum_{i=j}^{i=k} n_i$$

$$\exists j, k \text{ s.t. } (\forall l, m \text{ s.t. } 1 \leq l \leq m \leq n \ (S_{j,k} \geq S_{l,m})) \ \wedge \ \max = S_{j,k}$$

error:

    input: s {}

    output: undefined

    input: $\{n_1, ..., n_k\}$

    output: $\exists j, k$ s.t. $S_{j,k} > maxnumber \rightarrow error (overflow)$

a)   the maximal subsequence in sub-array A

b)   the maximal subsequence in sub-array B

c)   a subsequence that overlaps the boundary between A and B. This latter is itself formed of a maximal subsequence of A that is constrained to lie at A's upper boundary, and a maximal subsequence that is constrained to lie at B's lower boundary.

The algorithm is stated as follows:

function $S(l,u)$

if $l > u$ then $S \leftarrow 0$
else if $l = u$ then $S \leftarrow a_i$
else
   $m \leftarrow (l+u)/2$

   { find max subsequence in A recursively }
   $sa \leftarrow S(l,m)$

   { find max subsequence in B recursively }
   $sb \leftarrow S(m+1,u)$

   { find max subsequence constrained to lie at
    the upper end of A }
   $sum \leftarrow 0$
   $ma \leftarrow 0$
   for $i \leftarrow m$ downto l do
    $sum \leftarrow sum + a_i$
    if $sum > ma$ then $ma \leftarrow sum$

   { find max subsequence constrained to lie at
    the lower end of B }
   $sum \leftarrow 0$
   $mb \leftarrow 0$
   for $i \leftarrow m+1$ to u do
    $sum \leftarrow sum + a_i$
    if $sum > mb$ then $mb \leftarrow sum$

   { find maximum overall }
   $sc \leftarrow ma + mb$
   $S \leftarrow maxof(sa,sb,sc)$

Finally Bentley presents a $O(n)$ algorithm:

function S

$sa \leftarrow 0$
$sb \leftarrow 0$

{ this loop scans the array once only }
for $i \leftarrow 1$ to n do

    { sb is the sum of the current subsequence }
    { sum is a trial extension of sb by a single element }
    $sum \leftarrow sb + a_i$
    if $sum > 0$ then $sb \leftarrow sum$ else $sb \leftarrow 0$

    { sa is the value of the largest maximal subsequence found thus far }
    if $sb > sa$ then $sa \leftarrow sb$

$S \leftarrow sa$

4.3 Algorithm Design for Specific Target Architectures

Quinn states that when programming for parallel processors it is essential that "the algorithm must fit the architecture". Below three different algorithms are developed for three different categories of parallel processors:

Array processor

Pipeline processor

MIMD

The first algorithm presented is for an array processor:

for all j $\leftarrow$ 1 to n do

    for all k $\leftarrow$ j to n do

$$\text{sum } (j,k) = \sum_{j}^{k} a\ (i)$$

maximal subsequence = maximum of sums

This algorithm assumes:

(i) $n^2 / 2$ processing elements

(ii) processing elements not involved in summations can be masked

(iii) maximum across processing elements is an in-built function.

The second algorithm is for a pipeline processor, which we assume has n̄ processors.

Initially assume that each value in the sequence is allocated to a processor (i.e. a(1) to processor 1, a(2) to processor 2 ... a(i) to processor i ... a(n) to processor n.

Then each processor performs the following actions:

(i) read value from the processor on the left (a large negative number in the case of processor 1),

add in own value to form own sum,

compare the result with own value and pass the larger on.

(ii) read maximum from the left (a large negative number in the case of processor 1),

compare against own sum and pass the larger on.

for j ← 1 to n

left ? sum (j-1)

sum (j) = maximum of (sum (j-1) + a (j) , a(j))

right ! sum (j)

left ? max

max = maximum of (max , sum (j))

right ! max

Assuming both (i) and (ii) take T seconds, then a single result will be produced in $(n+1) \times T$ seconds. If there were a number of sequences that required their maximal subsequence to be calculated a new result could be produced every $2 \times T$ seconds. Further refinements could include allocating parts (i) and (ii) of each process to separate processors, this would however increase the pipe priming time and thus not be suited to a few sequences.

The final algorithm in this section is suited to a general class of loosely coupled multiple processors, known as MIMD processors.

Quinn has identified a number of approaches to solving such combinatorial problems on MIMD architectures, including building a tree of real or virtual processors to match the problem.

A divide and conquer technique, using n processors, may be well suited to this problem.

a (n+1) = large negative number

par i ← 1 to n

    oldsum (i) = large negative number

    newsum (i) = a(i)

    j = 0

    while (newsum (i) > oldsum (i)) and (i+j < n+1)

        j = j+1

        oldsum (i) = newsum (i)

        newsum (i) = newsum (i) + a (i+j)

    maximum of oldsum

maximum of can also be run in parallel, as processes finish their oldsum is checked for been maximum. If many processes finish together a comparison tree can be constructed.

4.4 Comparison of the Algorithms

The first algorithm of Bentley's is very similar to the algorithm chosen for the array processor. For a sequential model, however, this algorithm is considered inefficient. The second algorithm presented in section 4.2 is based on the divide and conquer technique, as was the algorithm for the MIMD architecture. However in the sequential case the divide was only by 2, whereas in the MIMD case, the constraint was dictated by the problem space (i.e. the number of elements in the sequence). Bentley's final algorithm and the pipeline algorithm only scan the sequence once, but there their approaches diverge.

The algorithm presented for the pipeline processor would be very inefficient if it were to be executed on either of the other parallel architectures presented; similarly their algorithms could not be efficiently executed on a pipeline architecture. The earlier stages of the algorithms for the array processor and MIMD processor could be more readily interchanged, but the subsequent extraction of the maximum would cause inefficiencies to be introduced if the algorithms were to be interchanged.

All the parallel algorithms presented would have been considered inefficient for execution on a conventional architecture, but they take advantage of the well known features of their parallel architectures and for these will be considered efficient (Quinn shows how timings can be calculated for such algorithms). It follows that the choice of algorithm (with implicit determination of parallelism) should be deferred as long as possible, until the choice of architecture is pressing.

5. Conclusions

A specification expresses the functional and nonfunctional requirements of a piece of software, and we have tried to show why a functional specification should not dictate any unnecessary ordering constraints. It should be independent of implementation details, and thus realisable on a variety of architectures. Ordering constraints and explicit concurrency should be introduced as part of the refinement phase as it is needed. Only the essential temporal ordering should be specified.

If an application calls for a high performance system, it may be more difficult to separate the problem from a solution, e.g. it may be more tempting to specify concurrency. In addition, the realization that a variety of parallel architectures are becoming available again may lure a specifier into making premature design decisions and enforcing them as requirements.

Keeping in mind the nature of a specification, analysis should uncover what needs to be done to solve the user's problem. Is concurrency **what** must be done, or **how** it will be done? If concurrency is a matter of choice then it should not be specified because it is a design decision. If there is no choice, then concurrency is a requirement and must be specified. More often than not, however, it is not concurrency that is required, but performance.

Incidentally the answer would be the same if the question were changed to:
   "Should Sequentiality be Specified?"

References

Alford, M., "SREM at the Age of Eight: the Distributed Computer Design System," *Computer* **18**, pp.36 - 44 (1985).

Ambler, L. et al, "GYPSY a Language for Specification and Implementation of Verifiable Programs," *SIGPLAN Notices* **12**, pp.1 - 10 (1977).

Ashcroft, A. E. and Wadge, W. W., "LUCID, A Non-Procedural Language with Iteration," *CACM* **20**, pp.519 - 526 (1979).

Bentley, J., "Algorithm Design Techniques," *CACM*, pp.865 - 871 (1984).

Booch, G., "Object-Oriented Development," *Transactions on Software Engineering* **SE-12**, pp.211 -221 (1986).

Flynn, M. J., "Some Computer Organisations and Their Effectiveness," *IEEE Transactions on Computers* **C-21**, pp.948 - 960 (1972).

Gane, C. and Sarson, T., *Structured Systems Analysis*, Prentice Hall (1979).

Gomaa, H., "A Software Design Method for Real-time Systems," *CACM* 27, pp.938 - 949 (1984).

Handler, W., "Innovative Computer Architecture," in *Parallel Processing Systems*, ed. D. J. Evans, Cambridge University Press (1982).

Hockney, R. W. and Jesshope, C. R., *Parallel Computers 2*, Adam Hilger Ltd, Bristol (1988).

Ibrahim, R. L., "Specifying the Maximal Subsequence Problem," Working Paper, University of Reading (1988).

INMOS, *Occam Programming Manual*, Prentice Hall (1984).

ISO, *Information Processing Systems - OSI LOTOS A Formal description Technique based on Temporal Ordering of Observational Behaviour*, 1986.

Meyer, B., *Object-oriented Software Construction*, Prentice Hall (1988).

Mulley, G. P., "CORE - A Method for Controlled Requirement Specification," pp. 24 - 320 in *Proceedings of the 4th International Conference on Software Engineering* (1980).

Peters, L. J., *Software Design: Methods and Techniques,* Yourdon Press (1981).

Quinn, M., *Designing Efficient Algorithms for Parallel Computers,* Mc Graw Hill (1987).

Reisig, W., *Petri Nets an Introduction,* Springer Verlag (1985 (English Edition)).

Ross, D., "Applications and Extensions of SADT," *Computer* **18**, pp.25 - 35 (1985).

Shore, J. E., "Second Thoughts on Parallel Processing," *Computers and Electrical Engineering* **1**, pp.95 - 109 (1973).

Stevens, W. P., Meyers, G. J., and Constantine, L. L., "Structured Design," *IBM Systems Journal* **13**, pp.115 - 139 (1974).

Swartout, W. and Balzer, R., "On the Inevitable Interwining of Specification and Implementation," *CACM* **25** (1982).

Williams, S. A., "Mapping Parallel Programs on to Distributed Architectures," Invited paper submitted to Advances in Parallel Computing (1988).

Wirth, N., *Programming in Modula 2,* Springer (1982).

Department-of-Defence, U.S., "Programming Language ADA : Reference Manual," *L.N.C.S.* **106**, Springer Verlag (1981).

Zave, P., "An Operational Approach to Requirements Specification for Embedded Systems," *IEEE Transactions on Software Engineering* **SE 8**, pp.250 - 269 (1982).

Semantics for Specifying Real-time Systems

Mathai Joseph, Asis Goswami*

University of Warwick

EXTENDED ABSTRACT

Introduction

Real-time programs are the least well understood of concurrent programs: apart from being subject to all the usual problems associated with concurrency and communication they must interact with a variety of agents[1] at points in the execution that have a specified ordering relation with time. So, for example, the familiar partial order over the execution of the statements of a concurrent program must be made more elaborate to accommodate time-ordering, and this requires a semantics that can account for programs whose execution may be constrained by limitations on the availability of resources.

Consider the network $P :: P_1\|P_2\|P_3$ where the processes P_1 and P_2 are

$$P_1 :: *[... P_2!e ...], \quad P_2 :: *[P_3?x \rightarrow P_1?y; \; delay \; d; \; P_3!(f(x,y))]$$

and P_3 is a physical device with the behaviour

$$P_3 :: *[true \rightarrow P_2!a; \; ... ; P_2?b]$$

where the command *delay d* has the effect of suspending the execution of a process for exactly d time units. Assume that the program has been designed for a machine with two identical processors. Under what conditions can the correctness of the program be guaranteed if only one processor is available for execution? With the existing techniques for analysis and verification, answering this question would require a complete re-examination of the program even though it is not the program but the execution environment that has changed.

In practice, real-time programs are of course far more complicated than this simple example and have many more processes. A variety of questions may be raised about such programs: e.g., what is the minimum number of processors for ensuring the correct

---

*Address for correspondence: Department of Computer Science, University of Warwick, Coventry CV4 7AL. This work was supported by research grant GR/D 73881 from the Science and Engineering Research Council.

[1]An *agent* may be another program, a person, or part of a physical system

implementation of a particular real-time program, what kinds of process structure will ensure that a program is insensitive to bounded variations in the numbers of processors, etc. Our concern has been to develop a model of real-time programs which addresses these questions. More generally, the semantics can be viewed as modelling the execution of programs in a domain of limited resources of which the processor, which controls the execution speed of the program, is merely one.

Time in Programs

Real-time programs can be distinguished from other programs because they operate in an explicit framework of time. Time progresses during the execution of each command and when a program is waiting for some external or internal action (such waiting can of course be represented as the execution of a *delay* command). An important question then concerns the values needed to represent time. Since programs execute with fixed precision, it can be argued that time can be represented by values from the natural numbers. Moreover, time is only observable within a program at the start and end of any command. On the other hand, since in a distributed system independent actions can take place arbitrarily close together, time should perhaps be represented by real values.

In this model, time has values from the real domain but there is no single system-wide clock. Instead, time is treated as a local property of each program unit: it can be observed only at the start or end of an action and it increases monotonically. When program units are composed together, a new time frame is established which is consistent with the actions of each separate unit. There is thus no inherent notion of 'global' time, though synchrony with any particular clock can be maintained by composing the program with a process representing that clock.

Semantics of primitive commands and combinators

The behaviour of a primitive command is characterized by observations made at the start and end of the execution of the command. So, there will be one observation for a non-terminating command and two observations for a terminating command. The set of behaviours of a communication command must take into account all the possible outcomes of communication and failure. The behaviour of a composite command is derived from the behaviours of its constituent (possibly primitive) commands and the nature of their combination.

Let $\diamond$ be an n-ary command combinator, and C be a composite command $\diamond(C_1, \ldots, C_n)$. The semantics of the combinator $\diamond$ defines the behaviour of C in terms of the behaviours of $C_1, \ldots, C_n$. This can be done uniformly for *any* combinator.

The semantics of $\diamond$ is given by the formula

$$\beta(C) = \{Combine_\diamond (F_1, \ldots, F_n) \mid [\forall i \in [1, n] : \\ F_i \in \beta(C_i)] \wedge Consistent_\diamond(F_1, \ldots, F_n)\}$$

The predicate *Consistent₀* tests whether the behaviours F_i describe only those observed values of the variables of the components C_i of C that are consistent with respect to the computation mechanism of $\diamond$. If so, then these behaviours can be combined to produce a behaviour of C. *Consistent₀* is a relation on the observed values (and their times) of the variables in a subset $VAR_{\diamond}(C)$ of $VAR(C)$; only variables in $VAR_{\diamond}(C)$ participate in the operation of $\diamond$. In the case of parallel composition, for example, $VAR_{\parallel}(C_1 \parallel \cdots \parallel C_n)$ is the set of all variables used for communication in $C_1, \ldots, C_n$. In the trivial case where $VAR_{\diamond}(C) = \emptyset$, the behaviours $F_1, \ldots, F_n$ are consistent with respect to $\diamond$.

In general, a combinator provides mechanisms for the transfer of values between its constituent commands. For any command C_i in C and combinator $\diamond$, $Input_{\diamond}(C_i)$ is the set of variables of C_i in which it expects to receive values from its environment, for example $C_1, \ldots, C_{i-1}, C_{i+1}, \ldots, C_n$, by the communication mechanisms of $\diamond$. Similarly, $Output_{\diamond}(C_i)$ is the subset of $VAR(C_i)$ for the output of values by C_i through $\diamond$. The sets $Input_{\diamond}(C_i)$ and $Output_{\diamond}(C_i)$ need not be disjoint and

$$VAR_{\diamond}(C) = \bigcup_{i=1}^{n}(Input_{\diamond}(C_i) \cup Output_{\diamond}(C_i))$$

The environment of C_i *sends* data to a variable $v \in Input_{\diamond}(C_i)$ through the variables in a nonempty set $Complement_{\diamond}(v)$. If v is an output variable of A, the environment of C_i *receives* data from v in the set of variables in $Complement_{\diamond}(v)$.

In the formula for $\beta(C)$, the predicate *Consistent₀* checks, for each $v \in VAR_{\diamond}(C)$, whether the observed values of v in F_i are consistent (with respect to $\diamond$) with the observation times and the observed values of the variables of the environment of C_i in C which are in $Complement_{\diamond}(v)$.

The function *Combine₀* reorders the observed values of its argument commands into a new total order of time values which is obtained from the observation times and the observed values of the arguments.

Limited Resource Semantics

Each command requires a number of resources such as processors, memory, communication channels etc. for its execution and the time taken by a command often depends on the resources provided for its execution. Assume that there are bounds, $Lower(C)$ and $Upper(C)$ to the resources needed for any command C. If the available resources exceed the upper bound, let the execution time of C be the same as that with the maximal resource. Let execution of a command be impossible if the available resource is smaller than the lower bound for the command.

If a sequential component S in C is suspended in an interval I, then the resources allocated to S for the interval I can be re-assigned to another command running concurrently with C.

Let the command C' be the *functional equivalent* of C, i.e. $VAR(C) = VAR(C')$ and for all $F' \in \beta(C')$ there is some $F \in \beta(C)$ with the same observations and the same temporal ordering (but possibly different times). For any component of C, a $< command, wait >$ pair is an interval of time during which the component waits for resources. Then C' is an$R$-*representation* of C and the semantics $\Gamma(C)$ of a command C is a set of quadruples $< R, C', F, W >$, where $F \in \beta(C)$, W is the set of (command,wait) pairs associated with F, R is a description of the resource availability and C' is a functional equivalent of C for which R is the upper bound of the resource requirement.

If C is a primitive command, then *Lower*(C) and *Upper*(C) are the same, and the C' is C. The set W may be nonempty even for a primitive command if its termination depends on cooperation from other commands.

Combinators and limited resources

Let $\diamond$ be an n-ary combinator and

$$C \triangleq C_1 \diamond \cdots \diamond C_n$$

The R-*semantics* of C is obtained by combining

$$< R_i, C'_i, F_i, W_i > \in \Gamma(C_i)$$

for $i \in [1, n]$, such that the F_i's are consistent with respect to $\diamond$ and an execution of C with R can be obtained by executing C_1 with R_1, C_2 with R_2, and so on.

Conclusions

In this paper we have considered two features that distinguish real-time programs from other programs: the relation between time and computations and the relation between command executions and the availability of resources. We have outlined how both of these features can be represented in a semantics for real-time programs. These two themes are considered elsewhere [1,2] in greater detail.

References

[1] A. Goswami and M. Joseph. A semantic model for the specification of real-time processes. In *CONCURRENCY 88, Lecture Notes in Computer Science 335*, pages 292–306, Springer-Verlag, Heidelberg, 1988.

[2] M. Joseph and A. Goswami. What's real about real-time systems? In *Procedings of the 9th IEEE Real-Time systems Symposium*, pages 78–85, Huntsville, Alabama, 1988.

Specifying Processes in Terms of Their Environments

Wang Yi

Department of Computer Science

Chalmers University of Technology

S-41296 Göteborg, Sweden

June 15, 1988

Abstract

A simple approach to specification and verification of processes is suggested. Instead of *process equivalence* as usual in process algebra, the correctness of an implementation w.r.t a specification is determined by a relation called *conformance*. A process is supposed to work in some *environment* in which a sought implementation is going to be embedded. The behaviour description of an environment is viewed as a specification, which makes implicit requirements on the potential implementations. According to the behaviour descriptions, we are able to decide if there are any *deadlocks* in the whole system consisting of the environment and the final implementation. If the whole system is *deadlock-free*, the implementation is the desired one. It is shown that the equivalence on processes, induced by the comformance relation coincides with failure equivalence.

Keywords: Formal Description, Transition system, CCS, Specification, Implementation, Correctness, Deadlock, Process Equivalence.

1 Motivation

Computer systems are often studied on two abstract levels: *specification* which describes a collection of requirements on a sought system and *implementation* which describes the behaviour of the final system. Thus we can reason about a system in terms of its specification and its behaviour's description. This creates a need for *proper correctness criteria* of an implementation w.r.t a specification.

Over the years, the research on process algebra has been focussed on process equivalences [Mi80,NH84,HBR84,La85,Ni87]. This allows the description languages such as CCS,CSP to be used as both   programming language and specification language. The correctness of an implementation w.r.t a specification is simply determined by their equivalence. Thus the potential implementations is restricted to a single equivalence class. In this paper, another suggestion is put forward. We allow the CCS like languages still to be both   programming language and specification language. Instead of equivalence as usual in process algebra, the notion of correctness is another relation called *conformance*. The conformance relation is formalized based on de Nicola and Hennessy's testing semantics for processes [NH84,Ni87].

A process (or system) is supposed to work in some environment in which a sought implementation is going to be embedded. The operational behaviour of such an environment makes implicit requirements to potential implementations. Indeed the requirements are often described by a specification written in some logical language, such as temporal logic, Hennessy-Milner logic. We shall use the behaviour description of an environment, written in process algebra, as specification.

Actually, a communication protocol usually goes through *conformance testing* before it is put into practical use. The goal of conformance testing is to determine whether the final system satisfies the requirements of its environment (or user). However a *collection* of tests, *which indeed reflect the behaviour of the concerned environment*, are often necessary. Hence to make a specification, we consider the behaviour (possibly *infinite*) of the environment instead of a set of tests. According to the behaviour descriptions, we are able to decide whether there are any *deadlocks* in the whole system consisting of the environment and the final implementation. If the whole system is *deadlock-free*, the implementation is the desired one.

Processes and their environments are modelled by *labelled transition system*. Actually, a number of so-called process description languages such as CCS, SCCS, LOTOS can be given an operational interpretation in terms of labelled transition systems. The notion of *labelled transition system* is described in **section 2**. Moreover, to have some formal notation for processes, a simple description language is presented, which is a variant of the standard CCS. In **section 3**, we formalize the notion, *conformance* in terms of the operational interpretation for processes and their environments. **Section 4** shows that the equivalence on processes, induced by the conformance relation coincides with failure equivalence. **section 5** discusses the differences between our work and the related work, particularly de Nicola and Hennessy's testing semantics for processes.

2   Labelled Transition Systems

Definition 1 *A labelled transition system S is a structure:*

$$< S, Act, \longrightarrow, \Phi >$$

where

- *S is a set of states or processes.*
- *Act is a set of atomic actions.*
- $\longrightarrow$: *$S \times Act \times S$ is the transition relation. With each $u \in Act$, we shall write $p \xrightarrow{u} q$ if $(p, u, q) \in \longrightarrow$ to mean p may become q by performing u.*
- *$\Phi \subseteq S$ is a set of accept states, modelling those states which are of particular interests, such as successful termination and unsuccessful termination.*

Note that the above definition is a variant of the standard one with extension of the forth component, i.e. the set of accept states. See bellow for motivation.

Now let $Act = L \cup \{\tau\}$, where $\tau \notin L$. τ is regarded as invisible. Generally we shall use the following notation.

Definition 2
Let L^* denote the set of finite strings over L, with ϵ as the empty string.
    For $s = a_1 a_2 ... a_n \in L^*$ where $n \geq 0$,
$p \xRightarrow{s} q$ iff $p(\xrightarrow{\tau})^* \xrightarrow{a_1} (\xrightarrow{\tau})^* ... (\xrightarrow{\tau})^* \xrightarrow{a_n} (\xrightarrow{\tau})^* q$
In particular, $p \xRightarrow{\epsilon} q$ if $n = 0$.

The juxtaposition is relation composition and $*$ is transitive closure. The transition relation $\Longrightarrow$ is called the experiment relation in [Mi80].

We will consider a particular transition system **Pr**, which is described by a CCS like description language. It has a set of prefixing operators $A \cup \{\tau\}$, a nondeterministic operator +, parallel operator and recursion. These operators are well motivated in [Mi80]. Instead of *nil*, we have two constants $\top, \bot$ to model *successful termination* and *unsuccessful termination*. As specification, $\top$ is the weakest specification and $\bot$ is the self-contradictionary specification which has no implementations at all. As implementations, the two types of terminations have no special meanings. $\top$ and $\bot$ just indicate that the physical system has stopped interacting with its environment. We omit restriction and renaming for convenience only. They can be added just by giving the computational rules. We shall see that our framework does not depend on language structure anyway.

Let X be a set of variables ranged over by x and $A \cup \{\tau\}$ a set of actions or communications. The elements of A are visible actions and $\tau \notin A$ is internal or invisible action. As in [Mi80], we let A have the structure $\Lambda \cup \bar{\Lambda}$, where Λ is a primitive set and $\bar{\Lambda} = \{\bar{a} | a \in \Lambda\}$. We say that action $\bar{a}$ is the complement of action a. We shall also view a as the complement of $\bar{a}$, i.e. $a = \bar{\bar{a}}$. Visibility of an action a of process p means that a requires a similar participation of the environment of p, i.e. its complement $\bar{a}$. We use u, v to range over $A \cup \{\tau\}$ and a, b to range over the set of visible actions of A.

Then the language is formlly defined by the following grammar:

$$t = \top \mid \perp \mid x \mid u.t \mid t + t \mid t|t \mid \mathbf{fix}_i \vec{x} : \vec{t}$$

where $\mathbf{fix}_i \vec{x} : \vec{t}$ stands for the ith component of the mutually recursive definition $\vec{x} \Leftarrow \vec{t}$. Closed terms are defined as usual with $\mathbf{fix}_i$ binding variables. We use $\mathbf{Pr}$ to denote the collection of closed terms called processes. We use p, q etc., to range over $\mathbf{Pr}$. Moreover, we shall use the notation $\sum_{i \in I} E_i$ to denote $E_1 + E_2 + ... + E_i + ...$ ($i \in I$).

An operational semantics for the language can be given in the usual way. Let $\xrightarrow{u}$ be the least relation over $\mathbf{Pr}$ that satisfies the following rules:

Definition 3

$$u.p \xrightarrow{u} p \tag{1}$$

$$\frac{p \xrightarrow{u} p'}{p + q \xrightarrow{u} p'} \qquad \frac{p \xrightarrow{u} p'}{q + p \xrightarrow{u} p'} \tag{2}$$

$$\frac{p \xrightarrow{u} p'}{p|q \xrightarrow{u} p'|q} \qquad \frac{p \xrightarrow{u} p'}{q|p \xrightarrow{u} q|p'} \tag{3}$$

$$\frac{p \xrightarrow{a} p' \quad q \xrightarrow{\bar{a}} q'}{p|q \xrightarrow{\tau} p'|q'}$$

$$\frac{t_i(\mathbf{fix}\vec{x} : \vec{t}/\vec{x}) \xrightarrow{u} t'}{\mathbf{fix}_i \vec{x} : \vec{t} \xrightarrow{u} t'} \tag{4}$$

where $t_i(\mathbf{fix}\vec{x} : \vec{t}/\vec{x})$ is the i'th component of $\vec{t}$ in which every x_j has been substituted by $\mathbf{fix}_j \vec{x} : \vec{t}$.

Note that we have no computational rules for $\top, \perp$, which means that they will not perform anything at all.

Clearly, the computational rules give rise to a transition system

$$< \mathbf{Pr}, A \cup \{\tau\}, \longrightarrow, \{\top, \bot\} >$$

where

- $\longrightarrow$ is induced by the computational rules and

- The set of accept states has two elements $\top, \bot$, namely, *successful termination* and *unsuccessful termination*.

3   The Correctness Criteria

In this section, we formalize the notion of *conformance*. The elements of **Pr**, which are considered as processes and environments will be used as both   implementations and specifications. *conformance* is a relation defined on **Pr**, which shall be taken as the satisfication relation between implementations and specifications. Similar formalizations can be found in de Nicola and Hennessy's works on testing equivalence [NH84,Ni87]. However our goal here is to apply these theories to describe and verify processes.

3.1   Deadlock, Termination and Conformance

Assume that we have an environment described by $e \in \mathbf{Pr}$ and an embedded system described by $p \in \mathbf{Pr}$.

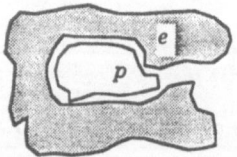

Fig 3.1.1

We view the whole as another system in which e interacts with p by message passing. e may terminate successfully after finite time. But the whole system may be deadlocked in the sense that there will be no communication between p and e. Suppose that a super-observer, to whom every communication between p and e is visible, is observing the behaviour of the whole system. Thus the possible observations will be the following:

- e and p keep communicating forever.

- e reach an expected state (*successful termination*).

- e reach an unexpected state (*unsuccessful termination*).

- e and p reach a state in which they can not communicate and e is not in any accept state.

In the first two cases, we shall say that p conforms to e. In other words, the implementation p satisfies the specification e.

To model these combined systems, we introduce another labelled transition system:

$$< \mathbf{Pr} \times \mathbf{Pr}, A \cup \{\tau\}, \longrightarrow, \emptyset >$$

where

1. $\mathbf{Pr} \times \mathbf{Pr}$ is the cartesian product of $\mathbf{Pr}$ with itself.

2. The set of atomic actions is $A \cup \{\tau\}$ as usual.

3. The transition relation $\longrightarrow$ is defined as follows:

$$\frac{p \xrightarrow{\tau} p'}{< p, e > \xrightarrow{\tau} < p', e >} \tag{5}$$

$$\frac{e \xrightarrow{\tau} e'}{< p, e > \xrightarrow{\tau} < p, e' >} \tag{6}$$

$$\frac{p \xrightarrow{\bar{a}} p' \quad e \xrightarrow{a} e'}{< p, e > \xrightarrow{a} < p', e' >} \tag{7}$$

where $a \in A$

4. The accept set is empty.

The transition relation $\longrightarrow$ can be extended to $\Longrightarrow$ in the same way as defining $\Longrightarrow$.

Intuitively the symbol a over the arrow $\longrightarrow$ is an event ocurring on the communication channel between p and e, which is observable to the *super-observer*. $< p, e >$ is the parallel process consisting of p and e. $< p, e >$ will evolve to another state $< p', e' >$ by internal moves of p and e or inter-communication between p and e.

We shall say that a state $< p, e >$ is deadlocked if p and e will never communicate. Formally, we define a predicate **deadlock** on $\mathbf{Pr} \times \mathbf{Pr}$.

Definition 4 *Let* $t \in \mathbf{Pr} \times \mathbf{Pr}$.
deadlock(t) *iff there is no* t' *such that* $t \xRightarrow{a} t'$ *for every* $a \in A$.

Here *deadlock* is defined in terms of communication capability. Namely if there exist any possible communications between the environment and the embedded system, then the state is not deadlocked, otherwise it is deadlocked.

Note that we use the double arrow $\Longrightarrow$ instead of $\longrightarrow$ in the definition of deadlock. This is because we view processes which can perform any number of internal moves but will not leave the possibility of sending out results, as *alive*. This is different from the view of [NH84, Ni87], where this type of processes are considered as *divergent*, i.e. the *undefined processes*. We shall see that our view of deadlock will cause processes p and $\tau^\omega|p$ to be equivalent. But they are distinguished by de Nicola-Hennessy's testing equivalence. Another point is that a deadlocked state may be capable of performing internal moves, but no external communication. For instance, $\neg\mathbf{deadlock}(< p, e >)$, but $\mathbf{deadlock}(< q, e >)$. p, q and e are as follows.

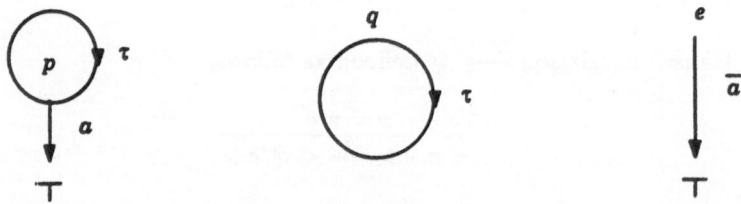

Fig 3.1.2

However the notion of deadlock above only reflects the communication possibility of the whole system. We have a particular case for those deadlocked states, i.e. successful termination of the whole system. Naturally, if the environment becomes a *successful termination*, the whole system does the same. This is determined by the predicate **successful**.

Definition 5 $\mathbf{successful}(< p, e >)$ *iff* $e \overset{\varepsilon}{\Longrightarrow} \top$

It is simply motivated by the fact that if the environment is *willing* to terminate the computation process, it does so, which is totally independent of whatever the involved system is doing. Even though the involved system has reached an unexpected state, i.e $\perp$ or whatever else, the state of the whole system is recognized as a successful termination as long as the environment is *satisfied*. A trivial example is that $\mathbf{successful}(< \perp, \top >)$ holds.

Now we are ready to define the conformance relation on **Pr**.

Definition 6
p *comforms to* e, *denoted by* $p \models e$, *iff* $< p, e > \overset{s}{\Longrightarrow} < p', e' >$ implies
$\neg\mathbf{deadlock}(< p', e' >)$ or $\mathbf{successful}(< p', e' >)$ *for all* s, p', e'.

Intuitively, if every possible computation of the system $< p, e >$ that leads to deadlock is successful, then p comforms to e (or p satisfies e). For a trivial example, let $p_s <= \bar{a}.p_s$ and $A_s <= a.A_s$. Using the computational rules, we get

$$\frac{p_{\scriptscriptstyle \delta} \xrightarrow{a} p_{\scriptscriptstyle \delta} \quad A_{\scriptscriptstyle \delta} \xrightarrow{a} A_{\scriptscriptstyle \delta}}{< p_{\scriptscriptstyle \delta}, A_{\scriptscriptstyle \delta} > \xrightarrow{a} < p_{\scriptscriptstyle \delta}, A_{\scriptscriptstyle \delta} >}$$

Since there exist no p', A' such that $< p_{\scriptscriptstyle \delta}, A_{\scriptscriptstyle \delta} > \xrightarrow{b} < p', A' >$ for every $b \neq a$ and $\neg\textbf{deadlock}(< p_{\scriptscriptstyle \delta}, A_{\scriptscriptstyle \delta} >)$, we can conclude $p_{\scriptscriptstyle \delta} \models A_{\scriptscriptstyle \delta}$. See section 3.2.

Before showing more examples, we give another formalization of the relation $\models$. Hopefully the new one will be easier to use than the old one. Let s be a string and its complement $\bar{s}$ is defined by taking the complement of every element in s. For instance, let $s = a_1 a_2 ... a_n$. Then $\bar{s} = \bar{a}_1 \bar{a}_2 ... \bar{a}_n$. But $\bar{\varepsilon} = \varepsilon$. Moreover let $\textbf{alive}(< p, e >)$ denote that $\neg\textbf{deadlock}(< p, e >)$ or $\textbf{successful}(< p, e >)$.

Proposition 1 $p \models e$ iff

1. $\forall s, p'.p \xRightarrow{s} p'$ implies $(\forall e'.e \xRightarrow{\bar{s}} e'$ implies $\textbf{alive}(< p', e' >))$
2. $\forall s, e'.e \xRightarrow{s} e'$ implies $(\forall p'.p \xRightarrow{\bar{s}} p'$ implies $\textbf{alive}(< p', e' >))$

3.2 Examples

3.2.1 a-transmitter and a-sender

a-transmitter a-sender

Fig 3.2.1.1 Specifications

The *a-transmitter* only requires that the potential implementations are able to perform a at least one time. However it does not matter if an implementation can perform an infinite number of a's or whatever else. Whereas the *a-sender* do demand that an implementation is able to perform an infinite number of a's.

Fig 3.2.1.2 Potential Implementations

Clearly, $p_t \models A_t$, $p_{\scriptscriptstyle \delta} \models A_{\scriptscriptstyle \delta}$ and $p_{\scriptscriptstyle \delta} \models A_t$, but $p_t \not\models A_{\scriptscriptstyle \delta}$. For instance, by the computational rules for pairs we can deduce that $< p_t, A_{\scriptscriptstyle \delta} >$ may be deadlocked after performing some sequence of a's.

$$\frac{p_t \xrightarrow{a} p_t \quad A_s \xrightarrow{a} A_s}{< p_t, A_s > \xrightarrow{a} < p_t, A_s >}$$

$$\frac{p_t \xrightarrow{a} T \quad A_s \xrightarrow{a} A_s}{< p_t, A_s > \xrightarrow{a} < T, A_s >}$$

Thus $< p_t, A_s > \overset{a^n}{\Longrightarrow} < T, A_s >$ for every $n \geq 1$. Therefore $p_t \not\models A_s$, since deadlock($< T, A_s >$) and ¬successful($< T, A_s >$). Actually we have the following diagrams.

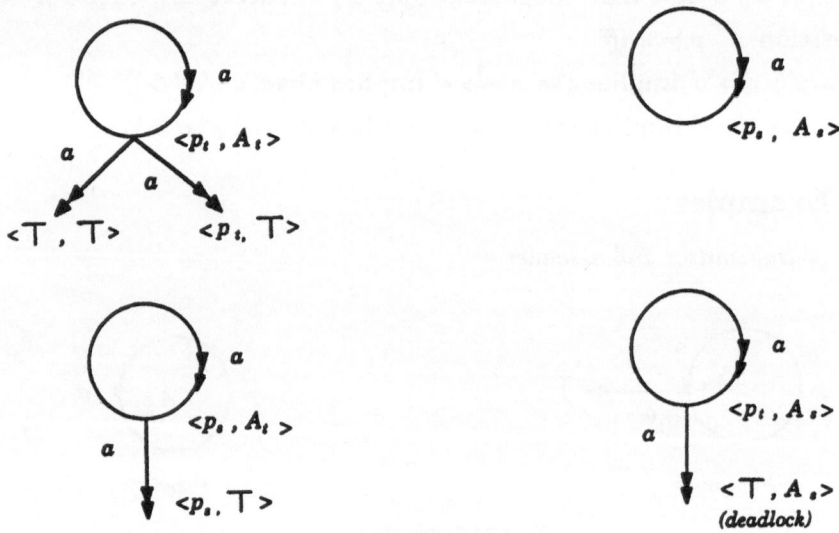

Fig 3.2.1.3

3.2.2 one-place buffer

$B_0 <= input.B_1 + output.\bot$
$B_1 <= input.\bot + output.B_0$

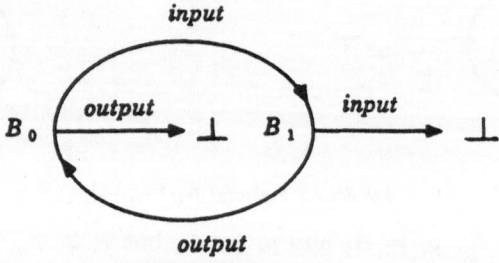

Fig 3.2.2.1 Specification

Observe that here we are just interested in the synchronization signals *input* and *output* but not the contents of the messages the buffer stores and forwards. A one-place buffer is required to be able to perform *input* and reach a state where it can perform *output*, and so forth. Moreover, the one-place buffer should refuse *output* at starting state (as the buffer is empty) and refuse *input* after performing *input* (as the buffer is occupied). Otherwise, it will lead the environment to *unsuccessful termination*. On the other hand, any other actions (perhaps housekeeping) except *input* and *output* are allowed in any states.

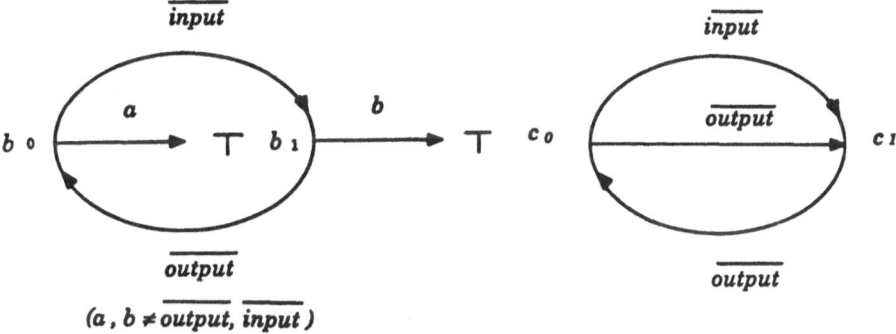

$(a, b \neq \overline{output}, \overline{input})$

Fig 3.2.2.2 Potential Implementations

It is trivial to check that $b_0 \models B_0$ and $c_0 \not\models B_0$.

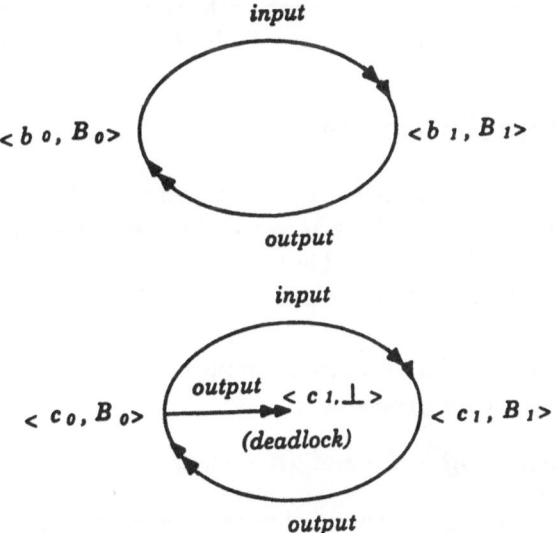

Fig3.2.2.3

3.2.3   *n-place buffer*

The specification B_0 for a one-place buffer can be extended in various ways to allow n-place buffers.

$$B_0 <= input.B_1 + output.\perp$$
$$B_1 <= input.B_2 + output.B_0$$
$$B_2 <= input.B_3 + output.B_1$$

......

$$B_{n-1} <= input.B_n + output.B_{n-2}$$
$$B_n <= input.\perp + output.B_{n-1}$$

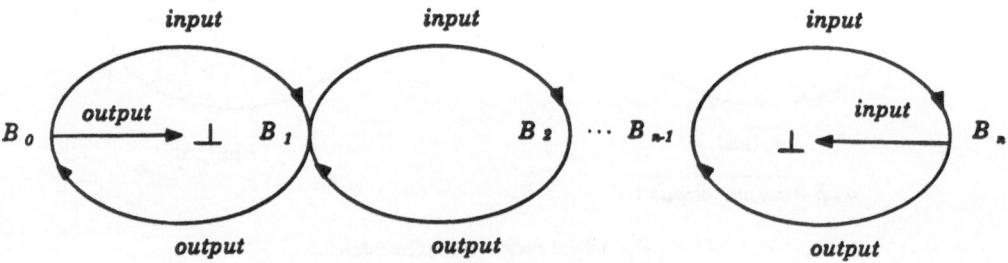

Fig 3.2.3.1 Specification

With this specification, all those buffers with at most n places are desired implementations. For instance, the one place buffer b_0 is one of them.

4   Process Equivalences

The equivalences over processes, such as observational equivalence, testing equivalence and failure equivalence have been recognized as fundamental notions in the study on processes. It would be desirable for a notion of correctness to respect some equivalence. Namely, if p satisfies e and p and q are equivalent then q satisfies e.

Let R be a satisfication relation between implementations and specifications, and $\approx$ be some equivalence for processes.

Definition 7 *R respects $\approx$ iff p R e and p $\approx$ q implies q R e.*

We shall show that the conformance $\models$ has this desired property w.r.t. failure equivalence.

4.1 The Equivalence $\approx_e$

Informally, two processes are equivalent iff they can not be distinguished in any environment. Naturally we can define our equivalence here by means of $\models$. Let $[p]$ be the set of environments to which p conforms.

Definition 8 $[p] = \{e \mid p \models e\}$

This immediately leads to an ordering on processes. Let $\sqsubseteq_e$ denote this ordering.

Definition 9 $p \sqsubseteq_e q$ iff $[p] \subseteq [q]$

It is easy to check that $\sqsubseteq_e$ is a preorder. The equivalence related to the preorder is induced in the usual fashion.

Definition 10 $p \approx_e q$ iff $p \sqsubseteq_e q$ & $q \sqsubseteq_e p$

Example: $p \approx_e p|\tau^\omega$ for every p.
    Note that the two processes are not testing equivalent in [NH84, Ni87].

Proposition 2 $\models$ *respects* $\approx_e$

This result follows directly from the definition of $\approx_e$. However we are able to show that $\models$ respects also Hoare-Brooks failure equivalence [HBR84, B83], $\approx_f$.

4.2 Failure Equivalence $\approx_f$

Various equivalences for processes have been defined according to different intuitions on process behaviour. Failure equivalence is based on the idiea that two processes are equivalent if and only if they refuse the same sets of actions after performing a finite sequence of actions. Originally the equivalence is proposed in [HBR84] for the theory of CSP. We start with the failure set for processes.

Definition 11 *Let* $p \overset{x}{\not\Longrightarrow}$ *denote that there is no* p' *such that* $p \overset{x}{\Longrightarrow} p'$
    $\mathbf{failures}(p) = \{(s, X) \mid \exists p' : p \overset{s}{\Longrightarrow} p' \ \& \ \forall x \in X : p' \overset{x}{\not\Longrightarrow}\}$

A natural equivalence immediately suggests itself.

Definition 12

1. $p \sqsubseteq_f q$ iff $\mathbf{failures}(q) \subseteq \mathbf{failures}(p)$
2. $p \approx_f q$ iff $p \sqsubseteq_f q$ & $q \sqsubseteq_f p$

Proposition 3 $\models$ *respects* $\approx_f$

It is sufficient to show that $\approx_e$ coincides with $\approx_f$.

4.3 $\approx_e = \approx_f$

For a detailed comparision of different process equivalences, we refer to [Ni87], where the author has introduced those equivalences proposed so far in the literature to transition systems. Here we just discuss the relationship between $\approx_f$ and $\approx_e$. It can be shown that for strongly divergent processes, i.e. those which can not perform infinite number of internal actions in any states, the equivalence $\approx_e$ is the same as de Nicola's testing equivalence $\approx_{test}$ [Ni87]. Thus it coincides with $\approx_f$ for the strongly convergent processes, as it has been shown in [Ni87] that testing equivalence is equal to failure equivalence on this restricted domain of processes, but not on the whole domain of processes. A trivial example is as follows.

Fig 4.3.1

The two processes p, q are not testing equivalent since the left one is divergent. For more examples, we refer to [Ni87]. However, they are failure equivalent and also equivalent in our definition. Moreover, they are observational equivalent too. We shall show that $\approx_e$ coincides with $\approx_f$ on the whole domain of processes (including those which may have possibly divergent states).

Theorem 1 $p \approx_e q$ iff $p \approx_f q$

Proof.

1. $\Longrightarrow$

    $[p] \subseteq [q]$ implies failures(q) $\subseteq$ failures(p).

    For every failure $(s, X) \in$ failures(p), where $s = a_1 a_2 \ldots a_n$ and $X = \{x_i | i \leq m\}$, we shall show that
    $[p] \subseteq [q]$ & $(s, X) \notin$ failures(p) implies $(s, X) \notin$ failures(q).

    We construct e in terms of (s, X) as follows:

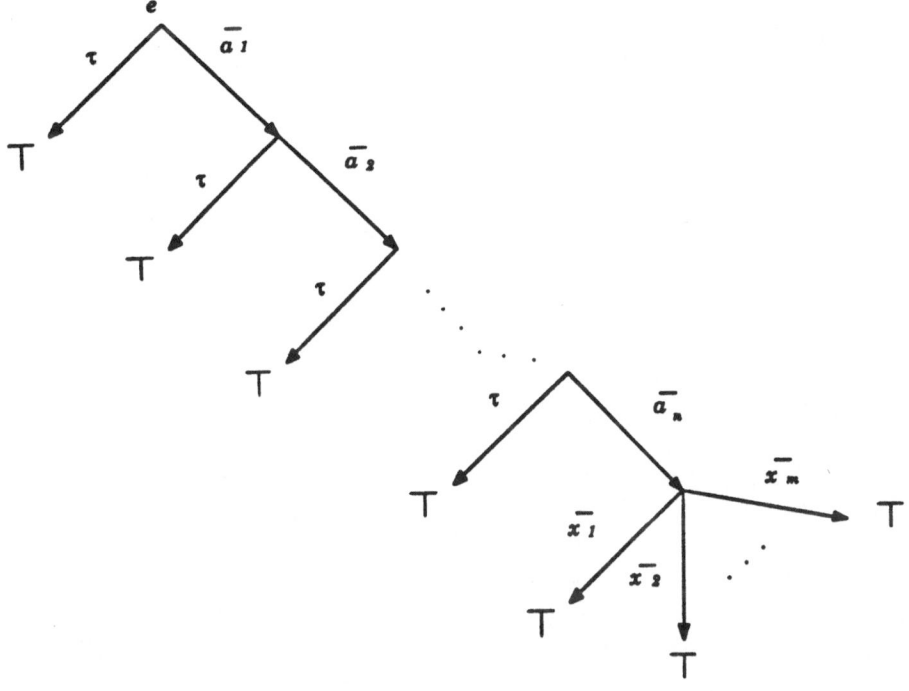

Fig 4.3.2

Since $(s, X) \notin \textbf{failures}(p)$, we have

(a) p rejects trace s or

(b) $\forall p'.p \overset{s}{\Longrightarrow} p'$ implies $\exists x_i \in X, p''.p' \overset{x_i}{\Longrightarrow} p''$

Thus we can conclude that $p \not\models e$.
By the hypothesis, $q \not\models e$, which clearly implies $(s, X) \notin \textbf{failures}(q)$.

2. $\Longleftarrow$

$\textbf{failures}(q) \subseteq \textbf{failures}(p)$ implies $[p] \subseteq [q]$, i.e.
$\textbf{failures}(q) \subseteq \textbf{failures}(p)$ implies ($q \not\models e$ implies $p \not\models e$) for an arbitrary e.

Let $\textbf{initial}(e) = \{a | \exists e'.e \overset{a}{\Longrightarrow} e'\}$

Suppose that there exists e such that $q \not\models e$.
Then there is a string s, p' and e' such that
$< q, e > \overset{s}{\Longrightarrow} < q', e' >$, $\textbf{deadlock}(< q', e' >)$ and $\neg\textbf{successful}(< q', e' >)$.

Then $q' \overset{x}{\nRightarrow}$ for every $x \in \mathbf{initial}(e')$.

This simply means $(s, \mathbf{initial}(t')) \in \mathbf{failures}(q)$.

By the hypothesis, we have $(s, \mathbf{initial}(t')) \in \mathbf{failures}(p)$.

Namely there exists p' such that

$p \overset{s}{\Longrightarrow} p'$ and $p' \overset{x}{\nRightarrow}$ for every $x \in \mathbf{initial}(e')$.

By the definition, we have $\mathbf{deadlock}(<p', e'>)$.

Moreover, $\neg\mathbf{successful}(<q', e'>)$ implies $\neg\mathbf{successful}(<p', e'>)$.

Therefore $p \not\models e$.

∎

5   Related Work and Concluding Remarks

The aim of this paper is to use process algebra to specify processes implicitly in order to allow a wide collection of implementations for a given specification. To some extent, our approach has fulfilled this requirement. Another direction is to develop specification languages separately from process algebra. In [Hol88, La87], the authors have successfully extended Hennessy-Milner logic [HM82] with recursion. These works have developped Hennessy-Milner logic as a specification language for processes. For detailed review along this line, we refer to [La87]. However, all these approaches demand both a specification language and a programming language. This often increases the complexity of the specification and verification procedures. At least, we can say that the *system designers* need to understand the semantics for both languages. Since the problems concerning concurrency are already so complicated, a simple approach is absolutely desired. One of the advantages of our approach is that it is simple. It allows the description language like CCS still to be used as both of specification language and programming language.

On the other hand, the environments are not trivial in general. They may include various technical components or they may be humans. Therefore, one needs to have a precise understanding of the environment's behaviour. However to write a *correct* specification, the *specifiers* in other approaches to system specification need to know well about the environment's behaviour too, perhaps in some *model* built in her/his mind, and then describe it using the logical language.

For a non-trivial example, we refer to a forthcoming paper [W88] in which the alternating bit protocol is verified using the approach presented here.

We shall point out that the formalization of *conformance* seems to be very close to the work of Nicola and Hennessy on testing equivalence reported in [NH84, Ni87]. However, we have different goals. Our goal here is to apply the testing theory to describe and verify systems. Whereas the works on equivalence is simply to establish a proper semantic foundation for such systems.

Technically, the works are different in the following aspects. First, we allow a process to satisfy an *infinite* test (if we do not distinguish tests from environments), but not in the formalization of testing equivalence. This allows us to specify infinite behaviour. So we use the notion of *environment* instead of *test*. This also has been motivated in section 1. For instance, we have $p \models e$, but p **must** e.

Fig 5.1

Moreover, the induced equivalence relation, $\approx_e$ on processes is not the same as testing equivalence. Consider the following examples.

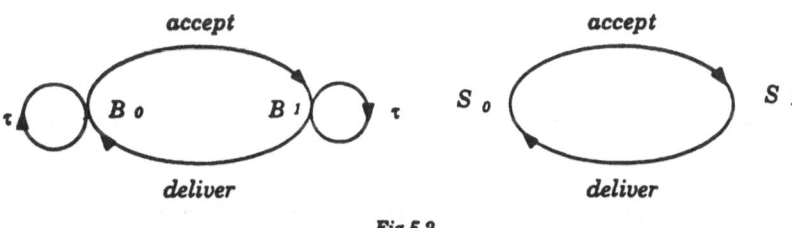

Fig 5.2

The first one B_0 may be the external behaviour of some protocol realization. The second one S_0 which has abstracted away the internal moves can be regarded as a specification for protocols in the usual sense, where equivalence determines correctness. Thus from a practical point of view, it would be desirable to treat them as equivalent or let the implementation satisfy the specification. But the testing equivalence distinguishes them.

Another example is $p \not\approx_{test} \tau^\omega | p$ and $p \approx_e \tau^\omega | p$!. Theoretically this is reasonable, since the process, B_0 on the left hand side is *divergent*, which is known as *undefined* generally. However in practice, it might be difficult to imagine that these two processes are not *equivalent*. The difference is caused by the different views on *divergence* and *deadlock*. See section 3. Actually the equivalence $\approx_e$ has been shown to coincide with failure equivalence.

The work of Larsen [La85] on context dependent equivalence over processes is about the equivalence problem for processes in a given environment (Larsen call it context). Whereas here we want to study the relation between processes and their environments.

Currently we are working on *submodule construction* which has been solved by Parrow [P87] in terms of observational equivalence. The problem is as follows. Find a CCS agent, X such that the equation $(X|p)\setminus R \approx S$ holds where p, R, S are given and $\setminus R$ and $|$ are the restriction and composition in CCS respectively. This obviously depends much on the equivalence $\approx$. We will give a solution in terms of the conformance $\models$. We want to find a construction C such that $(X|p)\setminus R \models S$ iff $X \models C(p, R, S)$. The result looks very much like Dijkstra's weakest pre-condition. This naturally leads to the problem of *program derivation*: given an environment e, find p such that $p \models e$. Thus if X is unknown and p, R, S are given, hopefully, it can be constructed from $C(p, R, S)$.

6 Acknowledgement

The author wishes to thank his advisor Sören Holmström and Uno Holmer, K.V.S Prasad and Sven Westin for valuable comments on this paper, and other members of the PMG group at Chalmers for their support and encouragement.

References

[Ab86]   Abramsky S. *Observation equivalence as a testing equivalence*, Dept. of computing, Imperial college, London, UK.

[B83]   Brooks S. *On the relationship of CCS and CSP*, LNCS 154,1983.

[Br87]   Brinksma E. *On the existence of canonical testers*, University of Twente, Box 217-9700 AE Enchede,The Netherlands, 1987.

[HBR84]   Hoare C., Brookes S. and Rounds A. *A theory of communicating sequential processes*, JACM, 1984.

[Ho84]   Hoare C. *Communicating sequential processes*, Prentice-Hall, 1985.

[Hol88]   Holmstrom S. *Reasoning about the alternating bit protocol*, Dept. of computer sciences, Chalmers university of technology, Sweden, 1988.

[K82]   Kennaway D. *A theory of nondeterminism and parallelism*, Ph.D thesis, Oxford university, UK. 1981.

[La85]   Larsen K. *A context dependent equivalence between processes*, LNCS 194, 1985.

[La87]   Larsen K. *Proof systems for Hennessy-Milner Logic with recursion*, to appear in CAAP'88.

[LM87]   Larsen K. and Milner R. *A complete protocol proof*, LNCS 226, 1987.

[LT87]   Larsen K. and Thomsen B. *A process modal logic*, Aalborg University, Denmark, 1987.

[Mi80]   Milner R. *A calculus of communicating systems*, LNCS 92, 1980.

[NH84]   de Nicola R. and Hennessy M. *Testing equivalences for processes*, TCS 34, 1984.

[Ni87]   de Nicola R. *Extentional Equivalences For Transition Systems*, Acta Informatia 24, 1987.

[OH83]   Olderog E. and Hoare C. *Specification oriented semantics for communicating processes*, LNCS 154, 1983.

[P87]   Parrow J. *Submodule construction as equation solving in CCS*, Laboratory for Foundations of Computer Science, University of Edinburgh 1987.

[W88]   Wang Y. *A simple protocol proof*, to appear in ISIIS'88, Second international symposium on interoperable information systems, 1988.

Hennessy-Milner Logic with recursion as a specification language, and a refinement calculus based on it

Sören Holmström[1]

ABSTRACT

This paper is about specification and verification of processes, modeled as CCS-agents.We show, by means of examples that Hennessy-Milner Logic (HML) with recursion is a suitable language for expressing *implicit* specifications. By extending this specification language with *refinement operators*, i.e. operators that describe the internal structure of a system, we obtain a calculus for stepwise refinement of agents from a specification in HML to a realization in CCS. The method is demonstrated by proving the alternating-bit protocol under weak assumptions about the unreliable media.

1 Introduction

Hennessy-Milner Logic (HML) is a modal-logic language in which properties of processes can be expressed. Originally it was introduced by Hennessy and Milner [HM85] to motivate their definition of observational equivalence based on ω-chains of equivalence relations. They proved that two agents are observationally equivalent iff they satisfy the same set of HML-formulas. Subsequently, other researchers have investigated correspondences between various sublanguages of HML and coarser equivalence relations [BR83, GS84, Lars86].

In this paper we shall use HML for a more practical purpose. By means of examples, we shall demonstrate that HML extended with *recursion* is adequate as a language for implicit specification of agents. We show that recursion makes many properties easier to express and, furthermore, it makes the logic strong enough to deal with bisimulation equivalence. (Recursion has been added to HML in other papers [Koze82, Lars87], but without exploring its relation to bisimulation equivalence on image infinite processes.)

Our approach to specification and verification is informal (though hopefully rigorous). Instead of giving a formal syntax and semantics of our specification language, we shall introduce it as a set theoretic notation where a specification is a set of agents and an agent satisfies a specification if it is a member of this set. (This approach has two advantages, first, we can rely on the reader's knowledge of set theory and second, it relieves us from a rather boring semantical description that merely translates HML into set theory.) Furthermore, we

---

[1] Authors address: Programming Methodology Group, Dept. of Computer Science, Chalmers University of Technology, S-412 96 Göteborg, Sweden. E-mail: holmstrom@cs.chalmers.se

shall not bother about completeness of the proof rules. No formal system of this kind can be complete in the usual sense.

We also introduce a collection of operators, with which one can specify the internal structure of a system. For example, an agent satisfying P|Q is of the form p|q where p satisfies P and q satisfies Q. We shall call these operators *refinement operators* because they are useful when expressing refinement steps. The refinement operators obey a collection of laws that forms a *refinement calculus*. Using this calculus we prove the correctness of the alternating-bit protocol. A more theoretical work that contains similar ideas for SCCS is [Wins85].

The paper is organized as follows. The next section contains a short summary of the syntax and semantics of CCS and the related notions of actual and observable (invisible τ) behaviour. In section 3 we show how actual behaviour can be specified and we also introduce recursion. Section 4 is devoted to specification of observable behaviour, and methods for simplifying these specifications. As a preparation for the proof of the alternating bit protocol we specify an unreliable medium in section 5. In section 6 we introduce the refinement operators and give their corresponding laws or "proof rules". Section 7 contains the proof of the alternating-bit protocol. In this proof, we have borrowed some ideas and notation from [LM86] but the proof itself is completely different. The unreliable medium is not specified explicitly as an agent but implicitly by an expression in HML. The consequence is that the correctness result is more general. Section 8 presents some conclusions.

2 CCS

In this section we recapitulate the basic definitions of CCS. For a full account see e.g. [Miln80,Miln84].

Assume a set Δ of *names* and a disjoint and equivalent set Δ^c of *conames*. Let $\alpha,\beta,\gamma,...$ range over names and let α^c be the coname corresponding to α. We extend the function ^{c} to also denote its inverse, i.e. let $(\alpha^c)^c = \alpha$. Let A range over sets of names. The names and conames form the set of *labels* $\Lambda = \Delta \cup \Delta^c$ ranged over by λ. The label λ^c is referred to as the *complement* of λ. The labels together with the distinguished element τ form the set of *actions* $Act = \Lambda \cup \{\tau\}$ ranged over by μ. The special action τ will as usual model "silent" or invisible transitions in the system, transitions that typically occur when two subsystems do an internal communication.

Define a *relabelling* to be a function S: Act→Act such that $S(\tau) = \tau$, $S(\lambda) \neq \tau$ and $S(\lambda^c) = (S(\lambda))^c$. We shall use the notation $\lambda_1...\lambda_n/\alpha_1...\alpha_n$ (all $\alpha_1,...,\alpha_n$ distinct) for the relabelling that maps α_i to λ_i and α^c_i to λ^c_i ($1 \leq i \leq n$) and every other action to itself.

Assume a set of agent *variables* ranged over by x,y,... and let e,f,... range over the set of *agent expressions* generated by the grammar below. We shall often let I denote a set understood from the context and ranged over by i. In such cases we let **e** abbreviate the I-indexed family $\langle e_i \rangle_{i \in I}$. Similarly we let **x** abbreviate $\langle x_i \rangle_{i \in I}$ an I-indexed family of distinct variables. In fact, we shall adopt the convention that a symbol in boldface denotes an I-indexed family of objects. The set of agent expressions is generated by the grammar

$$e ::= \ x \ | \ \mu e \ | \ \Sigma\mathbf{e} \ | \ e_1 | e_2 \ | \ e\backslash A \ | \ e[S] \ | \ \text{fix}_i \mathbf{x}.\mathbf{e}$$

All variables in **x** become bound in $\text{fix}_i \mathbf{x}.\mathbf{e}$.

Some abbreviations of $\Sigma\langle e_i \rangle_{i \in I}$ are standard: We write NIL and $e_0 + e_1$ when I is $\emptyset$ and $\{0,1\}$ respectively and we write fix x.p instead of $\text{fix}_0 \mathbf{x}.\mathbf{p}$ when I = $\{0\}$. Furthermore, we let fixx.e abbreviate the family $\langle \text{fix}_i \mathbf{x}.\mathbf{e} \rangle_{i \in I}$. Instead of introducing a family **k** of constants by defining $k_i = \text{fix}_i \mathbf{x}.\mathbf{e}$ we shall sometimes use the equivalent recursion equation syntax $k_i \Leftarrow e_i\{\mathbf{k}/\mathbf{x}\}$ ($i \in I$). For example we shall write $k \Leftarrow \alpha l + \beta k$, $l \Leftarrow \alpha k + \beta l$ instead of $\langle k,l \rangle = \text{fix}\langle x,y \rangle.\langle \alpha y + \beta x, \alpha x + \beta y \rangle$.

An *agent* is a closed agent expression and we shall let p,q,... range over agents. The operational semantics of CCS is given as a family of binary action relations $\langle \xrightarrow{\mu} \rangle_{\mu \in \text{Act}}$ over agents. The relation $p \xrightarrow{\mu} p'$ means that the agent p can do the action μ and become p' in doing so. The family $\langle \xrightarrow{\mu} \rangle_{\mu \in \text{Act}}$ is the least family of relations that satisfies the following rules

(i) $\quad \mu p \xrightarrow{\mu} p$

(ii) $\quad \dfrac{p_j \xrightarrow{\mu} p'}{\Sigma\mathbf{p} \xrightarrow{\mu} p'} \ (j \in I)$

(iii) $\quad \dfrac{p \xrightarrow{\mu} p'}{p|q \xrightarrow{\mu} p'|q} \quad \dfrac{q \xrightarrow{\mu} q'}{p|q \xrightarrow{\mu} p|q'} \quad \dfrac{p \xrightarrow{\lambda} p' \quad q \xrightarrow{\lambda^c} q'}{p|q \xrightarrow{\tau} p'|q'}$

(iv) $\quad \dfrac{p \xrightarrow{\mu} p'}{p\backslash A \xrightarrow{\mu} p'\backslash A} \ (\mu, \mu^c \notin A)$

(v) $\quad \dfrac{p \xrightarrow{\mu} p'}{p\,[S] \xrightarrow{S(\mu)} p'[S]}$

(vi) $\quad \dfrac{e_i\{\text{fixx.e}/\mathbf{x}\} \xrightarrow{\mu} p'}{\text{fix}_i \mathbf{x}.\mathbf{e} \xrightarrow{\mu} p'}$

where $e_i\{\text{fix } \mathbf{x}.\mathbf{e}/\mathbf{x}\}$ is the expression e_i where $\text{fix}_j \mathbf{x}.\mathbf{e}$ has been substituted for every free occurrence of x_j.

When p does an action sequence, for example $\tau\alpha\tau\tau\beta^c\gamma\tau$ and becomes p', it is only the visible actions $\alpha\beta^c\gamma$ that can be observed. We say that the label sequence $\alpha\beta^c\gamma$ is the *visible content* of the action sequence $\tau\alpha\tau\tau\beta^c\gamma\tau$. In other words, the visible content of an action sequence is what remains when the invisible τ-actions are removed. Let an *experiment* be a label sequence and let s range over experiments. We shall write $p \stackrel{s}{\Longrightarrow} p'$ if p can do actions whose visible content is the experiment s and become p'. More formally:

$$p \stackrel{\lambda_1\lambda_2...\lambda_n}{\Longrightarrow} p'$$
$$\text{iff} \quad p(\stackrel{\tau}{\rightarrow})^*\stackrel{\lambda_1}{\rightarrow}(\stackrel{\tau}{\rightarrow})^*\stackrel{\lambda_2}{\rightarrow}(\stackrel{\tau}{\rightarrow})^*... (\stackrel{\tau}{\rightarrow})^*\stackrel{\lambda_n}{\rightarrow}(\stackrel{\tau}{\rightarrow})^*p'$$

(where juxtaposition is composition of relations and * is transitive closure as usual). The relation $\stackrel{s}{\Longrightarrow}$ is called the experiment relation [Miln80]. Given the two families $\langle\stackrel{\mu}{\rightarrow}\rangle_{\mu\in Act}$ and $\langle\stackrel{s}{\Longrightarrow}\rangle_{s\in Act*}$ of relations on agents we can define two equivalences on agents using the notion of *bisimulation* which we now define.

Definition: A binary relation R on agents is a *strong bisimulation* if whenever pRq

(i)   if $p \stackrel{\mu}{\rightarrow} p'$ then $\exists q'.\ q \stackrel{\mu}{\rightarrow} q'$ and p'Rq'.
(ii)   if $q \stackrel{\mu}{\rightarrow} q'$ then $\exists p'.\ p \stackrel{\mu}{\rightarrow} p'$ and p'Rq'.

R is a *weak bisimulation* if whenever pRq

(i)   if $p \stackrel{s}{\Longrightarrow} p'$ then $\exists q'.\ q \stackrel{s}{\Longrightarrow} q'$ and p'Rq'.
(ii)   if $q \stackrel{s}{\Longrightarrow} q'$ then $\exists p'.\ p \stackrel{s}{\Longrightarrow} p'$ and p'Rq'.

Moreover, p is *strongly (weakly) equivalent* to q and we write p~q (p≈q) iff there is a strong (weak) bisimulation R such that pRq. ☐

It is not difficult to prove that the relation ~ (≈) is itself a strong (weak) bisimulation, namely the greatest strong (weak) bisimulation and, furthermore, that ~ and ≈ are equivalence relations. It is more complicated to show that ~, extended to agent expressions in the natural way, is a *congruence*. (See e.g. [Miln83] for a proof of this result for the similar calculus SCCS.) We shall not make use of this fact here, however.

If we can see every action, including τ, two agents are indistinguishable iff[1] they are strongly equivalent. We shall say that strongly equivalent agents have the same *actual behaviour*. On the other hand, an observer who is unable to

---

[1]This depends strongly on what kinds of experiments an observer is able to perform. Among other things, it must be possible to restart an agent from an intermediate state.

see τ-actions cannot distinguish between weakly equivalent agents. Weakly equivalent agents are therefore said to have the same *observable behaviour*.

3 Specifying actual behaviour

By a *specification* we shall understand an expression denoting a set of agents. An agent *satisfies* a specification if it is a member of the set denoted by the specification. In principle we could use set theoretic notation directly as our specification language. However, there is good reason to restrict ourselves to a weaker language built from a few well understood constructs. By doing so, we can derive useful "laws" or "proof rules" that turn reasoning about behaviour into routine activity. The ultimate goal is to develop a formal calculus for this reasoning but this is outside the scope of this paper.

In this section we shall derive a specification language along these lines. We shall show that specifications in this language respect strong equivalence, i.e. two strongly equivalent agents either both satisfy the specification or fail to satisfy it. Hence, a specification in this language specifies actual behaviour. We cannot within the language express non-behavioural properties, i.e. that an agent is composed from two subagents working in parallel. Later we shall see how the language can be extended with refinement operators, which makes it possible to describe internal structure of an agent.

Let P,Q,R range over sets of agents and let 1 be the set of all agents. Thus, 1 is the specification that is satisfied by every agent and $\emptyset$ is the specification that is satisfied by no-one. The operators $\cup$ and $\cap$ give two ways of composing specifications. For example, an agent satisfies $P \cup Q$ if it satisfies P or if it satisfies Q.

To be able to express something about the possible actions of an agent, we need two new operators.

Definition:

$$\langle \mu \rangle P = \{p \mid \exists p'.\ p \xrightarrow{\mu} p' \wedge p' \in P\}$$
$$[\mu]P = \{p \mid \forall p'.p \xrightarrow{\mu} p' \Rightarrow p' \in P\} \qquad \qquad \square$$

The intuition is that an agent satisfies $\langle \mu \rangle P$ iff it can do α and become one satisfying P and an agent satisfies $[\mu]P$ iff it cannot do α without becoming one satisfying P. With these operators and $\cap$, $\cup$, $\emptyset$ and 1 we can express non-trivial properties of agents. Here are some examples. (Let the unary operators $\langle \mu \rangle$ and $[\mu]$ have higher priority than the binary operators $\cup$ and $\cap$.)

$\langle\alpha\rangle 1$	the agent can do α
$[\beta]\varnothing$	the agent cannot do β.
$\langle\alpha\rangle([\beta]\varnothing\cap[\gamma]\varnothing)$	the agent can do α and become an agent that can do neither β nor γ.
$\langle\alpha\rangle 1\cap[\alpha]\varnothing$	the agent can do α and cannot do α.

The last specification is clearly self-contradictory and impossible to satisfy. It is easy to prove that $\langle\alpha\rangle 1\cap[\alpha]\varnothing$ is equivalent to the empty set by the $\langle\alpha\rangle\cap$-*law* below.

Proposition 3.1: $\langle\mu\rangle$ and $[\mu]$ interact with $\cup$ and $\cap$ in the following way. (Note the special cases when $I = \varnothing$.)

$\langle\mu\rangle\cup$-*distributivity*: $\langle\mu\rangle\cup_{i\in I}P_i = \cup_{i\in I}\langle\mu\rangle P_i$ spec. case: $\langle\mu\rangle\varnothing = \varnothing$

$[\mu]\cap$-*distributivity*: $[\mu]\cap_{i\in I}P_i = \cap_{i\in I}[\mu]P_i$ spec. case: $[\mu]1 = 1$

$[\mu]\cup$-*law*: $[\mu](P\cup Q) \subseteq [\mu]P \cup \langle\mu\rangle Q$

$\langle\mu\rangle\cap$-*law*: $\langle\mu\rangle P \cap [\mu]Q \subseteq \langle\mu\rangle(P\cap Q)$

Proof: Straightforward from the definitions. $\qquad\qquad\qquad\qquad\qquad\qquad$ ☐

More interesting is specification of infinite behaviour. Let an α-*sender* be an agent that can do α any number of times without risk of deadlock. Examples of α-senders are fix x.αx and fix x. $\alpha\alpha$x+βNIL. A counter-example is the agent (fix x.αx)+αNIL which can deadlock after doing the first α. How do we form the set of all α-senders? One possibility is to take the limit of a sequence of successively smaller sets. Let A_n be the set of agents that can do α at least n times without risk of deadlock. It can be inductively defined by

$$A_0 = 1$$
$$A_{n+1} = \langle\alpha\rangle 1\cap[\alpha]A_n$$

The set $\cap_n A_n$ is now the set of α-senders since it is the set of agents that can do α any number of times (without risk of deadlock).

Another way of specifying infinite behaviour is by recursion. After some thought, we realize that α-sender is an agent that (i) can do α and (ii) cannot do α without being an α-sender again. Letting A be the set of α-senders we get the following inequation for A:

$$A \subseteq \langle\alpha\rangle 1\cap[\alpha]A$$

This inequation has several solutions, e.g. the trivial least solution $A = \emptyset$. Which one should we choose for A? A likely candidate is the greatest solution, which exists if the right hand side fulfils a certain condition. To show this we need some standard set-theoretical results.

Let F,G range over functions on sets. An expression $F(X)$ is said to be *monotonic in* X if $F(P) \subseteq F(Q)$ whenever $P \subseteq Q$ and we say that a function F is *monotonic* if $F(X)$ is monotonic in X.

Proposition 3.2: If F is monotonic then

(i) $\quad \cup \{X \mid X \subseteq F(X)\}$ is the greatest solution of $X \subseteq F(X)$.

(ii) $\quad \cup \{X \mid X \subseteq F(X)\}$ is a fixed point of F.

Proof: (i) We first show that $\cup \{X \mid X \subseteq F(X)\} \subseteq F(\cup \{X \mid X \subseteq F(X)\})$ by showing that $Y \subseteq F(\cup \{X \mid X \subseteq F(X)\})$ for every Y such that $Y \subseteq F(Y)$. Assume $Y \subseteq F(Y)$. We then have $Y \subseteq \cup \{X \mid X \subseteq F(X)\}$. Now

$$Y \subseteq F(Y) \qquad \text{by assumption}$$
$$\subseteq F(\cup \{X \mid X \subseteq F(X)\}) \qquad \text{by monotonicity of F}$$
$$\text{since } Y \subseteq \cup \{X \mid X \subseteq F(X)\}$$

Hence $\cup \{X \mid X \subseteq F(X)\}$ is a solution of $X \subseteq F(X)$. Furthermore it is a superset of every solution, so it must be the greatest solution.

(ii) It remains to prove that $\cup \{X \mid X \subseteq F(X)\}$ is a fixed point to F. By monotonicity of F and (i) above we have $F(\cup \{X \mid X \subseteq F(X)\}) \subseteq F(F(\cup \{X \mid X \subseteq F(X)\}))$. Hence $F(\cup \{X \mid X \subseteq F(X)\})$ is also a solution of the equation $X \subseteq F(X)$ and since $\cup \{X \mid X \subseteq F(X)\}$ is the greatest solution, we have $F(\cup \{X \mid X \subseteq F(X)\}) \subseteq \cup \{X \mid X \subseteq F(X)\}$. Hence $\cup \{X \mid X \subseteq F(X)\} = F(\cup \{X \mid X \subseteq F(X)\})$ by (i). $\quad \square$

The set $\cup \{X \mid X \subseteq F(X)\}$ will appear so frequently that it is worthwhile to introduce special syntax for it.

Definition: $\nu X.F(X) = \cup \{X \mid X \subseteq F(X)\}$ $\qquad\qquad \square$

It is not difficult to show that the expression $\langle \alpha \rangle 1 \cap [\alpha] X$ is monotonic in X so the inequation $A \subseteq \langle \alpha \rangle 1 \cap [\alpha] A$ has the greatest solution $\nu X.\langle \alpha \rangle 1 \cap [\alpha] X$. In order to show that, for example, the agent fix $x.\alpha \alpha x + \beta \text{NIL}$ satisfies the specification $\nu X.\langle \alpha \rangle 1 \cap [\alpha] X$, it is enough to find a set P such that $P \subseteq \langle \alpha \rangle 1 \cap [\alpha] P$ and P contains fix $x.\alpha \alpha x + \beta .\text{NIL}$. Take, for example, $\{\text{fix } x.\alpha \alpha x + \beta \text{NIL}, \alpha(\text{fix } x.\alpha \alpha x + \beta \text{NIL})\}$.

The reasoning about greatest fixed points is often simplified by the following two laws which follow directly from the definition and proposition above.

Proposition 3.3:

(i)   *Park's rule*: if $P \subseteq F(P)$ then $P \subseteq \nu X.F(X)$

(ii)   *Unfolding*: $\nu X.F(X) = F(\nu X.F(X))$ provided F is monotonic. $\square$

Instead of introducing a constant K by declaring $K = \nu X.F(X)$ where F is monotonic, we shall at times define K by the inequation $K \subseteq F(K)$, where it is implicit that we mean the greatest solution. Furthermore, when the right hand side $F(K)$ is an intersection, we shall often increase readability by splitting the inequation into several inequations, where every inequation correspond to one desired property. For example, the inequation defining the set A of α-senders, can be written

$A \subseteq \langle \alpha \rangle 1$ 　　　　　　　an α-sender can do α

$A \subseteq [\alpha]A$ 　　　　　　　an α-sender cannot do α without becoming
　　　　　　　　　　　　　　　an α-sender

In the above example we defined a set of agents by single recursion but in general we must use mutual recursion. Let a *pulse buffer* be a buffer that holds synchronization pulses, i.e. messages without content. A pulse buffer can accept a pulse if it is not full and it can deliver a pulse if it is not empty. Let {accept,deliver} be the set of actions and let B_n be the set of agents that are pulse buffers holding n pulses. The family $\langle B_n \rangle_{n \in N}$ of sets satisfy:

B_0 　$\subseteq \langle accept \rangle 1 \cap [accept]B_1 \cap [deliver]\emptyset$
$B_{n+1} \subseteq [accept]B_{n+2} \cap \langle deliver \rangle 1 \cap [deliver]B_n$ 　　$(n \in N)$

Splitting the right hand sides gives the more readable set of inequations

(B.1)　B_0　$\subseteq \langle accept \rangle 1$ 　　　　　　　an empty buffer can accept a pulse

(B.2)　B_0　$\subseteq [deliver]\emptyset$ 　　　　　　　an empty buffer cannot deliver a pulse

(B.3)　B_n　$\subseteq [accept]B_{n+1}$ $(n \in N)$ 　　　a buffer holding n pulses cannot accept a pulse
　　　　　　　　　　　　　　　　　　　　　　　without becoming a buffer holding n+1 pulses

(B.4)　$B_{n+1} \subseteq \langle deliver \rangle 1$ 　$(n \in N)$ 　　　a non-empty buffer can deliver a pulse

(B.5)　$B_{n+1} \subseteq [deliver]B_n$ 　$(n \in N)$ 　　　a buffer holding n+1 pulses cannot deliver a
　　　　　　　　　　　　　　　　　　　　　　　pulse without becoming a buffer holding n
　　　　　　　　　　　　　　　　　　　　　　　pulses

Fortunately, propositions 3.2 and 3.3 generalize easily to systems of several inequations. Since the right hand sides are monotonic in the variables B_i there exists a greatest solution which defines the family of pulse buffers. In order to formalize this we need some more notation.

Let P, Q, R range over I-indexed families of sets and extend the meaning of $\subseteq$ by writing $P \subseteq Q$ when $P_i \subseteq Q_i$ for every $i \in I$. Analogously, we write $P \cup Q$ for the tuple R, such that $R_i = P_i \cup Q_i$ for every $i \in I$ (and similarly for $P \cap Q$). The definition of monotonicity extends also in the obvious way to families. Moreover, let $F, G, \ldots$ range over functions returning I-indexed families of sets and let $F_i(P)$ be the i-th component of $F(P)$. With this new notation we have:

Proposition 3.4: If F is monotonic then the inequation system $X \subseteq F(X)$ has the greatest solution $\bigcup\{X \mid X \subseteq F(X)\}$ which is also a fixed point of F. $\qquad\square$

Definition:

$$vX.F(X) \;\; = \bigcup\{P \mid P \subseteq F(P)\}$$
$$v_iX.F(X) \;\; = Q_i \text{ where } Q = \bigcup\{P \mid P \subseteq F(P)\} \qquad\qquad\square$$

Proposition 3.5:

(i)   *Park's rule*: if $P \subseteq F(P)$ then $P \subseteq vX.F(X)$.

(ii)   *Unfolding*: $vX.F(X) = F(vX.F(X))$ and $v_iX.F(X) = F_i(vX.F(X))$
             provided F is monotonic. $\qquad\qquad\qquad\qquad\qquad\qquad\qquad\square$

Using v, we can write the set of empty pulse buffers

$$v_0X.F(X) \quad \text{where}$$
$$F_0(X) = \langle accept\rangle 1 \cap [accept]X_1 \cap [deliver]\varnothing$$
$$F_{n+1}(X) = [accept]X_{n+2} \cap \langle deliver\rangle 1 \cap [deliver]X_n \qquad (n \in N)$$

but we shall often prefer the inequation system above as a definition of the family $\langle B_n\rangle_{n \in N}$ of constants.

The specification of the pulse buffer above is *implicit*. It only expresses a collection of properties that a pulse buffer should have. The specification admits several (inequivalent) implementations, for example, buffers of different capacities. It is of course important that the specification is *complete* in the sense that it expresses every expected property. (Unfortunately, this requirement is often difficult to check.)

Let p be a one place buffer defined by p $\Leftarrow$ accept.deliver.p. (We shall sometimes separate action names by dots.) In order to prove that p is an empty pulse buffer it is sufficient to find a family **P** of sets such that P_0 contains p and $\mathbf{P} \subseteq$ F(**P**). Take, for example, $P_0 = \{p\}$, $P_1 = \{deliver.p\}$ and $P_{n+2} = \emptyset$. It is trivial to see that P_0 contains p and it is not difficult to prove that $\mathbf{P} \subseteq$ F(**P**) using the operational semantics of agents.

When specifying an agent by recursion, it is important that the corresponding functional F is monotonic, otherwise vX.F(X) is not necessarily a solution to $X \subseteq$ F(X). Therefore the following result is essential.

Proposition 3.6: $\langle\mu\rangle$, $[\mu]$, $\cup$ and $\cap$ are monotonic functions. Moreover, vX.F(X,Y) and vX.G(X,Y) are monotonic in **Y** if F and G are monotonic.

Proof: It is easy to show monotonicity of $\langle\mu\rangle$, $[\mu]$, $\cup$ and $\cap$ from the definitions. The only non-trivial part is recursion. To prove that vX.F(X,Y) is monotonic in **Y**, assume $\mathbf{P} \subseteq \mathbf{Q}$ and show vX.F(X,**P**) $\subseteq$ vX.F(X,**Q**). Now

$$vX.F(X,\mathbf{P}) = F(vX.F(X,\mathbf{P})) \qquad\qquad \text{by unfolding}$$
$$\subseteq F(vX.F(X,\mathbf{P}),\mathbf{Q}) \qquad\qquad \text{by monotonicity of F}$$

Hence vX.F(X,**P**) $\subseteq$ vX.F(X,**Q**) by Park's rule. $\qquad\qquad\qquad\qquad$ $\square$

Monotonicity is preserved by composition so every expression F(**X**) composed from 1, $\emptyset$, $\langle\mu\rangle$, $[\mu]$, $\cup$, $\cap$, v and variables in **X** is monotonic in **X**. Apart from ensuring existence of greatest solutions to our inequations, monotonicity provides a convenient method of inequational reasoning as we shall see later.

We shall now turn to the question of how powerful the language generated by $\emptyset$, 1, $\langle\alpha\rangle$, $[\alpha]$, $\cup$, $\cap$, and v is and prove that it is possible to specify an agent up to (but not beyond) strong equivalence. It is standard to say that P *respects* $\sim$ if p $\sim$ q and p$\in$ P implies q$\in$ P. We generalize this notion to tuples and functions by saying that **P** *respects* $\sim$ if P_i respects $\sim$ for every i$\in$ I and an agent-set function F *respects* $\sim$ if F(**P**) respects $\sim$ whenever **P** respects $\sim$.

Proposition 3.7: 1, $\emptyset$, $\langle\mu\rangle$, $[\mu]$, $\cup$ and $\cap$ respect $\sim$. Furthermore, vX.F(X,**P**) and vX.G(X,**P**) respect $\sim$ provided F, G and **P** respect $\sim$ and F and G are monotonic.

Proof: We shall show that (i) $\langle\alpha\rangle$ respects $\sim$ and that (ii) vX.F(X,**P**) respects $\sim$ if F and **P** respect $\sim$ and F is monotonic. The other cases are similar.

(i) Assume **P** respects $\sim$. We shall prove that $\langle\mu\rangle$**P** respects $\sim$. Assume p$\in \langle\mu\rangle$**P** and p $\sim$ q. It remains to show q$\in \langle\mu\rangle$**P**. Since p$\in \langle\mu\rangle$**P** there is a p' s.t. p$\xrightarrow{\mu}$p' and

$p' \in P$. Using the fact $p \sim q$ we know that there is a q' s.t. $q \xrightarrow{\mu} q'$ and $p' \sim q'$. But P respects $\sim$, hence $q' \in P$ and we finally get $q \in \langle \mu \rangle P$.

(ii) Let $P^\sim$ be P closed under $\sim$ (i.e. $P^\sim = \{q \mid \exists p \in P \wedge p \sim q\}$). Thus, $P^\sim$ is the smallest superset of P that respects $\sim$. Assume F and $\mathbf{P}$ respect $\sim$. It remains to show that $(\nu X.F(X,\mathbf{P}))^\sim \subseteq \nu X.F(X,\mathbf{P})$

$$
\begin{aligned}
(\nu X.F(X,\mathbf{P}))^\sim &= (F(\nu X.F(X,\mathbf{P}),\mathbf{P}))^\sim && \text{unfolding} \\
&\subseteq (F((\nu X.F(X,\mathbf{P}))^\sim,\mathbf{P}))^\sim && \text{monotonicity of } F \\
&\subseteq F((\nu X.F(X,\mathbf{P}))^\sim,\mathbf{P}) && F, \mathbf{P} \text{ and } (\nu X.F(X,\mathbf{P}))^\sim \text{ respect } \sim.
\end{aligned}
$$

Hence $(\nu X.F(X,\mathbf{P}))^\sim \subseteq \nu X.F(X,\mathbf{P})$ by Park's rule. $\qquad \square$

From this proposition we can conclude that a closed expression built from $\varnothing$, 1, $\langle \mu \rangle$, $[\mu]$, $\cup$, $\cap$, ν respects $\sim$. Hence, such a specification cannot distinguish between agents with the same actual behaviour. The reverse, i.e. that inequivalent agents can be distinguished, is easy to see. Let the family $\langle A_p \rangle_{p \in 1}$ be the greatest solution to

$$
A_p \subseteq \bigcap_{\mu \in \mathrm{Act}} \left(\bigcap_{p':p\xrightarrow{\mu}p'} \langle \mu \rangle A_{p'} \cap [\mu] \bigcup_{p':p\xrightarrow{\mu}p'} A_{p'} \right) \qquad (p \in 1)
$$

(where $p':p\xrightarrow{\mu}p'$ is short for $p' \in \{q \mid p\xrightarrow{\mu}q\}$). A_p is the set of agents that behave like p. This is formalized by the following proposition:

Proposition 3.8: $p \sim q$ iff $q \in A_p$

Proof: $\Rightarrow$: Let $P_p = \{p\}$ (for every p) and show that $\langle P_p \rangle_{p \in 1}$ satisfies the equation that defines $\langle A_p \rangle_{p \in 1}$. Park's rule then gives $P_p \subseteq A_p$, hence $p \in A_p$ and since A_p respects $\sim$ we have $q \in A_p$.

$\Leftarrow$: We show that $\{(p,q) \mid q \in A_p\}$ is a bisimulation. Assume $p\xrightarrow{\mu}p'$ and $q \in A_p$. Using the first half of the equation for A_p and $q \in A_p$ we know that there is a q' such that $p\xrightarrow{\mu}q'$ and $q' \in A_{p'}$. For the other direction, assume $q\xrightarrow{\mu}q'$. From the second half of the equation for A_p we know that $q' \in \bigcup_{p':p\xrightarrow{\mu}p'} A_{p'}$. Hence, there is a p' such that $p\xrightarrow{\mu}p'$ and $q' \in A_{p'}$. $\qquad \square$

Since A_p is defined solely in terms of $\varnothing$, 1, $\langle \mu \rangle$, $[\mu]$, $\cup$, $\cap$ and recursion we see that the specification language generated by $\varnothing$, 1, $\langle \mu \rangle$, $[\mu]$, $\cup$, $\cap$, and ν matches strong bisimulation equivalence. In the rest of the paper we shall call this language *strong HML*.

That recursion is essential for the power of the specification language can be demonstrated by an example. Let $\alpha^0 x = x$, $\alpha^{n+1}x = \alpha\alpha^n x$ and $\alpha^\omega = \mathrm{fix}\ x.\alpha x$ and let $p = \Sigma_n \alpha^n \text{NIL}$ and $q = p+\alpha^\omega$. Now $p \not\sim q$ and it is easy to see that $q \in vX.\langle\alpha\rangle X$ and $p \notin vX.\langle\alpha\rangle X$. But p and q are *strongly congruent* as defined in [Miln80] and HML without recursion respects strong congruence. (For a proof, see [HM85]).

We finish this section by showing that the recursive specification of the α-sender is at least as strong as the non-recursive one, i.e. we shall prove that $vX.\langle\alpha\rangle 1\cap[\alpha]X \subseteq \cap_n A_n$. This is done by showing $vX.\langle\alpha\rangle 1\cap[\alpha]X \subseteq A_n$ by induction on n.

Base: $vX.\langle\alpha\rangle 1\cap[\alpha]X \subseteq 1 = A_0$ $\qquad\qquad$ immediate

Step: $vX.\langle\alpha\rangle 1\cap[\alpha]X$
$\qquad = \langle\alpha\rangle 1\cap[\alpha](vX.\langle\alpha\rangle 1\cap[\alpha]X)$ $\qquad\qquad$ unfolding
$\qquad \subseteq \langle\alpha\rangle 1\cap[\alpha]A_n$ $\qquad\qquad$ ind. hyp. and monotonicity
$\qquad = A_{n+1}$ $\qquad\qquad$ def of A_{n+1}

Note the importance of monotonicity in the reasoning above. From now on we use monotonicity implicitly. The two specifications of α-senders are in fact equivalent. The proof in the other direction, i.e. that $\cap_n A_n \subseteq vX.\langle\alpha\rangle 1\cap[\alpha]X$ is left to the reader. (Hint: Use Park's rule, $[\alpha]\cap$-distributivity and simple set theoretic laws.)

4 Specifying observable behaviour

Clearly, strong HML is too powerful if one is only concerned with observable behaviour. An observer cannot see τ-actions and therefore it is not meaningful for him to write a specification like $\langle\tau\rangle 1$. He cannot possibly see whether an agent satisfies his specification or not. Therefore we shall define a weaker variant of HML that matches weak bisimulation equivalence. The definitions and propositions are as in the previous section but with actions replaced by experiments.

Definition:

$\qquad \langle\langle s\rangle\rangle P = \{p \mid \exists p'. \ p\overset{s}{\Longrightarrow}p' \wedge p' \in P\}$
$\qquad [[s]]P = \{p \mid \forall p'. \ p\overset{s}{\Longrightarrow}p' \Rightarrow p' \in P\}$ $\qquad\qquad\qquad$ □

An agent satisfying $\langle\langle s\rangle\rangle P$ can do the experiment s and become one satisfying P and an agent satisfying $[[s]]P$ cannot do the experiment s without becoming one satisfying P. When s is empty we write $\langle\langle\rangle\rangle P$ and $[[]]P$ instead of $\langle\langle\varepsilon\rangle\rangle P$ and $[[\varepsilon]]P$. So $\langle\langle\rangle\rangle P$ is the set of agents that can silently enter the set P and $[[]]P$ is the set of agents that cannot silently leave the set P.

Proposition 4.1: $\langle\langle s \rangle\rangle$ and $[[s]]$ interact with $\cup$ and $\cap$ in the following way.

$\langle\langle s \rangle\rangle\cup$-*distributivity*: $\langle\langle s \rangle\rangle\cup_{i\in I}P_i = \cup_{i\in I}\langle\langle s \rangle\rangle P_i$ spec. case: $\langle\langle s \rangle\rangle\emptyset = \emptyset$

$[[s]]\cap$-*distributivity*: $[[s]]\cap_{i\in I}P_i = \cap_{i\in I}[[s]]P_i$ spec. case: $[[s]]\,1 = 1$

$[[s]]\cup$-*law*: $[[s]]\,(P\cup Q) \subseteq [[s]]P \cup \langle\langle s \rangle\rangle Q$

$\langle\langle s \rangle\rangle\cap$-*law*: $\langle\langle s \rangle\rangle P \cap [[s]]Q \subseteq \langle\langle s \rangle\rangle(P\cap Q)$ □

Proposition 4.2: $\langle\langle s \rangle\rangle$ and $[[s]]$ are monotonic.   □

Proposition 4.3: $\langle\langle s \rangle\rangle$, $[[s]]$, $\cup$ and $\cap$ respect $\approx$. Furthermore, $\nu X.F(X,P)$ and $\nu X.G(X,P)$ respect $\approx$ if F, G and P respect $\approx$ and F and G are monotonic.   □

We shall call the specification language generated $\langle\langle s \rangle\rangle$, $[[s]]$, $\cup$, $\cap$ and ν *weak HML*. In the light of the previous section it should be obvious that every specification in weak HML respects weak equivalence and that two agents that are not weakly equivalent can be distinguished by weak HML. In other words, weak HML is a specification language for *observable behaviour*.

The pulse buffer specification from the previous section must be revised if we only want to express observable behaviour. Let C_n be the set of buffers that hold n pulses and are allowed to make τ-transitions. This family can be defined as the greatest solution to:

(C.1) C_0 $\subseteq \langle\langle \text{accept} \rangle\rangle 1$ an empty buffer can accept a pulse

(C.2) C_0 $\subseteq [[\text{deliver}]]\emptyset$ an empty buffer cannot deliver a pulse

(C.3) C_n $\subseteq [[\text{accept}]]C_{n+1}$ $(n\in N)$ a buffer holding n pulses cannot accept a pulse without becoming a buffer holding n+1 pulses

(C.4) $C_{n+1} \subseteq \langle\langle \text{deliver} \rangle\rangle 1$ $(n\in N)$ a non-empty buffer can deliver a pulse

(C.5) $C_{n+1} \subseteq [[\text{deliver}]]C_n$ $(n\in N)$ a buffer holding n+1 pulses cannot deliver a pulse without becoming a buffer holding n pulses

(C.6) C_n $\subseteq [[]]C_n$ $(n\in N)$ a buffer cannot lose or duplicate pulses silently

(C.7) C_n $\subseteq [[\lambda]]\emptyset$ $(n\in N, \lambda \notin \{\text{accept,deliver}\})$ a buffer cannot do anything else than accept or deliver a pulse.

The inequation C.6 is essential though it has no counterpart in the previous definition of B_n. Removing it would make the specification too liberal accepting agents like (fix x. accept.deliver.x) + τNIL. By adding C.7 we can allow Act to contain other actions than accept and deliver.

Weak HML has the advantage of expressing only what can be observed, but strong HML is easier to use when reasoning about agents since it is based on the more primitive action relation. Fortunately, a specification in weak HML can easily be translated to a specification in strong HML using:

Proposition 4.4 *(weak operator elimination laws)*:

(a) $\langle\langle st\rangle\rangle P = \langle\langle s\rangle\rangle \langle\langle t\rangle\rangle P$

(b) $\langle\langle\lambda\rangle\rangle P = \langle\langle\rangle\rangle \langle\lambda\rangle \langle\langle\rangle\rangle P$

(c) $\langle\langle\rangle\rangle P = \bigcup_n \langle\tau\rangle^n P$

(d) $\langle\langle\rangle\rangle 1 = 1$

(e) $[[st]]P = [[s]] [[t]] P$

(f) $[[\lambda]]P = [[]] [\lambda] [[]]P$

(g) $[[]] P = \bigcap_n [\tau]^n P = \nu X.P \cap [\tau]X$

(h) $[[]] \emptyset = \emptyset$

(i) $[[]] P \subseteq P \subseteq \langle\langle\rangle\rangle P$

Proof: All equalities except (g) follow straightforwardly from the definitions of $\langle\langle s\rangle\rangle$, $[[s]]$ and $=\overset{s}{=}>$. For g, show $\bigcap_n [\tau]^n P \subseteq \nu X.P\cap[\tau]X$ by Park's rule and $\nu X.P\cap[\tau]X \subseteq [\tau]^n P$ by induction on n. $\square$

However, with these laws alone we cannot get the simplest possible equivalent specification in strong HML. We must also use some facts about *invariants*.

Definition: P is an *invariant* if $P \subseteq [\tau] P$. $\square$

Saying that P is an invariant is the same as saying that P is preserved by silent actions. Invariants can be characterized in several ways.

Proposition 4.5: The following assertions about P are equivalent

(i) P is an invariant.

(ii) $P \subseteq [[]] P$

(iii) $P = [[]] P$

Proof: (i) $\Rightarrow$ (ii): Assume $P \subseteq [\tau] P$. Show $P \subseteq [\tau]^n P$ for every n by induction on n and hence $P \subseteq \bigcap_n [\tau]^n P = [[]] P$. (ii) $\Rightarrow$ (iii): Follows easily from the fact that $[[]]P \subseteq P$ for every P. (iii) $\Rightarrow$ (i): Assume $P = [[]] P$. Now $P = [[]]P = \bigcap_n [\tau]^n P \subseteq [\tau]P$. $\square$

Using propositions 4.4 and 4.5 it is easy to show that C is equal to C' which is the maximal solution to

(C'.1) C'_0 $\subseteq \langle\langle\rangle\rangle \langle accept\rangle 1$

(C'.2) C'_0 $\subseteq [deliver]\varnothing$

(C'.3) C'_n $\subseteq [accept] C'_{n+1}$ $(n \in N)$

(C'.4) $C'_{n+1} \subseteq \langle\langle\rangle\rangle \langle deliver\rangle 1$ $(n \in N)$

(C'.5) $C'_{n+1} \subseteq [deliver] C'_n$ $(n \in N)$

(C'.6) C'_n $\subseteq [\tau] C'_n$ $(n \in N)$

(C'.7) C'_n $\subseteq [\lambda]\varnothing$ $(n \in N, \lambda \notin \{accept, deliver\})$

thus getting a simpler definition of the set of pulse buffers.

The proof is an easy application of Park's rule and of proposition 4.4 and 4.5. For example, to prove that $C'_n \subseteq C_n$, we show that C'_n satisfies the equations that defines C_n. In the case of equation (C.3) we have:

$C'_n = [[]] C'_n$ C'_n is invariant (from prop 4.5)

$\subseteq [[]] [accept] C'_{n+1}$ C'.3

$\subseteq [[]] [accept] [[]] C'_{n+1}$ C'_{n+1} is invariant

$= [[accept]]C'_{n+1}$ prop 4.4(f)

The other inequations are proved analogously.

We end this section by introducing some new notation. If A is a set of experiments then $\langle\langle A\rangle\rangle P$ abbreviates $\bigcup_{s \in A}\langle\langle s\rangle\rangle P$ and $[[A]]P$ abbreviates $\bigcap_{s \in A}[[s]]P$. The sets of experiments will often be expressed as regular expressions. For example $[[(\alpha+\beta)^*]] P$ is the set of agents that cannot leave the set P by doing any sequence of α or β. We also have a generalization of proposition 4.1:

Proposition 4.6: $\langle\langle A\rangle\rangle$ and $[[A]]$ interact with $\cup$ and $\cap$ in the following way.

$\langle\langle A\rangle\rangle\cup$-*distributivity*: $\langle\langle A\rangle\rangle\bigcup_{i \in I}P_i = \bigcup_{i \in I}\langle\langle A\rangle\rangle P_i$

$[[A]]\cap$-*distributivity*: $[[A]]\bigcap_{i \in I}P_i = \bigcap_{i \in I}[[A]]P_i$

$[[A]]\cup$-*law*: $[[A]] (P\cup Q) \subseteq [[A]]P \cup \langle\langle A\rangle\rangle Q$

$\langle\langle A\rangle\rangle\cap$-*law*: $\langle\langle A\rangle\rangle P \cap [[A]]Q \subseteq \langle\langle A\rangle\rangle(P\cap Q)$

Proof: Immediate from the definitions and the distributivity laws for $\cup$ and $\cap$. $\square$

5 Specifying an unreliable medium.

The alternating-bit protocol is a distributed system that offers reliable one-way communication given a pair of unreliable communication lines (or media), one in each direction. In section 7 we shall prove the correctness of this protocol assuming certain properties of an unreliable medium. Here we shall discuss these properties and formalize them in weak HML.

(It is of course possible to specify an unreliable medium simply by saying that it should make the alternating-bit-protocol work. This specification is short, simple and can easily be made precise but it has a serious disadvantage: The specification of the unsafe medium involves the whole protocol so it becomes difficult to prove that a given medium satisfies it. Instead, we want to specify the unreliable medium without referring to the alternating-bit protocol at all. It is then possible to prove that the alternating-bit protocol is correct provided the unreliable media satisfy their specifications.)

An *unreliable medium* has the following properties. It transfers messages from one place to another. It can lose and duplicate messages typically due to noise or buffer overflow but never change their order. Furthermore, a message will eventually be delivered if it is sent repeatedly. It is surprisingly hard to formally specify this simple component if we want a general specification that allows for every thinkable implementation.

Let m range over the set Msg of *messages* and let s,t range over sequences of messages. As usual, we let ε be the empty sequence and st be the concatenation of s and t. We also use the notation #s for the length of s. Here are some examples of unreliable media:

$$\text{one-place-buffer} \quad \Leftarrow \sum_m \text{accept}_m.\text{deliver}_m.\text{one-place-buffer}$$

$$\text{unbounded-buffer}_\varepsilon \quad \Leftarrow \sum_{m'} \text{accept}_{m'}.\text{unbounded-buffer}_{m'}$$
$$\text{unbounded-buffer}_{sm} \Leftarrow \sum_{m'} \text{accept}_{m'}.\text{unbounded-buffer}_{m'sm}$$
$$+ \text{deliver}_m.\ \text{unbounded-buffer}_s$$

$$\text{n-bounded-buffer}_\varepsilon \quad \Leftarrow \sum_{m'} \text{accept}_{m'}.\text{n-bounded-buffer}_{m'} \qquad (n{>}0)$$
$$\text{n-bounded-buffer}_{sm} \Leftarrow \sum_{m'} \text{accept}_{m'}.\text{n-bounded-buffer}_{m'sm} \quad (\#sm{<}n)$$
$$+ \ \text{deliver}_m.\text{n-bounded-buffer}_s$$
$$\text{n-bounded-buffer}_{sm} \Leftarrow \text{deliver}_m.\ \text{n-bounded-buffer}_s \qquad (\#sm{=}n)$$

$$\text{register}_m \qquad\qquad \Leftarrow \sum_{m'} \text{accept}_{m'}.\ \text{register}_{m'}$$
$$+ \text{deliver}_m.\ \text{register}_m$$

$$\text{shift-register}_{sm} \quad <= \Sigma_m \text{'accept}_m\text{'. shift-register}_{m's}$$
$$+ \text{deliver}_m. \text{shift-register}_{sm}$$

$$\text{lossy-buffer} <= \Sigma_m \text{'accept}_m\text{'. (deliver}_m\text{'.lossy-buffer} + \tau.\text{lossy-buffer})$$

$$\text{every-2nd-buffer} <= \Sigma_m \text{'accept}_m\text{'.}\Sigma_m \text{ accept}_m.\text{deliver}_m.\text{every-2nd-buffer}$$

A buffer (one-place, bounded or unbounded) is a perfect medium in the sense that it does not lose or duplicate messages. It is a special case of an unreliable medium. A shift register, of which an ordinary register is a special case, loses and duplicates messages if reading and writing does not alternate. A lossy buffer is similar to a one-place buffer but can silently lose its contents. The lossy-buffer is perhaps the most typical of the unreliable media shown here. Another imperfect medium is the every-2nd-buffer which loses every second message. All these agents are unreliable media according to the informal characterization we started from. How do we write a specification that embodies them all without being too weak?

We write $s \leq t$ if we can get the message sequence s by removing and duplicating messages in t. When $s \leq t$ we say that s contains *no more information than* t. If $s \leq t$ and $t \leq s$ then we write $s \cong t$,and if $s \leq t$ but not $t \leq s$ then we write $s < t$. For example, let $s = m_1 m_2 m_2 m_4$, $t = m_1 m_1 m_2 m_3 m_4$ and $u = m_1 m_2 m_3 m_3 m_4$ (all m_i distinct). Now $s < t$ (s contains strictly less information than t since m_3 is lost) and $t \cong u$ (t and u contains the same information).

Proposition 5.1:

(i) $\leq$ is a preorder, i.e. $\leq$ is reflexive and transitive.

(ii) $\leq$ has the least element ε.

(iii) $\leq$ is preserved by concatenation, $stu \leq st'u$ if $t \leq t'$

(iv) $m \cong m^n$ (n>0) $\square$

An unreliable medium holds a sequence s of messages. Since the medium can duplicate and lose messages we do not know s exactly but we have an upper bound on the amount of information in s. For example, if we feed the sequence $m_1 m_2 m_3$ of messages into an empty medium (m_3 first) then we know that the medium can contain, for example, $m_1 m_3 m_3$ but never $m_2 m_1 m_3$ or any message that is not in $m_1 m_2 m_3$. If the medium delivers m_2 then we know that it cannot hold a sequence that contains more information than $m_1 m_2$.

Let M_s be the set of agents that are unreliable media containing at most the information in s. First we note that a medium cannot silently increase the amount of information in the sequence it holds. If it holds at most the information in s then it still holds at most the information in s after some silent actions. So we have:

(M.1) $M_s \subseteq [[]]\ M_s$

Second, if a medium that holds at most the information in s accepts the message m then it holds at most the information in ms.

(M.2) $M_s \subseteq [[accept_m]]\ M_{ms}$

Third, if a medium that holds at most the information in s delivers the message m then there must be a sequence tm such that (i) tm contains no more information than s and (ii) the resulting agent holds at most the information in t.

(M.3) $M_s \subseteq [[deliver_m]]\ \bigcup_{t:tm\leq s} M_t$

Note the special case when m does not occur in s:

$$M_s \subseteq [[deliver_m]]\ \bigcup_{t:tm\leq s} M_t$$
$$= [[deliver_m]]\ \varnothing \qquad \text{since } \{t \mid tm\leq s\} = \varnothing$$

This is exactly what we want; a medium cannot deliver a message that it does not hold. A second example: assume that the medium holds at most the information in $s=m_1 m_2 m_3$ (all m_i distinct). If m_2 is delivered then this formula says that the resulting medium holds at most the information in $t=m_1 m_2$, since t is the most informative sequence for which $tm_2\leq s$.

Fourth, a medium cannot do anything else than accepting and delivering messages.

(M.4) $M_s \subseteq [[\lambda]]\ \varnothing$ $\qquad (\forall \lambda \notin \{accept_m, deliver_m \mid m \in Msg\})$

All the assertions we have seen so far have been safety requirements, i.e. requirements saying what the medium *cannot* do. This is clearly not sufficient since, e.g. NIL satisfies M.1-4. We also need some positive assertions stating what it *can* do.

It is important that the medium is not completely broken, i.e. we demand that a message is eventually delivered if sent persistently. Unfortunately this *liveness* condition is not expressible in weak HML. The best we can do is to demand that it should be *possible* for a medium to deliver a message m provided it is given m persistently.

(M.5) $M_s \subseteq \langle\langle (\text{accept}_m + \Sigma_{m' \neq m}\text{deliver}_{m'})^* \text{ deliver}_m \rangle\rangle 1$ $\qquad\qquad (\forall m \in \text{Msg})$

Put in other words, every agent should be able to do a sequence of accept_m and deliver actions finished by deliver_m.

The following result will be needed later. It says that the information ordering on message sequences induces a subset ordering on sets of unreliable media.

Proposition 5.2: If $t \leq s$ then $M_t \subseteq M_s$ (and if $s \cong t$ then $M_s = M_t$).

Proof: (outline.) Show that $\cup_{t \leq s} M_t \subseteq M_s$ by Park's rule, i.e. show that $\cup_{t \leq s} M_t$ satisfies the inequations that define M_s. $\qquad\qquad\qquad\qquad\qquad\qquad\qquad\square$

It remains to show that the proposed implementations above satisfy the given specification. Using the technique demonstrated in the previous section we can first show that the family **M'** defined as the greatest solution to

(M'.1) $M'_s \subseteq [\tau] M'_s$

(M'.2) $M'_s \subseteq [\text{accept}_m]M'_{ms}$

(M'.3) $M'_s \subseteq [\text{deliver}_m] \cup_{t:tm \leq s} M'_t$

(M'.4) $M'_s \subseteq [\lambda]\emptyset$ $\qquad\qquad\qquad (\forall \lambda \notin \{\text{accept}_m, \text{deliver}_m \mid m \in \text{Msg}\})$

(M'.5) $M'_s \subseteq \langle\langle (\text{accept}_m + \Sigma_{m' \neq m}\text{deliver}_{m'})^* \text{ deliver}_m \rangle\rangle 1$ $\qquad (\forall m \in \text{Msg})$

is equal to **M**. With this equivalent specification of an unreliable medium, it is not difficult to find a family $\mathbf{P} \subseteq \mathbf{M'} = \mathbf{M}$ of sets containing the proposed implementations. For example, to see that the shift register satisfies the specification, let $P_s = \{\text{shift-register}_t \mid t \leq s\}$ and show that $\mathbf{P} \subseteq \mathbf{M'}$ using Park's rule. The details are left to the reader.

6 A Refinement Calculus

To decompose a problem is to reduce it to a number of simpler subproblems. When the problem is to find a program satisfying a specification, problem decomposition goes under the name *stepwise refinement*. Rigorous stepwise refinement starts from a rigorous specification and every refinement step is justified by showing that the solutions to the subproblems compose into a solution to the whole problem.

In the context of CCS/HML a typical refinement step can look like this: Let the problem be to find an agent p satisfying P. Let us assume that we believe that p can be composed from two agents q and r working in parallel where q satisfies Q and r satisfies R. By proving $\forall q,r.\ q \in Q \wedge r \in R \Rightarrow (q|r) \in P$, we *justify* the refinement step. It now remains to solve the two subproblems of finding $q \in Q$ and $r \in R$. The important information in the refinement step is (i) the specification P, (ii) the subspecifications Q and R and (iii) the composition method I. (The names p, q, and r of the agents are bound and not relevant.)

By introducing *refinement operators* into the specification language we achieve three things: (i) refinement steps can be succinctly formulated, (ii) we do not have to introduce names for the subagents and (iii) refinement steps can be justified using convenient "proof rules". The refinement operators are defined to be the familiar agent operators, action, sum, parallel composition, etc., lifted to sets. For example $P|Q$ is the set of agents of the form $(p|q)$ where p satisfies P and q satisfies Q. With this notation we write $Q|R \subseteq P$ instead of the longer (and less perspicuous) $\forall q,r.\ q \in Q \wedge r \in R \Rightarrow (q|r) \in P$.

Definition *(refinement operators)*:

(i) $\alpha P = \{\alpha p \mid p \in P\}$

(ii) $\Sigma P = \{\Sigma p \mid \forall i \in I.p_i \in P_i\}$

(iii) $P|Q = \{\ p|q \mid p \in P, q \in Q\}$

(iv) $P \backslash \alpha = \{p \backslash \alpha \mid p \in P\}$

(v) $P[S] = \{p[S] \mid p \in P\}$ $\square$

We shall abbreviate $\Sigma_{i \in \{0,1\}} P_i$ by $P_0 + P_1$ and $\Sigma_{i \in \varnothing} P_i$ by NIL. Let the binary refinement operators + and I have lower priority than $\cup$ and $\cap$.

To complete the refinement language we also want a refinement operator that corresponds to the recursion operator on agent expressions. Allowing expressions like fix $x.\langle\alpha\rangle x$, where set expressions occur within fix, leads to problems, however. Instead, we shall use fix in a more restrictive way by letting an *agent* of the form fix$_i$x.e be a nullary refinement operator that denotes the singleton set {fix$_i$x.e}.

By a *refinement expression* we shall understand an expression built from HML operators and refinement operators. For example, the refinement expression $\langle\alpha\rangle P + Q$ I $[\beta]$(fix $x.\alpha x$) denotes the set of agents of the form $p + q | r$ where p can do α and become an agent satisfying P, q satisfies Q and r cannot do β without becoming an agent satisfying fix $x.\alpha x$, i.e. identical to fix $x.\alpha x$.

This confusion of programming language and specification language has been advocated by Hoare, Hehner and others under the slogan "Programs are

Predicates". The method is convenient for stepwise refinement since it replaces the satisfaction relation ($\in$) between programs and specifications by a refinement relation ($\subseteq$) between more or less refined specifications.

Fortunately, monotonicity is preserved by the refinement operators so inequational reasoning is still allowed.

Proposition 6.1: The refinement operators α, Σ, $|$, $\backslash\alpha$, [s] are monotonic. $\quad\Box$

Proposition 6.2: The refinement operators α, Σ, $|$, $\backslash\alpha$ and [s] are *additive*, i.e. they distribute over union, so

(i) $\alpha(P \cup Q) = \alpha P \cup \alpha Q$ and $\alpha\varnothing = \varnothing$

(ii) $P \cup Q + R = (P+R) \cup (Q+R)$ and $\varnothing + R = \varnothing$

(iii) $P \cup Q \mid R = (P \mid R) \cup (Q \mid R)$ and $\varnothing \mid R = \varnothing$

(iv) $(P \cup Q) \backslash A = P \backslash A \cup Q \backslash A$ and $\varnothing \backslash A = \varnothing$

(v) $(P \cup Q)[s] = P[s] \cup Q[s]$ and $\varnothing[s] = \varnothing$

(This result generalizes in the natural way to infinite sums and unions.) $\quad\Box$

From the operational semantics of CCS is it straightforward to show the following rules that forms the basis for our refinement calculus.

Proposition 6.3:

$act\langle\mu\rangle$: $\mu.P \subseteq \langle\mu\rangle P$

$act[\mu]$: $\mu.P \subseteq [\mu]P$

$act[\nu]$: $\mu.P \subseteq [\nu]\varnothing$ $(\nu \neq \mu)$

$sum\langle\mu\rangle$: $\dfrac{P_i \subseteq \langle\mu\rangle Q}{\Sigma P \subseteq \langle\mu\rangle Q}$

$sum[\mu]$: $\dfrac{\forall i \in I.\ P_i \subseteq [\mu]Q}{\Sigma P \subseteq [\mu]Q}$

$par\langle\mu\rangle$:
$$\frac{P\subseteq\langle\mu\rangle P'}{P\,|\,Q\subseteq\langle\mu\rangle(P'\,|\,Q)} \qquad\qquad \frac{Q\subseteq\langle\mu\rangle Q'}{P\,|\,Q\subseteq\langle\mu\rangle(P\,|\,Q')}$$

$par\langle\tau\rangle$:
$$\frac{P\subseteq\langle\lambda\rangle P' \quad Q\subseteq\langle\lambda^c\rangle Q'}{P\,|\,Q\subseteq\langle\tau\rangle(P'\,|\,Q')}$$

$par[\lambda]$:
$$\frac{P\subseteq[\lambda]P' \quad Q\subseteq[\lambda]Q'}{P\,|\,Q\subseteq[\lambda]\,((P'\,|\,Q)\cup(P\,|\,Q'))}$$

$par[\tau]$:
$$\frac{\forall\mu.\ P\subseteq[\mu]P_\mu \quad Q\subseteq[\mu]Q_\mu}{P\,|\,Q\subseteq[\tau]\,((P_\tau\,|\,Q)\cup(P\,|\,Q_\tau)\ \cup\ \bigcup_\lambda(P_\lambda\,|\,Q_{\lambda c}))}$$

$res\langle\mu\rangle$:
$$\frac{P\subseteq\langle\mu\rangle Q}{P\backslash A\subseteq\langle\mu\rangle\,(Q\backslash A)} \qquad (\mu\notin A,\ \mu^c\notin A\ )$$

$res[\mu]$:
$$\frac{P\subseteq\langle\mu\rangle Q}{P\backslash A\subseteq[\mu](Q\backslash A)}$$

$res[\alpha]$:
$$P\backslash A\subseteq[\alpha]\varnothing \qquad (\alpha\in A)$$

$res[\alpha^c]$:
$$P\backslash A\subseteq[\alpha^c]\varnothing \qquad (\alpha\in A)$$

$ren\langle\mu\rangle$:
$$\frac{P\subseteq\langle\mu\rangle Q}{P[S]\subseteq\langle S(\mu)\rangle(Q[S])}$$

$ren[\mu]$:
$$\frac{\forall\mu.\ P\subseteq[\mu]P_\mu}{P[S]\subseteq[\mu]\bigcup_{\nu:S(\nu)=\mu}(P_\nu[S])}$$

$fix\langle\mu\rangle$:
$$\frac{e_i\{\text{fix }\mathbf{x}.e/\mathbf{x}\}\subseteq\langle\mu\rangle Q}{\text{fix}_i\mathbf{x}.e\subseteq\langle\mu\rangle Q}$$

$fix[\mu]$:
$$\frac{e_i\{\text{fix }\mathbf{x}.e/\mathbf{x}\}\subseteq[\mu]Q}{\text{fix}_i\mathbf{x}.e\subseteq[\mu]Q}$$

Proof: we show $par\langle\mu\rangle$, $par[\lambda]$, $fix\langle\mu\rangle$. The others are similar or trivial.

$par\langle\mu\rangle$: Assume $P\subseteq\langle\mu\rangle P'$ and $r\in P\,|\,Q$. Then $r=(p\,|\,q)$ where $p\in P$ and $q\in Q$. From the assumption we know that there is a p' s.t. $p\overset{\mu}{\to}p'$ and $p'\in P'$. From the operational semantics we then have $r=(p\,|\,q)\overset{\mu}{\longrightarrow}(p'\,|\,q)$. Hence $r\in\langle\mu\rangle(P'\,|\,Q)$.

par[μ]: Assume $P \subseteq [\lambda]P_\lambda$, $Q \subseteq [\lambda]Q_\lambda$, $r \in P \mid Q$ and $r \overset{\lambda}{\to} r'$. Under these assumptions we shall show that $r' \in (P' \mid Q) \cup (P \mid Q')$, i.e. that $r' \in (P' \mid Q)$ or $r' \in (P \mid Q')$. From the assumption $r \in P \mid Q$ we know that $r = (p \mid q)$ where $p \in P$ and $q \in Q$. From the operational semantics and the fact $r \overset{\lambda}{\to} r'$ we also know that either (i) $p \overset{\lambda}{\to} p'$ and $r' = (p' \mid q)$ or (ii) $q \overset{\lambda}{\to} q'$ and $r' = (p \mid q')$. In the first case we know from $P \subseteq [\lambda]P_\lambda$ that $p' \in P'$, hence $r' = (p \mid q') \in (P' \mid Q)$. The other case is symmetric.

fix⟨μ⟩: Assume $e_i\{fix\ \mathbf{x}.e/\mathbf{x}\} \subseteq \langle\mu\rangle Q$ and $p \in fix_i\mathbf{x}.e$. Then $p = fix_i\mathbf{x}.e$ and, moreover, there is a p' such that $e_i\{fix\ \mathbf{x}.e/\mathbf{x}\} \overset{\mu}{\to} p' \in Q$. From the operational semantics we then know that $fix_i\mathbf{x}.e \overset{\mu}{\to} p' \in Q$. Hence $p \in \langle\mu\rangle Q$. $\qquad\square$

Most of these rules can be simplified and more adapted to inequational reasoning. The following proposition displays a collection of derived rules that are more convenient to use in most cases.

Proposition 6.4:

sum'⟨μ⟩: $\langle\mu\rangle P + Q \subseteq \langle\mu\rangle P$

$\qquad\qquad P + \langle\mu\rangle Q \subseteq \langle\mu\rangle Q$

sum'[μ]: $[\mu]P + [\mu]Q \subseteq [\mu] (P \cup Q)$

par'⟨μ⟩: $\langle\mu\rangle P \mid Q \subseteq \langle\mu\rangle (P \mid Q)$

$\qquad\qquad P \mid \langle\mu\rangle Q \subseteq \langle\mu\rangle (P \mid Q)$

par'[λ]: $P \cap [\lambda]P' \mid Q \cap [\lambda]Q' \subseteq [\lambda]((P' \mid Q) \cup (P \mid Q'))$

par'[τ]: $P \cap \bigcap_\mu[\mu]P_\mu \mid Q \cap \bigcap_\mu[\mu]Q_\mu \subseteq [\tau]((P_\tau \mid Q) \cup (P \mid Q_\tau) \cup \bigcup_\lambda(P_\lambda \mid Q_{\lambda c}))$

res'⟨μ⟩: $\langle\mu\rangle P \setminus A \subseteq \langle\mu\rangle (P \setminus A)$ $\qquad\qquad\qquad (\mu \notin A, \mu^c \notin A)$

res'[μ]: $[\mu]P \setminus A \subseteq [\mu] (P \setminus A)$

ren'⟨μ⟩: $(\langle\mu\rangle P) [S] \subseteq \langle S(\mu)\rangle (P [S])$

ren'[μ]: $(\bigcap_\nu[\nu]P_\nu)[S] \subseteq [\mu] \bigcup_{\nu:S(\nu)=\mu}(P_\nu[S])$ $\qquad\qquad\qquad\square$

Here is a trivial example of the use of the refinement calculus. Recall the specification $\nu X.\langle\alpha\rangle 1 \cap [\alpha]X$ of the α-sender. We show that fix x.ααx is a correct refinement of this specification.

(1) $\qquad \alpha\alpha(fix\ x.\alpha\alpha x) \subseteq \langle\alpha\rangle\alpha(fix\ x.\alpha\alpha x)$ $\qquad\qquad\qquad\qquad$ by act⟨α⟩

(2) $\qquad fix\ x.\alpha\alpha x \subseteq \langle\alpha\rangle\alpha(fix\ x.\alpha\alpha x)$ $\qquad\qquad\qquad\qquad$ by (1) and fix⟨α⟩

(3) $\qquad \subseteq \langle\alpha\rangle 1$ $\qquad\qquad\qquad\qquad$ by set theory

(4) $\qquad \alpha(fix\ x.\alpha\alpha x) \subseteq \langle\alpha\rangle(fix\ x.\alpha\alpha x)$ $\qquad\qquad\qquad\qquad$ by act⟨α⟩

(5) $\qquad \subseteq \langle\alpha\rangle 1$ $\qquad\qquad\qquad\qquad$ by set theory

(6) $(\text{fix } x.\alpha\alpha x)\cup\alpha(\text{fix } x.\alpha\alpha x)\subseteq\langle\alpha\rangle 1$ by (3), (5) and set theory

(7) $\alpha\alpha(\text{fix } x.\alpha\alpha x)\subseteq[\alpha]\alpha(\text{fix } x.\alpha\alpha x)$ by act$[\alpha]$

(8) $\text{fix } x.\alpha\alpha x\subseteq[\alpha]\alpha(\text{fix } x.\alpha\alpha x)$ by (7) and fix$[\alpha]$
(9) $\subseteq[\alpha]((\text{fix } x.\alpha\alpha x)\cup\alpha(\text{fix } x.\alpha\alpha x))$ by set theory

(10) $\alpha(\text{fix } x.\alpha\alpha x)\subseteq[\alpha](\text{fix } x.\alpha\alpha x)$ by act$[\alpha]$
(11) $\subseteq[\alpha]((\text{fix } x.\alpha\alpha x)\cup\alpha(\text{fix } x.\alpha\alpha x))$ by set theory

(12) $(\text{fix } x.\alpha\alpha x)\cup\alpha(\text{fix } x.\alpha\alpha x)\subseteq[\alpha]((\text{fix } x.\alpha\alpha x)\cup\alpha(\text{fix } x.\alpha\alpha x))$

          by (9), (11) and set theory

(13) $(\text{fix } x.\alpha\alpha x)\cup\alpha(\text{fix } x.\alpha\alpha x)\subseteq\nu X.\langle\alpha\rangle 1\cap[\alpha]X$ by Park's rule

(14) $\text{fix } x.\alpha\alpha x\subseteq\nu X.\langle\alpha\rangle 1\cap[\alpha]X$ by (13) and set theory

More examples of use of the refinement calculus will appear in the next section.

There is also a collection of simple laws that show how the weak operator $\langle\langle s\rangle\rangle$ interacts with the refinement operators. These are completely analogous to the laws for the strong modal operator $\langle\mu\rangle$. Unfortunately , no such collection of simple laws for $[[s]]$ can be found.

Proposition 6.5:

$sum\langle\langle s\rangle\rangle$: $\langle\langle s\rangle\rangle P + Q \subseteq \langle\langle s\rangle\rangle P$
          $P + \langle\langle s\rangle\rangle Q \subseteq \langle\langle s\rangle\rangle Q$

$par\langle\langle s\rangle\rangle$: $\langle\langle s\rangle\rangle P \mid Q \subseteq \langle\langle s\rangle\rangle (P\mid Q)$
          $P \mid \langle\langle s\rangle\rangle Q \subseteq \langle\langle s\rangle\rangle (P\mid Q)$
$par\langle\langle\rangle\rangle$: $\langle\langle s\rangle\rangle P \mid \langle\langle s^c\rangle\rangle Q \subseteq \langle\langle\rangle\rangle (P\mid Q)$
$res\langle\langle s\rangle\rangle$: $(\langle\langle s\rangle\rangle P) \setminus A \subseteq \langle\langle s\rangle\rangle (P\setminus A)$ $(\lambda\notin A,\lambda^c\notin A,(\forall\lambda\in s))$

$ren\langle\langle s\rangle\rangle$: $(\langle\langle s\rangle\rangle P) [S] \subseteq \langle\langle S(s)\rangle\rangle (P[S])$

where $\lambda\in s$ means that λ is a label in the sequence s. (The relabelling function S and the complement function ^{c} are here extended to sequences of labels in the obvious way, so $S(\lambda_1\lambda_2...\lambda_n) = S(\lambda_1)S(\lambda_2)...S(\lambda_n)$ and $(\lambda_1\lambda_2...\lambda_n)^c = \lambda^c_1\lambda^c_2...\lambda^c_n.)$ ☐

7 The alternating-bit protocol

The alternating-bit protocol is a well known solution to the problem of obtaining reliable communication on unreliable communication lines. Its correctness has been proved in various formalisms and has become a benchmark for formal verification methods for concurrency. The protocol is a distributed system with four components connected in the following way

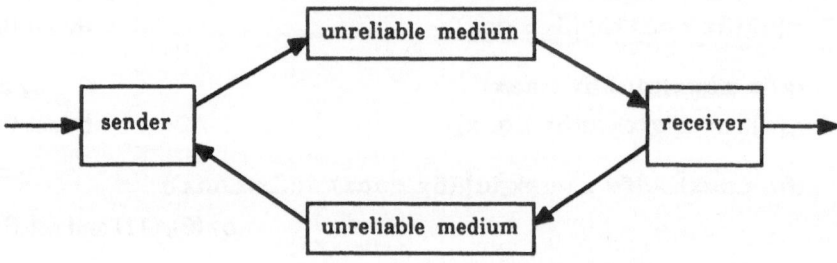

The protocol provides reliable communication and externally it behaves as a one place buffer. It works by repeated transmission as follows: The sender accepts a message and this message together with a one bit sequence number is repeatedly sent to the receiver through the unreliable medium. The receiver will eventually receive the packet, deliver it and repeatedly send it back to acknowledge that the packet has been delivered. When the sender eventually receives the packet it is currently sending, i.e. the *acknowledgement*, it knows that the message has been delivered. So it ceases sending and becomes ready to accept the next message. The receiver, on the other hand, stops acknowledging the packet as soon as a packet with the next message arrives. The sender toggles the one bit sequence number for every new message it accepts (hence the name of the protocol.) The receiver uses this bit to decide whether a packet contains a fresh message or is just a resent copy of the previous one.

This is a simplified version of what is normally implemented. In practice, the sender waits a specified time interval, the *time out* before it sends the packet anew. The time out prevents the lines to be flooded by copies of the same packet. For the same reason, the receiver returns just one acknowledgement for each incoming packet.

Without losing any interesting aspect of the problem we shall simplify it more by assuming that messages are always empty, i.e. the messages are only synchronization pulses. This means that the packets contain only the one bit sequence number.

We shall model the protocol by the agent $\text{Sys}(s,m_1,r,m_2)$ where s is the sender, r the receiver and m_1 and m_2 are the two unreliable media. Let b range over the set $\{0,1\}$ and let $\neg b = (b+1) \bmod 2$. Formally, $\mathrm{Sys}(s,m_1,r,m_2)$ is defined by

$$\mathrm{Sys}(s,m_1,r,m_2) = (\ \ s$$
$$| \ \ m_1[\mathrm{send}^c{}_b,\mathrm{receive}^c{}_b/\mathrm{accept}_b,\mathrm{deliver}_b \ | b \in \{0,1\}]$$
$$| \ \ r$$
$$| \ \ m_2[\mathrm{reply}^c{}_b,\mathrm{ack}^c{}_b/\mathrm{accept}_b,\mathrm{deliver}_b \ | b \in \{0,1\}]$$
$$) \setminus \{\mathrm{send}_b,\mathrm{receive}_b,\mathrm{reply}_b,\mathrm{ack}_b \ | b \in \{0,1\}\}$$

but the structure of the agent is maybe easier to understand from a diagram.

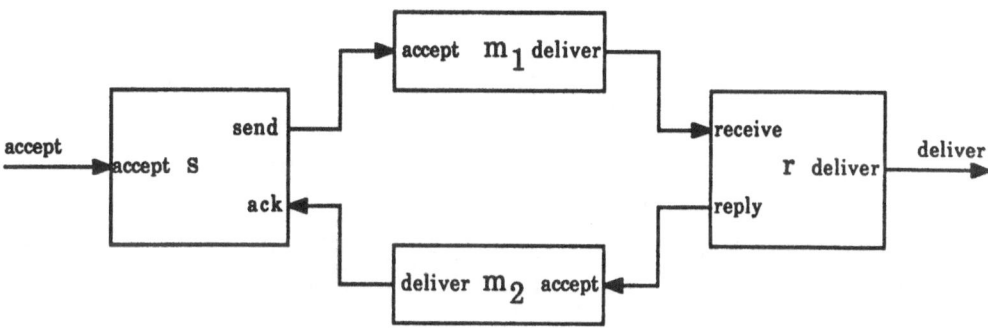

This configuration is straightforward to create using renaming, parallel composition and restriction. Only the labels accept_b and $\mathrm{deliver}_b$ ($b \in \{0,1\}$) are visible from outside. The "channels" send, trans, reply and ack are internalized by the restriction.

We define the sender and the receiver explicitly as agents. The sender is in one of four states, each represented by the agents: $\mathrm{acknowledged}_0$, $\mathrm{sending}_1$, $\mathrm{acknowledged}_1$ and $\mathrm{sending}_0$. The agent $\mathrm{acknowledged}_0$ models a sender that has just been acknowledged a 0 and is prepared to accept the next pulse and start sending a 1. The agent $\mathrm{sending}_1$ models a sender that keeps sending a 1 through the medium until it is acknowledged a 1.

$$\mathrm{acknowledged}_b \ \Leftarrow \ \mathrm{accept}.\mathrm{sending}_{\neg b}$$
$$+ \ \mathrm{ack}_b.\mathrm{acknowledged}_b$$
$$\mathrm{sending}_b \ \ \ \ \Leftarrow \ \mathrm{send}_b.\mathrm{sending}_b$$
$$+ \ \mathrm{ack}_{\neg b}.\mathrm{sending}_b$$
$$+ \ \mathrm{ack}_b.\mathrm{acknowledged}_b$$

Sender:

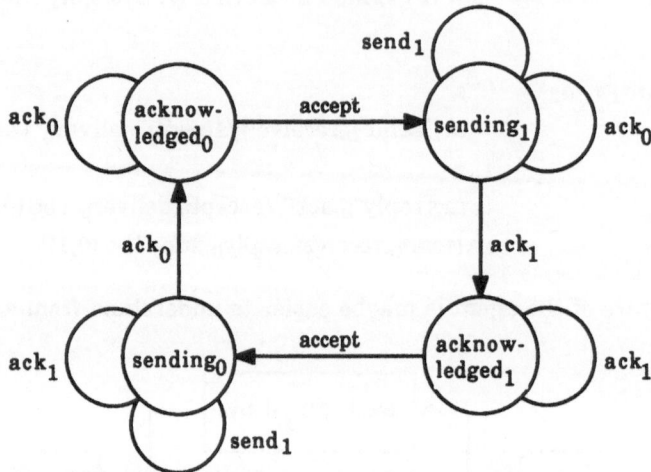

The receiver has also four states: replying$_0$, received$_1$, replying$_1$ and received$_0$. When in the state replying$_0$, it keeps acknowledging that it has received a 0 and delivered the corresponding pulse. When it receives a 1 it moves to the state received$_1$ and becomes ready to deliver a new pulse.

$$\text{replying}_b \Leftarrow \text{reply}_b.\text{replying}_b$$
$$+ \text{receive}_b.\text{replying}_b$$
$$+ \text{receive}_{\neg b}.\text{received}_{\neg b}$$
$$\text{received}_b \Leftarrow \text{receive}_b.\text{received}_b$$
$$+ \text{deliver}.\text{replying}_b$$

Receiver:

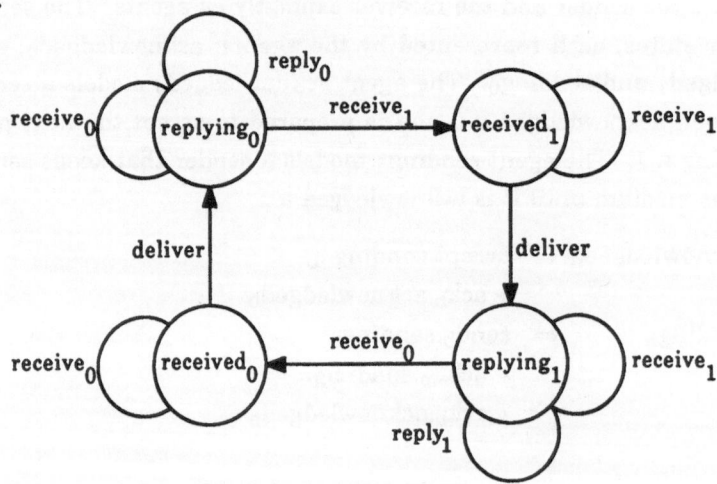

We shall not give explicit specifications (agents) for the two media m_1 and m_2. Instead we shall use the implicit specification from section 5, thus getting a more general result.

We assume that the protocol starts in a state where the sender is in the state acknowledged$_0$, the receiver is in the state replying$_0$ and the two media contain nothing but 0:s. We shall prove that this start state behaves as an empty pulse buffer. So we shall show that the agent Sys(acknowledged$_0$,m_1,replying$_0$,m_2) satisfies the specification C_0 in section 4 provided m_1 and m_2 are taken from the set M_0 of unreliable media that hold a sequence of 0:s. Using refinement operators we can write:

Theorem 7.1: Sys(acknowledged$_0$, M_0, replying$_0$, M_0) $\subseteq C_0$. $\quad\square$

Before we prove the theorem we shall explain how the protocol works once again, using the new notation and some pictures. After this we state some lemmas whose meaning and relevance should then be obvious.

Let's represent an unsafe medium that holds a (possibly empty) sequence of 0:s by the picture

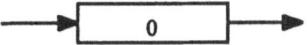

This represents a typical member of M_0. A typical member of M_{10} is represented by

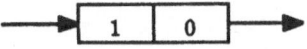

or by

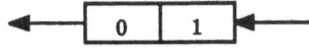

Note that both the sequence of 1:s and the sequence of 0:s can be empty. (One must always be very careful when using pictures; it is easy to read out information that is not there.)

For every pulse that is transferred, the protocol goes through four phases and returns to the start phase but with complemented bits. So altogether there are eight phases that are repeated cyclically for every two pulses. Let's assume that the protocol starts in the phase Sys(acknowledged$_0$, M_0, replying$_0$, M_0):

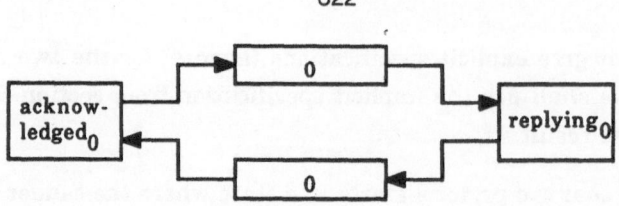

A protocol in this phase is waiting for input. It cannot do anything else than accept a pulse and it cannot leave the phase by internal communication. Formally, this means that the $Sys(acknowledged_0, M_0, replying_0, M_0)$ is an invariant. If the protocol accepts a pulse it moves to the next phase which is $Sys(sending_1, M_{10}, replying_0, M_0)$:

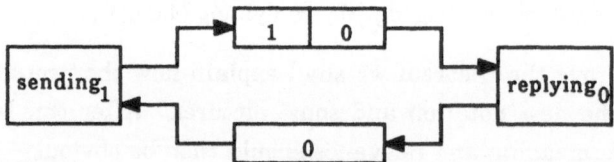

Here the sender starts filling the forward medium with 1:s. The protocol cannot do any external communication during this phase. It can, however, silently move to the next phase which is $Sys(sending_1, M_1, received_1, M_0)$:

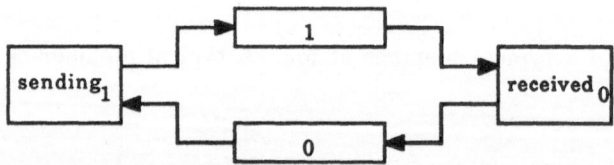

In this phase, the protocol waits until it can deliver the message. The protocol cannot leave this phase in any other way. After delivery the protocol enters the phase $Sys(sending_1, M_1, replying_1, M_{10})$:

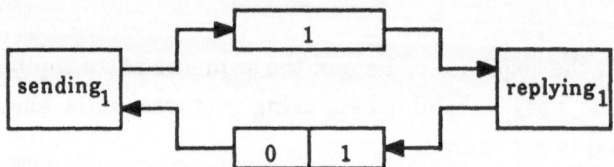

Here the protocol starts sending acknowledgements. It is unable to do anything but silently move to the phase $Sys(acknowledged_1, M_1, replying_1, M_1)$ which is the complement of the start phase.

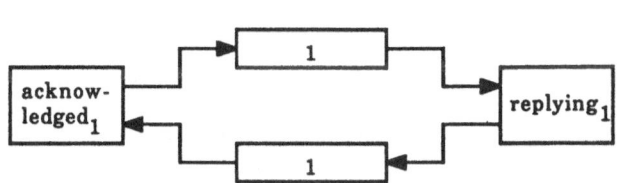

The protocol is now prepared to accept the next pulse and then the pattern is repeated but with all bits complemented.

When understanding this protocol it is important to note that it is not the individual packages (bits) that carry the pulse through the medium but rather a *shift* from 0:s to 1:s or from 1:s to 0:s. One shift at a time can safely be passed through an unsafe medium and the protocol ensures that there is at most one shift on its way at any time.

We now state two lemmas about unsafe media that we will need later.

Lemma 7.2.1:

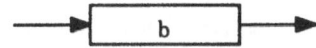

(a) $M_b \subseteq [\tau]M_b$
(b) $M_b \subseteq [accept_b]M_b$
(c) $M_b \subseteq [deliver_b]M_b$
(d) $M_b \subseteq [deliver_{\neg b}]\emptyset$
(e) $M_b \subseteq M_{\neg bb}$
(f) $M_b \subseteq [\lambda]\emptyset$ $\qquad$ $(\forall \lambda \notin \{accept_b, deliver_b \mid b \in \{0,1\}\})$

Proof: Follows easily from the fact that **M** is the a solution to the equations (M'.1-4) and proposition 5.2 and the techniques in section 4. $\qquad\square$

Lemma 7.2.2:

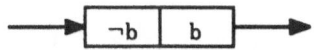

(a) $M_{\neg bb} \subseteq [\tau]M_{\neg bb}$
(b) $M_{\neg bb} \subseteq [accept_{\neg b}]M_{\neg bb}$
(c) $M_{\neg bb} \subseteq [deliver_b]M_{\neg bb}$
(d) $M_{\neg bb} \subseteq [deliver_{\neg b}]M_{\neg b}$
(e) $M_{\neg bb} \subseteq \langle\langle (accept_{\neg b} + deliver_b)^* \rangle\rangle \langle\langle deliver_{\neg b} \rangle\rangle M_{\neg b}$
(f) $M_{\neg bb} \subseteq [\lambda]\emptyset$ $\qquad$ $(\forall \lambda \notin \{accept_b, deliver_b \mid b \in \{0,1\}\})$

Proof: Most of the assertions follow easily from the specification of **M** using the techniques in section 4 to simplify specifications. To show (e), first prove

$$M_{\neg bb} \subseteq [[(accept_{\neg b} + deliver_b)^* \; deliver_{\neg b}]] \; M_{\neg b}$$

and then use (M.5) and the $\langle\langle A\rangle\rangle\cap$-law (Proposition 4.6). ☐

Using lemma 7.2.1 and 7.2.2 we can prove four lemmas that state important properties of the eight phases, informally described above. The proofs use the laws from the previous section. The proof of lemma 7.3.2(a,d) is given in an appendix. The other proofs are similar.

Lemma 7.3.1: $Sys(acknowledged_b, M_b, replying_b, M_b)$

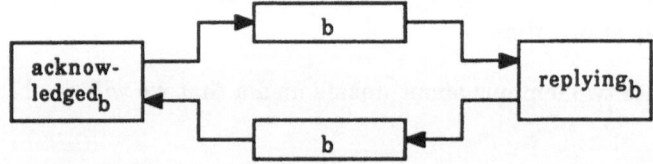

(a) $Sys(acknowledged_b, M_b, replying_b, M_b)$ is an invariant.
(b) $Sys(acknowledged_b, M_b, replying_b, M_b)$
        $\subseteq [accept] \; Sys(sending_{\neg b}, M_{\neg bb}, replying_b, M_b)$
(c) $Sys(acknowledged_b, M_b, replying_b, M_b) \subseteq [deliver]\emptyset$
(d) $Sys(acknowledged_b, M_b, replying_b, M_b) \subseteq \langle accept\rangle 1$
(e) $Sys(acknowledged_b, M_b, replying_b, M_b) \subseteq [\lambda]\emptyset$ $(\forall \lambda \notin \{accept, deliver\})$ ☐

Lemma 7.3.2: $Sys(sending_{\neg b}, M_{\neg bb}, replying_b, M_b)$

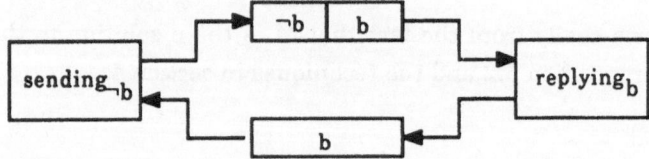

(a) $Sys(sending_{\neg b}, M_{\neg bb}, replying_b, M_b)$
        $\subseteq [\tau] \; (Sys(sending_{\neg b}, M_{\neg bb}, replying_b, M_b)$
                $\cup Sys(sending_{\neg b}, M_{\neg b}, received_{\neg b}, M_b))$
(b) $Sys(sending_{\neg b}, M_{\neg bb}, replying_b, M_b) \subseteq [accept] \; \emptyset$
(c) $Sys(sending_{\neg b}, M_{\neg bb}, replying_b, M_b) \subseteq [deliver] \; \emptyset$
(d) $Sys(sending_{\neg b}, M_{\neg bb}, replying_b, M_b)$
        $\subseteq \langle\langle\rangle\rangle \; Sys(sending_{\neg b}, M_{\neg b}, received_{\neg b}, M_b)$
(e) $Sys(sending_{\neg b}, M_{\neg bb}, replying_b, M_b) \subseteq [\lambda]\emptyset$ $(\forall \lambda \notin \{accept, deliver\})$ ☐

Lemma 7.3.3: Sys(sending$_{\neg b}$,M$_{\neg b}$,received$_{\neg b}$,M$_b$)

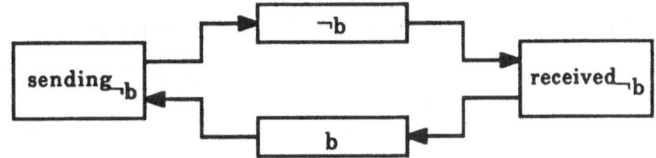

(a)   Sys(sending$_{\neg b}$,M$_{\neg b}$,received$_{\neg b}$,M$_b$) is an invariant

(b)   Sys(sending$_{\neg b}$,M$_{\neg b}$,received$_{\neg b}$,M$_b$)
       $\subseteq$ [deliver] Sys(sending$_{\neg b}$,M$_{\neg b}$,replying$_{\neg b}$,M$_{\neg bb}$)

(c)   Sys(sending$_{\neg b}$,M$_{\neg b}$,received$_{\neg b}$,M$_b$) $\subseteq$ [accept] $\varnothing$

(d)   Sys(sending$_{\neg b}$,M$_{\neg b}$,received$_{\neg b}$,M$_b$) $\subseteq$ $\langle$deliver$\rangle$1

(e)   Sys(sending$_{\neg b}$,M$_{\neg b}$,received$_{\neg b}$,M$_b$) $\subseteq$ [λ]$\varnothing$ ($\forall \lambda \notin$ {accept,deliver}) $\square$

Lemma 7.3.4: Sys(sending$_{\neg b}$,M$_{\neg b}$,replying$_{\neg b}$,M$_{\neg bb}$)

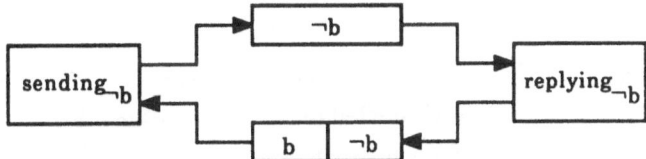

(a)   Sys(sending$_{\neg b}$,M$_{\neg b}$,replying$_{\neg b}$,M$_{\neg bb}$)
       $\subseteq$ [τ] (Sys(sending$_{\neg b}$,M$_{\neg b}$,replying$_{\neg b}$,M$_{\neg bb}$)
                    $\cup$ Sys(acknowledged$_{\neg b}$,M$_{\neg b}$,replying$_{\neg b}$,M$_{\neg b}$))

(b)   Sys(sending$_{\neg b}$,M$_{\neg b}$,replying$_{\neg b}$,M$_{\neg bb}$) $\subseteq$ [accept] $\varnothing$

(c)   Sys(sending$_{\neg b}$,M$_{\neg b}$,replying$_{\neg b}$,M$_{\neg bb}$) $\subseteq$ [deliver] $\varnothing$

(d)   Sys(sending$_{\neg b}$,M$_{\neg b}$,replying$_{\neg b}$,M$_{\neg bb}$)
       $\subseteq$ $\langle\langle\rangle\rangle$ Sys(acknowledged$_{\neg b}$,M$_{\neg b}$,replying$_{\neg b}$,M$_{\neg b}$)

(e)   Sys(sending$_{\neg b}$,M$_{\neg b}$,replying$_{\neg b}$,M$_{\neg bb}$) $\subseteq$ [λ]$\varnothing$ ($\forall \lambda \notin$ {accept,deliver}) $\square$

Let P$_n$ be the set of alternating-bit protocols where n pulses are on their way.
Formally:

$$P_0 = \bigcup_b(\text{Sys(sending}_b\text{,M}_b\text{,replying}_b\text{,M}_{b\neg b})$$
$$\cup \text{Sys(acknowledged}_b\text{,M}_b\text{,replying}_b\text{,M}_b))$$

$$P_1 = \bigcup_b(\text{Sys(sending}_{\neg b}\text{,M}_{\neg bb}\text{,replying}_b\text{,M}_b)$$
$$\cup\text{Sys(sending}_{\neg b}\text{,M}_{\neg b}\text{,received}_{\neg b}\text{,M}_b))$$

$$P_{n+2} = \varnothing$$

Lemma 7.4: $P_n \subseteq C_n$ $(n \in N)$.

Proof: According to Park's rule it is enough to show that P is a solution to the equation defining C' which is equal to C, i.e. we shall show

(1)	P_n	$\subseteq [\tau] P_n$	$(n \in N)$
(2)	P_0	$\subseteq \langle \langle \rangle \rangle \langle \text{accept} \rangle 1$	
(3)	P_0	$\subseteq [\text{deliver}] \varnothing$	
(4)	P_n	$\subseteq [\text{accept}] P_{n+1}$	$(n \in N)$
(5)	P_{n+1}	$\subseteq \langle \langle \rangle \rangle \langle \text{deliver} \rangle 1$	$(n \in N)$
(6)	P_{n+1}	$\subseteq [\text{deliver}] P_n$	$(n \in N)$
(7)	P_n	$\subseteq [\lambda] \varnothing$	$(\forall n \in N, \forall \lambda \notin \{\text{accept}, \text{deliver}\})$

We shall only prove (1) and (2). The other inequations are similar or trivial.

(1):
$P_0 = \bigcup_b (\ \text{Sys}(\text{sending}_b, M_b, \text{replying}_b, M_{b \neg b})$
$\qquad\qquad \cup \text{Sys}(\text{acknowledged}_b, M_b, \text{replying}_b, M_b))$

$\subseteq \bigcup_b (\ [\tau] (\ \text{Sys}(\text{sending}_b, M_b, \text{replying}_b, M_{b \neg b}) \qquad$ (lemma 7.3.4(a) and
$\qquad\qquad \cup \text{Sys}(\text{acknowledged}_b, M_b, \text{replying}_b, M_b))$
$\qquad \cup [\tau] \ \text{Sys}(\text{acknowledged}_b, M_b, \text{replying}_b, M_b)) \qquad$ lemma 7.3.1(a))

$\subseteq \bigcup_b [\tau] (\ \text{Sys}(\text{sending}_b, M_b, \text{replying}_b, M_{b \neg b})$
$\qquad\qquad \cup \text{Sys}(\text{acknowledged}_b, M_b, \text{replying}_b, M_b))$

$\subseteq [\tau] \bigcup_b (\ \text{Sys}(\text{sending}_b, M_b, \text{replying}_b, M_{b \neg b})$
$\qquad\qquad \cup \text{Sys}(\text{acknowledged}_b, M_b, \text{replying}_b, M_b)))$

$= [\tau] P_0$

$P_1 = \bigcup_b (\ \text{Sys}(\text{sending}_{\neg b}, M_{\neg bb}, \text{replying}_b, M_b)$
$\qquad\qquad \cup \text{Sys}(\text{sending}_{\neg b}, M_{\neg b}, \text{received}_{\neg b}, M_b))$

$\subseteq \bigcup_b (\ [\tau] (\ \text{Sys}(\text{sending}_{\neg b}, M_{\neg bb}, \text{replying}_b, M_b) \qquad$ (lemma 7.3.2(a) and
$\qquad\qquad \cup \text{Sys}(\text{sending}_{\neg b}, M_{\neg b}, \text{received}_{\neg b}, M_b))$
$\qquad \cup [\tau] \text{Sys}(\text{sending}_{\neg b}, M_{\neg b}, \text{received}_{\neg b}, M_b)) \qquad$ lemma 7.3.3(a))

$\subseteq \bigcup_b (\ [\tau] (\ \text{Sys}(\text{sending}_{\neg b}, M_{\neg bb}, \text{replying}_b, M_b)$
$\qquad\qquad \cup \text{Sys}(\text{sending}_{\neg b}, M_{\neg b}, \text{received}_{\neg b}, M_b)))$

$\subseteq [\tau] \bigcup_b (\ \text{Sys}(\text{sending}_{\neg b}, M_{\neg bb}, \text{replying}_b, M_b)$
$\qquad\qquad \cup \text{Sys}(\text{sending}_{\neg b}, M_{\neg b}, \text{received}_{\neg b}, M_b))$

$\subseteq [\tau] P_1$

$P_{n+2} = \varnothing \subseteq [\tau] \varnothing = [\tau] P_{n+2}$ $\qquad\qquad\qquad (n \in N)$

(2): $P_0 = \cup_b ($ $Sys(sending_b, M_b, replying_b, M_{b \to b})$

$\cup Sys(acknowledged_b, M_b, replying_b, M_b))$

$\subseteq \cup_b (\langle\langle\rangle\rangle Sys(acknowledged_b, M_b, replying_b, M_b))$ (lemma 7.3.4(d))

$\cup Sys(acknowledged_b, M_b, replying_b, M_b))$

$\subseteq \cup_b \langle\langle\rangle\rangle Sys(acknowledged_b, M_b, replying_b, M_b)$ (prop 4.4.(i))

$\subseteq \cup_b \langle\langle\rangle\rangle \langle accept \rangle\, 1$ (lemma 7.3.1(d))

$= \langle\langle\rangle\rangle \langle accept \rangle\, 1$ $\square$

Proof of theorem 7.1: Immediate from lemma 7.4:

$Sys(acknowledged_0, M_0, replying_0, M_0) \subseteq P_0$ (def of P_0)

$\subseteq C_0$ (lemma 7.4)

$\square$

8 Conclusions and further work.

We have shown how to implicitly specify agents using Hennessy-Milner logic with recursion and how the correctness of the alternating-bit protocol can be proved using the refinement operators and the laws for them. We hope that the reader finds the proof natural and not too difficult to follow.

Some questions remain:

–   Some "relative" completeness result of the laws for the refinement operators are desired. How do we now that we have not "forgotten" some rule?

–   Many steps in the proof are boring and this indicates that some kind of computer aid could pay off. How should one formalize the refinement calculus so it can be implemented on a computer?

Acknowledgements

The author wants to thank Kent Petersson, K.V.S. Prasad, Wang Yi, Sven Westin and other members of the Programming Methodology Group for valuable comments on earlier versions of this paper.

References

[BR83]   S. Brookes and W. Rounds, Behavioural equivalences induced by programming logics, ICALP'83, LNCS 154, 1983.

[GS84]   S. Graf and J. Sifakis, A modal characterization of observational congruence on finite terms of CCS, ICALP'84, LNCS 172.

[HM85]   M. Hennessy and R. Milner, Algebraic laws for Nondeterminism and Concurrency, JACM 32 (1), (1985).

[Koze82] D. Kozen, Results on the Propositional μ-calculus, ICALP'82, LNCS 140.

[Lars86] K.G. Larsen, *Context-Dependent Bisimulation between processes*, Ph. D. thesis CST 37-86, University of Edinburgh, 1986.

[LM86]   K. G. Larsen and R. Milner, A Complete Protocol Verification using Relativized Bisimulation, R 86-12, Institute of Electronic Systems, Aalborg University Center.

[Lars87] K.G. Larsen, Proof Systems for Hennessy-Milner Logic with Recursion, to appear in CAAP'88.

[Miln80] R. Milner, *A Calculus of Communicating Systems*, LNCS 92.

[Miln83] R. Milner, Calculi for Synchrony and Asynchrony, *Theoretical Computer Science* 25 (1983) 267-310.

[Miln84] R. Milner, The calculus CCS and its evaluation rules, Seminar on Concurrency, CMU, LNCS 197.

[Wins85] G. Winskel, A Complete Proof System for SCCS with Modal Assertions, Cambridge Computer Lab., Techn. Rep. 78, September 1985.

APPENDIX: Proof of lemma 7.3.2(a,d).

Lemma 7.3.2: $Sys(sending_{\neg b}, M_{\neg bb}, replying_b, M_b)$

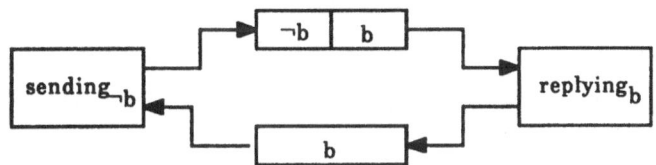

(a) $Sys(sending_{\neg b}, M_{\neg bb}, replying_b, M_b)$
 $\subseteq [\tau] (Sys(sending_{\neg b}, M_{\neg bb}, replying_b, M_b)$
   $\cup Sys(sending_{\neg b}, M_{\neg b}, received_{\neg b}, M_b))$

(d) $Sys(sending_{\neg b}, M_{\neg bb}, replying_b, M_b)$
 $\subseteq \langle\langle\rangle\rangle Sys(sending_{\neg b}, M_{\neg b}, received_{\neg b}, M_b)$

Proof:

(a) Let $N^1_s = M_s [send^c_b, receive^c_b/accept_b, deliver_b \mid b \in \{0,1\}]$ ($s \in \{0,1\}^*$) and
$N^2_s = M_s [reply^c_b, ack^c_b/accept_b, deliver_b \mid b \in \{0,1\}]$. We have

 $Sys(sending_{\neg b}, M_{\neg bb}, replying_b, M_b)$
 $=(\quad sending_{\neg b} \cap [\tau]\emptyset \cap [send_{\neg b}]sending_{\neg b} \cap [ack_b]sending_{\neg b}$

   $\cap [ack_{\neg b}]acknowledged_{\neg b} \cap \bigcap_{\lambda \notin \{send \neg b, ackb, ack \neg b\}} [\lambda]\emptyset$

 $\mid\quad N^1_{\neg bb} \cap [\tau] N^1_{\neg bb} \cap [send^c_{\neg b}] N^1_{\neg bb} \cap [receive^c_b] N^1_{\neg bb}$

   $\cap [receive^c_{\neg b}] N^1_{\neg b} \cap \bigcap_{\lambda \notin \{send^c b', deliver^c b' \mid b' \in \{0,1\}\}} [\lambda]\emptyset$

 $\mid\quad replying_b \cap [\tau]\emptyset \cap [receive_{\neg b}]received_{\neg b} \cap [receive_b]replying_b$

   $\cap [reply_b]replying_b \cap \bigcap_{\lambda \notin \{receive \neg b, receiveb, replyb\}} [\lambda]\emptyset$

 $\mid\quad N^2_b \cap [\tau] N^2_b \cap [reply^c_b] N^2_b \cap [ack^c_b] N^2_b$

   $\cap \bigcap_{\lambda \notin \{reply^c b, reply^c \neg b, ackb\}} [\lambda]\emptyset$
 $) \setminus \{send_b, receive_b, reply_b, ack_b \mid b \in \{0,1\}\}$ (by lemma 7.3.1)

 $\subseteq [\tau]((sending_{\neg b} \mid N^1_{\neg bb} \mid replying_b \mid N^2_b)$
    $\setminus \{send_b, receive_b, reply_b, ack_b \mid b \in \{0,1\}\}$
   $\cup (sending_{\neg b} \mid N^1_{\neg b} \mid received_{\neg b} \mid N^2_b)$
    $\setminus \{send_b, receive_b, reply_b, ack_b \mid b \in \{0,1\})$ (by the law *par[τ]*)

(d) From lemma 7.3.2(e) we have $M_{\neg bb} \subseteq \langle\!\langle(\text{accept}_{\neg b}+\text{deliver}_b)^*\text{deliver}_{\neg b}\rangle\!\rangle M_{\neg b} = \bigcup_{s \in A} \langle\!\langle s \rangle\!\rangle \langle\!\langle \text{deliver}_{\neg b} \rangle\!\rangle M_{\neg b}$ where $A = (\text{accept}_{\neg b}+\text{deliver}_b)^*$. Hence

$\quad$ Sys(sending$_{\neg b}$,$M_{\neg bb}$,replying$_b$,M_b)

$\quad \subseteq$ Sys(sending$_{\neg b}$,$\bigcup_{s \in A} \langle\!\langle s \rangle\!\rangle\langle\!\langle \text{deliver}_{\neg b} \rangle\!\rangle M_{\neg b}$,replying$_b$,$M_b$)

$\quad = \bigcup_{s \in A}$Sys(sending$_{\neg b}$,$\langle\!\langle s \rangle\!\rangle\langle\!\langle \text{deliver}_{\neg b} \rangle\!\rangle M_{\neg b}$,replying$_b$,$M_b$)

$\hfill$ (by monotonicity)

It remains to show $\bigcup_{s \in A}$Sys(sending$_{\neg b}$,$\langle\!\langle s \rangle\!\rangle\langle\!\langle \text{deliver}_{\neg b} \rangle\!\rangle M_{\neg b}$,replying$_b$,$M_b$) $\subseteq$ $\langle\!\langle\rangle\!\rangle$Sys(sending$_{\neg b}$,$M_{\neg b}$,received$_{\neg b}$,$M_b$). So we prove

$\quad$ Sys(sending$_{\neg b}$,$\langle\!\langle s \rangle\!\rangle\langle\!\langle \text{deliver}_{\neg b} \rangle\!\rangle M_{\neg b}$,replying$_b$,$M_b$)

$\quad \subseteq \langle\!\langle\rangle\!\rangle$Sys(sending$_{\neg b}$,$M_{\neg b}$,received$_{\neg b}$,$M_b$)

by induction on s.

Base $s = \varepsilon$:

$\quad$ Sys(sending$_{\neg b}$,$\langle\!\langle \text{deliver}_{\neg b} \rangle\!\rangle M_{\neg b}$,replying$_b$,$M_b$)

$\quad \subseteq$ Sys(sending$_{\neg b}$,$\langle\!\langle \text{deliver}_{\neg b} \rangle\!\rangle M_{\neg b}$,$\langle\!\langle \text{receive}_{\neg b} \rangle\!\rangle$ received$_{\neg b}$,M_b)

$\quad \subseteq \langle\!\langle\rangle\!\rangle$Sys(sending$_{\neg b}$,$M_{\neg b}$,received$_{\neg b}$,$M_b$) $\hfill$ (prop 6.5)

Step $s = \text{accept}_{\neg b}s'$:

$\quad$ Sys(sending$_{\neg b}$,$\langle\!\langle \text{accept}_{\neg b}\ s' \rangle\!\rangle\langle\!\langle \text{deliver}_{\neg b} \rangle\!\rangle M_{\neg b}$,replying$_b$,$M_b$)

$\quad \subseteq$ Sys($\langle\!\langle \text{send}_{\neg b} \rangle\!\rangle$sending$_{\neg b}$,$\langle\!\langle \text{accept}_{\neg b}s' \rangle\!\rangle\langle\!\langle \text{deliver}_{\neg b} \rangle\!\rangle M_{\neg b}$,replying$_b$,$M_b$)

$\quad \subseteq \langle\!\langle\rangle\!\rangle$ Sys(sending$_{\neg b}$,$\langle\!\langle s' \rangle\!\rangle\langle\!\langle \text{deliver}_{\neg b} \rangle\!\rangle M_{\neg b}$,replying$_b$,$M_b$) $\hfill$ (prop 6.5)

$\quad \subseteq \langle\!\langle\rangle\!\rangle$ Sys(sending$_{\neg b}$,$M_{\neg b}$,received$_{\neg b}$,M_b) $\hfill$ (by ind. hyp.)

Step $s = \text{deliver}_b s'$:

$\quad$ Sys(sending$_{\neg b}$,$\langle\!\langle \text{deliver}_b s' \rangle\!\rangle\langle\!\langle \text{deliver}_{\neg b} \rangle\!\rangle M_{\neg b}$,replying$_b$,$M_b$)

$\quad \subseteq$ Sys(sending$_{\neg b}$,$\langle\!\langle \text{deliver}_b s' \rangle\!\rangle\langle\!\langle \text{deliver}_{\neg b} \rangle\!\rangle M_{\neg b}$,$\langle\!\langle \text{receive}_b \rangle\!\rangle$replying$_b$,$M_b$)

$\quad \subseteq \langle\!\langle\rangle\!\rangle$ Sys(sending$_{\neg b}$,$\langle\!\langle s' \rangle\!\rangle\langle\!\langle \text{deliver}_{\neg b} \rangle\!\rangle M_{\neg b}$,replying$_b$,$M_b$) $\hfill$ (prop 6.5)

$\quad \subseteq \langle\!\langle\rangle\!\rangle$ Sys(sending$_{\neg b}$,$M_{\neg b}$,received$_{\neg b}$,M_b) $\hfill$ (by ind. hyp.)

$\hfill \square$

A Functional Programming Approach
to the Specification and Verification
of Concurrent Systems

Peter Dybjer *   Herbert Sander *

Abstract

We propose a collection of principles for reasoning about lazy functional programs. These include rules for predicate logic, rules of conversion of programs, and rules for predicates defined in terms of least and greatest fixed points. These rules can be used for reasoning about networks of deterministic computing agents in view of Kahn's principle. We also propose that certain networks which include non-deterministic agents can be treated in a similar way by viewing such agents as incompletely specified deterministic agents. To illustrate this technique we discuss the specification and correctness of the alternating bit protocol.

1   Introduction

The functional programming approach to concurrency can be traced back to Landin's remark [19]:

> It appears that in stream transformers we have a functional analogue of what Conway calls coroutines.

Kahn [15] and Kahn and MacQueen [16] demonstrated the power of this idea. They showed how networks of *deterministic* computing agents can be described as systems of mutually recursive stream transformers and thus be given denotational semantics. As a consequence, proof techniques from domain theory [13] can be applied, see also [1].

Extending these ideas to networks with *non-deterministic* agents has been considered problematic. A naive approach may give rise to anomalies, such as Keller's [18] and Brock and Ackerman's [4]. Even though such anomalies can be overcome, it is at a certain cost of abstractness, see Park [23], Kearney and Staples [17], and Broy [6].

In this paper we reconsider the functional programming (or stream) approach to concurrency in the light of recent developments in the field of programming logic. Firstly, the logic of finitely terminating functional programs has been studied extensively. Several formal systems have been proposed, for example, Manna and Waldinger's deductive

---
*Authors' address: Programming Methodology Group, Department of Computer Sciences, Chalmers University of Technology and University of Göteborg, S-412 96 Göteborg, Sweden. Electronic mail: peterd@cs.chalmers.se, herbert@cs.chalmers.se.

synthesis [21], Martin-Löf's type theory [22], Coquand and Huet's calculus of constructions [7], and Aczel's logical theory of construction [2]. Secondly, it has been argued that greatest fixed points give natural rules for reasoning about infinitely proceeding programs such as streams, see de Roever [8]. Such rules have been formalised in the μ-calculus of Scott and de Bakker, see Park [24]. Greatest fixed points have also been used in the context of concurrency by Park and others for formalising concepts such as observational equivalence and fairness.

Our favorite framework for reasoning about functional programs is based on Aczel's logical theory of constructions. One may consider an untyped [10] or a typed [9] version of such a theory. Its essential components are:

- natural deduction rules for predicate logic;

- rules of conversion for lazy functional programs;

- rules for inductively defined predicates.

Here we shall outline a kind of μ-calculus extension of this framework, where we have access to predicates defined by least as well as greatest fixed points. This logic gives a formal context for reasoning about lazy functional programs and thus for reasoning about Kahn-networks.

One can also reason about networks which include non-deterministic agents in this way. The technique is basically Park's [23]: non-deterministic agents are represented as incompletely specified deterministic ones. So non-deterministic agents are not represented on the program level, but on the specification level. Having access to a powerful specification language (the predicates representable in the logic) may make this approach feasible. Observe that we end up with a logic for concurrent systems which has no features particular to concurrency. There are no primitives for the parallel composition of networks for example. (Such constructs could perhaps be derived, however.)

We are not in a position to make any claims about the generality and feasibility of this approach - our experiments have just begun. At present we think of it as yet another application of functional programming and logics of functional programs. It should be worth pursuing however, not only for logical reasons, but also for practical reasons: lazy functional languages are now being implemented efficiently [14,3] and implementations of their logics are improving in quality.

The rest of this paper is organised as follows:

- In section 2 we discuss the notions of specification and correctness of functional programs and concurrent systems.

- In section 3 we describe some proof rules and introduce auxiliary notation.

- In section 4 we specify the task of a communication protocol.

- In section 5 we program the protocol in a lazy functional language using the technique with an alternating bit. We also show the correctness of this implementation with respect to the specification.

- In section 6 we conclude and compare our approach to related approaches.

2 What is a specification of a concurrent system?

Let us first recall what we mean by a specification of a single program. Girard [11] distinguished between two approaches: *integrated logic* and *external logic*.

In integrated logic, which we shall not be concerned with here, a specification is a proposition (formula) and a program satisfying the specification is a constructive proof of the proposition. Examples are Martin-Löf's type theory and the calculus of constructions for functional programming.

In external logic a specification is a property of a program. Usually, such a specification is divided into an input condition P and an input-output relation R. The specification then has the form

$$Spec\ f \equiv \forall x.Px \supset x\,R(f\,x).$$

(If we work in a typed framework, we first specify a type $A \to B$ of f - A is the type of inputs and B is the type of outputs. The specification is then a property of programs of type $A \to B$: $Spec\ f \equiv \forall x : A.Px \supset x\,R(f\,x)$).

Can this notion of specification be generalised to concurrent systems in a natural way? The idea is to model a system as a Kahn-network or, if there are non-deterministic agents, as an incompletely specified Kahn-network. Assume that a system satisfies the following criteria:

- The topology of the system is fixed. Input and output channels are determined. Streams of messages are communicated between the agents.

- Agents are divided into those which we wish to program and those which are predetermined. The latter kind may be non-deterministic.

- The purpose of the system is to realise a certain relation R between inputs and outputs, provided that the inputs satisfy certain conditions P, and the predetermined agents satisfy certain other conditions Q.

For a given topology one can associate functions $\overline{trans}$ by Kahn's principle. (There is one function for each output channel - the overbar suggests that there may be more than one.) These are higher-order functions and Kahn-network analogues of function application. If $\overline{h}$ are the stream transformers associated with the agents (one for each output of an agent), $\overline{is}$ the input streams, and $\overline{js}$ the output streams, then we have

$$\overline{js} = \overline{trans}\ \overline{h}\ \overline{is}.$$

By the second assumption, agents (and thus stream transformers) are divided into those which we wish to program and those which are predetermined. If we use $\overline{f}$ for the former and $\overline{g}$ for the latter, we can write

$$\overline{js} = \overline{trans}\ \overline{f}\ \overline{g}\ \overline{is}.$$

By our third assumption, the intended behaviour of the network is specified by the input conditions P, by the conditions on the predetermined agents Q, and by the input-output relation R. Thus the specification of our task, that is, to program $\bar{f}$, is

$$Spec\ \bar{f} \equiv \forall \bar{g}.Q\bar{g} \supset \forall \overline{is}.P\overline{is} \supset \overline{is}\ R(\overline{trans\ \bar{f}\ \bar{g}\ is}).$$

(We could of course make the corresponding analysis in a typed framework. Then we would have to associate types to stream transformers, etc.)

We have thus shown the analogy between the specification of a concurrent system, which satisfies the criteria above, and the specification of a single program.

The specification of a communication protocol satisfies these criteria:

- It has a fixed topology:

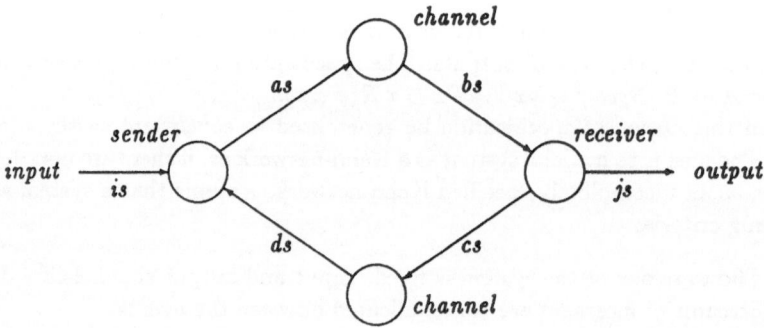

- Two of the four agents (the channels) are predetermined and our task is to program the other two (the sender and the receiver).

- The purpose of the system is to produce output which is extensionally equal (or observationally equivalent) to the input, provided the input is a stream of data items. In our proof below we assume that the input stream is infinite. We know that the channels are unreliable transmitters, that is, that each item is either correctly or erroneously transmitted. We also assume that they are fair in the sense that for each 'time' there is a later 'time' when they transmit an item correctly.

Let *trans* be the network transfer function, such that

$$js = trans\ f_1\ f_2\ f_3\ g_1\ g_2\ is\ ,$$

where *is* is the input stream, *js* the output stream, f_1 the function associated with the sender, f_2 and f_3 the functions associated with the receiver, and g_1 and g_2 the functions associated with the the channels. We can determine *js* from *is* by using the following system of equations:

$$as\ =\ f_1\ is\ ds\ ,$$

$$
\begin{aligned}
bs &= g_1 \, as \,, \\
cs &= f_2 \, bs \,, \\
ds &= g_2 \, cs \,, \\
js &= f_3 \, bs \,.
\end{aligned}
$$

If we use $\approx$ for the relation of extensional equality, *Inf* for the property of being an infinite stream of data, and *Futc* for the property of being a fair unreliable transmission channel, then the specification of our programming task is

$$
\textit{Protocol } f_1 \, f_2 \, f_3 \; \equiv \; \forall g_1, g_2. \textit{Futc } g_1 \wedge \textit{Futc } g_2 \supset \forall is. \textit{Inf } is \supset is \approx \textit{trans } f_1 \, f_2 \, f_3 \, g_1 \, g_2 \, is \,.
$$

We have thus reduced the protocol problem to a pure functional programming problem.

3 Some formal rules for reasoning about functional programs

We shall now outline the formal system we use for reasoning about lazy functional programs. We are not going to give a complete description here, since this is work in progress. The reader is referred to [10] and [9] for detailed descriptions of two closely related formal systems. The present system can be characterised as an extension of these two systems where we have access to predicates defined in terms of greatest fixed points. We achieve this by assuming classical logic (the previous systems are both constructive) and a general least fixed point operator μ. The greatest fixed point operator ν can then be defined in terms of μ.

We shall present the three components of the system: predicate logic, conversion rules, and rules for fixed points.

Predicate logic. We assume a natural deduction formulation of a classical predicate logic with equality. We choose an untyped logic as in [10]. This is for notational convenience only (see [9] for a typed version). The individuals of the logic are untyped functional programs, that is, terms of an untyped λ-calculus with constants.

It is convenient to introduce auxiliary notation, such as bounded quantifiers

$$
\begin{aligned}
\forall x \in P.Q &\equiv \forall x.P \, x \supset Q, \\
\exists x \in P.Q &\equiv \exists x.P \, x \wedge Q,
\end{aligned}
$$

'set notation' for unary predicates

$$
\begin{aligned}
a \in P &\equiv P \, a \,, \\
\{x \mid P\} &\equiv x.P \,, \\
\{a_1, \ldots, a_n\} &\equiv x.x = a_1 \vee \cdots \vee x = a_n \,, \\
P \cup Q &\equiv \{x \mid x \in P \vee x \in Q\} \,,
\end{aligned}
$$

etc.,

and infix notation for binary predicates such as $=$.

The constants of the λ-calculus are divided into three kinds: constructors, selectors (case analysis functions), and recursive functions. Which particular constants to choose depends on the problem at hand.

When we discuss the alternating bit protocol we need for example the constructors 0 and s (for natural numbers), $\langle -, - \rangle$ (for pairs), *nil* and $::$ (for lists and streams), O and L (for bits), and *error* and *ok* (for corrupted messages). We also need the selectors *natcases*, *split*, *listcases*, *streamcases*, *if*, and *errcases*, and a number of constants for recursively defined functions.

Conversion rules. The equality of the system is convertibility, so the basic rule is β-conversion, but there are also conversion rules relating the constructors and selectors:

$$
\begin{aligned}
natcases\ 0\ d\ e &= d, \\
natcases\ (s\ a)\ d\ e &= e\ a, \\
listcases\ nil\ d\ e &= d, \\
listcases\ (a::as)\ d\ e &= e\ a\ as, \\
streamcases\ (a::as)\ d &= d\ a\ as, \\
split\ \langle a, b \rangle\ d &= d\ a\ b, \\
if\ O\ d\ e &= d, \\
if\ L\ d\ e &= e, \\
errcases\ error\ d\ e &= d, \\
errcases\ (ok\ a)\ d\ e &= e\ a.
\end{aligned}
$$

Moreover, for every recursive function constant c, there is a rule of conversion of the form

$$c = e,$$

where e is a closed λ-term possibly containing c and other recursive function constants. Mutual recursion is possible.

For example, we can introduce a new constant *append* with the conversion rule

$$append = as.bs.listcases\ as\ bs\ (a.as'.a::append\ as'\ bs).$$

(We follow the type theoretic tradition and distinguish between abstractions $x.a$ and λ-abstractions $\lambda x.a$. In this way λ is considered as a constructor among others.) From this rule we can derive the recursion equations

$$
\begin{aligned}
append\ nil\ bs &= bs, \\
append\ (a::as')\ bs &= a::append\ as'\ bs,
\end{aligned}
$$

by using substitution and conversion rules. It is often convenient to write the recursion equations directly, and view them as a definition of the constant in question, provided they have such a form that they can be derived from a proper definition. It is essential that we have this restriction to 'complete pattern matching' in order to get a smooth logic.

Rules for fixed points. Let T be a monotonic function which maps n-ary predicates to n-ary predicates. Then a new n-ary predicate $\mu_n T \equiv \mu_n X.T\,X$ (or μT for short), the least fixed point of T, can be introduced. It satisfies a rule

$$\frac{T(\mu T)\,\bar{a}}{\mu T\,\bar{a}}$$

which expresses that μT is a prefixed point of T, and a rule

$$\frac{\mu T\,\bar{a} \qquad \begin{array}{c}(T\,P\,\bar{x})\\ P\,\bar{x}\end{array}}{P\,\bar{a}}$$

which expresses that μT is the least prefixed point of T. The former rule gives rise to introduction rules and the latter to elimination or induction rules.

We also introduce a new n-ary predicate $\nu_n T$ (or νT) for the greatest fixed point of T. It satisfies a rule

$$\frac{\nu T\,\bar{a}}{T(\nu T)\,\bar{a}}$$

which expresses that νT is a postfixed point of T, and a rule

$$\frac{P\,\bar{a} \qquad \begin{array}{c}(P\,\bar{x})\\ T\,P\,\bar{x}\end{array}}{\nu T\,\bar{a}}$$

which expresses that νT is the greatest postfixed point of T. The latter rule has been called 'greatest fixed point induction' or 'Park's rule'.

In the next section we show examples of predicates defined in terms of μ and ν.

4 The specification of a protocol

In section 2 we arrived at the following form for the specification of a communication protocol:

$$Protocol\ f_1\ f_2\ f_3 \equiv \forall g_1, g_2 \in Futc.\forall is \in Inf.is \approx trans\,f_1\,f_2\,f_3\,g_1\,g_2\ is.$$

The predicates Inf, $\approx$, and $Futc$ can be defined as greatest fixed points in the following way.

$Inf\ as$ is true provided as is an infinite stream of data. Let D be a unary predicate such that $D\,a$ is true provided a is a data item. Then we define

$$Inf \equiv \nu X.D::X,$$

where $::$ is used as a map on unary predicates:

$$(P::Q)\ as \equiv \exists a, as'.P\,a \wedge Q\,as \wedge as = a::as'.$$

$as \approx bs$ is true provided as and bs are extensionally equal infinite streams. We define

$$\approx \; \equiv \; \nu X. =::X,$$

where $=$ is the convertibility relation and $::$ is used as a map on binary relations:

$$as\,(R::S)\,bs \; \equiv \; \exists a, as', b, bs'.a\,R\,b \wedge as'\,S\,bs' \wedge as = a::as' \wedge bs = b::bs'.$$

Futc g is true provided g is a fair unreliable transmission channel. We express this formally by introducing an auxiliary function *corrupt* of two arguments: an oracle stream (of bits) and a stream of messages. The bits in the oracle stream determine whether a message is corrupted or not. We have the following recursion equations:

$$
\begin{aligned}
corrupt\,(O::os)\,(x::xs) &= error::corrupt\,os\,xs, \\
corrupt\,(L::os)\,(x::xs) &= (ok\,x)::corrupt\,os\,xs.
\end{aligned}
$$

These channels are non-deterministic in the sense that we do not know in advance when messages are corrupted. By using oracles we isolate this non-determinism, and the fairness of a channel is reduced to the fairness (or more properly, the *L*-fairness) of an oracle:

$$Futc\,g \; \equiv \; \exists os \in Fair.g = corrupt\,os,$$

where

$$Fair \; \equiv \; \nu X.append\,O^{*}L\,X.$$

A $::$ before *append* is used as a map on unary predicates here. $O^{*}L$ is a unary predicate such that $O^{*}L\,al$ is true provided al is a list of zero or more O's followed by a final L. It can be defined either as a least fixed point

$$O^{*}L \; \equiv \; \mu X.\{L::nil\} \cup O::X,$$

or in terms of a function $O^{-}L$ such that $O^{n}L$ is a list of n O's followed by a final L:

$$O^{*}L\,al \; \equiv \; \exists n \in Nat.al = O^{n}L,$$

where

$$
\begin{aligned}
O^{0}L &= L::nil, \\
O^{sn}L &= O::O^{n}L.
\end{aligned}
$$

5 A proof of correctness of the alternating bit protocol

We shall now program the alternating bit protocol in our lazy functional language by writing down the recursion equations for some new constants. Our programs differ only marginally from the corresponding programs written in Miranda, in appendix A. The sender is programmed with the two mutually recursive functions *abpsend* and *await*.

The receiver is programmed with the two functions *abpack*, which sends acknowledgement bits, and *abpout*, which produces output. The functions satisfy the following recursion equations:

$$abpsend\ b\,(i::is)\ ds\ =\ \langle i,b\rangle :: await\ b\,(i::is)\ ds,$$

$$await\ b\,(i::is)\,((ok\ b_0)::ds)\ =\ if\,(eq\ b\ b_0)\,(abpsend\,(not\ b)\ is\ ds)\,(\langle i,b\rangle :: await\ b\,(i::is)\ ds),$$
$$await\ b\,(i::is)\,(error::ds)\ =\ \langle i,b\rangle :: await\ b\,(i::is)\ ds,$$

$$abpack\ b\,((ok\ \langle i,b_0\rangle)::bs)\ =\ if\,(eq\ b\ b_0)\,(b:: abpack\,(not\ b)\ bs)\,((not\ b):: abpack\ b\ bs),$$
$$abpack\ b\,(error::bs)\ =\ (not\ b):: abpack\ b\ bs,$$

$$abpout\ b\,((ok\langle i,b_0\rangle)::bs)\ =\ if\,(eq\ b\ b_0)\,(i:: abpout\,(not\ b)\ bs)\,(abpout\ b\ bs),$$
$$abpout\ b\,(error::bs)\ =\ abpout\ b\ bs.$$

Formally, the alternating bit protocol gives two implementations - one for each possible start bit. We shall prove the correctness of both implementations simultaneously, that is, that

$$Protocol\,(abpsend\ b)\,(abpack\ b)\,(abpout\ b)$$

for $b = O, L$. If we unfold the definition of *Protocol* and introduce the auxiliary program

$$abptrans\ b_0\ b_1\ os_0\ os_1\ is\ =\ trans\,(abpsend\ b_0)\,(abpack\ b_1)$$
$$(abpout\ b_1)\,(corrupt\ os_0)\,(corrupt\ os_1)\ is,$$

then we can rewrite the correctness proposition:

$$\forall os_0, os_1 \in Fair.\forall is \in Inf.is \approx abptrans\ b\ b\ os_0\ os_1\ is$$

for $b = O, L$.

Since $\approx$ is defined as a greatest fixed point we can use an instance of Park's rule:

$$\frac{is\ R\,(abptrans\ b\ b\ os_0\ os_1\ is)\qquad \dfrac{(is\ R\ js)}{is\,(=\,::R)\ js}}{is \approx abptrans\ b\ b\ os_0\ os_1\ is}.$$

R is an auxiliary relation:

$$is\ R\ js\ \equiv\ Inf\ is \wedge \exists b \in Bit.\exists os_0, os_1 \in Fair.js = abptrans\ b\ b\ os_0\ os_1\ is.$$

$is\ R\ js$ is true provided *is* is infinite and *js* is a possible output from the protocol (for some start bit *b* and some fair oracles os_0 and os_1) when the input stream is *is*.

In CCS jargon we would say that we prove that *is* and $abptrans\ b\ b\ os_0\ os_1\ is$ are in the largest bisimulation $\approx$ by finding another bisimulation R such that it is easy to see that $is\ R\,(abptrans\ b\ b\ os_0\ os_1\ is)$.

The proof of the major premise of Park's rule is trivial. We proceed to prove the minor premise which states that R is a bisimulation (of infinite streams). So assume that $is\ R\ js$, that is, that

$$Inf\ is$$

and

$$\exists b \in Bit.\exists os_0, os_1 \in Fair.js = abptrans\,b\,b\,os_0\,os_1\,is.$$

Since $Inf \equiv \nu X.D :: X$, it follows by the first rule for ν that

$$\exists i \in D.\exists is' \in Inf.is = i :: is'.$$

Since $Fair \equiv \nu X.append\,O^*L\,X$, it also follows that

$$\exists os_0' \in Fair.os_0 = \underbrace{O :: \cdots :: O}_{m} :: L :: b_{01} :: \cdots :: b_{0n} :: os_0'$$

and

$$\exists os_1' \in Fair.os_1 = b_{11} :: \cdots :: b_{1m} :: \underbrace{O :: \cdots :: O}_{n} :: L :: os_1'.$$

for some natural numbers m and n.

Finally, we deduce by induction the following conversions:

$$abptrans\,b\,b\,(\underbrace{O :: \cdots :: O}_{m} :: L :: b_{01} :: \cdots :: b_{0n} :: os_0')\,(b_{11} :: \cdots :: b_{1m} :: \underbrace{O :: \cdots :: O}_{n} :: L :: os_1')\,(i :: is')$$

$$= \quad abptrans\,b\,b\,(L :: b_{01} :: \cdots :: b_{0n} :: os_0')\,(\underbrace{O :: \cdots :: O}_{n} :: L :: os_1')\,(i :: is).$$

If $n > 0$ we continue with

$$= \quad i :: abptrans\,b\,(not\,b)\,(b_{01} :: \cdots :: b_{0n} :: os_0')(\underbrace{O :: \cdots :: O}_{n-1} :: L :: os_1')\,(i :: is')$$

$$= \quad i :: abptrans\,(not\,b)\,(not\,b)\,os_0'\,os_1'\,is'.$$

If $n = 0$ we arrive at the last line directly.

By putting all this together we conclude that

$$is\,(= :: R)\,js,$$

and the proof is complete.

Obviously, to get a formal proof we have to fill in many details. We intend to present some of these details in an expanded version of this paper.

6 Conclusion and Related Approaches

This paper describes the synthesis of two quite separate suggestions:

- that we should reduce problems in concurrency to pure functional programming problems;

- and that we should use a certain logic for reasoning about functional programs.

Both suggestions follow old traditions.

As regards the functional programming approach to concurrency we referred to the ideas of Kahn and Park, for example. What we have described is a pure version of this approach: no features particular to concurrency are introduced. An alternative approach is 'applicative multiprogramming' which differs from the present approach in that non-deterministic constructs are added to a deterministic language, see for example Broy [5].

In hardware verification a similar pure approach has been proposed by Gordon [12]. His logic HOL is an extension of Church's simple type theory and is a powerful logic with no features particular to hardware. Hardware modelling in HOL is based on using streams in a similar way as here, but an important difference is that these streams are represented as functions.

Our logic is a kind of μ-calculus. Another μ-calculus, which has been proposed recently, is Larsen's 'Hennessy-Milner logic with recursion' [20]. This logic can be used for reasoning about CCS-agents.

Finally, we should say something about LCF, since it is a logic which can be used in a similar way as ours. The basic difference is that LCF reflects the domain theoretic model it is based on. For example, there is a least element $\perp$ of each type. There are two relations, $\sqsubseteq$ and $=$, which mean that denotations are ordered and equal (respectively) in the model. There is also the principle of Scott induction. In our logic there is no primitive order relation or least element. LCF's denotational equality (of infinite streams) corresponds to our $\approx$ rather than to our $=$ which means convertibility. See also de Roever [8] for a discussion of greatest fixed point induction and Scott induction.

References

[1] S. Abramsky. Reasoning about concurrent systems. 1983.

[2] P. Aczel. Frege structures and the notions of proposition, truth and set. In *The Kleene Symposium*, pages 31–59, North-Holland, 1980.

[3] L. Augustsson. *Compiling Lazy Functional Languages, Part II*. PhD thesis, Chalmers University of Technology, 1987.

[4] J. D. Brock and W. B. Ackerman. Scenarios: a model of non-determinate computation. In J. Diaz and I. Ramos, editors, *Formalisation of Programming Concepts*, Springer-Verlag, LNCS 107, 1981.

[5] M. Broy. A fixed point approach to applicative multiprogramming. In M. Broy and E. M. Schmidt, editors, *Theoretical Foundations of Programming Methodology*, pages 565–623, Reidel, Dordrecht, 1982.

[6] M. Broy. Nondeterministic data flow programs: how to avoid the merge anomaly. *Science of Computer Programming*, 10:65–85, 1988.

[7] T. Coquand and G. Huet. Constructions: a higher order proof system for mechanizing mathematics. In B. Buchberger, editor, *EUROCAL '85: European Conference on Computer Algebra, Volume 1: Invited Lectures*, pages 151–184, Springer-Verlag, LNCS 203, 1985.

[8] W. P. de Roever. On backtracking and greatest fixed points. In E. J. Neuhold, editor, *Formal Description of Programming Concepts*, pages 621–639, North-Holland, 1978.

[9] P. Dybjer. Comparing integrated and external logics of functional programs. 1988.

[10] P. Dybjer. Program verification in a logical theory of constructions. In J. Jouannaud, editor, *Functional Programming Languages and Computer Architecture*, pages 334–349, Springer-Verlag, LNCS 201, September 1985. Appears in revised form as Programming Methodology Group Report 26, June 1986.

[11] J. Y. Girard. *Linear Logic and Parallelism.* September/October 1986.

[12] M. Gordon. How to specify and verify hardware using higher order logic. 1984. Lecture Notes, University of Cambridge.

[13] M. Gordon, R. Milner, and C. Wadsworth. *Edinburgh LCF.* Springer-Verlag, LNCS 70, 1979.

[14] T. Johnsson. *Compiling Lazy Functional Languages.* PhD thesis, Chalmers University of Technology, 1987.

[15] G. Kahn. The semantics of a simple language for parallel processing. In *Information Processing 74*, pages 471–475, North Holland, 1974.

[16] G. Kahn and D. MacQueen. Coroutines and networks of parallel processes. In *Information Processing 77*, pages 993–998, North Holland, 1977.

[17] P. F. Kearney and J. Staples. An extensional fixed point semantics for nondeterministic data flow.

[18] R. M. Keller. Denotational models for parallel programs with indeterminate operators. In E. J. Neuhold, editor, *Formal Description of Programming Concepts*, pages 377–366, North-Holland, 1978.

[19] P. Landin. A correspondence between ALGOL 60 and Church's lambda notation: part I. *CACM*, 8(2):89–101, 1965.

[20] K. G. Larsen. Proof systems for Hennessy-Milner logic with recursion. In *CAAP '88*, Springer-Verlag, LNCS 299, 1988.

[21] Z. Manna and R. Waldinger. A deductive approach to program synthesis. *ACM TOPLAS*, 2(1):92–121, 1980.

[22] P. Martin-Löf. Constructive mathematics and computer programming. In *Logic, Methodology and Philosophy of Science, VI, 1979*, pages 153–175, North-Holland, 1982.

[23] D. Park. The 'fairness' problem and nondeterministic computing networks. In *Foundations of Computer Science IV.2*, pages 133–161, Mathematical Centre Tracts 159, Amsterdam, 1983.

[24] D. Park. Fixpoint induction and proofs of program properties. In *Machine Intelligence 5*, pages 59–78, Edinburgh University Press, 1970.

A The alternating bit protocol written in Miranda[1]

```
|| Type-definitions
stream * == [*]
bit ::= 0 | L
err * ::= Error | Ok *
|| Function definitions
not x = 0 , x = L
      = L , otherwise


abp_send :: bit -> stream * -> stream (err bit) -> stream (*,bit)
abp_send b (i:is) ds = (i,b):await b (i:is) ds

await :: bit -> stream * -> stream (err bit) -> stream (*,bit)
await b (i:is) ((Ok b0):ds) = abp_send (not b) is ds,    b = b0
await b (i:is)     (d:ds)   = (i,b):await  b (i:is) ds

abp_out :: bit -> stream (err (*,bit)) -> stream *
abp_out b ((Ok (i,b0)):bs) = i:abp_out (not b) bs , b=b0
abp_out b        (c:bs)    = abp_out b bs

abp_ack :: bit -> stream (err (*,bit)) -> stream bit
abp_ack ·b ((Ok (i,b0)):bs) = b:abp_ack (not b) bs , b=b0
abp_ack b       (c:bs)     = (not b):abp_ack b bs

corrupt (0:os)  (x:xs)  = Error:corrupt os xs
corrupt (L:os)  (x:xs)  = (Ok x):corrupt os xs

abp_trans b0 b1 os0 os1 is = abp_out b1 bs
                        where
                        as = abp_send b0 is ds
                        bs = corrupt os0 as
                        cs = abp_ack b1 bs
                        ds = corrupt os1 cs
```

---

[1]Miranda is a trademark of Research Software Ltd.

Synchronization in Network Protocols

J. D. Parker
Department of Computer Science
Boston College
Chestnut Hill, Mass

Revised June 2, 1989

§0. Introduction   As the number and diversity of networks increases, the need for sophisticated network protocols has grown. Because network protocols must deal with distant parties, they are prey to all the well-known problems faced by distributed systems. This paper addresses one of these problems: synchronization. Though connectionless protocols require little knowledge of the other party's state, in connection-oriented protocols, we hope that the parties remain synchronized. Thus, for example, if one party disconnects, we expect the other to disconnect. If one party thinks that a connection has been established, we expect the other to agree. In this paper we address two issues: the loss of synchronization itself, and detection of this loss.

Our underlying model uses communicating finite-state machines (see, for example, [1]), discussed in section 2. We use added transitions to capture interactions brought about by channel errors. This model captures the major attributes of network protocols. Its richness implies that many questions of interest are undecidable. We show that the problem is undecidable in general, and we suggest alternative approaches to characterizing problematic states. We characterize synchronization loss in a manner that allows us to deal with the possibility of channel errors. We describe techniques that may be used to detect such problems, and illustrate their use in finding and solving the loss of synchronization in the data transfer portion of the alternating bit protocol.

§1. Previous Work   A variety of models have been used to discuss network protocols. One uses the notion of synchronous communication, as seen in CSP [2] or in the Ada rendezvous statement. While this model is well suited for closely-coupled systems, it makes little provision for modeling a distributed network. We will use communicating finite-state machines ([1]) instead.

One technique that can detect a wide variety of problems is *state expansion*, or *reachability analysis*, as was discussed by   Merlin [3] and Postel and Farber [4]. This technique searches the space of reachable global states, looking for problematic states. Though we can combine instances of the same state and use a preferred representative for each search path, systematic expansion cannot be guaranteed to terminate, as a two-party network with unbounded channels can be used to simulate a Turing machine (see [5] for references). Thus deciding if a state is reachable is equivalent to solving the halting problem.

Another fruitful technique focuses on one party of the protocol, as was done by Brand and Zafiropulo [6]. Rather than looking at global state expansion, they look at *tree expansions*: the set of all paths from some vertex x. A state in the tree gives the machine's history since leaving x. In this manner, they project the global state and global history onto trees associated with each process. The authors assume that messages are unique, so that receipt of a message tells the recipient where his correspondent was. They define a set of functions that capture what each state "knows" about the others, allowing them to reconstruct the global state. Our technique also focuses on one party, though we do not attempt to retain the global state and we do not assume unique labels.

We may attempt to *synthesize protocols*, hoping to design only correct protocols. To guarantee correctness, we need either an underlying search or a synthesis based on proof techniques. Gouda [7] has addressed the loss of synchronization in synthesizing protocols, and proposed a strategy he calls "winners and losers". Miller [8] notes that loss of synchronization in send-receive symmetric protocols is due to mixed states, and proposes abolishing them through the use of dummy messages. Both ideas are discussed in section 3.

§2. **Definitions**   We model a pair of communicating processes by a finite-state machine (FSM) for each process, and two unbounded FIFO queues to hold messages being sent in each direction. In each machine, the transitions between states represent the sending and receiving of messages. Each queue represents a one directional, unbounded FIFO channel that holds the messages sent by one process, but not yet read by the other.

A *communicating finite-state machine* P is a labelled directed graph with two types of edges, *sending* and *receiving*. A sending (respectively receiving) edge is labelled $-m$ ($+m$) for some *message* m in an alphabet M of symbols. A message is an indivisible token. When we don't need to specify if a message has been sent or received, we will write it as an unsigned symbol. The vertex p_0 in P is identified as its *initial vertex*, and each vertex in P is reachable by a directed path from p_0.

A vertex whose outgoing edges are all sending (receiving) is called a *sending (receiving) vertex*. A vertex whose outgoing edges include both sending and receiving edges is called a *mixed vertex*.

Let P and Q be two communicating finite-state machines with the same set M of messages. Let (P, Q) denote the network consisting of machines P and Q connected by two FIFO channels in opposite directions. A *state* of network (P, Q) is a four-tuple $[p, q, \sigma, \tau]$, where p and q are vertices in P and Q respectively, and σ and τ are strings over the messages in M. Informally, a state $[p, q, \sigma, \tau]$ means that the executions of P and Q have reached vertices p and q respectively, while the input channels of P and Q store the strings σ and τ respectively. The *initial state* of network (P, Q) is $[p_0, q_0, E, E]$, where E denotes the empty string. Our model assumes that the channels are ideal, so we will introduce channel events (see, for example, [9]) to represent the loss, modification, or duplication of messages.

We define a function *succ* that tells what happens to the state of an FSM, such as P, upon reading or sending a message. If x is a state of P, then $succ(x, +m) = u$ means that if P receives m while in state x, it enters state u. We also define $succ(x, -m) = v$ if P enters state v after sending message m while in state x. We extend the definition of *succ* to include sequences of messages $\sigma = m_1 m_2 \ldots m_k$ in the natural way:

$$succ(x, E) = x, \text{ and } succ(x, m_1 m_2 \ldots m_k) = succ(succ(x, m_1), m_2 \ldots m_k).$$

The sequence of edges traversed in P while sending and receiving the messages m_i in σ is called the *path* defined by σ, and we say that a path in the graph P generates the message sequence obtained by stringing together the labels on the path edges.

Let $s = [p, q, \sigma, \tau]$ be a state of the network (P, Q) and let e be an outgoing edge of vertex p. A state s' is said to *follow* s *over* e if and only if one of the following conditions is satisfied:

- e is a sending edge, labelled $-m$, from p to p' in P, and $s' = [p', q, \sigma, \tau m]$

- e is a receiving edge, labelled $+m$, from p to p' in P, and $s = [p', q, \sigma', \tau]$ where $\sigma = m\sigma'$.

If s and s' are two states of network (P, Q), we say s' *follows* s if and only if there is a directed edge e in P or in Q such that s' follows s over e.

If s and s' are two states of (P, Q), we say s' *is reachable from* s if $s = s'$ or there exist states $s_1, \ldots s_k$ such that $s = s_1, s' = s_k$ and s_{i+1} follows s_i for $i = 1, 2, \ldots, k-1$. A state s of network (P, Q) is said to be *reachable* if it is reachable from the initial state of (P, Q).

A *reception* is a pair (x, m) for a vertex x and a message m. A reception (x, m) is *specified* if and only if $succ(x, m)$ is defined. For a message $+m$, a reception $(x, +m)$ is *executable* if there is a reachable global state $s = [x, y, m\sigma, \tau]$ (or $[y, x, \sigma, m\tau]$). For a message $-m$, the reception $(x, -m)$ is executable if there is a reachable state $[x, y, \sigma, \tau]$ (or $[y, x, \sigma, \tau]$). A reachable state $[p, q, \sigma, \tau]$ of a network (P, Q) is an *unspecified reception state* if one of the following two conditions is satisfied:

- $\sigma = m_1 m_2 \ldots m_k$ $(k \geq 1)$, and p is a receiving vertex and none of its outgoing edges is labelled $+m_1$.

- $\tau = m_1 m_2 \ldots m_k$ $(k \geq 1)$, and q is a receiving vertex and none of its outgoing edges is labelled $+m_1$.

If the global state $s = [p, q, \sigma, \tau]$ is reachable, we say that p and q are *adjoint states*, and that p is *adjoint* to q. A reachable state $s = [p, q, E, E]$, where $\sigma = \tau = E$ (the empty string) is said to be *stable*, and we say that p and q are *stably adjoint*. We will abuse the notation by saying that p is a stable state of P. We say that p is an *ambiguous state* if p is stably adjoint to two distinct states q and q'.

A common design technique is to specify that P and Q are *dual machines*, or *send-*

receive symmetric. That is, there is a one to one and onto function d from P to Q, called the *dual function*, that maps each vertex $p \in P$ to a unique vertex $d(p) \in Q$, and each edge e from p to p' labeled with message $+m$ $(-m)$ is mapped to a unique edge $d(e)$ from $d(p)$ to $d(p')$ labeled $-m$ $(+m)$. We extend d to apply to sequences of messages $\sigma = m_1 m_2 \ldots m_k$ by defining $d(\sigma)$ to be the sequence of messages with signs reversed.

One definition of synchronization loss relies on ambiguous states. If (P, Q) is dual, and $q = d(p)$, then when (P, Q) is in state $[p, q', E, E]$ with $q \neq q'$, we might say that *the protocol has lost synchronization.* There are two problems with this definition. First, our FSM models will not always be dual. Second, this definition is too restrictive. If the channel can lose messages, then every stable receiving state is ambiguous, even in dual protocols. That is, if $succ(q, -m) = q'$ and $[p, q, E, E]$ is reachable, then if the channel loses the message m, $[p, q', E, E]$ is reachable, so q and q' are stable adjoint. We suggest below an alternative definition that singles out a subset of states.

For a protocol to synchronize, we must assume that there are dual pairs of states, that we will informally call "synchronization points". Let $S_P = \{p_1, p_2, \ldots p_k\}$ and $S_Q = \{q_1, q_2, \ldots q_k\}$ be subsets of P and Q such that (p_i, q_i) are stably adjoint states for $1 <= i <= k$. We define a protocol to be *ambiguous* if there is a pair of indices i, j, $i \neq j$ such that p_i, q_j are stably adjoint. The key point is not that p_i is ambiguous, but that these states have been singled out as "synchronization points".

§3. Synchronization in Dual Machines   Although the loss of synchronization has long been seen as a problem, (see, for example, [1]), two recent papers have addressed the problem directly. Miller discusses the issue in [8], and characterizes loss of synchronization in dual protocols. He shows that loss of synchronization occurs only at mixed nodes and proposes that mixed states be removed by the introduction of *dummy messages*. For each mixed state $x \in P$, he alters the state x by splitting the sending and receiving edges, and introducing a dummy message $-D$, as show in Figure 1. The mixed node x becomes a sending node x' followed by a receiving node x'', while the dual node y becomes a receiving node y' followed by a sending node y''.

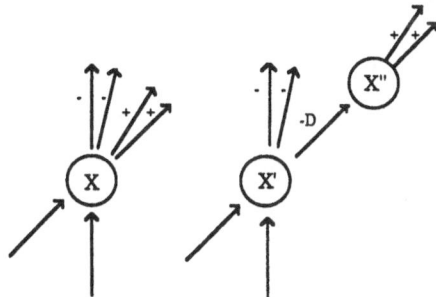

Figure 1. Mixed Node x is split into x' and x''

However, mixed states are required to mediate between error-prone machines and channels. For example, if Q enters a receiving state y' and the channel loses P's dummy message $-D$ (or P fails), then Q is forced to wait for a message that will never arrive. This problem is usually addressed with timeouts. If Q waits a fixed time without receiving a message, Q sends a message to P requesting a retransmission. But this turns the receiving state y' into a mixed state, and defeats the proposed solution.

Another contribution has been made by Gouda [7] who synthesizes protocols based on dual machines, with provisions made for loss of synchronization due to mixed states. Gouda's solution is to define one of the parties to be a *loser* and the other a *winner*. When the two parties detect a loss of synchronization, the loser then enters the state that the winner "expects" him to be in, while the winner stays in place and discards messages until synchronization has been restored.

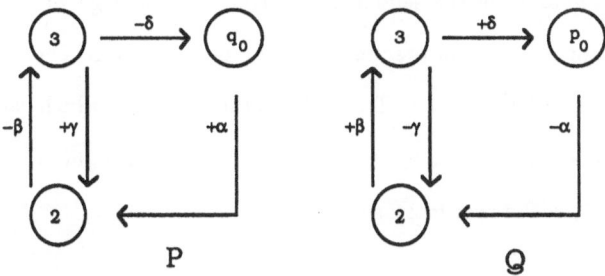

Figure 2. Mixed nodes lead to loss of synchronization

In his example, given in Figure 2, Q starts by sending $-\alpha$. When P reads $+\alpha$ and sends $-\beta$, the parties enter a pair of dual mixed states: $[3, 3, E, E]$. If Q now sends $-\gamma$ while P sends $-\delta$, we have a loss of synchronization which leads, in this example, to undefined receptions. In Gouda's sample solution, we define Q to be the "loser," and P the "winner", giving the protocol described by Figure 3. Upon the receipt of the message $+\delta$, Q "gives up" his progress, and enters state q_0, the dual to P's state p_0. When P reads the message $+\gamma$, it loops in place.

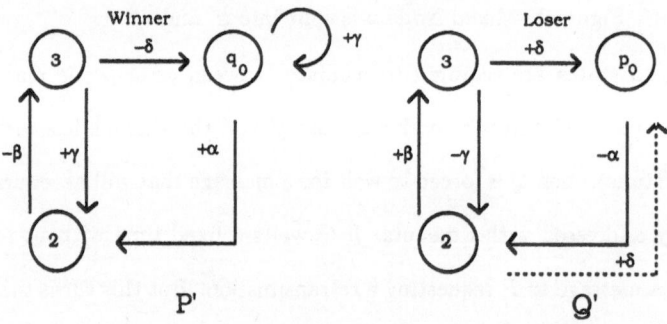

Figure 3. Loser renounces progress: winner waits

This presents a new problem: Can P tell where Q is? How can P detect that the message $+\gamma$ is to be discarded? Had the message $-\delta$ in Figure 2 been called $-\beta$ instead, the P could not detect the loss of synchronization. This problem is exacerbated when the channel can lose messages.

At issue is the uniqueness of messages. If each edge has a unique label, then any message implies the state of the sender at the time the message was sent, and Gouda's solution may be applied. But in practice, protocol designers will reuse messages, sending the same token from different states. This reduces the size of M, which has two advantages: it reduces the bandwidth required by the protocol, and it can simplify the logic of the FSM, which can now deal with fewer messages.

§4. The Abracadabra Protocol   To illustrate the issues with a more realistic protocol, we introduce the data transfer portion of the alternating bit protocol, as described in a working paper by Turner [8] on the Abracadabra protocol. The Abracadabra protocol is a version of the alternating bit protocol that includes provisions for channel errors by including timers that force retransmission of messages. Errors that arise in practice include lost messages, garbled messages, delayed messages, and messages that arrive out of order. We will assume a checksum large enough to detect garbled messages, which are then discarded and treated as lost messages. The protocol assumes that messages arrive in order. If the two parties are on the same token ring or broadcast medium, this is a valid assumption. If the topology of the network does not guarantee this, sequence numbers can allow the receiver to discard elements that arrive after a previous message, so that messages that arrive out of order may be treated as lost messages. This reduces the messages that we consider to lost messages and delayed messages. Our goal is to show that one party might not know the other party's state, to illustrate the consequences of this lack of knowledge, and to show how our analysis can lead to a solution.

There are several ways to simulate errors within the model. One technique is to introduce an additional FSM to model each direction of the channel. Thus when the channel receives a message, it can discard it, modeling lost messages, or pass it on. This has the advantage that it is simple to produce a description of the channel mechanically from a set M of messages.

Instead, we will introduce additional messages to M and additional interactions to the FSM's to model channel errors. Thus we only need to deal with two FSM's, though we need to represent a transition for each instance of a lost message. Though much of this could be done mechanically, given the original protocol, we may also use knowledge of the protocol to reduce the size of the FSM.

The protocol given by Turner is partially represented by the FSM in Figure 4, though this initial model does not include any indications of time-outs. A standard interpretation of the protocol requires $\{p_1, p_3\}$ and $\{q_1, q_3\}$ to be synchronization states. Thus the protocol is in error if it reaches $[p_1, q_3, E, E]$. The receptions (p_2, A_0) and (p_4, A_1) are included to signal a loss of synchronization, and lead P to an error state in Figure 4. In Turner's version, P disconnects without sending a message to Q.

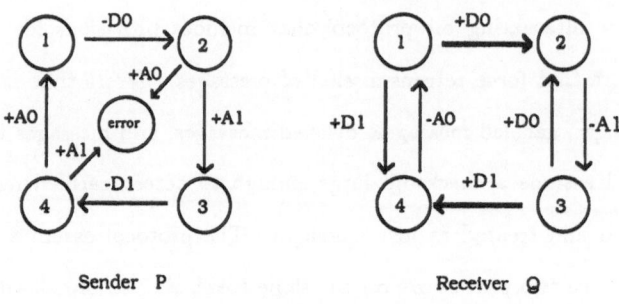

Sender P   Receiver Q

Figure 4. Abracadabra Protocol

§5. Equivalence Relations   If we are to adopt Gouda's technique, we must ask "If Q is in an ambiguous state, does P recognize it?" We identify potentially ambiguous states, and modify the messages they send to allow the other party to determine their state. If there are n states that can be confused, we need only add $log(n)$ bits to messages from these states to resolve the problem and attendant transitions to deal with them.

To begin the discussion, we initially make three assumptions.

(5.1) We assume that communication is synchronous,

(5.2) We ignore the interaction with P, and simply analyze Q,

(5.3) We assume that any two vertices $x, y \in Q$ are potentially ambiguous: we do not ask if $[p, x, E, E]$ and $[p, y, E, E]$ are reachable. At worst, this provides excess information.

If x and y are states of Q, we say that x is *absolutely equivalent to y over strings of length k* (written $x \simeq_k y$) if all paths in Q of length k starting at x produce message sequences valid at y and vice versa. In other words, no exchange of k messages is enough to allow P to decide if Q started in state x or y. If $x \simeq_k y$ for all finite k, then we say $x \simeq y$, or x and y are *absolutely equivalent*. This notion of absolute equivalence is akin to the issue of *observational congruence* as defined by Milner [10].

For each state $x \in Q$ we define the equivalence classes $A_k(x)$ to be the set of all states y such that $x \simeq_k y$. For any states $x, y \in Q$, $x \simeq_0 y$. First, observe that $x \simeq_1 y$ if and only if every message that is valid at x is also valid at y. By definition, if $x \simeq_k y$ then $x \simeq_{k-1} y$. This gives us a natural way to compute the equivalence classes A_k. Next, observe that if $x \simeq_k y$, then $x \simeq_{k+1} y$ if and only if for every message m valid at x, $succ(x, m) \simeq_k succ(y, m)$.

Lemma 1.   If $A_k(x) = A_{k-1}(x)$ for all $x \in Q$ for some $k > 1$, then $A_t(x) = A_k(x)$ for all $t > k$.

Proof: Assume that $y \in A_k(x)$. We wish to show that $y \in A_{k+1}(x)$. Let m be any message valid at x, and thus at y, as $x \simeq_1 y$. Let $succ(x, m) = u$ and $succ(y, m) = v$, so $u \simeq_{k-1} v$ by the definition of $\simeq_k$. Since $A_{k-1}(u) = A_k(u)$ by assumption, we have $u \simeq_k v$. Since m was an arbitrary message, $x \simeq_{k+1} y$. The lemma follows by induction. ∎

Our algorithm computes successive classes, until $A_k(x) = A_{k-1}(x)$ for all $x \in Q$. At this point, no further changes in the equivalence class are possible so our algorithm halts. The algorithm is presented in the appendix. It uses a 3 dimensional bit matrix *equiv* to store the equivalence classes. For each value of k and each vertex x, it inspects each column y in the row of the matrix $equiv[x, y, k]$. If the maximum out-degree for each vertex is d, and we store the graph using adjacency lists, this has a worst case running time of $O(n^2 d)$ for each stage k. We quit unless there is a change in some equivalence class, so there can be at most $n - 1$ stages, leading to a total of $O(n^3 d)$ time and $O(n^2 d)$ space in the worst case. It is easy to produce examples that take the algorithm this long. With a modest effort, we can obtain an algorithm that takes $O(n^2 d)$ time and $O(n^2)$ space.

To disambiguate a protocol, we examine each half of a protocol and identify all pairs of nodes that would be equivalent. By appending extra bits to any messages sent, we can make any ambiguity recognizable if communication is synchronous: that is, if both FSMs can empty their channels. In Figure 11 below we illustrate such a modification. Before discussing it, we will introduce a sequence of channel errors, and see how this affects the correctness of the protocol.

Applying these notions to the Abracadabra protocol, we find that P has no equivalent states, though $p_2 \simeq_1 p_4$. However, each state of Q is equivalent to another: $q_1 \simeq q_3$ and $q_2 \simeq q_4$. Thus if $[p_2, q_1, E, E]$ and $[p_2, q_3, E, E]$ were both reachable states, P would be unable to determine Q's state.

The protocol in Figure 4 has no mixed states. If we imagine that the channel never loses messages, then some receptions, such as $(q_3, +D_0)$ are non-executable, and could be removed. Once that has been done, we obtain an equivalent protocol that is dual. However, to prove that reception is non-executable is difficult in general, so we will add transitions rather than removing them.

We embed the protocol of Figure 4 into a larger dual protocol, given in Figure 5. Dotted lines represent transitions that are included for completeness, but would not be permitted to the party. Since each of these is a sending edge, this is a valid restriction. As Figure 5 has no mixed states, it has no ambiguous states, and cannot enter the error state (sending $-A_0$ from state q_2 or $-A_1$ from q_4 is prohibited.) Corollary 3 below demonstrates that the Abracadabra protocol is correct if there are no channel errors.

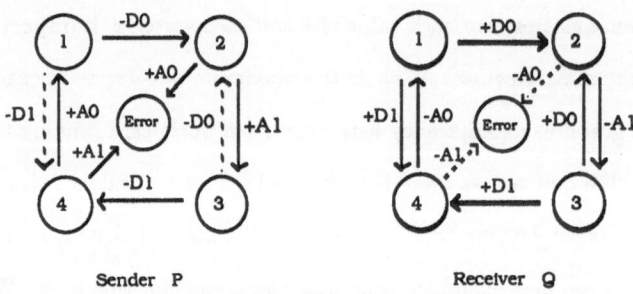

Figure 5. Dual Protocol equivalent to Figure 4

Theorem 2. *If (P, Q) is embedable in a protocol (P', Q') that is unambiguous, then (P, Q) is also unambiguous.*

Proof: Assume there are valid path pairs (σ, τ_1) and (σ, τ_2) such that $succ(q_0, \tau_1) \neq succ(q_0, \tau_2)$. Since these paths would also lie in the unambiguous protocol (P', Q'), they cannot exist. ∎

Corollary 3. *If we can embed a protocol (P, Q) inside a dual protocol with no mixed states, then (P, Q) is unambiguous.*

Proof: Miller demonstrates in [8] that dual protocols with no mixed states are unambiguous. ∎

We can use a similar technique to decide if Q will resynchronize. We say that a pair of vertices x, y in Q is *synchronizing over messages of length k*, written $x \cong_k y$, if for some $k \geq 1$, any message sequence $\sigma = m_1 m_2 .... m_k$ that is valid at x is valid at y and $succ(x, \sigma) = succ(y, \sigma)$. Message sequences of length k will force synchronization. If $x \cong_k y$, then $x \simeq y$. We may extend our algorithm to compute $\cong$ by allowing three possibilities for each pair of vertices: $x \not\cong_k y$, $x \simeq_k y$, and $x \cong_k y$.

The definition of $\simeq$ is quite strong: we assume that *every* legitimate sequence valid at x is also valid at y. It may be possible for an adversary who picks Q's transactions to fool us for an indefinite period. We define x to be *existentially similar to y over strings of length k* (written $x \sim_k y$) if there is some sequence of messages of length k that is valid at both x and y. Observe that $\sim_k$ is not an equivalence relation, as it is not transitive. We say that x is *existentially similar* to y, written $x \sim y$ if $x \sim_k y$ for all $k \geq 0$.

Our original algorithm to find absolute equivalence classes may be modified to decide which pairs are existentially similar, as well as which existentially similar pairs always resynchronize. We keep track of a pair of edges (the initial segment of equivalent paths) for each pair of equivalent vertices x and y.

§6. Errors in the Channel

Let us turn to problems that arise due to channel errors. The Abracadabra protocol uses retransmission to deal with lost packets. When the P sends D_0, it sets a timer, and enters state p_2 to wait for the acknowledgement A_1. If A_1 arrives, the timer is shut off, and P enters state p_3. However, if the timer fires, the sender returns to state p_1 and sends D_0 again. To model channel events that lose messages from the sender, we include a second transition from p_1 to p_2 : a transition to send the message L (for Lost). In our Figure, we use an additional label to describe a new interaction, so the edge from p_1 to p_2 now has two possible messages; $-D_0$ and $-L$. When q_1 receives L, it sends the message T (for TimeOut) to force P to return to state p_1. The protocol in Figure 6 models the behavior of the protocol in the presence of lost data packets. This technique introduces new states and transactions to reflect the state of the channel. Naturally, these will not be reflected in the implementation, and the designer will have little control over these messages and states. The protocol in Figure 6 can also be embedded in a larger dual machine, shown in Figure 7. By corollary 3, there can be no ambiguous states.

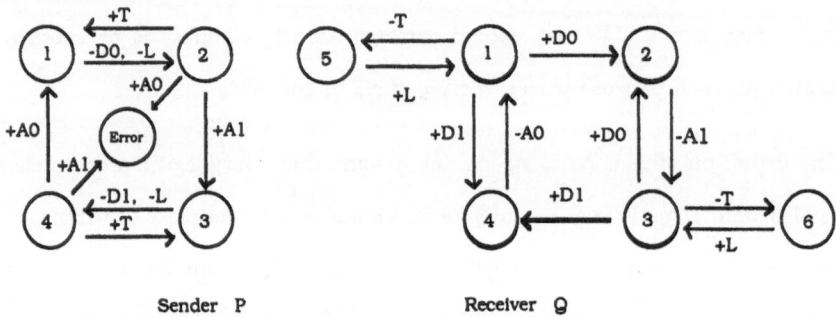

Figure 6. Abracadabra Protocol which may lose Data packets D_0 or D_1

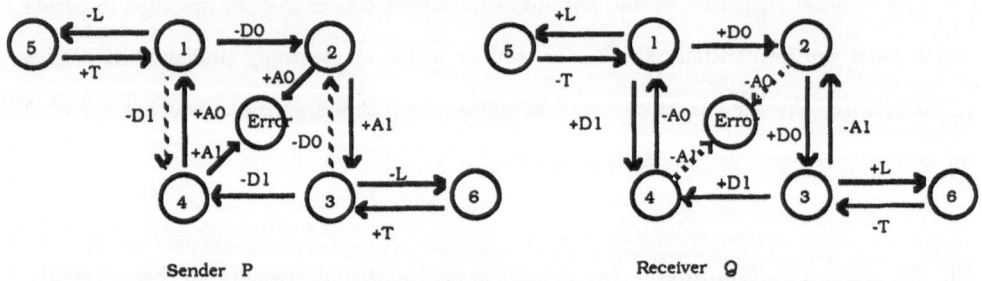

Figure 7. Unambiguous Protocol that contains Figure 6

While this models the loss of a data packet, it does not model the loss of an acknowledgement. In the Abracadabra protocol, the receiver does not use a timer to verify the arrival of an acknowledgement. Instead, it relies on the sender's timer. If the acknowledgement does not arrive, then the sender's timer will fire, and the data packet will be transmitted again. Thus we may simulate the loss of an acknowledgement by having the receiver send the message $-T$ to force a retransmission. In Figure 8, we only show transitions to model the loss of D_0 or A_1 to simplify the diagram. It is easy to see how to add transitions to deal with losing packets D_1 or A_0. Note that with this modification, the transaction $(q_3, +D_0)$, which provides for the receipt of a second copy of D_0, now becomes executable. We produce a larger dual protocol without mixed states in Figure 9 to show that the original was also free of ambiguous states.

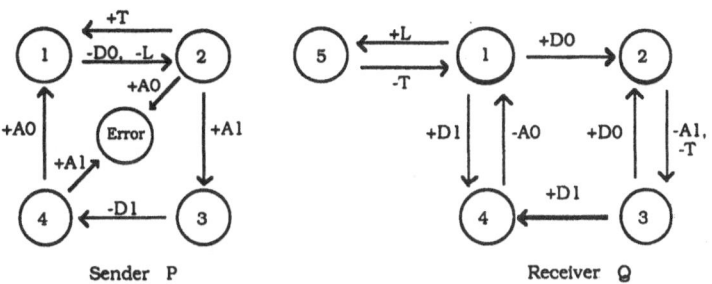

Figure 8. Either D_0 or A_1 can be lost

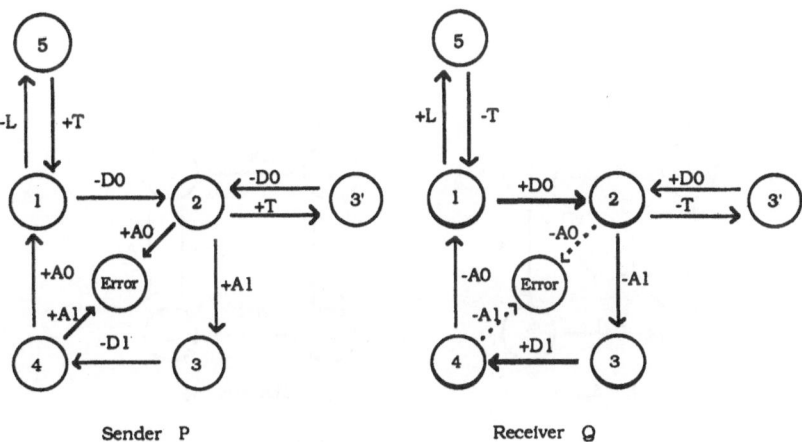

Figure 9. Dual Protocol that contains Figure 8

Our model now captures the behavior of the protocol upon the loss of any message. As seen, the protocol does not lose synchronization, though there is no progress unless the channel is fair and sends a non-zero percentage of the messages.

Finally we consider delayed messages. We still assume that messages arrive in order, though they may arrive after the timer fires. Referring to the original protocol, the reader may verify that if the sender sends D_0, times out, and then sends D_0 again, the receiver will send two copies of A_1. If both of these arrive, the sender is forced into the error state. We will first capture this behavior in our model. This causes us to enter into the ambiguous state discussed in section 5. Finally, we will use the notions of equivalence to provide a cure. One way to fix this protocol is simply to remove the error state, and force the reception $(p_4, +A_1)$ to loop in place. However, we assume that the designer wishes to retain this as a legitimate condition to test for severe errors.

To simulate slow messages, we introduce a new message $-S$ (for Slow), a new state p_6 and a new transition from p_1 to p_6. The receiver treats $+S$ as it would a lost message $+L$, sending a timeout $-T$ that forces P back to state p_1. This addition correctly models the behavior of Turner's protocol in the face of a delayed data packet. We treat a delayed acknowledgement as a delayed data packet, as the induced behavior is equivalent.

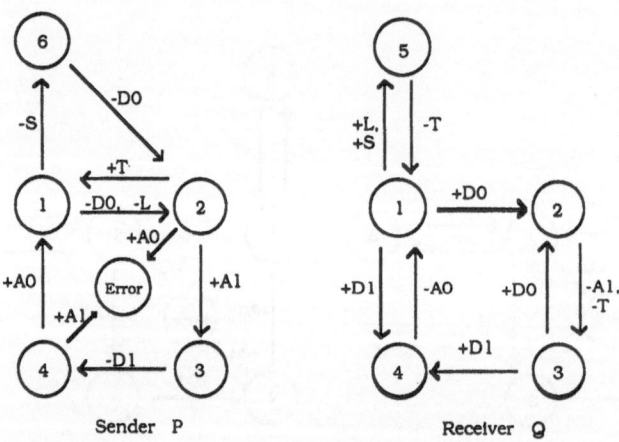

Figure 10. Abracadabra with slow message transitions

These new machines no longer need to alternate sending and receiving. Further, it is not possible to embed this new protocol in a dual machine without mixed states. When P consumes $-S - D_0 + T$, the FSM Q has consumed $+S - T + D_0$. These streams are not dual, as Q sends the timeout before reading D_0. (Note that this ambiguity arises without mixed states.) The problem is complicated by the fact that neither party can tell that there has been any loss of synchronization.

We see that P has an ambiguous state, but we do not have the liberty to change any of the messages sent so far to avoid this problem. The ambiguity is inherent in the protocol: any accurate model must be able to reach this state. If the protocol can resynchronize, however, the ambiguity will only be transient. We continue the search to see if the protocol recovers. However, the messages $\sigma = -S - D_0 + T - D_0 + A_1 - D_1 + A_1$ and $\tau = +S - T + D_0 - A_1 + D_0 - A_1$ force Q to enter state 3, while P enters the error state. In Turner's protocol, a slow ack forces P to disconnect while Q waits for data.

But we can control this exchange. We can modify the protocol to remove this ambiguity by distinguishing the first time an ack has been sent from later transmissions. The receipt of a second copy of an acknowledgement can be discarded, while the original copy received in the wrong state will still signal an error. This permits exactly enough information to allow us to disambiguate the loss of synchronization, while preserving the error condition of the original protocol. We present the new protocol in Figure 11.

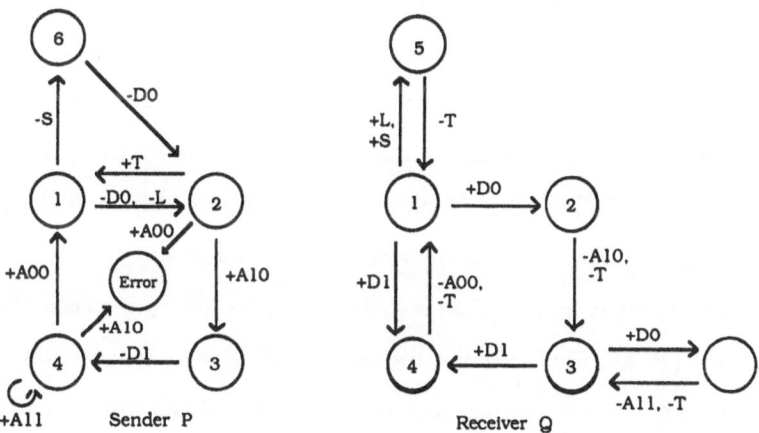

Figure 11. Abracadabra that deals correctly with slow messages D_0 and A_0

By comparing paths at p_1 and $d(p_1)$ in Figure 11, we can determine that all message exchanges will resynchronize. While this protocol is small enough to check by hand, we would like to have an algorithm that we can apply mechanically.

§7. Interaction with P   While we have illustrated the value of identifying problematic states, the forms of equivalence discussed in section 5 are can not deal with all the problems of real protocols. We may have $x \sim y$ via some interchange of messages that could not happen in practice, due to internal constraints not visible in the FSM. Likewise, we could fail to have $x \simeq y$, even though interchanges that arose in practice due to reordering would be indistinguishable. We might wish to consider another notion: *pragmatic equivalence*. Informally, we might say that two states $x, y \in Q$ are *pragmatically equivalent* if there is a state $p \in P$ that cannot distinguish between x and y. In fact, this is too much to ask. We will show that there are protocols (P, Q) and states x and y such that deciding if x and y are pragmatically equivalent is undecidable.

The problem is the deviation from the model of synchronously communicating machines. If the queues buffering communication have a fixed length, then the protocol has a finite number of states, and can be analyzed by exhaustive expansion. However, it is often difficult to design protocols with this property, or to prove that they meet this condition. Even with a protocol as simple as Abracadabra, it is easy to find interchanges that produce more tokens that they consume, forcing a steady increase in queue length.

If we allow unbounded queues in our model to simplify the design constraints, the model can be used to simulate an arbitrary Turing machine by using the channels to simulate the tape (see [5] or [11]). Given an arbitrary Turing Machine T, let (P, Q) be a protocol that models T, with initial state $[p_0, q_0, E, E]$ and an accept state $[p_F, q_F, E, E]$. Deciding if (P, Q) enters $[p_F, q_F, E, E]$ is equivalent to the halting problem. We construct a new protocol (P', Q') which contains two copies of P, $P \times \{0, 1\}$, as shown in Figure 5, and one copy of Q. We add different receptions $(p_F \times 0, -\alpha)$ and $(p_F \times 1, -\beta)$. Note that x and y are pragmatically equivalent if and only if (P, Q) never arrives at $[p_F, q_F, E, E]$.

§8. **Reordering message sequences**   Rather than restrict our search to states that are guaranteed to be confusable, we may wish to relax our criterion in hopes of obtaining an effective algorithm. Given that mixed nodes may differ in the order that they process messages, we define what it means for strings and vertices to be *abelian equivalent*. We say that $\sigma \Leftrightarrow \tau$ if

$$\sigma = m_1 m_2 \dots m_i m_{i+1} \dots m_k, \quad \tau = m_1 m_2 \dots m_{i+1} m_i \dots m_k$$

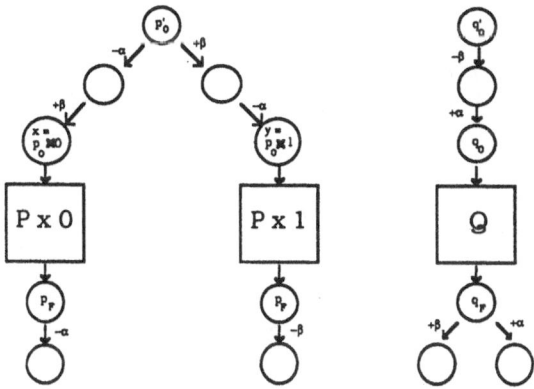

Figure 12. If x and y are not pragmatically equivalent, then (P, Q) accepts

and if m_i and m_{i+1} have different signs. We let $\Leftrightarrow^+$ be the transitive closure of $\Leftrightarrow$, and say that σ and τ are Abelianly equivalent strings. We say that x and y are Abelianly equivalent over strings of length k, written $x \bowtie_k y$, if there are messages σ and τ of length k such that $\sigma \Leftrightarrow^+ \tau$, σ is valid at x, and τ is valid at y. This relation allows messages of opposite sign to commute freely. Clearly this allows reorderings that could not happen in practice. However, the problem of computing this "simpler" equivalence is still hard.

Consider a protocol with an ambiguous states p_i, such as that in Figure 10. If $[p_i, q_j, E, E]$ is reachable, then there is a pair of dual vertices p, q and of paths (σ, τ) such that $succ(p, \sigma) = p_i$, $succ(q, \tau) = q_j$, and $\sigma \bowtie \tau$. We can always pick $p = p_0, q = q_0$. Thus p in P and $d(p)$ in the dual of Q have abelianly equivalent paths that lead to distinct endpoints. Returning to our example of Figure 10 we have the paths $\sigma = -S - D_0 + T$ and $\tau = +S - T + D_0$.

We have seen that it is useful to be able to find states that are abelianly equivalent. We would like to design an algorithm to compute $\bowtie$. Our algorithms for computing $\simeq$ and $\sim$ relied upon the decomposition of a pair of message sequences of length k into two strings: a common initial message and a pair of identical message sequences of length $k - 1$. This decomposition allowed a standard solution using dynamic programming. But when we allow reordering there is no similar decomposition. The intuition behind this is illustrated by the sequences $-\alpha - \beta + \gamma + \gamma \Leftrightarrow^+ +\gamma - \alpha + \gamma - \beta$, which do not allow any

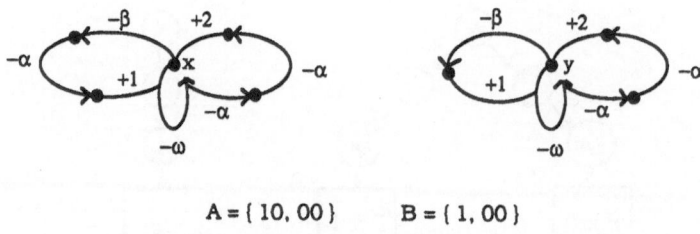

$$A = \{ 10, 00 \} \qquad B = \{ 1, 00 \}$$

Figure 13. Loops derived from Post correspondence problem.

Massey [15] for a discussion of covering spaces.) If $x \asymp y$ in our new graph, then there is a solution to the bounded PCP, so our problem is NP-hard.

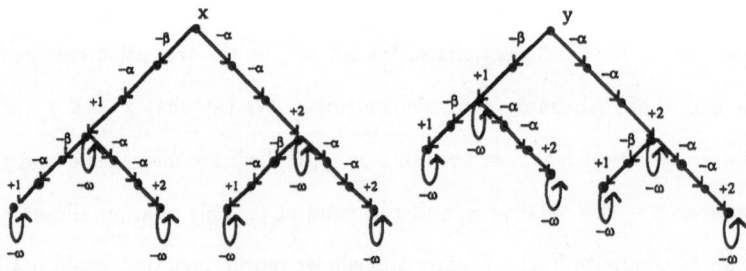

Figure 14. Network equivalent to instance of bounded PCP with $k = 2$

If we can make some assumptions about the protocol, we may be able to compute equivalence effectively. For example, if we know that in each FSM, each sending node is followed by a receiving node, then the search is no more difficult than searching for equivalence with synchronous communication, as this discipline forces each FSM to wait for a message before proceeding. If each message includes its state, equivalence classes are quite small. In general, any technique that helps a protocol remain in synchronization will also cut down on the search space.

§9. Conclusion   In section 3, we showed that ambiguous states are inevitable in the presence of errors. They can be exacerbated if the protocol has equivalent paths that make detection impossible.

There is a tension between the goals of a small message set and the desire for accurate information about a partner's state. We have investigated this tension by introducing the such decomposition.

We may still design search algorithms that hunt for problematic mixed nodes in this case. There is a simple representation of an Abelian equivalence class as a pair of strings over M. The first string holds the messages sent, while the second holds those messages received. We may use standard techniques to search for matching pairs of strings, and guarantee termination by only considering paths of length k or less, for some fixed k.

However, computing Abelian equivalence is NP-hard. We can transform an instance of the bounded correspondence problem (PCP) into the problem of deciding if $x \asymp y$. The classical PCP (see, for example Hopcroft and Ullman, [12]) starts with a finite alphabet Σ and two sequences $A = \{a_1, a_2, \ldots a_N\}$ and $B = \{b_1, b_2, \ldots b_N\}$ of strings from Σ^*. The PCP asks if there is a sequence $i_1, i_2, \ldots i_k$ of indices, not necessarily distinct, each between 1 and N, such that $a_{i_1} a_{i_2} \ldots a_{i_k}$ and $b_{i_1} b_{i_2} \ldots b_{i_k}$ are identical strings. Consider $\Sigma = \{0, 1\}, N = 4, A = \{1, 10111, 10, 1\}$ and $B = \{111, 10, 0, 11\}$. This instance of the PCP has a solution: the indices $(2, 1, 1, 3)$. In general, the PCP is undecidable.

The *bounded PCP* starts with all of the above as well as an integer $K \leq N$, and requires that $k \leq K$. As we bound the search space, we can guarantee termination. However, the problem is NP-complete as shown by Constable et. al [13], (or see [14].)

Our transformation begins by converting each string $a_i \in A$ into a loop at a vertex x, replacing 0 with $-\alpha$, 1 with $-\beta$, and appending the symbol $+i$ to the end of the loop. We perform the same transformation to turn strings from B into loops at y. Figure 13 illustrates this transformation on the following instance of the PCP. Let $\Sigma = \{0, 1\}, N = 2, A = \{10, 00\}$ and $B = \{1, 00\}$. We convert the sequence of strings into a bouquet of loops, and add a terminal loop labeled ω.

The transformation is not complete. We do not wish to allow solutions which consume only part of a loop. Thus, $x \asymp y$ in Figure 13 via the path $-\beta - \alpha + 1 \, (-\alpha - \alpha + 2)^+$ at x and $-\beta + 1 - \alpha - \alpha \, (+2 - \alpha - \alpha)^+$ at y. To assure that we have a finite correspondence, we "unwind" the loops representing strings in A and B and obtain Figure 14. (Figure 14 represents a finite portion of a covering space of our collection of loops in Figure 13. See

notion of equivalent states: states that can process similar message streams. This allows us to identify those states (p, q) that might lead to synchronization loss, and to those states that would be difficult to disambiguate. Identification can lead to prevention, by allowing the designer to alter the messages to force resynchronization. In our example, we would wish to distinguish paths out of q_1 and q_3 and those out of q_2 and q_4. This would require us to add, at most, an extra bit to each message. (Note that this requires us to add extra bits to the data messages, even though our analysis show that this was not needed to disambiguate the protocol.) Since the complete Abracadabra protocol has many states, this is less expensive than sending the states with each message, and simplifies the logic that would need to deal with a larger message set. We have shown that our techniques are compatible with channel events used to model imperfect channels.

If we can eliminate equivalent paths by redefining messages, we are assured that the protocol will not lose synchronization without detecting it. There are two obstacles to this goal. The first is that not all equivalent pairs of paths are ambiguous, due to reachability considerations. This may result in the detection of equivalent paths that will not arise in practice, and the consequent needless lengthening of messages. Second, the problem of finding all equivalent paths in a general setting is NP-complete. We can do better only if we can restrict our protocol to a tractable class.

We hope that this point of view will be useful, and will lead to other techniques that can identify problematic states without global state expansion. In the future, we hope to find realistic classes of protocols that are more amenable to analysis, and to find a better characterization of ambiguity.

§10. References

[1] Bochmann, G. v., "Finite state descriptions of communication protocols," IFIP, Liege, Belgium, Feb. 1978, pp. F3-1–F3-11.

[2] Hoare, C. A. R., "Communicating Sequential Processes," *Communications of the ACM*, vol. 21, No. 8, August 1978, pp. 666–677

[3] Merlin, P. M., "A Methodology for the Design and Implementation of Communication

Protocols," *IEEE Trans. on Communications*, vol. COM-24, No. 7, June 1976, pp. 614–621.

[4] Postel, J. B., D. Farber, "Graph Modeling of Computer Communications Protocols," Proc. 5th Texas Conf. on Computing Systems, Austin, Texas, October 1976, pp. 66-77.

[5] Brand, D. and P. Zafiropulo, "On communicating finite-state machines," Tech Report RZ-1053, IBM Zurich Research Lab, Ruschlikon, Switzerland.

[6] Brand, D. and P. Zafiropulo, "On communicating finite-state machines," *Journal of the ACM*, vol. 30, no. 2, April 1983, pp. 323–342.

[7] Gouda, M. G., "Synthesis of communicating finite state machines with guaranteed progress," *IEEE Trans. on Communications*, vol. COM-32, No. 7, July 1984, pp. 779–788.

[8] Miller, R. E., "The construction of self-synchronizing finite state protocols," *Distributed Computing*, vol 2, 1987, pp. 104–112.

[9] Chow, C.-H., M. G. Gouda, and S. S. Lam, "A discipline for constructing Multiphase Communication Protocols," *ACM Transactions on Computer Systems*, vol 3., No. 4, November 1985, pp. 315–343.

[10] Milner, R., "Calculi for Synchrony and Asynchrony", *Theoretical Computer Science*, vol 25, 1983, pp. 267–310.

[11] Gouda, M. G., C. H. Chow, and S. Lam, "On the decidability of livelock detection in networks of communicating finite state machines," IFIP, Liege, Belgium, 1985, pp. 47–56.

[12] Hopcroft, J. E., and J. D. Ullman, "Formal Languages and their relation to Automata," Addison-Wesley, Reading Mass 1969.

[13] Constable, R. L., H. B. Hunt, III, and S. Sahni, "On the computational complexity of scheme equivalence," Report No. 74-201, Dept. of Computer Science, Cornell University, Ithaca, NY, 1974.

[14] Garey, M. R., and D. S. Johnson, "Computers and Intractability: A guide to the

Theory of NP-Completeness," W. H. Freeman and Company, San Francisco, 1979.

[15] Massey, W. S., "Algebraic Topology: an Introduction," Harcourt, Brace & World, Inc, New York, 1967.

Algorithm to Compute Absolute Equivalence

```
type
      vertex = 1..n;
      stage = 0..n-2;
      link = ^node;
      node = record
              msg : message;
              dest : vertex;
              next : link;
      end;

var
      equiv = array [vertex, vertex, stage] of boolean;
      adj = array [1..n] of ^node;{ adjacency list}
      p, q : link;
      change : boolean;

                      {Initial conditions: equiv[x,y,0] := true for all x,y}
                      { and adjacency lists are sorted by message }

k := 0;
repeat
      k := k + 1; change := false;
      for x := 1 to n do
            for y := x+1 to n do
                  if  equiv[x,y,k-1] do
                          p := adj[x];
                          q := adj[y];
                          equiv[x,y,k] := true; { until proven otherwise }
                          while (p<>nil) and (q<>nil) do
                                if p^.msg <> q^.msg
                                then
                                        equiv[a,b,k] := false;
                                        p := nil; { break loop }
                                else
                                        p := p^.next; q := q^.next;
                          end; { while }
                          if equiv[x,y,k] <> equiv[x,y,k-1] then change := true;
            end; { for y }
      end; { for x }
      Print_Equivalence ( equiv, k ); { save the current work }
until not change;
```

From Synchronous to Asynchronous Communication

E. Pascal GRIBOMONT

Philips Research Laboratory

Av. Van Becelaere 2, Box 8

1170 Brussels, Belgium

Abstract. The design and the verification of distributed systems consisting of communicating processes can be difficult. The task is simpler when the rendezvous mechanism of CSP is used, but this mechanism often leads to rather inefficient implementations. For this reason, an asynchronous communication mechanism is preferred, provided that the correctness of the system can be preserved. A methodology is proposed to check whether a correct CSP system can be implemented with asynchronous communication or not. The methodology is illustrated by converting a pure-CSP solution to the mutual exclusion problem into a more realistic version.

1 Introduction

The design and the verification of a system of concurrent processes turns to be difficult. Several mechanisms are used to specify the interference; the most important ones are the sharing of memory and the communication by messages. Both mechanisms are used in many applications in the area of computer networks. In general, processes attached to distinct nodes communicate by asynchronous message passing; processes attached to a common node communicate by shared memory.

It is well known that the undisciplined use of interference mechanisms leads to clumsy, or incorrect, programs. A successful attempt to improve the clarity of concurrent systems is Hoare's language of CSP [5,6]. In this language, processes can communicate in a synchronous way only, which makes the design and the verification easier.

Nevertheless, the fair implementation of CSP can be inefficient, especially in computer networks, where synchronous communication is expensive and replaced by asynchronous communication. The favourable case occurs when this replacement does not destroy the correctness of the system.

A methodology is proposed to check the correctness of such a transformation. The principle is as follows. The starting point is a set of synchronously communicating processes, together with an invariant, which implies some specified safety properties. The couple (system, invariant) is incrementally transformed. A step consists in replacing a synchronous channel by an asynchronous one. Most of the time, this transformation destroys the invariant. More precisely, the invariant does not take into account states where a message has been sent but not yet received. However, it can be possible to transform the invariant in order to take such states into account. If it is possible, some weakest-precondition [1] and strongest-postcondition calculus leads to an adequate invariant of the new version of the system, which proves that the transformation is licit; other transformations can be attempted, until a fully asynchronous version of the system is obtained.

Another kind of transformation often turns out to be useful: the splitting of a process into a set of processes communicating by shared memory. Once again, such a transformation is not always licit and, therefore, it must be checked that the invariant can be adapted.

This paper goes on as follows. Some formal tools needed for the representation and the verification of concurrent processes are recalled in section 2. A simple example is introduced in section 3 to show that synchronous and asynchronous versions of a distributed algorithm may exhibit very different behaviours. The suggested methodology to convert a correct synchronous version into a correct asynchronous one is outlined in section 4. A simple CSP algorithm for implementing mutual exclusion in a computer network is described in section 5 and converted, in section 6, into an asynchronous version.

2 Formal tools

The formal tools involved by the methodology proposed here are elementary. First order logic, slightly enriched, is used as a specification language. Programs are represented by transition systems; the kind of transition system used in this paper is introduced below. A small subset of Hoare logic is also used. The predicate transformers wp and sp appear only in a restricted framework, which is detailed below.

2.1 Transition systems

Sets of distributed processes are represented by a variant of transition systems [7,8]. A *transition system* S consists of three components. The first one is a set $\mathcal{P}$ of processes. A *process* is a finite set of *labels*, or *control points*. The processes are pairwise disjoint. Second, the memory $\mathcal{M}$ of the system is modelled by a set of variables. Variables can be private or shared by two or several processes. The last component is a set $\mathcal{T}$ of *transitions*, which specifies the possible computations of the system. The description and the study of a transition system $S = \{\mathcal{P}, \mathcal{M}, \mathcal{T}\}$ involve the concepts defined below.

A *system state*, or simply a *state*, is a pair (control state, memory state). A *memory state* associates a value to each variable. A *control state* is a mapping which assigns to each process a label of this process.

Every transition specifies a conditional modification of the system state. A transition T may involve one process P or two processes P, Q; transitions involving one or two processes are represented as lists like

$$(l, C \longrightarrow A, l') \qquad\qquad l, l' \in P,$$
$$(lm, C \longrightarrow A, l'm') \qquad\qquad l, l' \in P, \ m, m' \in Q.$$

The *condition* C is a predicate on the set of memory states and the *assignment* A specifies a transformation on the same set.

Transition T can be executed in a system state $s = (N, \sigma)$ only if $N(P) = l$ (and $N(Q) = m$) and $C(\sigma)$ is true. The execution induces a new state $s' = (N', \sigma')$, called the *T-successor* of s. The control state N' is equal to N, except that $N'(P) = l'$ (and $N'(Q) = m'$). The memory state σ' is $A(\sigma)$. A state s' is a *valid successor* of a state s if there exists a transition T leading from s to s'. A state is *terminal* if it has no valid successor.

A *computation* is represented by a sequence of system states. The first state is the *initial* state; the last one (if any) is the *final* state; it must be a terminal state. Each state except the first one is a valid successor of its predecessor in the sequence.

Let us note that a transition is *deterministic*: there is at most one state σ' such that $\sigma' = A(\sigma)$. On the contrary, the whole system S is usually *non-deterministic*: a state can be the initial state of more than one computation.

2.2 A language for specification and proof

An element of a computation is a system state; in order to describe computations, it is necessary to write predicates over the set of system states. This set is the product of the set of control states and the set of memory states. The classical first order logic is convenient to write predicates on the set of memory states; it will be enriched to deal also with the control states.

A *place predicate* "*at l*" is associated with each label $l \in P \in \mathcal{P}$; it is true for the system state (N, σ) if and only if $N(P) = l$.

From now on, we suppose that a formula is written in the usual language of first order logic, augmented with place predicates. As a consequence, a formula appears as a predicate on the set of system states.

Let B be a formula and let l be a label of some process P. The formula $B_{[at\ l]}$ is obtained by replacing in B the occurrences of *at l* by *true*, and the occurrences of *at l'*, for all $l' \in (P \setminus \{l\})$, by *false*.

Formula $B_{[at\ l]}$ is characterized by two useful properties. First, its value on some system state (M, σ) is independent from $M(P)$, since $B_{[at\ l]}$ does no longer contain any occurrence of *at l'*, for any $l' \in P$. Second, the formula $B_{[at\ l]}$ is equivalent to B when *at l* is true. This can be stated more formally as follows.

$$(B_{[at\ l]})_{[at\ l']} \equiv B_{[at\ l]}, \quad \text{if } l, l' \in P;$$
$$at\ l \Rightarrow (B_{[at\ l]} \equiv B).$$

This concept will be illustrated later.

The language so defined allows to write predicates over the set of system states. It can be used for various purposes; the most important ones are listed below.

- Usually, only computations whose initial state satisfies some constraint are considered. This constraint is formalized into a formula, called the *initial condition*.
- A *safety property*, also called an *invariance property*, is a formula which must be true throughout the execution.
- An *invariant* is a formula I which is respected by all transitions. A state s is *respected* by a transition T if the following condition is satisfied: if $I(s)$ is true and if the T-successor s' of s exists, then $I(s')$ is also true.

Remarks. Many useful properties of concurrent systems are modelled by safety properties. Such properties assert that "something bad" never happens. Examples are freeness of deadlock (no state without valid successor can be reached) and partial correctness (no terminal state can be reached, unless with a memory part satisfying the program specification).

An invariant is a safety property if it is true initially.

It is rather easy to determine whether a formula is an invariant. The standard way to establish a safety property is to construct an invariant which implies it and which is true initially.

2.3 Weakest precondition, strongest postcondition

The notions of predicate transformers and, especially, of weakest precondition, have proved useful in the area of programming methodology (see [1,3]). The principle of the predicate transformers *wp* and *sp* is as follows. Let $\{P\}\ S\ \{Q\}$ be a classical Hoare triple. Formula P is the precondition, S is a program and Q is the postcondition. The triple is true if every finite computation of S, whose initial state satisfies P, terminates in a final state satisfying Q.

In this paper, only deterministic programs without infinite computations will appear in Hoare triples.[1]

An important question is as follows: if S and Q are fixed, what must be required about the precondition P in order to guarantee the validity of the triple $\{P\} S \{Q\}$? It can be proved that the set of acceptable preconditions form a boolean lattice; its minimum element is called the *weakest precondition* corresponding to S and Q, and is denoted $wp[S;Q]$. The dual concept is the *strongest postcondition*.

These results are summarized in asserting that the three formulas

$$\{P\} S \{Q\},$$
$$P \Rightarrow wp[S;Q],$$
$$sp[S;P] \Rightarrow Q.$$

are equivalent. Hoare axiom [4] is used to determine $wp[S;Q]$ and $sp[S;P]$ when S is an assignment. The most useful rule is

$$wp[x := f(x,y); \ Q(x,y,z)] \ = \ Q(f(x,y),y,z),$$

where x, y and z are disjoint subsets of the memory, and where f, P and Q are a function and two predicates of adequate types.

The use of predicate transformers is extended to transitions by the following identities.

$$
\begin{aligned}
\{P\} (L, C \rightarrow S, L') \{Q\} &=_{def} \{P_{[at \ L]} \wedge C\} S \{Q_{[at \ L']}\}, \\
wp[(L, C \rightarrow S, L'); Q] &=_{def} (at \ L \wedge C) \Rightarrow wp[S; Q_{[at \ L']}], \\
sp[(L, C \rightarrow S, L'); P] &=_{def} at \ L' \wedge sp[S; C \wedge P_{[at \ L]}].
\end{aligned}
\tag{1}
$$

("L" denotes a label l for a transition involving one process, and two labels lm for a transition involving two processes; "$at \ lm$" stands for "$(at \ l \wedge at \ m)$" and "$B_{[at \ lm]}$" stands for "$(B_{[at \ l]})_{[at \ m]}$".

Comment. The formulas $\{P\} T \{Q\}$, $(P \supset wp[T;Q])$ and $(sp[T;P] \supset Q)$ are equivalent for every transition.

2.4 Models of synchronous and asynchronous mechanisms

The formalism of transition systems has first been used to model the communication by shared variables, but is also convenient to model several kinds of message passing mechanisms.

In CSP, a matching pair $\{C?x, C!e\}$ of communication statements results in a distributed assignment $x := e$; as a consequence, such a pair will be modelled by a single transition involving both the sending and the receiving processes.[2] Examples are given in the following sections.

Let us now consider the case of asynchronous communication. It is supposed that channels are one-way and fully reliable (as in CSP); they will be modelled by linear buffers [9]. (Other models are possible; the methodology presented in section 4 can be adapted easily for most usual models.) If C is a channel, the statements "send e through C" and "receive x from C" are modelled respectively by the conditional statements

---

[1] As a consequence, the distinction between weakest precondition and weakest liberal precondition is not needed here.

[2] A transition can be introduced for all syntactically matching pairs, but only those corresponding to semantically matching pairs can be executed.

$$send(e, C): \qquad |C| < max_c \to C := C.e,$$

$$receive(x, C): \qquad |C| > 0 \to (x, C) := (hd(C), tl(C)),$$

with usual notations for linear buffers. The number max is the size of the buffer. Two frequent particular cases are $max = \infty$ (unbounded buffer) and $max = 1$ (one-place buffer). For a one-place buffer, the $send(e, C)$ and $receive(x, C)$ statements are rewritten respectively as

$$C = \emptyset \to C := e,$$

$$C \neq \emptyset \to (x, C) := (C, \emptyset).$$

With this classical communication model, communication statements are modelled by transitions involving only one process; the channel C appears as a variable shared by the sending and the receiving processes.

3 An explanatory example

An elementary example is introduced now, to illustrate the formalism of transition systems, and also to demonstrate that synchronous and asynchronous versions of the same algorithm may exhibit very different behaviours.

Let us consider an elementary CSP network consisting of two processes P and Q. A channel C is oriented from P to Q and a channel D is oriented from Q to P. These processes are

$$P ::= *[\ C!x;\ x := x|y$$
$$\square\ D?y \to skip\,],$$

$$Q ::= *[\ C?v \to u := u|v$$
$$\square\ D!u\,].$$

(2)

All variables are private; the symbol "|" denotes some binary operation. Every time the control is at the beginning of both loops, the relation

$$v|y = x \ \lor \ y = u \tag{3}$$

holds (provided it is satisfied initially); in fact, the network $(P\|Q)$ is equivalent to the cyclic, nondeterministic loop given below.

$$*[\ v := x;\ (x, u) := (x|y, u|v)$$
$$\square\ y := u\,]. \tag{4}$$

The network (2) can also be modelled by a transition system whose transitions are

$$(p_0q_0,\ v := x,\ p_1q_1),$$
$$(p_1,\ x := x|y,\ p_0),$$
$$(q_1,\ u := u|v,\ q_0),$$
$$(p_0q_0,\ y := u,\ p_2q_0),\ ^"$$
$$(p_2,\ skip,\ p_0). \tag{5}$$

As communicating sequential processes do not share memory, a more succint model can be obtained. A transition involving two processes can model not only a matching pair of communicating statements, but also one or more internal statements. The system (5) can be rewritten as

$$(p_0q_0, (v, x, u) := (x, x|y, u|x), p_0q_0),$$
$$(p_0q_0, y := u, p_0q_0).$$ (6)

This system admits the same invariant as the nondeterministic loop (4).

Let us now consider an asynchronous version of the network (2), where channels are modelled by one-place buffers. The corresponding transition system is:

$$(p_0, C = \emptyset \rightarrow C := x, p_1),$$
$$(p_1, x := x|y, p_0),$$
$$(p_0, D \neq \emptyset \rightarrow (y, D) := (D, \emptyset), p_2),$$
$$(p_2, skip, p_0),$$ (7)
$$(q_0, C \neq \emptyset \rightarrow (v, C) := (C, \emptyset), q_1),$$
$$(q_1, u := u|v, q_0),$$
$$(q_0, D = \emptyset \rightarrow D := u, q_0).$$

The system can be simplified into

$$(p_0, C = \emptyset \rightarrow (C, x) := (x, x|y), p_0),$$
$$(p_0, D \neq \emptyset \rightarrow (y, D) := (D, \emptyset), p_0),$$
$$(q_0, C \neq \emptyset \rightarrow (v, C, u) := (C, \emptyset, C|v), q_0),$$ (8)
$$(q_0, D = \emptyset \rightarrow D := u, q_0).$$

However, this asynchronous version is not a faithful implementation of the synchronous one. More precisely, the invariant of (6) is not respected by (8). To see why, let us consider the following symbolic computation.

$$\{ C = D = \emptyset \wedge x = x_0 \wedge y = y_0 \wedge u = u_0 \wedge v = v_0 \}$$
$$(C, x) := (x, x|y)$$
$$\{ C = x_0 \wedge D = \emptyset \wedge x = x_0|y_0 \wedge y = y_0 \wedge u = u_0 \wedge v = v_0 \}$$
$$D := u$$
$$\{ C = x_0 \wedge D = u_0 \wedge x = x_0|y_0 \wedge y = y_0 \wedge u = u_0 \wedge v = v_0 \}$$
$$(v, C, u) := (C, \emptyset, C|v)$$
$$\{ C = \emptyset \wedge D = u_0 \wedge x = x_0|y_0 \wedge y = y_0 \wedge u = u_0|x_0 \wedge v = x_0 \}$$
$$(y, D) := (D, \emptyset)$$
$$\{ C = D = \emptyset \wedge x = x_0|y_0 \wedge y = u_0 \wedge u = u_0|x_0 \wedge v = x_0 \}$$

The invariant (3) is not preserved; more specifically, the assertion

$$C = D = \emptyset \Rightarrow (v|y = x \vee y = u)$$

is not an invariance property of the asynchronous version. The symbolic computation listed above leads to a simple counterexample. If the initial values of the variables satisfy the identity $v_0|y_0 = x_0$, then the relation $(v_f|y_f = x_f \vee y_f = u_f)$, between the final values of the variables, is not guaranteed; it is true only if the initial values of the variables satisfy the relation $(x_0|u_0 = x_0|y_0 \vee u_0 = u_0|x_0)$.

As a conclusion, transforming a synchronous algorithm into an asynchronous one can introduce incorrect behaviour, even for elementary algorithms. However, the practical experience shows that, in many cases, correct synchronous networks can be implemented asynchronously without loss of correctness. The method proposed in the next section allows to detect such favourable cases in a systematic way.

4   A transformation method

If a synchronous network $\mathcal{N}$ is correct with respect to some specified invariance property A, then this can be established by an invariant I. Such a formula is a relation between the variables of the processes and their labels, which satisfies the following requirements.

$$\vDash\ Init\ \Rightarrow\ I\,,\quad \text{where } Init \text{ specifies the initial conditions;}$$
$$\vDash\ \{I\}\,T\,\{I\}\,,\quad \text{if } T \text{ is any transition;}$$
$$\vDash\ I\ \Rightarrow\ A\,.$$

The first requirement asserts that the invariant is true initially. The second one asserts that it is respected by all transitions (involving one or two processes); as a consequence, the invariant remains true throughout the execution. The last requirement guarantees that the property A also remains true throughout the execution.

Let us now consider an asynchronous version $\mathcal{N}'$ of $\mathcal{N}$. In order to prove the correctness of $\mathcal{N}'$, it is sufficient to design an invariant I' of $\mathcal{N}'$ such that

$$\vDash\ Init\ \Rightarrow\ (I'\wedge C_1 = \cdots = C_m = \emptyset)\,,$$
$$\vDash\ (C_1 = \cdots = C_m = \emptyset)\ \Rightarrow\ (I' \equiv I)\,,$$

where $\{C_1,\ldots,C_m\}$ is the whole set of channels.

The method proposed here for obtaining I' is incremental; a sequence $(N_0, N_1,\ldots,N_m)$ of "mixed" networks is considered, with $N_0 = \mathcal{N}$ and $N_m = \mathcal{N}'$. The network N_r is obtained from the network N_{r-1} by replacing the synchronous channel C_r by an asynchronous one. A corresponding sequence of invariants $(I = I_0, I_1,\ldots,I_n = I')$ is constructed in such a way that, for every r, the implication $[(I_r \wedge C_r = \emptyset)\ \Rightarrow\ I_{r-1}]$ is valid. If this sequence can be constructed, then the asynchronous network $\mathcal{N}'$ is a faithful implementation of the synchronous network $\mathcal{N}$.

This approach is acceptable only if I_r can be obtained (if it exists) from I_{r-1} in a rather systematic way. The relation between these formulas are investigated now.

Let J be an invariant of a mixed network $\mathcal{M}$, comprising a synchronous channel C, and let $\mathcal{M}'$ be the network obtained by replacing C by an asynchronous channel, that is, a linear buffer. An adequate invariant J' for $\mathcal{M}'$ (if it exists) would be a formula depending of the contents of channel C. We write $J'(C)$ to emphasize this dependency.

As $J'(C)$ must be determined from J, it will be postulated that $J'(C)$ reduces to J when C is empty. The invariant J' must be a solution of the system

$$\vDash\ \{J'\}\,T\,\{J'\}\,,\qquad \text{for each transition } T.$$

Most of the time, the replacement of a synchronous channel by an asynchronous one will enforce only a "small" modification of the invariant. More precisely, let us suppose that the invariant J is in conjunctive form, i.e., is the conjunction of a set of formulas (called *assertions*). Some of these formulas will be respected by the newly introduced send and receive statements, while other will not. The invariant J is rewritten as $(J^0 \wedge J^+)$, where J^0 contains the unaltered assertions and J^+ contains the altered ones. As a consequence, the new invariant $J'(C)$ will have the form $J'(C) =_{def} (J^0 \wedge K(C))$, where $K(C)$ is unknown, but reduces to J^+ when C is empty. The set of contraints about $K(C)$ is

$$J^0 \vDash \{K(C)\} T \{K(C)\}, \qquad \text{for each transition } T.$$

The postulate

$$J^0 \vDash K(\emptyset) \equiv J^+$$

allows to rewrite the system of constraints about $K(C)$ in a more operational way. These conditions can be partitionned into three sets, according to the nature of T. The set T of transitions is equal to $T_s \cup T_r \cup T_i$, where the elements of these subsets respectively are transitions containing a $send(e, C)$ statement, a $receive(x, C)$ statement, and no communication statement at all. The constraints can be rewritten as

$$J^0 \vDash \{K(c) \wedge C = c\} T_s \{K(c.e) \wedge C = c.e\}, \qquad \text{for all } T_s \in T_s;$$
$$J^0 \vDash \{K(a.c)_x^a \wedge C = a.c\} T_r \{K(c) \wedge C = c\}, \qquad \text{for all } T_r \in T_r;$$
$$J^0 \vDash \{K(C)\} T_i \{K(C)\}, \qquad \text{for all } T_i \in T_i.$$

In this system, c denotes an acceptable value for the buffer C and a, e denote acceptable values for a cell of this buffer. (NB. x denotes a program variable, but a, c and e do not.) This system can be explicited further into

$$J^0 \vDash sp[T_s; (J^+ \wedge C = \emptyset)] \Rightarrow [K(C) \wedge C = (e)],$$
$$J^0 \wedge c \neq \emptyset \vDash \{K(c) \wedge C = c\} T_s \{K(c.e) \wedge C = c.e\},$$
$$J^0 \vDash [K(C)_x^a \wedge C = (a)] \Rightarrow wp[T_r; (J^+ \wedge C = \emptyset)], \qquad (9)$$
$$J^0 \wedge c \neq \emptyset \vDash \{K(c)_x^a \wedge C = a.c\} T_r \{K(c) \wedge C = c\},$$
$$J^0 \vDash \{K(C)\} T_i \{K(C)\}.$$

Constraints in which K appears only once, outside the scope of any predicate transformer, can be solved explicitly. They determine a Boolean lattice, called the "candidate-set"; a solution for K cannot be selected outside of this set. The maximum and the minimum of the candidate-set respectively are

$$\bigwedge_{T \in T_r} wp[T; (J^+ \wedge C = \emptyset)] \qquad \text{and} \qquad \bigvee_{T \in T_s} sp[T_s; (J^+ \wedge C = \emptyset)].$$

The proposed stategy is two-phased. First, the minimum and the maximum of the candidate-set are evaluated; this can be done mechanically. The second phase consists in selecting within the candidate-set an adequate element, that is, an element which is also a solution of all the remaining constraints.

This strategy is sound since, if the transformation of $\mathcal{M}$ into $\mathcal{M}'$ is not correct, then no adequate element will be found. It is also complete; to see why, let us suppose that the asynchronous version $\mathcal{M}'$ is correct with respect to some safety property, and that this correctness is established by some invariant I'. Then, the synchronous version $\mathcal{M}$ is also correct and an adequate invariant is $(I' \wedge C = \emptyset)$. Nevertheless, a failure can be due not only to the incorrectness of the transformation, but also to an inadequate choice of the invariant J.[3]

It can be thought that the second phase of the method is practically infeasible when the candidate-set is large or infinite. This is not the case; the experience shows that, most of

---

[3]The same occurs, for instance, for the resolution method: this deduction technique is complete, provided that a "good" strategy of choices is used.

the time, if a solution exists, one of the bounds of the candidate-set is an adequate solution. Furthermore, as the invariant is intended to summarize the semantics of the algorithm, the strongest choice is also the best one; it will be adopted everytime it is possible.

When channels are one-place buffers, the system (9) has a simpler form, which is

$$
\begin{aligned}
J^0 &\models sp[T_s; (J^+ \wedge C = \emptyset)] \Rightarrow [K(C) \wedge C = e], \\
J^0 &\models [K(C)_x^a \wedge C = a] \Rightarrow wp[T_r; (J^+ \wedge C = \emptyset)], \\
J^0 &\models \{K(C)\} T_i \{K(C)\}.
\end{aligned}
\tag{10}
$$

The channel is now modelled as a store containing one value or no value (or some "dummy" value called "$\emptyset$"). A further simplification occurs when a one-place buffer is used for synchronization only. In this case, the message reduces to a signal and the asynchronous channel is modelled by a binary value; the constraints are therefore rewritten into

$$
\begin{aligned}
J^0 &\models sp[T_s; (J^+ \wedge C = 0)] \Rightarrow [K \wedge C = 1], \\
J^0 &\models [K \wedge C = 1] \Rightarrow wp[T_r; (J^+ \wedge C = 0)], \\
J^0 &\models \{K\} T_i \{K\}.
\end{aligned}
\tag{11}
$$

5 A simple solution to the mutual exclusion problem

5.1 Informal description

The classical problem of *mutual exclusion* between distributed computing stations can be stated as follows. A resource is shared between n computing stations, which can communicate by message passing. Only one station at a time may access the resource. The problem is how to implement this mutual exclusion.

A simple solution is informally as follows. The use of the resource is managed by a special process, called the controller. When a station p needs the access to the resource, it sends a request to the controller, waits until an authorization is received, and then accesses the resource. When the station has completed its task with the resource, it sends a release message to the controller, which can grant the access to another station. The controller uses a variable to record which station presently has access to the resource, and a waiting set (or queue) to record which stations are currently waiting for the access.

5.2 An abstract solution in CSP

The informal solution introduced above can easily be translated in the formalism of CSP. The following pieces of notation will be used.

Channels REQ_p and END_p are oriented from station p to the controller (a single channel could be used, but some clarity is gained by considering two channels). Channel OK_p is oriented from the controller to the station. Predicate crs_p denotes an internal condition of station p; the access to the resource is needed as long as the condition is true. When it is false, the station can perform internal computation which does not involve the shared resource. The truthvalue of the condition crs_p can be changed by internal computation only. The symbol [ ] denotes internal computation, with or without the resource.

A variable $INCS$ is used by the controller to record the number (or the name) of the station lying in its critical section. When the resource is idle, the value of $INCS$ is 0. The numbers of the waiting stations are stored in E; this structure is handled as a set, but it is also possible to handle E as a first-in-first-out queue, for instance.

The process executed by station p is

$$*[\,*[\neg crs_p \longrightarrow [\,]\,];$$
$$REQ_p!;$$
$$OK_p?;$$
$$*[crs_p \longrightarrow [\,]\,];$$
$$END_p!$$
$$]\,.$$

The process executed by the controller is

$$*[\,\square_{p=1}^{n}\,REQ_p? \longrightarrow \quad [\quad INCS = 0 \longrightarrow INCS := p;\ OK_p!$$
$$\square\quad INCS \neq 0 \longrightarrow E := E \cup \{p\}]$$
$$\square_{p=1}^{n}\,END_p? \longrightarrow \quad [\quad E = \emptyset \longrightarrow INCS := 0$$
$$\square\quad \square_{r=1}^{n}\,r \in E \longrightarrow OK_r!;\ (INCS, E) := (r,\ E \setminus \{r\})]$$
$$]\,.$$

The communication statements do not implement any actual value transmission, but only synchronization; that is the reason why no argument has been mentioned for input and output statements.

5.3   The transition system

The network is now rewritten as a transition system. Some control points are introduced now; they are defined as follows. Each station p can be at three control points. At the non-critical control point p_0, station p can perform internal computation, without using the resource. At the waiting point p_1, station p has requested the access and waits for the authorization. At the critical point p_2, station p effectively uses the resource. The idle state of the controller is modelled by a control point called C. If the controller has just received a request (a release) from station p, then it remains at the control point C_p (C'_p) until this request (this release) has been processed.

The transitions of the system are listed below (the channels do not appear explicitly).

1. $(p_0,\ \neg crs_p \longrightarrow [\,],\ p_0)$, $\forall p$
2. $(p_0 C,\ crs_p \longrightarrow skip,\ p_1 C_p)$, $\forall p$
3. $(p_1 C_p,\ INCS = 0 \longrightarrow INCS := p,\ p_2 C)$, $\forall p$
4. $(C_p,\ INCS \neq 0 \longrightarrow E := E \cup \{p\},\ C)$, $\forall p$
5. $(p_2,\ crs_p \longrightarrow [\,],\ p_2)$, $\forall p$
6. $(p_2 C,\ \neg crs_p \longrightarrow skip,\ p_0 C'_p)$, $\forall p$
7. $(q_1 C'_p,\ q \in E \longrightarrow (INCS, E) := (q, E \setminus \{q\}),\ q_2 C)$, $\forall p, q$
8. $(C'_p,\ E = \emptyset \longrightarrow INCS := 0,\ C)$, $\forall p$

The initial conditions are modelled by the formula

$$at\ C\ \wedge\ \forall p\,(at\ p_0)\ \wedge\ INCS = 0\ \wedge\ E = \emptyset,$$

which simply expresses that, initially, all stations are in their non-critical section; the controller and the resource are idle and the waiting set is empty.

Let us now recall a useful Boolean operation. If $A_1, \ldots, A_m$ are logical formulas and if N is a number, then the equality $(A_1 + \cdots + A_m = N)$, also written $(\sum_{i=1}^{m} A_i = N)$, is a logical formula. This formula is true if and only if exactly N elements of the bag $\{A_1, \ldots, A_m\}$ are true. This "Boolean addition" is natural if *false* is identified with 0 and if *true* is identified with 1.

Using Boolean addition strongly reduces the size of formulas. For instance, the fact that a station p is exactly in one state at a time is expressed by the formula

$$at\ p_0 + at\ p_1 + at\ p_2\ =\ 1.$$

The formula

$$at\ C + \sum_{p=1}^{n}[at\ C_p + at\ C'_p]\ =\ 1$$

expresses the same property for the controller.

5.4 The invariant

Let us now consider the formula I:

$$
\begin{array}{ll}
& \forall p\,[at\ p_1 = (at\ C_p + p \in E)] \\
\wedge & \forall p\,[(INCS = p) = (at\ C'_p + at\ p_2)] \\
\wedge & (INCS = 0 \Rightarrow E = \emptyset)
\end{array}
$$

This formula is an invariant of the system. Let us verify, for instance, that I is respected by transition $4(p)$. This amounts to check the validity of the triple

$$\{I\}\,(C_p,\ INCS \neq 0 \longrightarrow E := E \cup \{p\}\ ,C)\,\{I\}.$$

This triple is rewritten as

$$\{I_{[at\ C_p]} \wedge INCS \neq 0\}\ E := E \cup \{p\}\ \{I_{[at\ C]}\}.$$

which is expanded into

$$
\begin{array}{l}
\{\ at\ p_1 \wedge p \notin E \wedge INCS \neq 0 \\
\wedge\ \forall q \neq p\,[at\ q_1 = q \in E] \\
\wedge\ \forall q\,[(INCS = q) = at\ q_2]\ \}
\end{array}
$$

$$E := E \cup \{p\}$$

$$
\begin{array}{l}
\{\ \forall q\,[at\ q_1 = q \in E] \\
\wedge\ \forall q\,[(INCS = q) = at\ q_2] \\
\wedge\ (INCS = 0 \Rightarrow E = \emptyset)\ \}
\end{array}
$$

It is also equivalent to the implication

$$
\begin{array}{l}
\{\ at\ p_1 \wedge p \notin E \wedge INCS \neq 0 \\
\wedge\ \forall q \neq p\,[at\ q_1 = q \in E] \\
\wedge\ \forall q\,[(INCS = q) = at\ q_2]\ \}
\end{array}
$$

$$\Rightarrow$$

$$
\begin{array}{l}
\{\ \forall q\,[at\ q_1 = q \in E \cup \{p\}] \\
\wedge\ \forall q\,[(INCS = q) = at\ q_2] \\
\wedge\ (INCS = 0 \Rightarrow E \cup \{p\} = \emptyset)\ \}
\end{array}
$$

which can be rewritten into

$$\{ \ at \ p_1 \ \wedge \ p \notin E \ \wedge \ INCS \neq 0$$
$$\wedge \ \forall q \neq p \, [at \ q_1 = q \in E]$$
$$\wedge \ \forall q \, [(INCS = q) = at \ q_2] \ \}$$

$$\Rightarrow$$

$$\{ \ [at \ p_1 = p \in E \cup \{p\}]$$
$$\wedge \ \forall q \neq p \, [at \ q_1 = q \in E \cup \{p\}]$$
$$\wedge \ \forall q \, [(INCS = q) = at \ q_2]$$
$$\wedge \ \neg(INCS = 0) \ \}$$

This last implication is easily shown to be valid; this establishes the proposition.

The formula I is checked to be respected by the other transitions in a similar way. Such verifications can be tedious but never are difficult. They can be handled in a mechanical way.

Let us also obsrve that, at the initial state, the invariant reduces to the formula

$$\forall p \, [0 = (0 + 0)]$$
$$\wedge \quad \forall p \, [0 = (0 + 0)]$$
$$\wedge \quad (true \Rightarrow true)$$

which is identically true.

5.5 Some safety properties

A first consequence of the invariant is the mutual exclusion. Indeed, the mutual exclusion is expressed by the formula

$$\sum_{p=1}^{n} at \ p_2 \leq 1.$$

However, this formula is a logical consequence of the formula

$$\forall p \, [at \ p_2 \Rightarrow INCS = p],$$

which itself is a logical consequence of the invariant.
Other interesting properties can be deduced from the invariant. First, the formula

$$INCS = 0 \Rightarrow E = \emptyset$$

expresses that the resource is well managed: it is idle only when the waiting set is empty. Second, the formula

$$p \in E \Rightarrow at \ p_1$$

expresses that the resource can be granted only to stations which are waiting for it.

The discovery of an adequate invariant for a CSP network, or the design of a network respecting some safety property, is not always easy (see [2]).

6 A more realistic version

The methodology introduced in section 4 will now be used to obtain a correct asynchronous solution of the mutual exclusion problem.

6.1 Replacing one channel

As a first step, the (synchronous) channel REQ_p is tentatively replaced, for some fixed p, by an asynchronous one-place channel, also called REQ_p. This channel will transmit only synchronization bits; furthermore, as "send" and "receive" statements are executed strictly in turn, the "send" operation will be modelled by an non-conditional statement. Formally, this amounts to replace the transition $2(p)$ by the transitions

$$2a(p).\ \ (p_0,\ crs_p \ \longrightarrow\ REQ_p := 1\ ,\, p_1)\, ,$$
$$2b(p).\ \ (C,\ REQ_p = 1\ \longrightarrow\ REQ_p := 0\ ,\, C_p)\, .$$

Transition $2a(p)$ contains the "send" and transition $2b(p)$ contains the "receive".

Let us now check whether these new two transitions respect the $2n + 1$ terms of the invariant I. We see that transitions $2a(p)$ and $2b(p)$ do respect all but one assertion I^+ of the invariant, which is:

$$at\ p_1 \ =\ (at\ C_p + p \in E)\, .$$

Let I^0 be the conjunction of the remaining assertions. The new invariant I_1 will be a formula like

$$I^0 \wedge (REQ_p = 0 \Rightarrow I^+) \wedge (REQ_p = 1 \Rightarrow K),$$

where K satisfies the constraints listed below.

$$I^0 \ \vDash\ sp[2a(p);\ (I^+ \wedge REQ_p = 0)] \Rightarrow (K \wedge REQ_p = 1),$$
$$I^0 \ \vDash\ (K \wedge REQ_p = 1) \Rightarrow wp[2b(p);\ (I^+ \wedge REQ_p = 0)],$$
$$I^0 \ \vDash\ \{K\}\, T_i\, \{K\}\, .$$

As REQ_p is a one-place channel and is used for synchronization only, the system (11) has been considered. The identities (1) are used to expand the first two constraints of this system into

$$I^0 \ \vDash\ (at\ p_1 \wedge sp[REQ_p := 1;\ (crs_p \wedge I^+_{[at\ p_0]} \wedge REQ_p = 0)]) \Rightarrow (K \wedge REQ_p = 1),$$
$$I^0 \ \vDash\ (K \wedge REQ_p = 1) \Rightarrow ([at\ C \wedge REQ_p = 1] \Rightarrow wp[REQ_p := 0;\ (I^+_{[at\ C_p]} \wedge REQ_p = 0)])\, .$$

The identities

$$I^+ \ \equiv\ [at\ p_1 \ =\ (at\ C_p + p \in E)],$$
$$I^+_{[at\ p_0]} \ \equiv\ [\neg at\ C_p \wedge p \notin E],$$
$$I^+_{[at\ C_p]} \ \equiv\ [at\ p_1 \wedge p \notin E]$$

allows to rewrite the constraints into

$$I^0 \ \vDash\ (at\ p_1 \wedge sp[REQ_p := 1;\ (crs_p \wedge \neg at\ C_p \wedge p \notin E \wedge REQ_p = 0)]) \Rightarrow (K \wedge REQ_p = 1),$$
$$I^0 \ \vDash\ (K \wedge REQ_p = 1) \Rightarrow ([at\ C \wedge REQ_p = 1] \Rightarrow wp[REQ_p := 0;\ (at\ p_1 \wedge p \notin E \wedge REQ_p = 0)])\, .$$

Elementary calculus leads now to

$$I^0 \ \vDash\ (at\ p_1 \wedge crs_p \wedge \neg at\ C_p \wedge p \notin E \wedge REQ_p = 1) \Rightarrow (K \wedge REQ_p = 1),$$
$$I^0 \ \vDash\ (K \wedge REQ_p = 1) \Rightarrow ([at\ C \wedge REQ_p = 1] \Rightarrow [at\ p_1 \wedge p \notin E])\, .$$

As a result, the maximum and the minimum for the candidate-set corresponding to K are respectively

$$at\ p_1 \wedge crs_p \wedge \neg at\ C_p \wedge p \notin E,$$
$$at\ C \Rightarrow (at\ p_1 \wedge p \notin E).$$

It is checked easily that the strongest choice is adequate here; however, as the predicate crs_p is internal to station p, it is not useful to introduce it in the invariant, and this term can be dropped. The invariant I_1 of the new version is therefore

$$I^0 \wedge (REQ_p = 0 \Rightarrow [at\ p_1 = (at\ C_p + p \in E)]) \wedge (REQ_p = 1 \Rightarrow (at\ p_1 \wedge \neg at\ C_p \wedge p \notin E)),$$

which simplifies into $(I^0 \wedge [at\ p_1 = (at\ C_p + REQ_p + p \in E)])$. The full invariant I_1 is

$$
\begin{aligned}
& [at\ p_1 = (at\ C_p + REQ_p + p \in E)] \\
& \forall q \neq p\ [at\ q_1 = (at\ C_q + q \in E)] \\
\wedge\ & \forall q\ [(INCS = q) = (at\ C'_q + at\ q_2)] \\
\wedge\ & (INCS = 0 \Rightarrow E = \emptyset)
\end{aligned}
$$

This establishes that the channel REQ_p can be implemented as an asynchronous one-place buffer with maintaining the mutual exclusion and the other useful safety properties.

6.2 Replacing all channels

It can be thought that all channels REQ_q can be implemented as asynchronous one-place buffers, and an obvious guess for the adequate invariant I_2 is

$$
\begin{aligned}
& \forall p\ [at\ p_1 = (at\ C_p + REQ_p + p \in E)] \\
\wedge\ & \forall p\ [(INCS = p) = (at\ C'_p + at\ p_2)] \\
\wedge\ & (INCS = 0 \Rightarrow E = \emptyset)
\end{aligned}
$$

The validity of this guess is checked in a straightforward way.

As further steps, it is attempted to implement authorizations and releases channels in an asynchronous way. As the case is very similar to the case of requests channels, only the resulting transition system and the corresponding invariant are written here.

The transition system is[4]

1. $(p_0,\ \neg crs_p\ \longrightarrow\ [\ ],\ p_0),\ \forall p$
2a. $(p_0,\ crs_p\ \longrightarrow\ REQ_p := 1,\ p_1),\ \forall p$
3a. $(p_1,\ OK_p = 1\ \longrightarrow\ OK_p := 0,\ p_2),\ \forall p$
5. $(p_2,\ crs_p\ \longrightarrow\ [\ ],\ p_2),\ \forall p$
6a. $(p_2,\ \neg crs_p\ \longrightarrow\ END_p := 1,\ p_0),\ \forall p$
7a. $(q_1,\ OK_q = 1\ \longrightarrow\ OK_q := 0,\ q_2),\ \forall q$

2b. $(C,\ REQ_p = 1\ \longrightarrow\ REQ_p := 0,\ C_p),\ \forall p$
3b. $(C_p,\ INCS = 0\ \longrightarrow\ INCS := p;\ OK_p := 1,\ C),\ \forall p$
4. $(C_p,\ INCS \neq 0\ \longrightarrow\ E := E \cup \{p\},\ C),\ \forall p$
6b. $(C,\ END_p = 1\ \longrightarrow\ END_p := 0,\ C'_p),\ \forall p$
7b. $(C'_p,\ q \in E\ \longrightarrow\ (INCS, E) := (q, E \setminus \{q\});\ OK_q := 1,\ C),\ \forall p, q$
8. $(C'_p,\ E = \emptyset\ \longrightarrow\ INCS := 0,\ C),\ \forall p$

---

[4]Transitions 3a and 7a are the same; one of them may be omitted.

and an adequate invariant I_3 is

$$\forall p\,[\,at\ p_1 = (at\ C_p + REQ_p + OK_p + p \in E)\,]$$
$$\wedge\quad \forall p\,[\,(INCS = p) = (at\ C'_p + OK_p + END_p + at\ p_2)\,]$$
$$\wedge\quad (INCS = 0 \Rightarrow E = \emptyset)$$

6.3   Splitting the controller

A last transformation is now attempted. The controller consists of a single process which receives both requests and releases. Why not split it into two processes, one for the requests, the other for the releases? Naturally, these processes would share the variable $INCS$ and the queue E.

Formally, this transformation is achieved by replacing the identity

$$at\ C + \sum_{p=1}^{n} [at\ C_p + at\ C'_p] = 1$$

by the following two identities:

$$at\ C + \sum_{p=1}^{n} at\ C_p = 1 \qquad \text{and} \qquad at\ C' + \sum_{p=1}^{n} at\ C'_p = 1\,.$$

Furthermore, the second part of the transition system given in paragraph 6.2 is replaced by

2b. $(C,\ REQ_p = 1\ \longrightarrow\ REQ_p := 0\ ,C_p)$, $\forall p$
3c. $(C_p,\ <INCS = 0\ \longrightarrow\ INCS := p;\ OK_p := 1\ >,\ C)$, $\forall p$
4a. $(C_p,\ <INCS \neq 0\ \longrightarrow\ E := E \cup \{p\}\ >,\ C)$, $\forall p$

6c. $(C',\ END_p = 1\ \longrightarrow\ END_p := 0\ ,C'_p)$, $\forall p$
7c. $(C'_p,\ <q \in E\ \longrightarrow\ (INCS,E) := (q, E \setminus \{q\});\ OK_q := 1\ >,\ C')$, $\forall p,q$
8a. $(C'_p,\ <E = \emptyset\ \longrightarrow\ INCS := 0\ >,\ C')$, $\forall p$

As shared memory has been introduced, the granularity of parallelism becomes relevant; as usual, statements which must be implemented atomically are surrounded by angle brackets.

This transformation could possibly alter the invariant of the system, but it can be checked that, in this particular case, no further adaptation is needed. We only verify here that transition $6c(p)$ respects the assertion

$$A:\quad (INCS = p) = (at\ C'_p + OK_p + END_p + at\ p_2)$$

First, the condition $wp[6c(p);\ A]$ is evaluated:

$$wp[6c(p);\ A]\ =\ ([at\ C' \wedge END_p = 1]\ \Rightarrow\ wp[END_p := 0;\ A_{[at\ C'_p]}])\,,$$
$$=\ ([at\ C' \wedge END_p = 1]\ \Rightarrow\ [INCS = p \wedge OK_p = 0])\,.$$

Second, the condition $(I_3 \Rightarrow wp[6c(p);\ A])$ must be proved. This condition is equivalent to

$$(I_3 \wedge at\ C' \wedge END_p = 1)\ \Rightarrow\ (INCS = p \wedge OK_p = 0)\,,$$

which is identically true.

7 Conclusion

The methodology presented here allows the designer of a distributed system to work first at the abstract level of CSP and then to obtain a more efficient version in a systematic way. The development involves much symbolic computation, but this computation can be mechanized.

References

[1] E.W. DIJKSTRA, "A discipline of programming", Prentice Hall, New Jersey, 1976.

[2] E.P. GRIBOMONT, "Design and proof of communicating sequential processes", LNCS, vol. 259, pp. 261-276, 1987.

[3] D. GRIES, "The Science of Programming", Springer-Verlag, Berlin, 1981

[4] C.A.R. HOARE, "An axiomatic basis for computer programming", CACM, vol. 12, pp. 576-583, 1969

[5] C.A.R. HOARE, "Communicating Sequential Processes", CACM, vol. 21, pp. 666-677, 1978.

[6] C.A.R. HOARE, "Communicating Sequential Processes", Prentice Hall, New Jersey, 1985.

[7] R.M. KELLER, "Formal Verification of Parallel Programs", CACM, vol. 19, pp. 371-384, 1976.

[8] J. SIFAKIS, "A unified approach for studying the properties of transition systems", TCS, vol. 18, pp. 227-259, 1982.

[9] R.D. SCHLICHTING, F.D. SCHNEIDER, "Using Message Passing for Distributed Programming: Proof Rules and Disciplines", ACM Toplas, vol. 6, pp. 402-431, 1984.

Formal Specification and Verification

of Asynchronous Processes

in Higher-Order Logic

Jeffrey J. Joyce

University of Cambridge
Computer Laboratory
Cambridge, England

Abstract

We model the interaction of a synchronous process with an asynchronous memory process using a four-phase "handshaking" protocol. This example demonstrates the use of higher-order logic to reason about the behaviour of synchronous systems such as microprocessors which communicate requests to asynchronous devices and then wait for unpredictably long periods until these requests are answered. Experience with this example suggests that higher-order logic may also be a suitable formalism for reasoning about more abstract forms of concurrency.

1 Introduction

This paper describes the use of higher-order logic to model the interaction of a synchronous process with an asynchronous process using a four-phase "handshaking" protocol. Other formalisms, notably temporal logic, have also been used to reason about handshaking protocols. In our approach, universal and existential quantification are used in approximately the same roles as the operators $\Box$ ("forever") and $\Diamond$ ("eventually"). But unlike temporal logic, our specifications allow explicit references to time; for example, $f(t)$ and $f(t+1)$ denote the values of signal f at two successive points in discrete time. Higher-order functions are used to define predicates expressing conditions such a signal rising or the stability of a signal over a specified interval. The use of explicit time references and higher-order functions leads to convenient mechanisms for deriving abstract views of system behaviour where intervals of time are collapsed into single time steps.

Our use of higher-order logic to verify signalling protocols is based on experience with this formalism in verifying synchronous hardware. We have used higher-order logic to verify the correctness of a very simple microprocessor where the formal semantics of the instruction set have been formally derived from a switch-level model of MOS circuit behaviour [15]. The current design for this microprocessor assumes that every read and write request to external memory is completed in a single clock cycle. This assumption could be validated by a detailed analysis of timing characteristics for both the microprocessor and the external memory; for example, see [1]. We are now designing a new version of this microprocessor which has wait states where the microprocessor waits one or more clock cycles after issuing a read or write request to the external memory until the request is acknowledged. Wait states are implemented by while-loops in the microcode.

The example described in this paper was developed to investigate how interaction between the microprocessor and the asynchronous memory could be formally specified in higher-order logic. To focus more clearly on this problem, the microprocessor specification is replaced by an abstract state machine which models the interaction of the microprocessor with the asynchronous memory. The abstract state machine implements a synchronous interface which accepts synchronous read and write requests and communicates these asynchronously to the external memory. We develop formal specifications for both the synchronous process and the asynchronous memory and prove that these two processes correctly implement the four-phase handshaking protocol.

2   Related Work

Warren Hunt [12] also considers the formal specification and verification of a microprocessor system which communicates asynchronously with external memory. The absence of existential quantification in quantifier-free Boyer-Moore logic used by Hunt motivates the use of "oracles" to predict asynchronous responses from the memory. An oracle is a list of numbers which specifies the length of wait states for a particular execution sequence; Hunt's microprocessor is shown to be correct for all such oracles.

The VIPER microprocessor also uses four-phase handshaking to communicate asynchronously with external memory [20]. The formal proof described by Avra Cohn in [3] deals with higher level aspects of the VIPER implementation where the details of asynchronous communication are not considered. However, the proof models the possibility of a failed memory interaction which occurs when a wait state exceeds a specified "timeout" period.

The use of higher-order logic to reason about asynchronous behaviour in the verification of latches and flip-flops implemented by fixed delay NOR gates is described by John Herbert in [9]. Herbert reasons about asynchronous behaviour at the nanosecond time scale whereas this paper focuses on the behaviour of an asynchronous memory in terms of a much coarser grain of time corresponding to the behaviour of clocked sequential circuits. However, the main difference in this paper is that the delayed response of the asynchronous component, *i.e.* the asynchronous memory, is not fixed.

3 Specification in Higher-Order Logic

Functions which accept other functions as arguments or return functions as results are called "higher-order" functions. This property often leads to function definitions which are both simple and direct. Higher-order logic is a formalism for reasoning about functions; this includes higher-order functions as well as relations, *i.e.* functions that return Boolean values.

The behaviour of a hardware device can be described by a relation on input and output signals where signals are modelled by functions from discrete time to sampled values. Because signals are modelled by functions, these will be higher-order relations. This is one reason why higher-order logic is a convenient formalism for reasoning about functions and relations modelling hardware devices.

This approach, inspired by [6], can also be used to describe input-output behaviour at the more abstract level of concurrent processes. The synchronous interface and asynchronous memory correspond to physical devices but the discussion in this paper is only concerned with their external behaviour. Their behaviour is described by relations on a set of signals associated with each process. Some of these signals correspond to wires and busses; other signals represent internal states and do not have physical counterparts.

Thus, the behaviour of a single process is a set of relations, or constraints, on output signals in terms of input signals and internal state signals. The behaviour of a network of concurrent processes is obtained by composing the constraints imposed by each process. This is expressed formally by logical conjunction which serves as a behaviour composition operator. Another operator which appears in specification languages such as CCS [17] and CSP [11] is hiding (or restriction). In our higher-order logic approach, we use existential quantification to hide internal signals, *i.e.* signals which are not external ports of the network.

Much of our notation such as "∀" (*"for all"*) and "∃" (*"there exists"*) should be familiar from ordinary predicate calculus. Higher-order logic also includes function-denoting terms called λ-expressions as well as Hilbert's ε-operator. For example, the λ-expression "λx. x + 1" denotes the successor function. The term "εx. P x" denotes a value satisfying the predicate P if such a value exists; otherwise, it denotes an arbitrary value of the appropriate type. For example, "εx. x < 10" denotes some natural number less than ten but "εx. x < 0" denotes an arbitrary natural number since the predicate "less than zero" cannot be satisfied in the natural numbers.

In addition to standard logical connectives, the language provides conditional expressions, "b ⇒ t1 | t2", which may be read as *"if b then t1 else t2"* and *n*-tuples of the form "(t1, ... tn)". Certain features of the language have a special syntactic status for improved readability, *e.g.* the definition of infix functions. The language also includes basic arithmetic functions and relations such as +, -, < and ≤.

4 Signal Types

Every term in higher-order logic has a type which is either a primitive type (*e.g.* Booleans, natural numbers) or built-up from existing types using type constructors such as Cartesian product. Every compound term must correctly type-check; for example, a function from numbers to Booleans can only be applied to a number. This section describes function types used to represent various signals in the synchronous interface and asynchronous memory.

A single wire is modelled by a function which maps every point in discrete time to either Hi or Lo. For instance, the function f models a signal which is low when sampled at times t and $t + 3$ and high at times $t + 1$ and $t + 2$.

$$... \; f(t) = \text{Lo}, \; f(t + 1) = \text{Hi}, \; f(t + 2) = \text{Hi}, \; f(t + 3) = \text{Lo}, \; ...$$

Hi and Lo are logical constants denoting a pair of distinct values of a type val. These constants could be synonyms for true and false in a simple Boolean model of circuit behaviour or they could denote signal values in a more complex circuit model. For simplicity, we assume that Hi and Lo are the only two values of the type val; a similar effect can be achieved for a larger set of values with "well-definedness" assumptions; for example, see [4] or [15].

A simple model is also used for n-bit word values, namely natural numbers, ignoring the fact that n-bit words are finite which is not important in this example. Hence, a bus is modelled by a function from discrete time to the natural numbers. A third type of signal represents a memory state as it varies over time; this is modelled by a function from discrete time to a memory state which, in turn, is modelled by a function from addresses to n-bit word values (this is another use of higher-order functions). Addresses are also modelled by natural numbers ignoring the fact that memories are finite which, once again, is not important in our example. These conventions are summarized by the following type abbreviations where $ty1 \rightarrow ty2$ denotes the type of all total functions with arguments of type $ty1$ and results of type $ty2$.

$$\text{val} = \{\text{Hi},\text{Lo}\}$$
$$\text{wire} = \text{time} \rightarrow \text{val}$$
$$\text{bus} = \text{time} \rightarrow \text{number}$$
$$\text{memory} = \text{time} \rightarrow (\text{number} \rightarrow \text{number})$$

We also introduce a type abbreviation for Boolean signals which are functions from discrete time to Boolean values. These may be thought of as virtual signals testing for specific logical conditions. For example, a Boolean signal could be defined to test when a particular wire is high. In the formal specification of the four-phase handshaking protocol, Boolean signals are often used to detect synchronization points such as when a memory request is signalled. We sometimes refer to Boolean signals as event signals or sampling functions.

$$\texttt{boolsig = time} {\rightarrow} \texttt{boolean}$$

Some predicates such as Stable (see next section) are used to describe signals of more than one type. A wire, bus, memory or even a logical condition can all be described as being stable over a specified interval of time. Type variables provide the logic with a limited amount of polymorphism. The following type abbreviation is parameterized by a type variable α to describe the type of a polymorphic signal. Predicates such as Stable are defined in terms of this polymorphic type.

$$\alpha \ \texttt{sig = time} {\rightarrow} \alpha$$

5 Temporal Predicates

Reasoning about asynchronous behaviour involves reasoning about the behaviour of signals over intervals of time. This section introduces some higher-order predicates which make it easier to describe conditions which involve one or more instants of time.

IsHi and IsLo are curried functions, that is, functions which take their arguments 'one at a time'. When applied to a wire signal, e.g. "IsHi w", these functions return Boolean signals detecting when the wire is high (or low). When applied to both a wire signal and time value, e.g. "IsHi w t", these functions return Boolean values indicating whether the wire is high (or low) at the specified time.

Definition 5.1:
⊢ IsHi (w:wire) (t:time) = (w t = Hi)

Definition 5.2:
⊢ IsLo (w:wire) (t:time) = (w t = Lo)

Rises and Falls are also curried functions. These functions are used to define Boolean signals which detect transitions from low to high and *vice versa* as shown in Figure 1. For example, a wire rises at time t if it is low at time t and then high at time t+1.

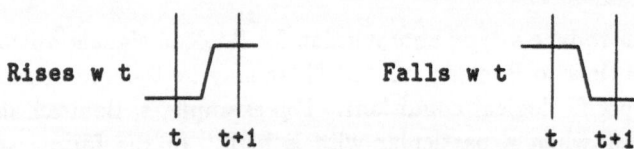

Figure 1: Rising and Falling Signals

Definition 5.3:
⊢ Rises (w:wire) (t:time) = ((w t = Lo) ∧ (w (t+1) = Hi))

Definition 5.4:
⊢ Falls (w:wire) (t:time) = ((w t = Hi) ∧ (w (t+1) = Lo))

Definition 5.5 shows the definition of stability during an interval. This definition takes advantage of polymorphic types provided in the HOL logic to define a predicate which will serve for all types of signal values.

Definition 5.5:
⊢ Stable (s:α sig,v:α,t1:time,t2:time) =
    ∀t. t1 ≤ t ∧ t < t2 ⟹ (s t = v)

Because asynchronous delays are not fixed, we often need to express the condition that a signal remains stable at a given value until a specified event occurs. If the event never occurs, then the signal remains stable forever. We define two predicates, StableUntil1 and StableUntil2, which capture slightly different interpretations of this informal notion. Polymorphic types are also used here so that these predicates can be used for all types of signals.

Definition 5.6:
⊢ StableUntil1 (s:α sig,v:α,t:time,b:boolsig) =
    ∀n. Stable (b,F,t,t+n) ⟹ (s (t+n) = v)

Definition 5.7:
⊢ StableUntil2 (s:α sig,v;α,t:time,b:boolsig) =
    ∀n. Stable (b,F,t,t+n) ⟹ (s ((t+n)+1) = v)

The choice between StableUntil1 and StableUntil2 is a matter of fine-tuning when these predicates are used to describe the behaviour of the synchronous in-

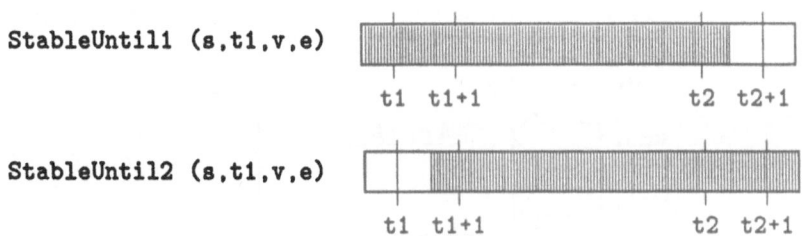

Figure 2: Intervals described by StableUntil1 and StableUntil2

terface and the asynchronous memory. The predicate StableUntil1 refers to an interval which begins immediately and ends exactly when the specified event occurs. StableUntil2 refers to an interval which begins and ends one time unit later than the interval described by StableUntil1. Figure 2 illustrates the intervals described by these two predicates where the specified event occurs at time t2. The shaded region indicates the stable interval in each case.

Interaction between the synchronous interface and the asynchronous memory process is often described in terms of the "first" or "next" occurrence of a specified event such as a particular signal rising or falling. The predicate First expresses the property that a specified event occurs for the first time after the start of an interval. The predicate Next is similar to the First except that the inequality "<" is used to ensure that the "next" time is strictly after the start of the interval.

Definition 5.8:
$\vdash$ First (t1:time,t2:time) (b:boolsig) =
   (t1 $\leq$ t2) $\wedge$ Stable (b,F,t1,t2) $\wedge$ (b t2)

Definition 5.9:
$\vdash$ Next (t1:time,t2:time) (b:boolsig) =
   (t1 < t2) $\wedge$ Stable (b,F,t1+1,t2) $\wedge$ (b t2)

6 Asynchronous Memory

In this section we develop the formal specification of an asynchronous memory based on the informal specification presented in [12]. As shown in Figure 3, the asynchronous memory has four inputs and two outputs. A read (write) request is signalled by a transition from low to high on the read (write) line. The request is eventually acknowledged by a transition from low to high on the dtack line. A request to end the cycle can then be signalled by a transition from high to low on the read (write) line which is acknowledged by a similar transition from high to low on the dtack line. This four-phase handshaking protocol is shown in Figure 4 for a read request.

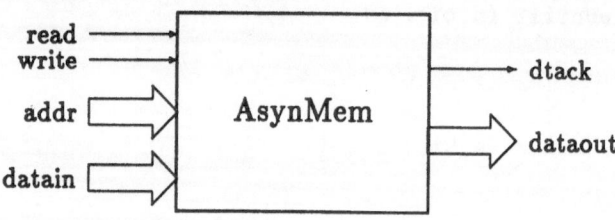

Figure 3: Asynchronous Memory Device

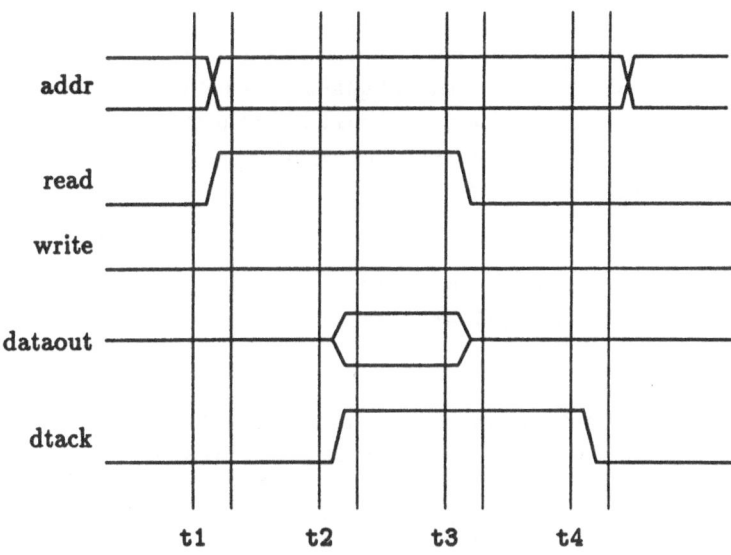

Figure 4: Read Cycle using a Four-Phase Handshaking Protocol. *This simplified timing diagram shows the relative order of events in a read cycle where a request occurs at t1 followed by an acknowledgement at t2. The acknowledgement may occur in the same interval as the request, i.e. t2 may be equal to t1. Similarly, t4 may be equal to t3. However, t3 must occur strictly after t1 and t4 strictly after t2 since signals are not allowed to rise and fall within the same interval.*

Once a read or write request has been signalled, several conditions must be maintained until the asynchronous memory responds or until the end of the cycle. After signalling a read request, the read line must remain high until the request is acknowledged. Furthermore, a write request cannot be signalled during a read cycle; that is, the **write** line must remain low until the end of the read cycle. Similarly, the address bus must remain stable throughout the cycle. During a write cycle, the **write** line must remain high until the request is acknowledged and the **read** line must remain low until the end of the cycle. In this case, the data bus as well as the address bus must remain stable throughout the write cycle. For both read and write cycles, a request to end the cycle is signalled by read or write falling. The next cycle is not allowed to begin until the end of the current cycle is acknowledged by the asynchronous memory; that is, the **read** and **write** lines must remain low until the **dtack** line also falls. Between read and write cycles, when the asynchronous memory is idle, the **dtack** line remains low and the internal state of the memory is stable.

Working from this informal description, the formal specification of the asynchronous memory process can be developed by outlining the overall structure of a memory cycle specification and then filling in precise details. We focus on the behaviour of the asynchronous memory during a read cycle, as illustrated in Figure 4, and begin with a pseudo-formal specification which shows the overall structure of the read cycle specification. Universally quantified time variables denote when the synchronous interface may initiate a request to begin or end a cycle. Existential quantification is used to specify when the asynchronous memory must eventually acknowledge one of these requests.

```
∀t1.
   asynchronous memory idle ∧                               (1)
   read cycle requested at time t1 ∧                        (2)
   ⟹
∃t2.
   read request acknowledged at time t2 ∧                   (3)
   fetched memory word available as output ∧               (4)
   ∀t3.
      end-of-cycle requested at time t3 ∧                   (5)
      ⟹
      ∃t4.
         end-of-cycle acknowledged at time t4 ∧             (6)
         asynchronous memory returns to idle state ∧        (7)
         internal memory state unchanged during cycle       (8)
```

In the terminology of [18], lines (1) to (2) and (5) specify "domain" constraints which are assumptions about inputs to the asynchronous memory. Lines (3) to (4) and (6) to (8) specify "functional" constraints, that is, outputs and internal state changes which the memory must produce in response to inputs.

Line (1) requires the asynchronous memory to be in an idle state when a read cycle is initiated (it cannot already be in the middle of another memory request). Unfortunately, the asynchronous memory device does not provide a "idle" flag as a physical output signal; while this would simplify the task of writing a formal specification, real memory devices do not usually provide such a signal. Clearly, the value of the dtack signal at any particular moment does not indicate when the memory is in an idle state since it can remain low for some time after a read or write cycle begins. For instance, the dtack signal is still low at time t1+1 in Figure 4 even though a read cycle has already begun and the memory is no longer idle. Instead, one of the conditions indicating that the memory is idle is that the dtack signal is low and will remain low until either read or write rises. The other condition is that the memory state is also stable until the next read or write request is signalled. These two conditions are easily expressed with the predicate StableUntil in Definition 6.2. We also define an infix function OrWhen which combines two event signals using logical "or".

Definition 6.1:
⊢ (b1:boolsig) OrWhen (b2:boolsig) = λt. b1 t ∨ b2 t

Definition 6.2:

⊢ MemIdle (read:wire,write:wire,dtack:wire,mem:memory) t =
    StableUntil1 (dtack,Lo,t,(Rises read) OrWhen (Rises write)) ∧
    StableUntil1 (mem,mem t,t,(Rises read) OrWhen (Rises write))

Line (2) refers to the initiation of a read cycle. A read request is signalled by a transition from from Lo to Hi on the read line. Once high, the read signal is required to stay high until the request is acknowledged; otherwise the behaviour of the asynchronous memory is undefined. More precisely, the read signal must remain stable up to and including some point when the dtack line is also high. Furthermore, the write signal must remain low and the address bus stable until the end of the read cycle, *i.e.* until dtack falls. These conditions are expressed formally by the following term.

    Rises read t1 ∧
    StableUntil2 (read,Hi,t1,Rises dtack) ∧
    StableUntil2 (write,Lo,t1,Falls dtack) ∧
    StableUntil2 (addr,address,t1,Falls dtack)

The eventual acknowledgement of the read request is signalled as soon as the dtack signal rises, line (3). Once the read request has been acknowledged, the dtack signal will remain high until the read signal falls. The acknowledgement of the request also means that the contents of memory at the given address are now available on dataout until the read signal falls, line (4). The function constant FETCH is used to represent this memory operation. Another constant, STORE, is used to specify the effect of a write cycle on the internal state of the asynchronous memory (see the Appendix for an example of its use). The memory operations denoted by these two constants are not formally defined here because we never need to reason about the effect of storing and later fetching a value from memory.

    First (t1,t2) (Rises dtack) ∧
    StableUntil1 (dtack,Hi,t2+1,Falls read) ∧
    StableUntil1 (dataout,FETCH (mem t1) address,t2+1,Falls read)

Line (5) refers to when read falls signalling a request to end the read cycle. Once low, the read signal must remain low until the request is acknowledged by the asynchronous memory.

    First (t2,t3) (Falls read) ∧
    StableUntil2 (read,Lo,t3,Falls dtack)

Lines (6) and (7) specify that the asynchronous memory will eventually acknowledge this request by resetting the dtack signal and entering the idle state. Finally, line (8) states that the internal state of the memory will be unchanged from its state at the beginning of the cycle.

```
        First (t3,t4) (Falls dtack) ∧
        MemIdle (read,write,dtack,mem) (t4+1) ∧
        (mem (t4+1) = mem t1)
```

The formal specification of the asynchronous memory behaviour during a read cycle is obtained by replacing the numbered lines of the pseudo-formal specification with these precise details. The predicate AsynMemRead is defined in terms of this specification. The variable address, which denotes the stable value of the addr signal during the read cycle, is included in the outermost quantification.

Definition 6.3:
```
⊢ AsynMemRead (
    read:wire,write:wire,addr:bus,datain:bus,
    dtack:wire,dataout:bus,
    mem:memory) =
    ∀ t1 address.
      Rises read t1 ∧
      MemIdle (read,write,dtack,mem) t1 ∧
      StableUntil2 (read,Hi,t1,Rises dtack) ∧
      StableUntil2 (write,Lo,t1,Falls dtack) ∧
      StableUntil2 (addr,address,t1,Falls dtack)
      ⟹
      ∃t2.
        First (t1,t2) (Rises dtack) ∧
        StableUntil1 (dtack,Hi,t2+1,Falls read) ∧
        StableUntil1 (
          dataout,FETCH (mem t1) address,t2+1,Falls read) ∧
        ∀t3.
        First (t2,t3) (Falls read) ∧
        StableUntil2 (read,Lo,t3,Falls dtack)
        ⟹
        ∃t4.
          First (t3,t4) (Falls dtack) ∧
          MemIdle (read,write,dtack,mem) (t4+1) ∧
          (mem (t4+1) = mem t1)
```

The formal specification of a write cycle is very similar to the above specification of the read cycle with the roles of read and write exchanged. There is also the extra condition that the data bus must remain stable throughout the write cycle. The formal specification uses the function constant STORE to specify that the memory state is updated as expected by the end of the write cycle. These two specifications are then combined to define a predicate, AsynMem, which denotes the behaviour of the asynchronous memory process. The formal specification also states that the memory will be initially idle at time 0. The complete specification is given in the Appendix.

7 Synchronous Interface

In this section we formally specify the behaviour of a synchronous interface to the memory device described in the previous section. To focus our discussion, we ignore implementation issues, *i.e.* the interconnection of clocked registers and control logic, and specify the behaviour of the synchronous interface in terms of an abstract state machine. The derivation of abstract state machine behaviour from an implementation has been demonstrated for several other examples including those described in [2] and [14].

As shown in Figure 5, the interface has four synchronous inputs and two synchronous outputs. The interface receives two asynchronous inputs from the memory and communicates requests and data to the memory through the four asynchronous outputs. To easily distinguish between synchronous and asynchronous signals with a common name, we adopt the naming convention of prefixing asynchronous signals with an "a", *e.g.* read and a_read.

The synchronous interface waits in an idle state until the read or write signals becomes high. As soon as this happens, the synchronous address and data busses are latched and the synchronous interface begins either a read or write cycle. After signalling a request to the asynchronous memory, the interface waits some number of clock ticks until the asynchronous memory responds. As soon as the request is acknowledged, the interface immediately resets the a_read or a_write line and waits one or more clock ticks until the asynchronous memory terminates the memory cycle. In the case of a read cycle, the value of a_dataout is latched just before resetting the a_read line. At the end of the memory cycle, the interface returns to the idle state.

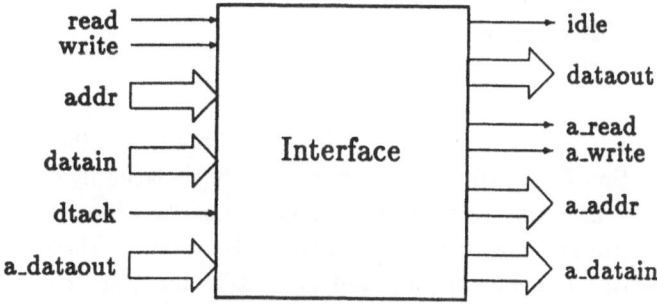

Figure 5: Synchronous Interface Device

We begin with a PASCAL-like notation to describe the implementation of the handshaking protocol in the synchronous interface. The wait states are implemented by repeat-until constructs where each iteration corresponds to a single clock tick. Assignment statements are used to specify both current outputs and updates to the internal state of registers in the interface. This algorithm is illustrated by the flow graph of Figure 6.

```
state0: idle := Hi;
        a_read := Lo;
        a_write := Lo;
        repeat_each_clock_tick {
          a_addr := addr;
          a_datain := datain
        } until ((read = Hi) or (write = Hi));
        if (read = Hi) goto state1 else goto state2;

state1: idle := Lo;
        a_read := Hi;
        repeat_each_clock_tick {
          dataout := a_dataout
        } until (dtack = Hi);
        goto state3;

state2: idle := Lo;
        a_write := Hi;
        repeat_each_clock_tick {
        } until (dtack = Hi);
        goto state3;

state3: a_read := Lo;
        a_write := Lo;
        repeat_each_clock_tick {
        } until (dtack = Lo);
        goto state0;
```

The behaviour implied by this algorithm is formally specified in higher-order logic by Definition 7.1. Assignment statements are replaced by output equations, e.g. "idle t = Hi", and next state equations, e.g. "a_addr (t+1) = addr t". Unlike assignment statements in the PASCAL-like notation, these equations must explicitly describe current outputs and next state for every clock tick. In the above "program" there is an implicit notion of state which is updated by assignment statements. This contrasts with our style of writing formal specifications in higher-order logic where state is made explicit. We are using the PASCAL-like notation here as an informal description; the formal semantics of this notation or its relationship to our formal specifications is not considered.

At this point we should clarify the distinction between two uses of the word "state". In the more general sense, we use this word to describe the current contents of all storage devices in the implementation of a process including both storage of data, *e.g.* contents of registers, and storage of control information, *e.g.* the program counter in a microprocessor. We also use the word "state" to refer very specifically to control points in a sequential process such as the four nodes of the flow graph in Figure 6. In the abstract state machine, the sequence of control points is represented by a function from discrete time to the numbers, 0, 1, 2 and 3, used to label the nodes of the flow graph. Even though we already have a type abbreviation for signals of this type, bus, we introduce a new type abbreviation, counter, to emphasis that the state signal does not necessarily have a physical counterpart.

$$\text{counter} = \text{time} \rightarrow \text{number}$$

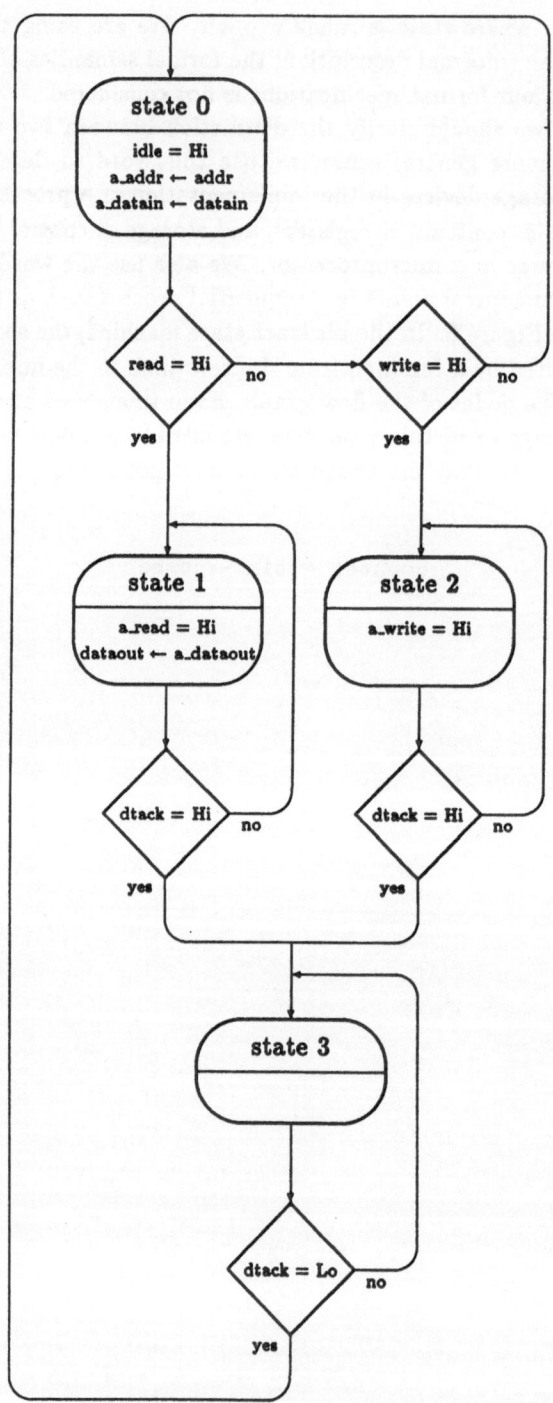

Figure 6: Flow Graph for the Synchronous Interface

Definition 7.1:
⊢ Interface (
    read:wire,write:wire,dtack:wire,
    addr:bus,datain:bus,a_dataout:bus,
    idle:wire,a_read:wire,a_write:wire,
    a_addr:bus,a_datain:bus,dataout:bus,
    state:counter) =
    (state 0 = 0) ∧
    ∀t.
      ((state t = 0) ⟹
        (idle t = Hi) ∧
        (a_read t = Lo) ∧
        (a_write t = Lo) ∧
        (a_addr (t+1) = addr t) ∧
        (a_datain (t+1) = datain t) ∧
        (dataout (t+1) = dataout t) ∧
        (state (t+1) =
          ((read t = Hi) ⇒ 1 | (write t = Hi) ⇒ 2 | 0))) ∧
      ((state t = 1) ⟹
        (idle t = Lo) ∧
        (a_read t = Hi) ∧
        (a_write t = Lo) ∧
        (a_addr (t+1) = a_addr t) ∧
        (dataout (t+1) = a_dataout t) ∧
        (state (t+1) = ((dtack t = Hi) ⇒ 3 | 1))) ∧
      ((state t = 2) ⟹
        (idle t = Lo) ∧
        (a_read t = Lo) ∧
        (a_write t = Hi) ∧
        (a_addr (t+1) = a_addr t) ∧
        (a_datain (t+1) = a_datain t) ∧
        (dataout (t+1) = dataout t) ∧
        (state (t+1) = ((dtack t = Hi) ⇒ 3 | 2))) ∧
      ((state t = 3) ⟹
        (idle t = Lo) ∧
        (a_read t = Lo) ∧
        (a_write t = Lo) ∧
        (a_addr (t+1) = a_addr t) ∧
        (a_datain (t+1) = a_datain t) ∧
        (dataout (t+1) = dataout t) ∧
        (state (t+1) = ((dtack t = Lo) ⇒ 0 | 3)))

Definition 7.1 also specifies that the initial state of the machine at time 0 is state 0 which is the idle state.

8 Formal Verification in Higher-Order Logic

The main emphasis of this paper is the use of higher-order logic as a specification language for asynchronous processes. However, higher-order logic is also a powerful means of reasoning about these specifications.

An interactive theorem-proving environment for higher-order logic is provided by the HOL system [7], a descendent of Edinburgh LCF [5]. A user of this system manipulates theorems as data objects in the meta-language. The meta-language is an interactive functional programming language called ML. Initially, only the axioms of higher-order logic exist as data objects. New theorems are generated by a small set of built-in ML functions which correspond to primitive inference rules of higher-order logic. Derived inference rules and powerful proof strategies can be programmed as ML functions to automate most of the minor steps, leaving the user to supply only the main steps in a proof [19]. Several large proofs involving more than a million primitive inference steps have been constructed in this system (for instance, see [3] or [15]).

The next two sections describe how higher-order logic can be used to reason about the specifications of the synchronous interface and the asynchronous memory, in particular, the interaction of these two processes. Our discussion focuses on theorems which highlight this process; we ignore details of their formal proof (although they have been generated as theorems by the HOL system).

9 Collapsing Wait States into Single Steps

States 1, 2 and 3 of the state machine implement wait states which occur during read and write cycles. In states 1 and 2 the machine waits until the dtack signal becomes high. In state 3 the machine waits until the dtack signal becomes low. During waits states, the outputs and internal state of the machine remain stable; hence, waits states can be viewed as single steps between synchronization points in the four-phase handshaking protocol.

Using induction on the length of a wait state, we can derive the behaviour of the machine in these wait states expressed in terms of StableUntil1. For example, Theorem 9.1 shows that the signals state, idle, a_read, a_write and a_addr remain stable while waiting in state 1 for an acknowledgement from the asynchronous memory.

Theorem 9.1:
```
⊢ Interface (
     read,write,dtack,addr,datain,a_dataout,
     idle,a_read,a_write,a_addr,a_datain,dataout,state)
   ⟹
   ∀t.
     (state t = 1)
     ⟹
     StableUntil1 (state,1,t,IsHi dtack) ∧
     StableUntil1 (idle,Lo,t,IsHi dtack) ∧
     StableUntil1 (a_read,Hi,t,IsHi dtack) ∧
     StableUntil1 (a_write,Lo,t,IsHi dtack) ∧
     StableUntil1 (a_addr,a_addr t,t,IsHi dtack)
```

Similar results can be derived for the behaviour of the machine in state 2 during a write cycle and for its behaviour in state 3 while waiting for the dtack signal to be reset. These theorems are used to reason about the interaction of the state machine with the asynchronous memory without further use of induction. Thus, we have factored out the inductive aspects of the verification task at this early stage in the formal proof.

10 Symbolic Simulation

This section explains how a proof technique called "symbolic simulation" is used to derive a more abstract view of interaction between the synchronous interface and the asynchronous memory where synchronization details of the handshaking protocol are hidden.

Figure 7 shows the interaction of the abstract state machine with the asynchronous memory during a read cycle. This timing diagram is consistent with the generalized read cycle shown in Figure 4 but shows how some of the unknowns in the generalized read cycle become fixed when the behaviour of the interacting process is known. In particular, Figure 7 shows that time t3 occurs immediately after time t2 (*i.e.* at time t2+1). Figure 7 also shows related activity in the synchronous interface such as the current state, the idle signal, and the latching of data from the asynchronous memory.

To formally reason about the interaction of the abstract state machine with the asynchronous memory process, we assume that a memory cycle is initiated at an arbitrary point in time, t1, and then use inference rules to derive the state of the two processes at each of the subsequent synchronization points, t2, t3 and t4. The asynchronous memory specification and the derived behaviour of the abstract state machine during wait states, *e.g.* Theorem 9.1, allows us to regard the intervals between synchronization points as single time steps. That is, the interaction of the two processes is deterministic for this abstract view of time where wait states are collapsed into single time steps.

This proof technique, called "symbolic simulation", has also been used to formally verify higher level aspects of microprocessor systems in [3] and [13], *e.g.* symbolic execution of microcode. The simulation takes the form of a sequence of inferences in higher-order logic. It is symbolic because variables are used in place of real data and because all possible asynchronous delays are considered at once. Even though we use the descriptive term "simulation", we emphasis that this technique is formal proof based on the inference rules of higher-order logic.

Symbolic simulation is used to reason about read and write cycles as well as idle cycles (idle cycles occur when there is no interaction between the two processes). The results of this major proof step are summarized in three theorems corresponding to the three types of memory cycles. Each of these theorems relates the state of the synchronous machine at the end of a memory cycle to its initial state at the beginning of the cycle. This is illustrated below for a read cycle using pseudo-formal notation.

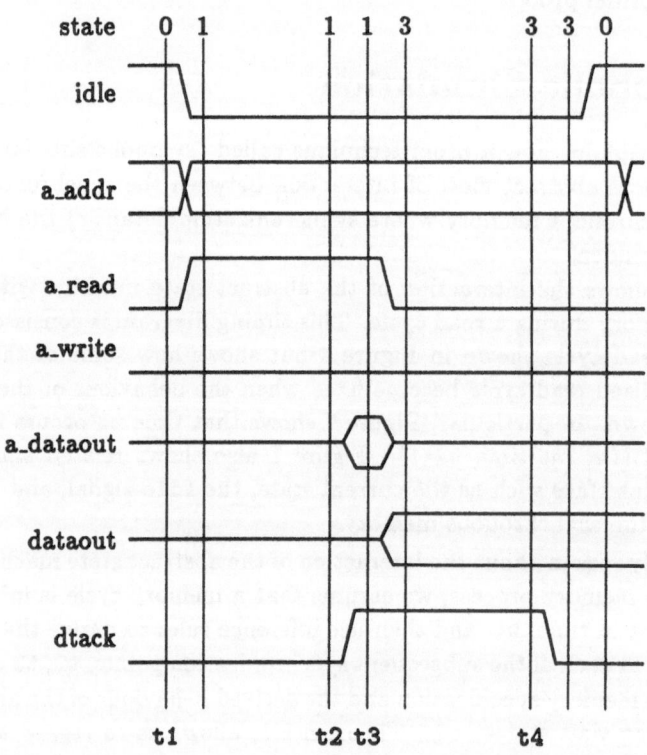

Figure 7: Interaction of State Machine with Asynchronous Memory

```
combined behaviour of interface and asynchronous memory     (1)
⟹
∀t1.
    synchronous read request at time t1 ∧                   (2)
    asynchronous memory idle                                (3)
    ⟹
∃t2.
      read cycle ends at t2 ∧                               (4)
      asynchronous memory returns to idle state ∧           (5)
      fetched memory word available as output ∧             (6)
      internal memory state unchanged during cycle          (7)
```

As shown above, the overall structure of these three theorems is an implication where the results of symbolic simulation are implied by the combined behaviours of the abstract state machine and the asynchronous memory. The behaviours of these two processes are combined by replacing the formal parameters of Interface and AsynMem with the names of interconnecting wires and buses and composing the two terms by logical conjunction, line (1).

```
Interface (
    read,write,dtack,addr,datain,a_dataout,
    idle,a_read,a_write,a_addr,a_datain,dataout,state) ∧
AsynMem (a_read,a_write,a_addr,a_datain,dtack,a_dataout,mem)
```

The right-hand side of the outermost implication, lines (2) to (7), states that a read cycle initiated at time t1 will eventually complete at time t2. The existentially quantified variable t2 denotes the end of the read cycle and should not be confused with time t2 in Figure 7.

Lines (2) and (3) refer to the initiation of a read cycle. This occurs when the synchronous read line is high, the abstract state machine is in state 0 and the asynchronous memory is idle.

```
            (state t1 = 0) ∧
            IsHi read t1 ∧
            MemIdle (a_read,a_write,dtack,mem) t1
```

If these conditions are satisfied, symbolic simulation shows that read cycle will eventually be completed at time t2. The end of the cycle is formally specified as the time when the idle signal of the state machine next becomes high after the start of the cycle at time t1, line (4). The predicate Next is used to formally state this condition.

```
            Next (t1,t2) (IsHi idle)
```

Symbolic simulation also shows that the asynchronous memory will return to its idle state by the end of the read cycle and that the fetched memory word will be available as output, lines (5) and (6). Finally, line (7) states that the internal state of the memory will be unchanged from the start of the read cycle.

```
MemIdle (a_read,a_write,dtack,mem) t2 ∧
(dataout t2 = FETCH (mem t1) (addr t1)) ∧
(mem t2 = mem t1)
```

Thus, the results of symbolic simulation for a read cycle are formally expressed by the following theorem.

Theorem 10.1:
```
⊢ ∀ read write dtack addr datain a_dataout
    idle a_read a_write a_addr a_datain dataout state mem.
    Interface (
      read,write,dtack,addr,datain,a_dataout,
      idle,a_read,a_write,a_addr,a_datain,dataout,state) ∧
    AsynMem (a_read,a_write,a_addr,a_datain,dtack,a_dataout,mem)
    ⟹
    ∀t1.
      (state t1 = 0) ∧
      IsHi read t1 ∧
      MemIdle (a_read,a_write,dtack,mem) t1
      ⟹
      ∃t2.
        Next (t1,t2) (IsHi idle) ∧
        MemIdle (a_read,a_write,dtack,mem) t2 ∧
        (dataout t2 = FETCH (mem t1) (addr t1)) ∧
        (mem t2 = mem t1)
```

The above theorem states that the synchronous interface and asynchronous memory correctly implement the four-phase handshaking protocol in the case of a read cycle. Similar theorems can be derived for a write cycle and for a idle cycle. These three theorems are relatively compact because they do not contain details of synchronization during memory cycles. This is desirable because we wish to present a more abstract view of interaction between the synchronous interface and the asynchronous memory where synchronization details are hidden.

11 Independent Process Specification

One of the main differences between the specifications described in this paper and the approach taken in [12] is that each process has an independent specification. In particular, the asynchronous memory specification is written as an independent definition in higher-order logic. Thus, it is possible to reason independently about each process; it is only necessary to compose specifications in order to reason about their interaction. This contrasts with Hunt's specification style where "the characterization of external devices are wrapped up in the same function which specifies the microprocessor" [1].

The relational style of specification used in our approach makes it easy to write independent specifications for each process and compose them using logical conjunction. The main problem with the functional style of specification is modelling bi-directional communication between two processes. [8] and [10] describe functional specification styles which solve this problem using lazy evaluation techniques.

12 Independent Time Scales

The formal specifications of the asynchronous memory process and the synchronous interface are expressed in terms of a common time scale, namely, the clocking rate of sequential components in the synchronous interface. This common time scale is an implicit form of synchronization which underlies the asynchronous interaction implemented by the four-phase handshaking protocol.

To clarify this situation, we distinguish between the time scale of the formal specification and the use of clocked components (if any) in an implementation of the asynchronous memory process. The formal specification describes the behaviour of the asynchronous memory process when observed in terms of the clocking rate of the synchronous interface. It is possible that the asynchronous memory is implemented by components clocked by the same clock used in the synchronous interface. However, the formal specification of the asynchronous memory process could also specify the observed behaviour of a memory with an independent clock or a memory implemented by unclocked circuits.

Hence, our specifications use a common time scale even though the corresponding devices may involve independent clocks. We have not used formal methods to relate specifications involving independent time scales. However, [16] explains that we have attempted to write the asynchronous memory specification in a "speed-independent" style where the underlying time scale is unimportant. Our use of the term "speed-independent" to describe this style of specification alludes to the use of the four-phase handshaking protocol (also called Muller signalling) in self-timed systems [18]. While the scale of time in the asynchronous memory specification is not important, the specification depends on "open-loop" relations which assume that address and data values are sent, not just concurrently, but in parallel with the request and acknowledgement signals.

---

[1][12], page 113.

13 Summary and Conclusion

We have modelled the interaction of a synchronous device with an asynchronous memory using a four-phase handshaking protocol. This example demonstrates the use of higher-order logic to reason about the behaviour of synchronous systems such as microprocessors which communicate requests to an asynchronous device and then wait for unpredictably long periods until these requests are answered. We showed how this behaviour can be formally related to more abstract time scale where wait states correspond to single abstract intervals. When viewed in terms of this abstract time scale, the behaviour of the synchronous device is entirely deterministic. We can then reason about higher level aspects of the synchronous behaviour, *e.g.* the correctness of microcode, without the extra complication of non-determinism. The main features of our approach are summarized in the following points:

- Existential quantification provides a straightforward means of specifying the finite but unknown length of a wait state.

- The use of explicit time references and higher-order functions leads to relatively simple and direct specifications.

- The relational specification style makes it easy to write independent specifications for processes.

In reasoning about the interaction of the synchronous interface with the asynchronous memory, inductive aspects were factored out of the verification procedure at an early stage; we obtained properties about the synchronous interface expressed in terms of the StableUntil predicate. We then used forward inference rules to symbolically simulate the interaction of the synchronous interface with the asynchronous memory.

The full-length version of this paper [16] describes the use of higher-order functions to directly relate different time scales. This leads to even simpler specifications of the interaction between the synchronous interface and asynchronous memory. The full-length paper also compares our model to real systems such as the M68000 microprocessor and describes how our model could be revised to include some of the detailed timing requirements found in these systems.

Although the example described in this paper is strongly oriented towards hardware, we believe that higher-order logic is also a suitable formalism for more abstract forms of concurrency. Furthermore, the expressive power of higher-order logic can be used to represent other formalisms such as temporal logic which are often used to specify concurrent systems. The readability of our specifications may be improved by borrowing notation from other formalisms and our proof strategies made more efficient by deriving special-purpose inference rules based on these formalisms.

Acknowledgements

Mike Gordon and Avra Cohn suggested many improvements to earlier versions of this paper. I am also grateful to Miriam Leeser for helpful discussions on handshaking protocols and their formal specification.

This research has been funded by the Cambridge Commonwealth Trust, the Canada Centennial Scholarship Fund, the Government of Alberta Heritage Fund, the Natural Sciences and Engineering Research Council of Canada and the UK Overseas Research Student Awards Scheme.

References

[1] A. Clements, *Microprocessor Systems Design*, PWS Publishers, Boston, 1987.

[2] A. Cohn and M. Gordon, "A Mechanized Proof of Correctness of a Simple Counter", Technical Report No. 94, Computer Laboratory, University of Cambridge, July 1986.

[3] A. Cohn, "A Proof of Correctness of the Viper Microprocessor: The First Level", *VLSI Specification, Verification and Synthesis*, Proceedings of the Workshop on Hardware Verification, Calgary, Canada, 12-16 January 1987, G. Birtwistle and P. Subrahmanyam, eds., 1987.

[4] I. Dhingra, "Formal Validation of an Integrated Circuit Design Style", *VLSI Specification, Verification and Synthesis*, Proceedings of the Workshop on Hardware Verification, Calgary, Canada, 12-16 January 1987, G. Birtwistle and P. Subrahmanyam, eds., 1987.

[5] M. Gordon, R. Milner and C. Wadsworth. *Edinburgh LCF: A Mechanised Logic of Computation*, Lecture Notes in Computer Science, Springer-Verlag, 1979.

[6] M. Gordon, "Why Higher-Order Logic is a Good Formalism for Specifying and Verifying Hardware", *Formal Aspects of VLSI Design*, Proceedings of the 1985 Edinburgh Conference on VLSI, G.J. Milne and P. Subrahmanyam, eds., North-Holland, Amsterdam, 1986.

[7] M. Gordon, "A Proof Generating System for Higher-Order Logic", *VLSI Specification, Verification and Synthesis*, Proceedings of the Workshop on Hardware Verification, Calgary, Canada, 12-16 January 1987, G. Birtwistle and P. Subrahmanyam, eds., 1987.

[8] P. Henderson, *Functional Programming*, Prentice-Hall, 1980.

[9] J. Herbert, "Application of Formal Methods to Digital System Design", Ph.D. Thesis, Computer Laboratory, Cambridge University, December 1986.

[10] S. Hill, "Simulating Digital Circuits in Miranda", University of Kent, 1986.

[11] C. Hoare, *Communicating Sequential Processes*, Prentice-Hall, 1985.

[12] W. Hunt, "FM8501: A Verified Microprocessor", PhD Thesis, Institute for Computer Science, University of Texas at Austin, 1986.

[13] J. Joyce, G. Birtwistle, and M. Gordon, "Proving a Computer Correct in Higher Order Logic", Technical Report No. 100, Computer Laboratory, University of Cambridge, December 1986.

[14] J. Joyce, Ph.D. Research Progress Report, Computer Laboratory, University of Cambridge, December 1987.

[15] J. Joyce, "Formal Specification and Verification of Microprocessor Systems", *EUROMICRO 88*, Proceedings of the 14th Symposium on Microprocessing and Microprogramming, Zurich, Switzerland, 29 August - 1 September, 1988, S. Winter and H. Schumny, eds., North-Holland, 1988.

[16] J. Joyce, "Formal Specification and Verification of Asynchronous Processes in Higher-Order Logic (Full-Length Version)", Technical Report No. 136, Computer Laboratory, University of Cambridge, 1988.

[17] R. Milner, *A Calculus of Communicating Systems*, Lecture Notes in Computer Science, Springer-Verlag, 1980.

[18] C. Seitz, "System Timing", Chapter 7 in *Introduction to VLSI Systems*, C. Mead and L. Conway, Addison-Wesley, Reading, Massachusetts, 1980.

[19] Paulson, L., *Logic and Computation*, Cambridge University Press, Cambridge, 1987.

[20] C. Pygott, "Electrical, Environmental and Timing Specification of the Viper Microprocessor", Memorandum No. 3753 (Unclassified), RSRE (Royal Signals and Radar Establishment), British Ministry of Defense, December 1984.

Appendix - Asynchronous Memory Specification

The definition of AsynMemRead in Section 6 specifies the behaviour of the asynchronous memory process during a read cycle. The corresponding specification for a write cycle is shown below. These two specifications are then combined to define AsynMem. In addition to the constraints imposed by AsynMemRead and AsynMemWrite during read and write cycles, AsynMem states that the memory is initially idle at time 0.

```
⊢ AsynMemWrite (
    read:wire,write:wire,addr:bus,datain:bus,
    dtack:wire,dataout:bus,
    mem:memory) =
  ∀ t1 address value.
    Rises write t1 ∧
    MemIdle (read,write,dtack,mem) t1 ∧
    StableUntil2 (read,Lo,t1,Falls dtack) ∧
    StableUntil2 (write,Hi,t1,Rises dtack) ∧
    StableUntil2 (addr,address,t1,Falls dtack) ∧
    StableUntil2 (datain,value,t1,Falls dtack)
    ⟹
    ∃t2.
      First (t1,t2) (Rises dtack) ∧
      StableUntil1 (dtack,Hi,t2+1,Falls write) ∧
      ∀t3.
        First (t2,t3) (Falls write) ∧
        StableUntil2 (write,Lo,t3,Falls dtack)
        ⟹
        ∃t4.
          First (t3,t4) (Falls dtack) ∧
          MemIdle (read,write,dtack,mem) (t4+1) ∧
          (mem (t4+1) = STORE (mem t1) address value))

⊢ AsynMem (
    read:wire,write:wire,addr:bus,datain:bus,
    dtack:wire,dataout:bus,
    mem:memory) =
  MemIdle (read,write,dtack,mem) 0 ∧
  AsynMemRead (read,write,addr,datain,dtack,dataout,mem) ∧
  AsynMemWrite (read,write,addr,datain,dtack,dataout,mem)
```

Temporal Specifications Directed by Grammar and Design of Process Networks

Francois D. CARREZ[1]
CNRS-CRIN, Université de Nancy I
B.P. 239
54506 Vandoeuvre-les-Nancy
France

Dominique MERY[2]
CNRS-CRIN, Université de Nancy I
B.P. 239
54506 Vandoeuvre-les-Nancy
France

15 juin 1988

Abstract

We present a framework of specification design and proof of processes networks. The design aspect is directed or aided by rules based on temporal specifications and grammar oriented specifications. A network is a set of processes communicating by using channels. The semantics of the network is defined with the help of histories of channel communications. But, more generally, a process is considered to be a set of possible operations on a set of channels. A channel is characterized by communication rules. These communication rules are in fact written as production rules of a *communication grammar*. We assume the existence of a global clock used to sort communications. Our semantics is based on the abstract data type approach and the temporal logics extended by specific operators (induced by the production rules). Inference rules are derived and allow to take into account communications.

[1]Supported by GRECO C3 and MRES grant
[2]Supported by GRECO C^3

1   Introduction

A network of processes is a set of processes that communicate using channels or other mediums. A process may be itself a network of hidden processes and may act on other unshared objects. We suggest to express the activity of processes on typed objects in a framework using abstract data types and temporal logics. Abstract data types express static properties of data and temporal logics express dynamic properties relative to message passing and channel traffic. This approach has been already followed by F.Kröger [KRO87]. In this paper, F. Kröger specifies the traffic in a filling station using abstract data types and temporal specifications. We choose the limits of our specification in order to obtain *manageable* specifications. It seems to be difficult to obtain a *synthetically built* station from the Kröger's specification. We concentrate our interest on the channels and try to generalize the method of Z. Manna and P. Wolper [MW84]. We give here some examples but don't propose any effective procedure of synthesis. Our current and future investigations concern this problem. S. Kaplan and A. Pnueli [KP87] suggest to use abstract data types and algebra of processes together. Their framework considers a process or a ground term in this calculus as a function acting on abstract data types. We think that this approach and the previous one are closely related. In fact temporal logics may be sufficiently extended to express what algebra of processes express, but in the terms of traces. It's clear that the logics become very complex. Now, another criticism that may be levelled at them is that the parallelism is reduced to non determinism. We think that some models like CCS [MIL86] may be generalized and take into account a synchronous aspect of execution. That is, during a computation step, one or more operations or actions are executed. This aspect corresponds to the simultaneous activity of some concurrent processes. Now, when we specify a process or a set of processes, we specify any process and after that, we specify the communications. The classical proof methods like S. Owicki and D. Gries [OG76], K. Apt, N. Francez and W. P De Roever [KAR80] have used this approach to generalize the method due to C. A. R. Hoare [HOA69]. But the manipulation of specifications must maintain some *compositional aspect*. What does it mean in our framework? It means that the semantics of two processes communicating by channels may be used to derive the semantics of the whole network of the two processes. This goal is achieved using the communication rules or the communication constraints. These communication constraints are more generally constraints on the operations. This means that we need to model or to interpret temporal specifications. A. Szalas [SZA88] gives some basis for such an approach and we exploit this idea in our work. The temporal framework is embedded in the algebraic one. The notion of Kripke model is used to interpret the specifications. What is a Kripke model in this framework? We define a sequence of states and an algebraic model namely the initial one. Usually, this model is needed to interpret objects typed by some sorts. We achieve some simplicity by choosing total operations on sorts but further works will explore the use of partial operations on sorts. So, we are able to interpret temporal and algebraic specifications in Kripke model. We obtain these properties but we are driven or guided by production rules. The communications are surely restricted but we limit our investigations to specific problems. We briefly present the content of the different parts in this paper. The next part presents some examples and a first approach relative to our study. We consider very simple examples and develop specifications associated to them. The third part contains precise explanations of the assertion language, specification language and model. We explain the design driven by some specific inference rules. In the fourth part, the design system is sketched. We derive new inference rules used for specific problems. We introduce a new temporal operator that seems to be more suited with respect to our approach. Finally, we conclude our paper and suggest the future works.

2 Specification of communications

A communication is an operation that includes a sender, a receiver and a medium. This operation can be algebraically specified. But some dynamic aspects must be included. Hence, the axioms may be temporally expressed but it is equivalent to describe algebraic Kripke models as A. Szalas [SZA88] has done. We use the word *operation* to define a communication but it is better to use the word *object*. So. a communication is an operation on objects that are typed or sorted. We can reason on algebraic structures in a uniform way. Let us consider a first example.

Example 1
A process P communicates by two channels I and O.

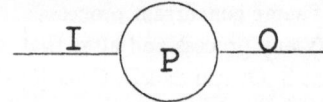

It produces on O the last read value on I. This property may be expressed in temporal logics but we must be able to express the notion of **last read value on I**. *We introduce a sort that allows to express it.*

<u>Sort</u> *Stack(Val)*
    <u>Used sorts</u> : *Boolean*
    <u>Constructors</u> :
        $\perp_{Stack}$: () → *Stack*
        Empty_stack : () → *Stack*
        Push : (*Val* × *Stack*) → *Stack*
    <u>Destructors</u> :
        Top : *Stack* → *Val*
        Pop : *Stack* → *Stack*

Observators :
  $Is\_empty : Stack \rightarrow Boolean$
  Axioms : { $v : Val, s : Stack$ }
  $Pop(Empty\_stack()) = \bot_{Stack}$
  $Pop(Push(v, s)) = s$
  $Is\_empty(Empty\_stack()) = True$
  $Is\_empty(Push(v, s)) = False$
  $Top(Empty\_stack()) = \bot_{Val}$
  $Top(Push(v, s)) = v$
End sort Stack

Now, we can specify the Process P by combining temporal logics and operations on an object s of sort Stack together :

New $s = Empty\_stack()$ $\land$ $\Box\Diamond$ $(Receive(I, v))$ $\land$
$\Box$ $(\neg$ $Is\_empty(s)$ $\leadsto$ $Send$ $(O, Top(s)))$ $\land$
$\Box$ $(Receive(I, v) \Rightarrow \bigcirc$ (**New** $s = Push(v, s))$ $\land$
  $Send(O, v) \Rightarrow \bigcirc$ (**New** $s = Pop(s)))$

where $\leadsto$ means Leads to.

We see that P is a process that use specific operations on sorts, namely a stack and its behaviour is equivalent to a stack. We don't specify the activity of P as CCS of R. Milner [MIL86] would have done.
New *s means that the value of s has been changed, and we are able to distinguish the old value s and the new value of s. Some authors use the notation s' and s for respectively the old and new value of s.*

Example 2
In the first example, we have shown how a process P behaving like a stack may be specified. Now, we illustrate a behaviour of communication using **grammar rules***. Let us consider two processes P and Q communicating by a channel J.*

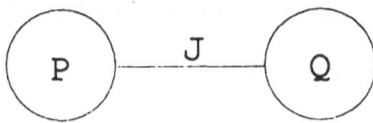

The communication is very special because P sends values or messages, but unfortunately, for two sent values, Q received only the second one. In fact, the channel J behaves like a filter which communicates only one value among two consecutive sent values. Now, the expression

of this situation is achieved using grammar rules.

$$P \longrightarrow ab \; Q$$
$$Q \longrightarrow \bar{b} \; P$$

The symbol $\longrightarrow$ *means that that the communication is acted by the left part and produces some values or effects on the right part. The letters a and b are two elements of an alphabet of communication. When we write a, we design a write operation and the dual operation (read operation) is denoted by* $\bar{a}$. *In a general framework, $ab\bar{b}$ may be replaced by algebraic operations on channels. Clearly, this means that the communication channel state is characterized by a behaviour language. The communication traffic on J may be characterized as follows :*

$$P \longrightarrow Send(J,m) \; Send(J,m') \; Q$$
$$Q \longrightarrow Receive(J,m') \; P$$

let us denote α *a communication assertion that allows to characterize the communication behaviour. This assertion* α *is built using an operator built from the grammar in an extended temporal framework as in [WOL83].*

$$\alpha \; (\underbrace{Send(J,m)}_{\alpha_1}, \underbrace{Send(J,m')}_{\alpha_2}, \underbrace{Receive(J.m')}_{\alpha_3}) = \alpha$$
$$\alpha = \alpha_1 \wedge \bigcirc \alpha_2 \wedge \bigcirc\bigcirc \beta$$
$$\beta = \alpha_3 \wedge \bigcirc \alpha$$

When we write $op(x_1,\dots,x_n)$, it means that the operator $op(x_1,\dots,x_n)$ is activated. The production rule is used to describe a part of a model.

The goal is to specify and to synthetize, but we use the axioms and rules on the new operator due to P. Wolper [WOL83]. Now, it is possible that P and Q have two specified behaviours with respect to I, J and J, K. At this stage, we consider very simple specifications for P and Q. The specification of P, namely ϕ may be written as :

$$P \; sat \; [(Read_P(I) = (New \; I,m)) \rightsquigarrow (New \; J = Write_P(J, Calcul_P(m)))]$$

The specification of Q, namely ψ is yet very simple :

$$Q \; sat \; [(Read_Q(J) = (New \; J,m)) \rightsquigarrow (New \; K = Write_Q(K, Calcul_Q(m)))]$$

Finally, we introduce a new kind of assertion denoted (ϕ, α, ψ) for the specification of the global network. The classical syntaxical construction has some semantical characterization (as the wp's ones for instance). What does it mean? If R is a network of the two processes P and Q, R *sat* (ϕ, α, ψ) means that ϕ and ψ hold in the connection satisfying α. The analysis of α suffices to infer a possible deadlock or transmission problem. But, α specifies the whole communication between P and Q. Moreover, if two different channels connect P and Q, we build a unique assertion to specify communications.

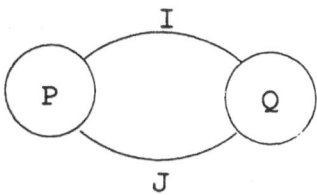

Hence, we obtain an assertion as $(\phi, \alpha(I, J), \psi)$. An assertion as α may be easily synthetised in the following sense : We obtain assertion of a decidable extended temporal logic (ETL) if the grammar is right regular. Else, a further study must be made. In the notation, we have three parts that are complementary. A first part that specifies the behaviours of P, a second one that specifies the behaviour of the channels between P and Q, and finally a third part that specifies the behaviours of Q. But it is possible that the three specifications are inconsistant together with respect to the execution model. For instance, if they lead to a deadlock, the assertion is equivalent to *false*. In fact, *false* means that the set of behaviours is empty and hence, the connection is not possible.

Example 3

Now, we consider a more sophisticated network. Let R be a network built from the following processes :

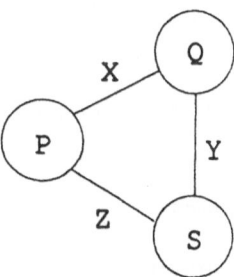

$$P \longrightarrow ab\ Q$$
$$Q \longrightarrow \bar{b}\ P$$
$$Q \longrightarrow a\ S$$
$$S \longrightarrow \bar{a}\ Q$$

$$S \longrightarrow a\ P$$
$$P \longrightarrow \bar{a}\ S \mid S$$

If we analyse the communication traffic. we see that :
let us assume that the first sent value is α_0. the second sent value is β_0. Then P will receive
in the best case the value β_0 else an other value sent after β_0.

We can imagine that the value a is a communication control value. In some protocol, this value ensures a good communication or reliable communication. The problem is now to refine the specification of the communication between P and Q. We propose to introduce a new process F (like Filter) that ensures the communication.

Example 4
Suppose that we want to specify an asynchronous channel between a sender P and a receiver
Q. To solve this problem, we introduce a third process F which behaves as a infinitary buffer.

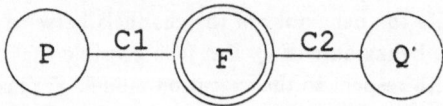

We define first the sort Buffer :

<u>Sort</u> *Buffer(Val)*
    <u>Used sorts</u> : *Boolean*
    <u>Constructors</u> :
        $\perp_{Buffer}$: $() \to Buffer$
        Empty_buffer : $() \to Buffer$
        Put : $(Val \times Buffer) \to Buffer$
    <u>Destructors</u> :
        Last : $Buffer \to Val$
        Kill : $Buffer \to Buffer$
    <u>Observators</u> :
        Is_empty : $Buffer \to Boolean$
    <u>Axioms</u> : { v : *Val*, b : *Buffer* }
        $Kill(Put(v, b)) = Put(v, Kill(b))$
        $Kill(Put(v, Empty\_buffer()))$
          $= Empty\_buffer()$
        $Kill(Empty\_buffer()) = \perp_{Buffer}$
        $Last(Put(v, b)) = Last(b)$

$$Last(Put(v, Empty\_buffer())) = v$$
$$Last(Empty\_buffer()) = \bot_{Val}$$
$$Is\_empty(Empty\_buffer()) = True$$
$$Is\_empty(Put(v, b)) = False$$

End sort Buffer

Then we associate to the process F an object b of sort $Buffer$ and we specify the behaviour of process F using temporal logics :

$\mathbf{New}\ b = Empty\_buffer() \wedge \Box\Diamond (Receive(C_1, v)) \wedge$
$\Box (\neg Is\_empty(b) \Rightarrow \Diamond Send (C_2, Last(b))) \wedge$
$\Box (Receive(C_1, v) \Rightarrow \bigcirc (\mathbf{New}\ b = Put(v, b)) \wedge$
$\quad Send(C_2, v) \Rightarrow \bigcirc (\mathbf{New}\ b = Kill(b)))$

where C_1 and C_2 are synchronous channels.
Afterwards we can specify the global communication between processes P, Q and F by the following grammar rules :

$$P \longrightarrow a\ F$$
$$F \longrightarrow \bar{a}\ P$$
$$F \longrightarrow a\ Q$$
$$Q \longrightarrow \bar{a}\ F$$

The process F recognizes a regular infinite word. Hence, it may be generated by the graph model principle as in [MW84].

Remark 1

If we consider a filter in Example 2 as in the last one, we can derive using histories of channels, that :
$\mathcal{H}(C_2) = Even(\mathcal{H}(C_1))$ i-e $\mathcal{H}(C_2) = \mathcal{H}(C_1)$ where all the odd messages are deleted.

Example 5

Now, we specify a process P that reads values on channel X and writes two copies of reading *values on channels Y_1 and Y_2.*

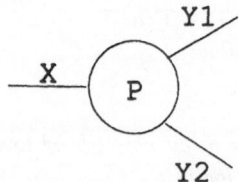

We suppose the channels to be synchronous with a Rendez-vous politics. There are two main manners for specifying such a process. We can use a one $-$ slot Buffer (synchronous approach) or a ∞ $-$ Buffer (asynchronous approach). We consider the second one, and we obtain, using an object b of sort Buffer :

P sat $\Box\Diamond$ $(Receive(X, v))$ $\wedge$
$\quad$ $\Box$ $(\neg$ $Is\_empty(b)$ $\rightsquigarrow$
$\qquad$ $\{(Send(Y_1, v)$ $\wedge$ $\bigcirc$ $Send(Y_2. v)$
$\qquad$ $\wedge$ $\bigcirc\bigcirc$ $(New$ $b = Kill(b)))$ $\oplus$
$\qquad$ $(Send(Y_2, v)$ $\wedge$ $\bigcirc$ $Send(Y_1, v)$
$\qquad$ $\wedge$ $\bigcirc\bigcirc$ $(New$ $b = Kill(b)))\}$
$\quad$ $\Box$ $(Receive(X, v) \Rightarrow \bigcirc(New$ $b = Put(v, b)))$

where $\oplus$ means Exclusive $-$ Or.
We may give a CCS specification of process P as follows :

$$P = (ab \ + \ ba) \ P$$

This specification is too "operational". We can refine this specification in another one. Let us consider that process P is in fact a network.

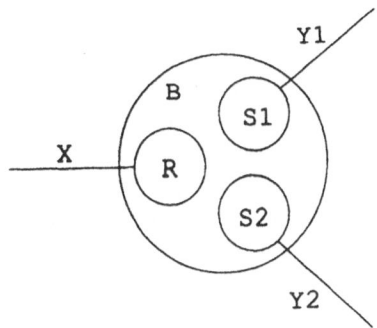

*It may be built of three main processes R, S_1 and S_2. The receiver R reads values on channel X then writes them on a buffer B. whereas both S_1 and S_2 read values on B then send them on Y_1 and Y_2 respectively. The problem which appears now is the shared access to a common object B. We can solve this problem by using an object S of type Semaphore with its well known primitives **wait** and **signal**, and a fourth process, the synchronizer, waiting for S_1 and S_2 having both read the same values before killing the last value of B.*

These examples are very simple but illustrate the power of this approach allowing to merge algebraic specifications, temporal ones and grammar based communication specifications. We limit the power of temporal language to obtain good procedures. It's clear that we use the temporal language as final specification language. Our thesis is that specifications are made easier if we use such and such a language. But at the final point of specification process, we must get a unified framework. in order to obtain a realistic synthesis. The algebraic specification process is a technical point and this process may be driven by some system like $SACSO$ [LPS87] for instance.

3   A   unified framework for the network specification

A process P is an entity characterized by its behaviours with respect to typed objects. A process P does actions that produce or induce reactions in others processes. We assume that there exists a class of basic processes able to perform simple or complex procedures. For instance, to transmit messages, to receive messages or to compute some functions . . . .

Definition 1 (Class of processes)
Let Π_0 be the class of basic processes. The class of processes Π contains Π_0 and is closed by

finite application of the following rules :

if $P_1, \ldots, P_n, c_1, \ldots, c_p, e_1, \ldots, e_q$ are respectively processes, internal channels and external channels, then $NET(P_1, \ldots, P_n, c_1, \ldots, c_p, e_1, \ldots, e_q)$ is a process or a network with internal and external channels.

A process P of Π is specified in an algebraic and temporal way. The semantics of a process P is made of a model for algebraic specifications and a sequence or trace of communications or operations. We consider a static part and a dynamic one. Such a structure is called an algebraic Kripke model, as [KRO87] and [SZA88]. A model for algebraic specifications is generally choosen as the initial one. We assume that the functions or operations are total, because it tells us some theoretical aspects not directly in our frame. A global clock is associated to any system or process and allows to arrange operations of any subprocess.

Definition 2 (Semantics)
Let P be a process of Π. A semantics for P is a couple $(\mathcal{J}, \sigma)$ where $\mathcal{J}$ is a initial model for specifications of P and σ is a sequence of algebraic operations. We denote $[\![P]\!]$, the set of semantics of process P.

Let us recall that if $(\mathcal{J}, \sigma) \in [\![P]\!]$ there exists a function $Date : \{\sigma_0, \ldots, \sigma_i, \ldots\} \longrightarrow N$. If P_1 and P_2 are two processes connected by a channel C satisfying the communication rule α such that : $(\mathcal{J}, \sigma_1) \in [\![P_1]\!]$ and $(\mathcal{J}, \sigma_2) \in [\![P_2]\!]$ then, a semantics of $P_1 -[C]- P_2$ is obtained from semantics of P_1 and P_2 by : $(\mathcal{J}, \sigma_1 \otimes_\alpha \sigma_2)$ where $\otimes_\alpha$ depends upon the rules of channel C.

Example 6
Let P and Q be two processes which communicate by a set of channels $C_1, \ldots, C_n$. We consider a Rendez-vous communication politics. Let Σ be the set of all matching pairs of Input/Ouput operations, for example : $New\, C_1 = Send(C_1, expr)$ and $New\, C_1 = Receive(C_1, var)$, that must occur at the same time. If $(\mathcal{J}, \sigma_1) \in [\![P]\!]$ and $(\mathcal{J}, \sigma_2) \in [\![Q]\!]$ then $[\![P \parallel_\Sigma Q]\!]$ using channels $C_1, \ldots, C_n$ is given by :
$(\mathcal{J}, \sigma_1 \otimes_\Sigma \sigma_2)$. *We define the operator $\otimes_\Sigma$ in a functional way by :*

$\sigma_1 \otimes_\Sigma \sigma_2 =$
if $((First(\sigma_1), First(\sigma_2)) \in \Sigma \wedge Date(First(\sigma_2)) = Date(First(\sigma_1)))\vee$
$\quad((First(\sigma_1), First(\sigma_2)) \notin \Sigma \wedge Date(First(\sigma_2)) = Date(First(\sigma_1)))$

$\quad then \begin{pmatrix} First(\sigma_1) \\ First(\sigma_2) \end{pmatrix} @ (Tail(\sigma_1) \otimes_\Sigma Tail(\sigma_2))$

else if $(\exists\ c_1,\ c_2 \in \Sigma\ s.t\ c_1 \neq c_2 \wedge$
$\quad First(\sigma_1)\ in\ c_1\ \wedge\ First(\sigma_2)\ in\ c_2)\ \vee$
$\quad ((First(\sigma_1), First(\sigma_2)) \in \Sigma\ \wedge$
$\quad Date(First(\sigma_2)) \neq Date(First(\sigma_1)))$
then $\perp$
else if $(\exists\ c \in \Sigma\ s.t\ First(\sigma_1)\ in\ c)\ \vee\ (Date(First(\sigma_2)) > Date(First(\sigma_1))$
$\quad\quad then\ First(\sigma_1)\ @\ (Tail(\sigma_1) \otimes_\Sigma \sigma_2$
$\quad\quad else\ First(\sigma_2)\ @\ \sigma_1 \otimes_\Sigma (Tail(\sigma_2))$

We see that the composition operator depends upon the communication and we must propose a hypothesis.

Hypothesis
Let be given a set of communication rules $\mathcal{R}$. Let $P_1,\ldots,P_n$ a set of processes communicating under $\mathcal{R}$. There exists an operation $\otimes_\mathcal{R}$ such that : $[\![P_1 \| \ldots \| P_n]\!] = \otimes_\mathcal{R}([\![P_1]\!],\ldots,[\![P_n]\!])$

In fact, our framework is restricted to this hypothesis. We do not know how to solve other problems. The problem of this hypothesis is now to verify if a given application satisfies it. Further investigations are needed in this way.

Some remarks must be done with respect to others approaches of semantics. In previous works, the second author D.Mery [MER86] [MER87] has investigated the problem of fairness in an interleaving semantics frame. But the questions of the assertion language and of the design where left unspecified. An assertion was a set of states. The goal was to prove eventuality properties under fairness hypothesis where a parallelism at one level was considered. Our current work is initially based on the proof system due to Van Nguyen et al [NDGO86] who use a trace model that records the communication between processes. The communication styles are restricted to synchronous and asynchronous communication politics. We have hence generalized this approach to general communications and used this general system as design oriented system. The use of abstract data types to specify communications has been soon proposed by G.R Perrin [PER85], but we have enriched here this approach because we include the temporal framework. Finally, the operator $\otimes_\mathcal{R}$ may be embedded in a frame as E. R. Olderog and C. A. R. Hoare [OH86] have proposed.

Definition 3 (Assertion language)
The assertion language $\mathcal{L}$ is a first order temporal language on the sorts of the processes. $\mathcal{L}$ may be extended if needed by operators derived from communication rules.

The communication rules are production rules that are expressed using some algebraic operations as elements of alphabet and using process names as nonterminal symbols. A communication contains a sender S, a receiver R and the message m which transits from S

to R.

$S \longrightarrow m\ R$ means that sender S is sending a message m to receiver R and $R \longrightarrow \overline{m}\ S$ means that receiver R has received a message m from S.

Definition 4 (Specification language)

Let P be a process of Π. Let $\tau_1, \ldots, \tau_p$ the sorts used by P. let $\lambda_1, \ldots, \lambda_p$ be the first order languages associated to $\tau_1, \ldots, \tau_p$.

The assertion language $\mathcal{P}$ of P is made of

- *temporal combinations*

- $\lambda_1, \ldots, \lambda_p$

A specification of P is a formula as P sat ϕ where $\phi \in \mathcal{P}$.

P sat ϕ means that any behaviour of $\mathcal{P}$ satisfies ϕ :

$\forall\ (\mathcal{J}, \sigma) \in [\![\mathcal{P}]\!],\ (\mathcal{J}, \sigma) \models \phi$

where $(\mathcal{J}, \sigma) \models \phi$ means that ϕ is satisfied in $(\mathcal{J}, \sigma)$

We don't more precisely specify the definition of $\models$ which is usual. We may refer to F. Kröger [KRO87] or A. Szalas [SZA88] for more details.

Our semantics is hence based on trace of operations. But, we may derive the states of objects. So, we may proceed as S. Kaplan and A. Pnueli [KP87] have done. They apply algebraic term of ACP on sorted objects. We may apply a sequence of algebraic operators σ on objects and derive the state at time t of objects.

Example 7 *Let P be a process using an object S of sort Stack and sending values on channel C.*

$$let\ \sigma = \underbrace{Empty\_stack()}_{\sigma_0}\ \underbrace{Push(m_0, S)}_{\sigma_1}\ \underbrace{Push(m_1, S)}_{\sigma_2}\ \underbrace{Send(C, Top(S))}_{\sigma_3}\ \underbrace{Pop(S)}_{\sigma_4}$$

$$\underbrace{Push(m_2, S)}_{\sigma_5}\ \underbrace{Send(C, Top(S))}_{\sigma_6}\ \underbrace{Pop(S)}_{\sigma_7}\ \underbrace{Send(C, Top(S))}_{\sigma_8}\ \underbrace{Pop(S)}_{\sigma_9} \ldots$$

be a behaviour of P.

We can compute the values of objects channel C and stack S and we obtain, using the axioms :

$State(\sigma, 0, S) = Empty\_stack()$

$State(\sigma, 1, S) = Push(m_0, Empty\_stack())$

$State(\sigma, 5, S) = Push(m_2, Push(m_0, Empty\_stack()))$
$State(\sigma, 8, C) = Send(Send(Send(Empty\_file(), m_1), m_2), m_0)$

We suppose here that C is of sort Channel with constructors : **Empty_channel, Send**
and with destructors : **Receive.**

The specification of a sort may be temporal. Szalas [SZA88] suggests to specify sorts in a temporal approach. In this new approach, we eliminate undesirable sequences : A *pop* operation may be made on a non empty stack. Hence, the algebraic specification axioms may contain temporal algebraic ones :
$\Box(pop(s) \Rightarrow s \neq ())$

4 Derivation of specifications

Our main goal is to design processes and networks of processes. The method is directed by production rules for the communication behaviour. In a synthesis framework, the choice of a concurrent solution depends upon some parameters as an explicit concurrency due to the specification or to an explicit choice during the resolution. We choose to decide the concurrency at the specification level : we indicate a set of processes acting on sorted data, and among the sorted data, there are specific communication ones. We notice that our framework is restricted to right regular grammars, because P. Wolper [WOL83] has proved the equivalence between right-regular grammar and decidable extended temporal logics. The stage of specification derivation is intermediate and must lead to an algorithmic solution. The key rule of our system is the network formation rule that has been generalized using communication rules. Let us recall that specification P *sat* ϕ of a process P uses temporal algebraic assertions. The first version of the rule is expressed in the following case : Two processes P and Q communicate using the channel C. The communication in C satisfies the communication rules $\mathcal{R}$. $\mathcal{R}$ allows to build a temporal assertion denoted ρ. This assertion uses the algebraic operations used in the specifications of P and Q. The rule can be expressed as follows :

If P *sat* ϕ. Q *sat* ψ, C *sat* ρ
then $P -[C]- Q$ *sat* (ϕ, ρ, ψ).

The expression (ϕ, ρ, ψ) is interpreted. if the channel C satisfies the hypothesis. This expression may be simplified to an other assertion as α. but we must be careful in the reduction.

Example 8

Process P receives values on I, computes a function f then sends results onto C. Process Q receives values v on C, but sends only the values v which verify assertion $q(v)$. C only transmits one among two consecutive read values. We have :

P sat $[(Read_P(I) = v) \rightsquigarrow (Write_P(C, f(v)))]$ $(is\ \phi)$
Q sat $[(Read_Q(C) = v \land q(v)) \rightsquigarrow (Write_Q(O, v) \land q(v)))]$ $(is\ \psi)$
$\mathcal{R}_C : P \longrightarrow Write_P(C, v_1)\ Write_P(C, v_2)\ Q$
$\quad\quad Q \longrightarrow (Read_Q(C) = v_2)\ P$

$\rho_C(Write_P(C, v_1), Write_P(C, v_2), (Read_Q(C) = v))$
$= Write_P(C, v_1) \land \bigcirc Write_P(C, v_2) \land$
$\quad \bigcirc\bigcirc (Read_Q(C) = v) \land \bigcirc\bigcirc\bigcirc \rho_C$

We derive the property : (ϕ, ρ_C, ψ). It is really difficult to give an equivalent assertion using classical temporal operators. In this case, we identify any occurence of $Read_P$ with an occurence of $Write_P$. We have relativized the assertion ρ_C.

We notice that we must be careful with respect to the local time and the global time for the next operator. But the hypothesis with respect to the composition operation leads to a sound rule. We see here that further developements are needed with respect to the system of simplification. It's clear that automatic techniques are better and currently explored. We give a general rule and interpret the new kind of assertion as (ϕ, ρ, ψ). In the expression of the rule above, we see that processes and communications are specified in the same way. In a design approach, processes and communications are not distinguished. But, we must say or express that there are n processes and p communications. The communications ensure a good global synchronization, but they are finally implemented by processes or synchronous/asynchronous channels. A generalized rule may be used as inference rule or as design rule. Let us asume that the network is made of n processes $P_1, \ldots, P_n$ and p communications $C_1, \ldots, C_p$. Any communication C_j is obtained using internal or external channels acting together. A communication C is specified by a set of communication rules. The set of communication rules $\mathcal{R}$ may be divided into independant communication rules set $\mathcal{R}_1, \ldots, \mathcal{R}_p$ that are associated to $C_1, \ldots, C_p$. In a first step of specification, we characterize $\mathcal{R}$. A process P_i is specified by a temporal ϕ_i. A communication C_j is specified by a temporal assertion α_j. The network satisfies the assertion $(\phi_1, \ldots, \phi_n, \alpha_1, \ldots, \alpha_p)$. If we refine C_j, we see that it is made of channels satisfying $\mathcal{R}_j$. We have $\alpha_j \stackrel{def}{=} \alpha(\mathcal{R}_j)$ where $\mathcal{R}_j$ is the set of rules of communication. An assertion of a network explicitly gives the architecture of this network. Let us justify this rule. Let σ be a sequence or a model of the network, σ is defined as : $\sigma_1 \otimes \ldots \otimes \sigma_n \otimes \tau_1 \ldots \tau_p$ or $\otimes(\sigma_1, \ldots, \sigma_n, \tau_1, \ldots, \tau_p)$. Any σ_i is a behaviour of P_i, it means that P_i satisfies ϕ_i. Any τ_i is a behaviour of C_j, it means that τ_j satisfies α_j. By the assumption of global clock, any sequence may be derived.

Definition 5 (Rule of formation)
Let P_i be a process specified by ϕ_i, $i \in \{1 \ldots n\}$.
Let C_j be a communication specified by α_j, $j \in \{1 \ldots p\}$.
Let $N = [P_1 \| \ldots \| P_n] \, (C_1, \ldots, C_p)$ be a network.
If P_i sat ϕ_i, C_j sat α_j Then N sat $(\phi_1, \ldots, \phi_n, \alpha_1, \ldots, \alpha_p)$

Now, using renaming operations, we may obtain equivalent assertions or weaker assertions. Let us consider a classical example :

Example 9
P sat $\Box \, (Read(In, v_1) \rightsquigarrow Write(C, f(v_1)))$ *(is ϕ)*
Q sat $\Box \, (Read(C, v_2) \rightsquigarrow Write(Out, g(v_2)))$ *(is ψ)*
C sat $\Box \, (Write(C, v_1) \Leftrightarrow Read(C, v_2) \, \wedge \, (v_1 = v_2))$ *(is α)*
$[P \| Q](C)$ sat (ϕ, α, ψ) .
We derive now the property :

$$[P \| Q](C) \text{ sat } \Box \, (Read(In, v_1) \rightsquigarrow Write(Out, g \circ f(v_1))$$

by using the property of Rendez-vous expressed in the specification of C.

Finally, we see that a metalanguage for derivation is needed. A derivation expresses the specification transformation. For instance, we use the metarules :

Channelization that decomposes a given assertion into some other assertions. It allows to choose the networks.

Distribution that decomposes a given assertion into specification processes and channel assertion

Finally, these two rules express a goal driven technique with respect to the given inference rules

5 Concluding remarks

The synthesis of networks is a very difficult task. The method due to Z. MANNA and P. WOLPER [MW84] is a possible way that may be followed in a richer framework. The combination of the algebraic approach. the temporal one and the language-based one allows to specify properties relative to networks in a *natural way*. It is clear that the underlying model is an operational one but it includes the algebraic aspect. We try to develop a general tool able to integrate the following characteristics:

- tool of specification aid as *SACSO* [LPS87].

- tool of simplification of temporal formulas.

- generalized version of WOLPER's synthesis algorithm.

This communication has presented the general framework. It is clear that examples play a central role in our approach but we restrict the class of programs in order to limit the complexity. The current works implement the general ideas of this paper and organize the approach. We may quicly describe it as follows:

1. Specification of objects.

2. Specification of constraints on the sorted objects.

3. Specification of communications.

References

[HOA69]   C. A. R. HOARE. An axiomatic basis for computer programming. *Communication of ACM*, 12:576–580, 1969.

[KAR80]   N. FRANCEZ K.R. APT and W.P. DE ROEVER. A proof system for communicating sequential processes. *ACM TOPLAS*, 2(3):359–385, 1980.

[KP87]   S. KAPLAN and A. PNUELI. Specification and implementation of concurrently accessed data structures : an abstract data type approach. *LNCS*, (247):220–244, 1987.

[KRO87]   F. KROGER. Abstract modules : combining algebraic and temporal logic specification means. *Technique et Science Informatiques*, 6(6):559–573, 1987.

[LPS87]   N. LEVY, A. PIGANIOL. and J. SOUQUIERES. Specifying in sacso. In *Fourth International Workshop On Software Engineering and Design*, 1987.

[MER86]   D. MERY. A proof system to derive eventuality properties under justice hypothesis. *LNCS*, (233), 1986. Bratislava, Tchecoslovaquie.

[MER87]   D. MERY. Méthode axiomatique pour les propriétés de fatalité des programmes parallèles. *RAIRO Informatique Théorique et Application*, 21(3):287–322, 1987.

[MIL86]   R. MILNER. *A Calculus of Communicating Systems*. Technical Report ECS-LFCS-87-7, Laboratory for Foundations of Computer Science, August 1986. First published by SPRINGER VERLAG as Vol 92 of LNCS.

[MW84]   Z. MANNA and P. WOLPER. Synthesis of communicating processes from temporal logic specifications. *ACM TOPLAS*, 6:68–93, 1984.

[NDGO86] Van N'GUYEN, A. DEMERS. D. GRIES, and S. OWICKI. A model and temporal proof system for networks of processes. *Distributing Computing*, 1:7–25, 1986.

[OG76]   S. OWICKI and D. GRIES. An axiomatic proof technique for parallel programs i. *Acta Informatica*, 6:319–340, 1976.

[OH86]   E.R OLDEROG and C.A.R HOARE. Specification-oriented semantics for communicating processes. *Acta Informatica*, 23:9–66, 1986.

[PER85]   G. R. PERRIN. *La Communication : Un Outil pour la Specification, la Construction et la Verification de Systemes Paralleles*. PhD thesis, Universite de Nancy I, 1985. FRANCE.

[SZA88]   A. SZALAS. Towards the temporal approach to abstract data type. *Fundamenta Informaticae*, XI(1):49–64. March 1988.

[WOL83]   P. WOLPER. Temporal logic can be more expressive. *Information and Control*, 56(1-2):72–99, 1983.

Analysis of Estelle Specifications

Udo Thalmann
Technische Universität Wien
Institut für Angewandte Informatik
Argentinierstrasse 8, 1040 Wien, Austria
e-mail: thalmann@slowhp.uucp

Abstract

This article presents an analysis method which evaluates an Estelle specification towards a temporal logic requirements specification. First, test data are derived from an Estelle specification. Then, the Estelle specification is executed according to these test data. During the execution of the Estelle specification, the temporal logic formulas are evaluated. If a temporal logic formula is injured, then the Estelle specification does not satisfy the requirements specification. This analysis method does not substitute verification methods, since it cannot assure the absence of errors. But it is possible to reveal errors in complex system specifications, which is a problem with todays verification methods.

I. Introduction

In this article I present an analysis method for specifications written in Estelle ([Lin86]). First, test data are derived from an Estelle specification. Then, the Estelle specification is executed according to these test data and checked towards a temporal logic specification. Four steps can be distinguished in this analysis approach:

1) Safety and liveness requirements of a system are specified in temporal logic. Then an algorithm is developed in Estelle which must fulfil the specified requirements. The formal specification of the system requirements in temporal logic is necessary, if the Estelle specification shall be automatically analyzed.

2) The data flow of the Estelle specification is analyzed. Data flow analysis serves three different purposes. Data flow analysis can be used to detect data flow anomalies, to construct data flow functions, and to derive test paths from the Estelle specification.

   - Data flow analysis is used to detect data flow anomalies in [Rap85] and [Ura87]. For example, the use of a variable which is not defined, is a data flow anomaly.

   - Data flow functions model the flow of functionally dependent data from an input interaction to an output interaction ([Sar87]). Test data can be derived separately for each data flow function. Thus a combinatorial explosion of test data can be avoided.

   - Test paths correspond to specific executions of the Estelle specification. Test data for a data flow function are derived such that they meet the path predicates of a certain test path.

3) Test data generation algorithms are used to derive test data for specific data flow functions and test paths.

4)   During the execution of the Estelle specification according to the derived test data, the behaviour of the Estelle specification is analyzed against the temporal logic specification.

This analysis method can be applied to more complex system specifications. This is a problem with other analysis methods, such as mechanical verification ([Hai82]), and model checkers ([Fer86]).

This article is organized as follows:
In chapter II I show, how to specify safety and liveness requirements of an unbounded buffer in temporal logic. Then I specify an algorithm in Estelle which implements the unbounded buffer.
In chapter III I apply data flow analysis to the unbounded buffer specification in Estelle. I explain, how to detect data flow anomalies, how to construct data flow functions, and how to derive test paths.
In chapter IV I describe two methods to derive test data such that they satisfy certain test path conditions.
In chapter V I show, how to analyze the behaviour of the Estelle specification against the safety and liveness requirements which are expressed in temporal logic formulas.

II. Specification with temporal logic and Estelle

System requirements such as safety and liveness requirements are expressed in temporal logic formulas, whereas a specific algorithm which hopefully satisfies the system requirements is expressed in Estelle. The semantics of temporal logic formulas are defined in terms of a structure which is explained below. I use branching time logic which is defined in [Ben83].

Semantics of branching time logic

To define the semantics, we must define how the temporal logic formulas are interpreted on an underlying structure. This structure is a triple (S,P,R) where

-    S is a set of states;

-    P is a function which assigns a set of predicates to each state. A predicate a, which is assigned to a certain state s by function P, is true in this state. Therefore we write $a \in P(s)$;

-    R is the successor relation between states. If the relation s R t holds, then t is a successor state of s.

A path b is a sequence of states $s = s_0, s_1, s_2,...$ for which the relation $s_i R s_{i+1}$ holds.
A ("for all path") and E ("for some path") are path quantifiers.
G ("always"), F ("sometimes"), and X ("nexttime") are path operators.
s |= p means that formula p is true in the structure at state s.
According to these definitions the semantics of temporal logic formulas are defined as follows:

s |= a	iff a∈ P(s)
s |= ~p	iff not (s |= p)
s |= p v q	iff s |= p or s |= q
s |= A p	iff for every path b, b |= p
s |= E p	iff for some path b, b |= p
b |= p	iff s |= p
b |= X p	iff for s R t, t |= p
b |= G p	iff for every state t∈ b, t |= p
b |= F p	iff for some state t∈ b, t |= p

The following mnemonics are usually used:
ALL (always) = AG,
INEV (inevitable) = AF,
SOME (sometimes) = EG,
POT (potentially) = EF.

The temporal logic formula s \|= ALL p means that the predicate p holds in state s, iff p holds in every state t on every path which emanates from s.

The temporal logic formula s \|= INEV p means that the predicate p holds in state s, iff p holds in some state t on every path which emanates from s.

I leave it to the reader to interpret the other temporal logic formulas accordingly. Now I will specify temporal logic formulas for an unbounded buffer.

Unbounded buffer example

Figure 1 outlines the structure of a message transfer system with an unbounded buffer. Messages are put into the buffer by a send operation. The variable in characterizes the sequence of messages which is sent to the buffer. The buffer delivers messages by a receive operation. The variable out characterizes the sequence of messages which is delivered by the buffer.

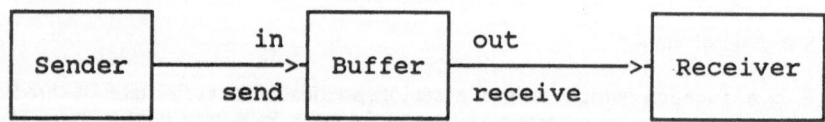

Figure 1: Structure of a message transfer system

Safety und liveness requirements

Safety and liveness requirements can be distinguished in a requirement specification. Safety requirements express the fact that "something bad never happens", whereas liveness requirements express the fact that "something good will happen".

Safety requirements can be further divided into invariants and pre- and postconditions.

Invariants are properties of the system which must always hold. The following buffer invariant says that the sequence of output messages must correspond to the sequence of input messages.

invariant: prefix (in, out) where prefix is a boolean function which yields true, if out is a prefix of in.

Pre- and postconditions are properties of a system which must hold before and after the execution of an operation. The following pre- and postcondition expresses a relation between the input message sequence in before and after the send operation. The auxiliary variable x represents the contents of the input message sequence before the send operation. The function concat adds the new input message mesg_in to the existing input message sequence in. AT(send) and AFTER(send) are action predicates which hold before respectively after the send operation is performed ([Fer86]).

pre: AT(send) => (in = x)
post: AFTER(send) => (in = concat(x,mesg_in))

A similar pre- and postcondition can be defined for the receive operation.

pre: AT(receive) => (out = y)
post: AFTER(receive) => (out = concat(y,mesg_out))

The auxiliary variable y is used to represent the contents of the output message sequence out before the receive operation.

The following liveness requirement for the send operation says that the send operation will terminate.

live: AT(send) => INEV AFTER(send)

A send operation can always be performed whereas a receive operation can only be performed, if the buffer is not empty. The first liveness requirement for the receive operation states that the buffer will be infinitely often not empty.

live: ALL POT ~ empty_buffer
where empty_buffer is an auxiliary boolean function which yields true, if the buffer is empty. The second liveness requirement for the receive operation states that the receive operation will terminate.

live: AT(receive) & ~ empty_buffer => INEV AFTER(receive)

Estelle specification

The underlying structure of an Estelle specification is a finite state machine (S,R) where S is a set of states and R is the successor relation between states. S and R have the same meaning as in the underlying temporal logic structure. The relation R between states is defined by transitions. Pascal statements are allowed within a transition body to express data manipulations. I assume that specifications are written in Estelle normal form in order to facilitate data flow analysis ([Ura87]). [Sar86] describes a method to obtain an Estelle normal form specification from an Estelle specification. Figure 2 shows the structure of the unbounded buffer where head references the last message sent to the buffer, and tail references the last message delivered by the buffer.

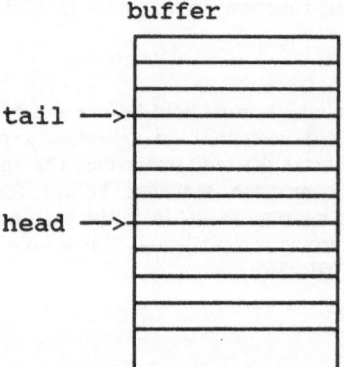

buffer

tail ——>

head ——>

Figure 2: Structure of the unbounded buffer

The unbounded buffer is specified in Estelle below. I assume that the reader is familiar with Estelle. The auxiliary variables in, out and the auxiliary function concat are only used for analysis purposes. They do not influence the behaviour of the buffer. Labels (t0, t1, t2) are not part of the Estelle syntax, but are convenient for data flow analysis.

```
VAR
      buf: array [1..n] of mesg_type
      {n is assumed to be infinite}
      head, tail: 0..n
      mesg_in, mesg_out: mesg_type
      in, out: array [1..n] of mesg_type

FUNCTION empty_buffer (head, tail): BOOLEAN
      empty_buffer := (head = tail)

FUNCTION concat ...

t0:   INITIALIZE
      head := 0
      tail := 0

t1:   FROM        s1
      TO          s1
      WHEN        send(mesg_in)
      BEGIN       head := head + 1
                  buf [head] := mesg_in
                  in := concat(in,mesg_in)
      END

t2:   FROM        s1
      TO          s1
      PROVIDED    NOT empty_buffer
      BEGIN       tail := tail + 1
                  mesg_out := buf[tail]
                  OUTPUT (mesg_out)
                  out := concat(out,mesg_out)
      END
```

III. Data flow analysis

Data flow analysis examines relations between variable definitions and variable uses. A variable definition assigns a value to a variable whereas a variable use references a variable. A variable cannot be used, if it has not been defined before. On the other hand a variable, which is defined without ever being used, does not serve any purpose. Before I apply data flow analysis to the Estelle buffer specification, I will review some data flow concepts. A detailed discussion can be found in [Rap85].

Data flow concepts

A definition of a variable is called def. A variable use is called c-use (computational use), if it occurs in the body of a transition. A variable use is called p-use (predicative use), if it occurs in a PROVIDED clause. Data flow analysis is based on def-, dcu-, and dpu sets of all variables.

def-, dcu-, und dpu-sets

The def set def(x) (definition set) of a variable x is the set of all transitions in which x is defined. The dcu set dcu(x,i) (definition c-use set) of a variable x is the set of all transitions with c-uses of x such that there exists a def-clear path w.r.t x from transition i, where x is defined, to each c-use of x in a transition of the dcu set. A def-clear path w.r.t. variable x is a path with no definition of x. The dpu set dpu(x,i) is defined accordingly. Table 1 shows def-, dcu-, and dpu sets of the Estelle buffer specification.

Table 1: def-, dcu-, dpu-sets

variable x	i = def(x)	dcu(x,i)	dpu(x,i)
head	t0 t1	t1 t1	t2 t2
tail	t0 t2	t2 t2	t2 t2
mesg_in	t1	t1	-
buf	t1	t2	-
mesg_out	t2	t2	-

In the remaining part of this chapter I will use the results of the above data flow analysis to detect data flow anomalies, to construct data flow functions, and to derive test paths from the Estelle buffer specification.

1) Detection of data flow anomalies

A data flow anomaly expresses the fact that one of the following three conditions is violated.

- Between two definitions of a variable x on a given path there exists at least one c-use or p-use of x.

- From each definition of a variable x emanates at least one path on which x is c-used or p-used.

- Each c-use or p-use of a variable x can be reached from the initial state of the Estelle specification by at least one path on which x is defined.

2) Construction of data flow functions

A data flow function represents the flow of functionally dependent information from input to output interactions in the Estelle specification. Thus to construct data flow functions requires to identify functionally dependant variables which are used to store information in the Estelle specification. Once data flow functions have been established, test data can be derived for each data flow function separately. This helps to avoid a combinatorial explosion of the number of test data.

[Sar87] examines three types of variables:

- Input variables are defined in Estelle input interactions (e.g. mesg_in). The set of input variables which belong to a data flow function is called SIN (set of input nodes).

- Output variables are used in Estelle output interactions (e.g. mesg_out). The set of output variables which belong to a data flow function is called SON (set of output nodes).

- Context variables are used to store data between input and output interactions (e.g. head, tail, buf). The set of context variables which belong to a data flow function is called SDN.

- The set of functions which belong to a data flow function is called SFN.

[Sar87] explains how to derive the above variable and function sets without using the data flow concepts described above. I will outline a similar approach which uses def-, dcu-, and dpu information to construct variable and function sets for data flow functions.

The sets SIN, SON, SDN, and SFN are constructed as follows:

-   Examine all context variables
    For each context variable examine all input-, output-, context variables, and functions as follows:
    If an input-, output-, or context variable, or a function occurs in a def, c-use, or p-use of the examined context variable, it is added to the set SIN, SON, SDN, or SFN respectively.

-   Input and output variables as well as functions are examined like context variables. Variables and functions which have already been assigned to a certain set are considered first.

There is only one data flow function in the Estelle buffer specification with the obvious sets:

SDN = {head, tail, buf}
SIN = {mesg_in}
SON = {mesg_out}
SFN = {empty_buffer}

3) Selection of test paths

A test path corresponds to a specific execution sequence of the Estelle specification. Test data for a specific data flow function are derived such that a selected test path can be executed. In order to execute a test path, the path predicates must be satisfied. The set of all paths in the buffer specification can be described by the following regular expression:

$$t0 \ (t1+t2)^*$$

where t0, t1, and t2 are the transition labels in the Estelle specification.

Test path selection criteria

Data flow analysis provides several criteria for the selection of test paths. These criteria are described in [Rap85]. I will use the all-uses criterion to select test paths. The all-uses criterion requires that there is at least one def-clear test path between each variable definition and variable use. The following test paths have been derived directly from the def-, dcu-, and dpu sets of table 1.

$$t0 \ (t1(t1+t2) + t2t2)$$

Row 1 in table 1 requires that transition t0 is followed by transition t1 or t2. Row 2 requires that transition t1 is followed by transition t1, or t2. Row 4 requires that transition t2 is followed by transition t2.

IV. Test data generation

Test data are derived for each data flow function such that the path predicates of a certain test path are satisfied. There are several methods to derive test data. It depends on the structure of the path predicates which test data generation method is used.

-   If the path predicates constitute a system of only linear equalities and/or linear inequalities, then this system can be solved with linear programming methods. The advantage of this method is that maximum and minimum test data are derived. Transformation rules for linear equalities and inequalities into a standard form are described in [Cla76].

-   If the path predicates do not constitute a linear equality or inequality system, then "try and error" methods can be used to derive test data. Try and error methods select test data from the domain of the variables randomly until the path predicates are satisfied. A try and error method is described in [Ram76].

So far I assumed that all paths are executable, if only the right test data are selected. Unfortunately not all paths are executable. What makes the problem even worse is the fact that it is not decidable whether a path is executable or not. The try and error method in [Ram76] informs the test person, if it cannot select test data which satisfy the path predicates.

V. Analysis of the Estelle specification

The underlying structure of the Estelle specification is a finite state machine (S,R), where S is a set of states and R is the successor relation between states. Since S and R have the same meaning in the underlying structure of the temporal logic specification, temporal logic formulas can be interpreted on the finite state machine model. Figure 3 outlines an analysis system for Estelle specifications. It consists of an Estelle interpreter and a temporal logic analyzer.

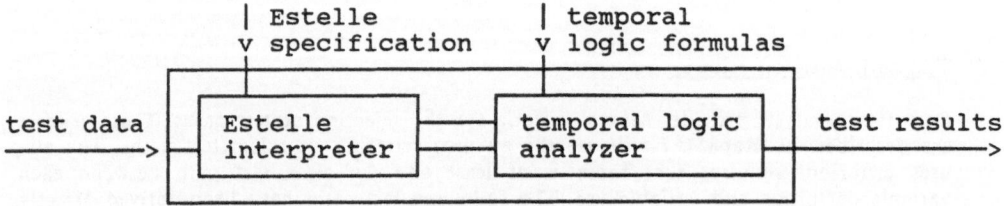

Figure 3: Analysis system for Estelle specifications

The Estelle interpreter selects a transition according to the derived test data. After each transition the temporal logic analyzer is activated which evaluates the temporal logic formulas according to the current values of the variables after the execution of a transition. If a temporal logic formula is evaluated to false, the behaviour of the Estelle specification does not satisfy the requirements. Not all temporal logic formulas can be evaluated (i.e. not all temporal logic formulas are decidable). [Kar86] distinguishes three kinds of decidability according to the structure of the temporal logic formula.

1) A logic formula f without temporal path operator or path quantifier is state decidable (SD). This means that f can be evaluated in the current state.

2) If f is state decidable, then Xf is next decidable (ND). This means that f can be evaluated in the next state. If f, f1, f2 are next decidable, so are ~f, f1&f2, Xf, Ef, and Af.

3) Every formula which is neither state decidable nor next decidable is nondecidable (UD).

Depending on the decidability property of the formula f, [Kar86] outlines an algorithm to evaluate a temporal logic formula.

- If formula f is SD, then evaluate f.

- If formula f is ND, then delay evaluation of f to the next state.

- If formula f is UD, then evaluate the subformula(s) of f in the following way:

operator	truth value of the subformula	truth value of the formula f
F,E	true	true
G,A	false	false

In any other combination truth values cannot be assigned. In this case the temporal logic analyzer stores information about undecided subformulas for a later decision.

VI. Conclusions

In this article I have presented an analysis method for Estelle specifications. An Estelle specification is executed according to a set of test data and checked whether it satisfies or injures the temporal logic requirements. This analysis method does not substitute verification methods, since it cannot assure the absence of errors. But it is possible to reveal errors in complex system specifications, which is a problem with todays verification methods.

A prototype of a data flow analyzer and a temporal logic analyzer will be implemented in the near future. When I elaborated this analysis approach, the question arose, whether it is useful to combine temporal logic with abstract data types. This can be an area of future research.

Acknowledgments

I thank professor Helmut Kerner and Thomas Mayr for their support and helpful discussions. Part of this work has been supported by Siemens Austria.

References

[Cla76]   L.A.Clarke;
          "A System to Generate Test Data and Symbolically Execute Programs";
          IEEE Tr.Soft.Eng., Vol.Se-2, No.3, September 1976.

[Ram76]   C.V.Ramamoorthy, S.F.Ho, W.T.Chen;
          "On Automated Generation of Program Test Data";
          IEEE Tr.Soft.Eng., Vol.Se-2, No.4, December 1976.

[Hai82]   B.Hailpern;
            "Verifying Concurrent Processes using Temporal Logic";
            Lecture Notes in Computer Science, Springer, 1982.

[Ben83]   M.B.Ari, A.Pnueli, Z.Manna;
            "The Temporal Logic of Branching Time";
            Acta Informatica 20, Springer, 1983.

[Rap85]   S.Rapps, E.Weyuker;
            "Selecting Test Data Using Data Flow Information";
            IEEE Tr.Soft.Eng., Vol.Se-11, No.4, April 1985.

[Fer86]   J.C.Fernandez, J.L.Richier, J.Voiron;
            "Verification of Protocol Specifications using the CESAR System";
            Protocol Specification, Testing, and Verification, V; North Holland; 1986.

[Lin86]   R.J.Linn;
            "The Features and Facilities of Estelle";
            Protocol Specification, Testing, and Verification, V, North Holland, IFIP, 1986.

[Sar86]   B.Sarikaya, G.v.Bochmann;
            "Obtaining Normal Form Specifications for Protocols";
            Computer Network Usage, North Holland, IFIP, 1986.

[Ura87]   H.Ural;
            "Test Sequence Selection based on Static Data Flow Analysis";
            Computer Communications, Vol.10, No.5, October 1987.

[Sar87]   B.Sarikaya, G.v.Bochmann, E.Cerny;
            "A Test Design Methodology for Protocol Testing";
            IEEE Tr.Soft.Eng., Vol.Se-13, No.5, May 1987.

Concurrency in Modula-2:
Properties of the Language Primitives

Robin A. Nicholl

Department of Computer Science
University of Western Ontario
London, Canada N6A 5B7

Abstract

The programming language Modula-2 provides two levels of support for concurrent programming, differing primarily in the level of abstraction at which programs can be written. We describe these two levels of concurrency in an algebraic style and show how this decription can be used to reason about the behaviour of concurrent programs in Modula-2. An intermediate level of "anonymous coroutines" is introduced to show the connection between these two levels. A number of alternative interpretations are given for those parts of the language where the precise semantics is not clear.

1   Introduction

The programming language Modula-2 [6] is intended to be useful for systems programming, as well as for a wide range of other applications. To enable the writing of process schedulers and device drivers, the language provides a small number of low level primitives for context switching. To enable quasi-parallel programming at a higher level the language provides a "standard module" Processes, which is responsible for sharing processor time among a number of "processes".

Modula-2's facilities for managing concurrent processes are oriented towards *multiprogramming* on a single processor machine. Hence it is always appropriate to talk about *the current process*, and this has a significant impact on the abstract notion of "process" presented to the programmer. This impact is so great that Wirth [6, p. 132] has suggested that Modula-2 "processes" should be termed *coroutines* instead, and recent revisions of the language definition have indeed eliminated the type PROCESS []. An extended version of the language has been used for programming multiprocessor machines [5].

To assist in the development of reliable Modula-2 programs which use several processes it is essential that we know how to reason about the behaviour of such programs. We should expect that the restricted form of concurrency provided in Modula-2 should be easier to work with than the more general forms of concurrency generally discussed in the literature, notably CSP [1]. To this end we will identify some properties of the form of concurrency found in Modula-2 and discuss the use of these properties in reasoning about program behaviour.

Modula-2 provides two levels of support for concurrent programming. At the lower level the programmer is responsible for sharing the processor among different processes. Since context switching is made explicit at this level, program behaviour is deterministic. It is at this level that processes may more reasonably be viewed as coroutines.

At the higher level the programmer is relieved of the responsibility for process scheduling. At this level program behaviour is no longer deterministic, since it now depends on the particular (and hidden) scheduling algorithm. The programmer is still able to identify the specific points at which context switching may occur, but can no longer determine which process will be dispatched. Programming at this level requires dealing with the concerns of mutual exclusion and process synchronization, although the ability to control when context switching can occur does eliminate some of the difficulties of concurrent programming.

We will describe these two levels of concurrency in terms of properties satisfied by the relevant operators. This style of definition is based on the ideas presented in [2].

2 Low-level Concurrency

At the lowest level it is quite clear that Modula-2 supports coroutines, and has no true concurrency. The lowest level appears as the following primitives, provided (EXPORTed) by the machine-dependent module SYSTEM:

```
TYPE PROCESS;
PROCEDURE NEWPROCESS (p: PROC; a: ADDRESS; n: CARDINAL;
                      VAR new: PROCESS);
PROCEDURE TRANSFER   (VAR old, new: PROCESS);
```

NEWPROCESS is used to create a new process to be identified by the parameter new, to occupy a block of memory of size n starting at address a, and to execute procedure p. We shall assume that the block of memory occupied by a process is chosen to ensure no interference with the correct behaviour of the process, and henceforth ignore the second and third parameters to NEWPROCESS.

TRANSFER is used to stop execution of the current coroutine, identify it by the parameter old and start execution of the coroutine identified by new. The stopped coroutine can be resumed at a later time, by another call to TRANSFER.

The Modula-2 programmer can also *declare* variables of type PROCESS and use these variables in *assignment* statements.

Hence the four methods of dealing with coroutines are:

1. Declaration: e.g. VAR p1, p2:   PROCESS;

2. Assignment: e.g. p1 := p2

3. Creation: e.g. NEWPROCESS (P, ..., p1)

4. Switching: e.g. TRANSFER (p1, p2)

Their behaviour, which has been informally described above, satisfies the following properties. These properties thus act as a *definition* of their behaviour. In expressing these properties we use the following abbreviations:

NEWPROCESS (P, ..., p1) is written $p1/P$

TRANSFER (p1, p2) is written $p1 \rightsquigarrow p2$.

These properties are expressed in a style similar to that used in describing CSP [1] and in stating some of the laws of programming [2].

1. The order in which two processes (coroutines) are created is not significant:

$$n/P; m/Q = m/Q; n/P$$

2. If two processes are created with the same name, the first is lost:

$$n/P; n/Q = n/Q$$

3. Switching to a newly created process causes that process to execute, and the current process to await execution:

$$n/P; m \rightsquigarrow n; Q = m/Q; P$$

4. If a statement does not refer to n and does not involve $\rightsquigarrow$ then it makes no difference if the statement is done before or after creating process n:

$$n/P; Q = Q; n/P$$

5. Assignments before context switches have the following effects:

   (a) $n := p; n \rightsquigarrow m = n \rightsquigarrow m$
   (b) $m := p; n \rightsquigarrow m = n \rightsquigarrow p$

2.1 Example

Based on these properties we can carry out the following derivation, based on the program in [3, pp. 68-69].

Suppose three coroutines Msg1, Msg2 and Msg3 as defined below. We have used $*P$ to indicate the repeated execution of P and the numbers 1, 2 and 3 to indicate the distinctive portion of each coroutine.

$$Msg1 = *(1; c1 \rightsquigarrow c2)$$

$$Msg2 = *(2; c2 \rightsquigarrow c3)$$

$$Msg3 = *(3; c3 \rightsquigarrow c1)$$

Thus each coroutine loops indefinitely, performing its distinctive task before switching context.

The main program creates these three coroutines and then switches to the first.

$$Main = c1/Msg1; c2/Msg2; c3/Msg3; main \rightsquigarrow c1$$

In explaining the resultant behaviour we use Π to denote the empty (or null) statement; we also use $<$ and $>$ to delimit the process of interest so that $/$ and $\rightsquigarrow$ may only be used between these symbols. The behaviour of $Main$ is:

$$
\begin{aligned}
&<; Main; > \\
={}& <; c1/Msg1; c2/Msg2; c3/Msg3; main \rightsquigarrow c1; > \\
={}& <; c2/Msg2; c3/Msg3; c1/Msg1; main \rightsquigarrow c1; > \\
={}& <; c2/Msg2; c3/Msg3; main/\Pi; Msg1; > \\
={}& <; c2/Msg2; c3/Msg3; main/\Pi; 1; c1 \rightsquigarrow c2; Msg1; > \\
={}& 1; <; c2/Msg2; c3/Msg3; main/\Pi; c1 \rightsquigarrow c2; Msg1; > \\
={}& 1; <; c3/Msg3; main/\Pi; c1/Msg1; Msg2; > \\
={}& 1; <; c3/Msg3; main/\Pi; c1/Msg1; 2; c2 \rightsquigarrow c3; Msg2; > \\
={}& 1; 2; <; main/\Pi; c1/Msg1; c3/Msg3; c2 \rightsquigarrow c3; Msg2; > \\
={}& 1; 2; <; main/\Pi; c1/Msg1; c2/Msg2; 3; c3 \rightsquigarrow c1; Msg3; > \\
={}& 1; 2; 3; <; main/\Pi; c2/Msg2; c3/Msg3; Msg1; > \\
={}& 1; 2; 3; <; Main; >
\end{aligned}
$$

Hence the program repeatedly writes the sequence of numbers 1,2,3 and may thus be rewritten as follows.

$$Main = *(1; 2; 3)$$

3 Anonymous Coroutines

We have now seen how to reason about coroutines at the lowest level. Before looking at the module **Processes** we will examine an intermediate level. At this intermediate level we will continue to have coroutines, but they are now *anonymous*. Coroutines are given no name when they are created, and context switching no longer identifies which coroutine will start execution. In this respect program behaviour is now nondeterministic.

We use the following notation, a slight modification of that used to describe low-level concurrency.

$/P$ - create a coroutine to execute P

$\odot$ - cause a context switch from the current coroutine to some coroutine, which could be the current one

These satisfy the following properties:

1. Context switching is symmetric, so the order in which anonymous coroutines are created does not matter.

$$/P; /Q = /Q; /P$$

2. With two coroutines the behaviour following a context switch does not depend on which of the two was executing.

$$/P; \odot; Q; >= /Q; \odot; P; >$$

3. When there is only one coroutine, context switches will not affect its execution.

$$<; \odot; P =<; P$$

4. If Q cannot cause a context switch (i.e. Q does not contain $\odot$), it does not matter whether it is executed before or after creating a coroutine.

$$/P; Q = Q; /P$$

The most important issue remains however: what behaviour is acceptable from $/P; \odot; Q; >$?

5. We will certainly accept that a context switch may allow the current coroutine to continue. Since other alternatives are also acceptable we state:

(a) A context switch is permitted to continue execution of the current coroutine:

$$/P; \odot; Q; > \supseteq /P; Q; >$$

(b) When this is taken together with 1, we can conclude:

$$/P; \odot; Q; > \supseteq /Q; P; >$$

(c) Letting $P \cup Q$ denote the non-deterministic choice between behaving like P or behaving like Q we have:

$$/P; \odot; Q; > \supseteq ((/P; Q) \cup (/Q; P)); >$$

In languages which permit a greater degree of concurrency than Modula-2 we may be unable to state any property stronger than this. In Modula-2 however, we find that we can replace "concurrency" by non-determinism.

(d) When only two coroutines P and Q exist, this describes the *only* acceptable behaviour:

$$<; /P; \odot; Q; > = <; (/P; Q) \cup (/Q; P); >$$

(e) We would like to extend this *equality* between behaviours to allow for more than two coroutines. We obtain the following property, of which those above form special cases:

$$/P; \odot; Q; > = \odot; (/P; Q) \cup (/Q; P); >$$

3.1 Example

Suppose we have two coroutines, one producing a stream of ones, the other a stream of zeros. Let each coroutine offer a context switch after producing each value, as defined by:

$$ONES = 1; \odot; ONES$$
$$ZEROS = 0; \odot; ZEROS$$

and suppose they are activated by the following main program:

$$MAIN = /ZEROS; ONES$$

Then the behaviour of this program is:

$$<; MAIN; >$$
$$= \ <; /ZEROS; ONES; >$$
$$= \ <; /ZEROS; 1; \odot; ONES; >$$
$$= \ 1; <; /ZEROS; \odot; ONES; >$$
$$= \ 1; <; \odot; (/ZEROS; ONES) \cup (/ONES; ZEROS); >$$
$$= \ 1; <; (/ZEROS; ONES) \cup (/ONES; ZEROS); >$$

Now $<; /ZEROS; ONES; >=<; MAIN; >$ and $<; /ONES; ZEROS; >= 0; <$ $; (/ONES; ZEROS) \cup (/ZEROS; ONES); >$.

So, substituting in the equation above, $MAIN = 1; (MAIN \cup X)$ where $X = 0; (MAIN \cup X)$ denotes $(/ONES; ZEROS)$.

Now we can use these equalities to define $MAIN \cup X$ directly:

$$MAIN \cup X$$
$$= \ (1; MAIN \cup X) \cup (0; MAIN \cup X)$$
$$= \ (1 \cup 0); (MAIN \cup X)$$
$$= \ *(1 \cup 0)$$

Then we have:

$$MAIN$$
$$= \ 1; ((1; (MAIN \cup X)) \cup (0; (MAIN \cup X)))$$
$$= \ 1; (1 \cup 0); (MAIN \cup X)$$
$$= \ 1; (1 \cup 0); *(1 \cup 0)$$
$$= \ 1; *(1 \cup 0)$$

4 High-level Concurrency

In Modula-2 the high level of concurrency is provided by the following primitives provided by module Processes as presented in [6].

```
TYPE SIGNAL;
PROCEDURE StartProcess (P: PROC; n: CARDINAL);
```

```
PROCEDURE Init (VAR s: SIGNAL);
PROCEDURE WAIT (VAR s: SIGNAL);
PROCEDURE SEND (VAR s: SIGNAL);
```

Informal descriptions of these primitives now follow.

StartProcess(P,n) creates and starts execution of a new process. The new process executes procedure P with a workspace of size n. We will ignore n and adopt the notation ↑P as an abbreviation for `StartProcess(P,...)`.

WAIT(s) suspends the current process until some other process sends the signal s. Abbreviated to ?s.

SEND(s) if any processes are waiting for signal s, one of them is allowed to continue execution; if no processes are waiting for s, there is no effect. Abbreviated to !s.

Init(s) initialises the signal s. Initially no processes are waiting for s. Abbreviated to ιs.

In order to define these primitives we have found it necessary to introduce one further primitive, not found in Modula-2. It indicates that a coroutine *is waiting on* a signal s, and is denoted by ▷s. These primitives possess the following properties. Note in particular that these primitives dictate the points at which rescheduling may occur.

1. A process starting P will itself be suspended while P executes.

$$\uparrow P; Q; >= /Q; P; >$$

2. A newly started process performing a **WAIT** operation can be replaced by a suspended process which is already waiting. Context switching determines which process will continue.

$$\uparrow(?s; P) = /(\triangleright s; P); \odot$$

3. A waiting process cannot execute, hence:

$$/P; \triangleright s; Q; >= /(\triangleright s; Q); P; >$$

4. Starting a process which then waits on some signal suspends the process until that signal is sent; hence a context switch takes place.

$$\uparrow(?s;P);Q;\, >= \odot;/(\triangleright s;P);Q;\, >$$

5. Sending a SIGNAL on which no process is waiting has the sole effect of allowing a context switch to occur.

$$\iota s;!s = \iota s;\odot$$

An interesting consequence of this property is that the context switch primitive introduced for discussing the behaviour of anonymous coroutines ($\odot$) can be written in Modula-2 as

$$\mathbf{var}s;\iota s;!s$$

which is

```
PROCEDURE ContextSwitch ;
   VAR s: SIGNAL ;
BEGIN
   Init(s) ; SEND(s)
END ContextSwitch ;
```

6. Initialisation of a signal is independent of process creation, unless the created process is already waiting on that signal. Hence provided P is not $\triangleright s$ we have:

$$\iota s;/(P;Q) = /(P;Q);\iota s$$

7. Initialisation of a signal is independent of any statement which does not involve a context switch. Hence provided P cannot involve rescheduling (i.e. no $\odot$, $\uparrow$, ?or !) we have:

$$\iota s;P = P;\iota s$$

In particular this includes moving initialisation passed the $<$ delimiter:

$$\iota s;\, <=<;\iota s$$

8. If a SIGNAL is sent to a process which is already waiting, the results are not clearly defined: does the waiting process become *current* or does it merely become *ready*?

  (a) If it becomes current we have

$$/(\triangleright s; P); !s; Q; >= /Q; P; >$$

  (b) If it becomes ready we have

$$/(\triangleright s; P); !s; Q; >= /P; \odot; Q; >$$

Note that 8a describes a possible behaviour of 8b since $/P; \odot; Q; > \sqsupseteq /Q; P; >$. Henceforth we shall use 8a. This reduces the degree of non-determinism involved and also seems more in keeping with Wirth's informal description [6].

4.1   Example

Suppose two processes

$$ZEROS \;=\; 0; ?z; ZEROS$$
$$ONES \;=\; 1; ?o; ONES$$

driven by a main program which constantly sends the relevant signals, after first starting the two processes:

$$MAIN = \iota z; \iota o; \uparrow ZEROS; \uparrow ONES; LOOP$$

where $LOOP = *(!z; !o)$.

Then the behaviour of $MAIN$ may be analyzed as follows:

$$<; MAIN; >$$
$$= \;\; <; \iota z; \iota o; \uparrow ZEROS; \uparrow ONES; LOOP; >$$
$$= \;\; <; \iota z; \iota o; 0; \uparrow(?z; ZEROS); \uparrow ONES; LOOP; >$$
$$= \;\; 0; <; \iota z; \iota o; /(\triangleright z; ZEROS); \odot; \uparrow ONES; LOOP; >$$
$$= \;\; 0; <; \iota z; \iota o; \odot; /(\triangleright z; ZEROS); \uparrow ONES; LOOP; >$$
$$= \;\; 0; <; \iota z; \iota o; /(\triangleright z; ZEROS); \uparrow ONES; LOOP; >$$

$$= \quad 0; <; \iota z; \iota o; /(\triangleright z; ZEROS); 1; \uparrow(?o; ONES); LOOP_i >$$
$$= \quad 0; 1; <; \iota z; \iota o; /(\triangleright z; ZEROS); /(\triangleright o; ONES); \odot; LOOP; >$$
$$= \quad 0; 1; <; \iota z; \iota o; /(\triangleright z; ZEROS); /(\triangleright o; ONES); !z; !o; LOOP; >$$
$$= \quad 0; 1; <; \iota z; \iota o; /(\triangleright o; ONES); /(\triangleright z; ZEROS); !z; !o; LOOP; >$$
$$= \quad 0; 1; <; \iota z; \iota o; /(\triangleright o; ONES); /(!o; LOOP); ZEROS; >$$
$$= \quad 0; 1; <; \iota z; \iota o; /(\triangleright o; ONES); /(!o; LOOP); 0; ?z; ZEROS; >$$
$$= \quad 0; 1; 0; <; \iota z; \iota o; /(\triangleright o; ONES); \uparrow(?z; ZEROS); !o; LOOP; >$$
$$= \quad 0; 1; 0; <; \iota z; \iota o; /(\triangleright o; ONES); /(\triangleright z; ZEROS); \odot; !o; LOOP; >$$
$$= \quad 0; 1; 0; <; \iota z; \iota o; /(\triangleright z; ZEROS); /(\triangleright o; ONES); !o; LOOP; >$$
$$= \quad 0; 1; 0; <; \iota z; \iota o; /(\triangleright z; ZEROS); /LOOP; ONES; >$$
$$= \quad 0; 1; 0; <; \iota z; \iota o; /(\triangleright z; ZEROS); \uparrow ONES; LOOP; >$$

From this derivation we can identify a repeated pattern

$$<; \iota z; \iota o; /(\triangleright z; ZEROS); \uparrow ONES; LOOP; >$$

Denote this expression by X (say) so that $MAIN = 0; X = 0; 1; 0; X$. Hence we deduce $X = *(1; 0)$ and so

$$MAIN = 0; *(1; 0)$$

4.2 Example

In this section we show some of the work required to analyse the behaviour of somewhat more realistic concurrent programs in Modula-2. Consider two processes communicating via a (shared) buffer of size 1. The producer process (PRO) puts the elements of a sequence s into the buffer; the consumer process (CON) appends values taken from the buffer to sequence t.

$$PRO = *(x, i := s_i, i + 1; PUT)$$
$$CON = *(GET; t_j, j := y, j + 1)$$

where we define PUT, GET and initialisation $INIT$ as follows, taking b as the buffer, e as a Boolean variable to indicate if the buffer is empty and T and F to represent TRUE and FALSE respectively.

$$PUT \;\; = \;\; (IF\;e\;THEN\;\Pi\;ELSE\;?s\;END); putx; !d$$
$$GET \;\; = \;\; (IF\;e\;THEN\;?d\;ELSE\;\Pi\;END); gety; !s$$
$$putx \;\; = \;\; b, e := x, F$$
$$gety \;\; = \;\; y, e := b, T$$
$$INIT \;\; = \;\; \iota s; \iota d; e := T$$

Then (assuming s is initialised elsewhere) the program is given by

$$MAIN = i, j := 1, 1; INIT; {\uparrow}PRO; CON$$

Since the derivation of this program is quite lengthy, consider instead some portions of the entire derivation. We begin with some derivations concerning PUT and state the corresponding (and symmetric) results for GET.

$$<; INIT; e := T; PUT$$
$$= \;\; <; INIT; PUT$$
$$= \;\; <; INIT; \Pi; putx; !d$$
$$= \;\; <; INIT; putx; !d$$
$$= \;\; <; INIT; putx; \iota d; !d$$
$$= \;\; <; INIT; putx; \odot$$
$$= \;\; <; INIT; putx$$
$$= \;\; putx; <; INIT; e := F$$

$$<; INIT; e := F; PUT$$
$$= \;\; <; INIT; e := F; ?s; putx; !d$$

Provided P is not already awaiting some signal (i.e. if it does not begin with ▷)

$$<; INIT; e := T; /P; PUT$$
$$= \;\; <; INIT; /P; PUT$$

$$
\begin{aligned}
&= \quad <; INIT; /P; \Pi; putx; !d \\
&= \quad putx; <; INIT; e := F; /P; !d \\
&= \quad putx; <; INIT; e := F; /P; \iota d; !d \\
&= \quad putx; <; INIT; e := F; /P; \odot
\end{aligned}
$$

$$
\begin{aligned}
&<; INIT; e := F; /P; PUT; Q; > \\
&= \quad <; INIT; e := F; /P; ?s; putx; !d; Q; > \\
&= \quad <; INIT; e := F; /(\triangleright s; putx; !d; Q); P; >
\end{aligned}
$$

$$
\begin{aligned}
<; INIT; e := T; GET &= \quad <; INIT; ?d; gety; !s \\
<; INIT; e := F; GET &= \quad gety; <; INIT \\
<; INIT; e := T; /P; GET &= \quad <; INIT; e := T; /(\triangleright d; gety; !d; Q); P; > \\
<; INIT; e := F; /P; GET &= \quad gety; <; INIT; e := T; /P; \odot
\end{aligned}
$$

5 Conclusions and Future Work

Using the properties given here, it is possible to analyse the behaviour and prove the correctness of small concurrent programs in Modula-2. The proofs are primarily concerned with ensuring that all non-deterministic choices associated with context switches continue to satisfy the correctness criterion. These proofs can be carried out at either the higher or lower level: the former involving less detail but more non-determinism than the latter.

Modula-2 also allows programmers to write device drivers, including clock handlers, which execute in response to hardware interrupts. When such processes are introduced into a program the degree of non-determinism is greatly increased and we can more closely approximate true concurrency. Work on identifying properties of Modula-2 processes in the face of interrupts is currently still in progress.

Acknowledgements

Financial support for this work was provided by the Natural Science and Engineering Research Council (NSERC) of Canada under Grant No. S265A1.

References

[1] C.A.R. Hoare, *Communicating Sequential Processes*, Prentice-Hall Int'l., 1985.

[2] C.A.R. Hoare *et al.*, "Laws of Programming", *Comm. ACM 30*, 8 (Aug. 1987), 672-686.

[3] *Modula-2/86 Software Development System, 3rd Ed.*, Logitech Inc., 1986.

[4] R.A. Nicholl, "A Specification of Modula-2 Process (Coroutine) Management", to appear in *Journal of Pascal, Ada and Modula-2*, 1988.

[5] P. Rovner, R. Levin and J. Wick, "On Extending Modula-2 for Building Large, Integrated Systems", Report No. 3, DEC Systems Research Center, 1985.

[6] N. Wirth, *Programming in Modula-2*, Springer-Verlag, 1982.

Appendix: Summary of Notation

For process identifiers $p1$, $p2$ and procedure identifiers P and Q:

Π the empty (null) statement

$P;Q$ do P, then do Q

$*P$ do P repeatedly; as `WHILE TRUE DO P END`

$P \cup Q$ do either P or Q; no Modula-2 counterpart

$p1/P$ create a new process identified by $p1$ and executing P;
    as `NEWPROCESS (P,...,p1)`

$p1 \rightsquigarrow p2$ switch the processor from $p1$ to $p2$; as `TRANSFER (p1,p2)`

$/P$ create a new *anonymous* process, executing P; no Modula-2 counterpart

$\odot$ cause a (possible) context switch; no Modula-2 counterpart

$\uparrow P$ create and start execution of a new process executing procedure P;
    as `StartProcess(P,...)`

ιs initialise SIGNAL s; as Init(s)

$?s$ (begin to) wait for SIGNAL s; as WAIT(s)

$!s$ send SIGNAL s; as WAIT(s)

$\triangleright s$ continue waiting for SIGNAL s; no Modula-2 counterpart

$P \supseteq Q$ all behaviours of Q are also possible behaviours of P

Specification and implementation of concurrent systems using PARLOG

David Gilbert

PARLOG Group
Department of Computing
Imperial College
180, Queens Gate
London SW7 2BZ
UK

email: drg@doc.ic.ac.uk

ABSTRACT

The specification and implementation of a class of concurrent systems is investigated in this paper using PARLOG as the implementation language, guided by specifications in CCS. This class of systems is restricted to an illustrative subset, including buffers and queues. A comparison is made between the computational models of concurrency that are expressed by each language. Illustrative programs are given which highlight the differences in operational behaviour between programs in the two languages. We investigate the ways in which equivalences in CCS programs may be used to compare PARLOG programs, and the areas of PARLOG programming which are not covered by this comparison. The advantages and disadvantages of CCS compared with PARLOG as a 'specification-implementation' language are discussed.

1. Introduction

Designers of concurrent systems are faced with the problems of firstly specifying the system and secondly implementing the design. There are several specification techniques available, including LOTOS, CCS, CSP and Petri Nets. However, the route from specification to implementation is not clear in many cases. The best that can be done is often proving that the implementation is in some way equivalent to the specification, rather than generating the implementation directly from the specification.

In this paper we wish to investigate the possibility of using the parallel logic programming language PARLOG both to specify and implement certain classes of communicating systems.

The PARLOG language has been described elsewhere in detail [Gregory87 , Gilbert87] and we will only describe its features relevant to this discussion. These are:

• asynchronous and synchronous stream based communication.

- committed choice and no output before commitment.

2. PARLOG

PARLOG as a Concurrent Logic programming language

PARLOG is a language which belongs to the family of committed choice parallel logic programming languages. It has the ability to explicitly express both OR and stream-AND parallelism. Committed choice non-determinism is implemented by the use of guards which ensure that committal is made to only one clause. These features enable the language to use non-determinism as an evaluation strategy if required by the programmer. They permit the writing of programs which require the separate use of both sequential and parallel evaluations.

The PARLOG syntax used in this article is given in the Table at the end of this paper. Note that in the examples in this paper, variable names start with a capital letter.

Mode declarations are used to specify communication restraints on shared variables, which are declared to be either ? ("input") or ^ ("output"), thus acting as communication channels. These declarations are made once for each relation, and each argument of the relation is annotated:

mode name(a1?,..,ak^,..)

Messages sent along these channels are incrementally constructed from partially determined data structures, usually consisting of lists of terms acting as message streams.

The general form of a PARLOG clause is:

<head> ← <guard> : <body> <or-op>

where <head> is in the form *name(a1,..,ak)* , **name** being the relation name, and **a1,..,ak** its arguments. The logical implication symbol is '←' and ':' the guard operator. Both the guard and the body can be a conjunction of calls, or empty, and the calls are separated by the sequential-AND or parallel-AND operators; an OR-operator terminates the clause. In the case that a guard is empty, it is omitted along with the guard operator; if a clause has a guard but no body, the body is replaced by **true**; a clause with no guard or body is represented by just the head. A clause is a candidate for evaluation if both input matching in the head and the evaluation of the guard succeeds, whereas in a non-candidate clause either of these fail. A clause can be suspended if either the input matching or guard evaluation suspend waiting for an input variable to become instantiated. A suspended call may eventually become either candidate or non-candidate. For example, given the following PARLOG program:

Example 1.

```
mode check(pattern?).
check([H|T]).
```

a call *? check(foo)* will **fail**, a call *? check([X/Y])* will **succeed**, and a call *? check(X)* will **suspend** (until X has become further instantiated). Note that no output bindings are made until committal has been made to a clause (ie the input and guard conditions are satisfied), and committal may be made to only one clause of a procedure.

Stream communication

The committed choice aspect of PARLOG give it the ability to express stream communication using **stream-AND-parallelism** in a very simple manner. This enables PARLOG to be used constructively in applications where concurrency and process state are of importance. Since PARLOG allows more than one call to be evaluated concurrently, the language permits computations which comprise the execution of several concurrent processes.

The basic paradigm of stream communication is that of producers and consumers. PARLOG's ability to perform **stream-AND-parallelism** gives the language the power to express stream communication in a very simple manner. Shared variables act as communication channels and messages can be sent by the incremental construction of partly instantiated data structures such as tuples or lists of terms. This incremental communication has an analogy with the lazy and eager parallel evaluation of functional programs. Output variables are produced in an asynchronous manner, but input variables are processed synchronously if an argument contains a partially instantiated term. Mode declarations mean that in a situation involving communication only one PARLOG process can be the producer, binding shared variables, whilst there may be one or more consumers of the communication stream.

An outline program which illustrates this form of stream communication is:

```
mode eager_producer(Stream^).
eager_producer([Item|Stream])←
     produce(Item) , eager_producer(Stream).


mode naive_consumer(Stream?).
naive_consumer([Item|Stream])←
     consume(Item) , naive_consumer(Stream).


mode produce(item^) , consume(item?).


call:
? eager_producer(Stream) , eager_consumer(Stream).
```

Note that we do not give the code for the procedures *produce* and *consume*, as these are not relevant to this particular algorithm; we assume that *produce* is 'eager' (asynchronous). The *eager_producer* process can run ahead of the consumer by an arbitrary amount, and we assume the existence of system buffers to permit this. An obvious question, to which there is no explicit answer since they are not formally part of the language, is what happens when these buffers become full.

Thus the naive mode of communication in PARLOG, which uses asynchronous sends and synchronous receives, makes reasoning about communicating systems difficult since the channels act as unbounded buffers in this case. The PARLOG programmer is, however, able to utilise the completely synchronous communication facilities offered by the language using *back-communication*, which can take two forms. These are firstly *mode reversal* (lazy evaluation) and secondly *cooperative construction of binding terms*

("Incomplete messages" [Shapiro83], or "back communication by cooperative construction of binding terms"[Gregory87]).

PARLOG in the light of Milner's interpretations

We note some correspondences between the concepts Milner expresses in his recent paper [Milner86] and those implemented in PARLOG. The interested reader is referred to [Ellis86 , Hussey87] for a discussion of the translation of PARLOG into CCS.

The "naive" form of communication in PARLOG (asynchronous producer, synchronous receiver) relies on buffering within the system, and does not conform to Milner's Principle 1 (all process interactions are atomic events); co-operative construction of binding terms is excluded for the same reason that Milner discounts the 'rendez-vous' of Ada. However, reverse-mode communication can be considered to conform to the first principle of Milner if we recognise that in a practical programming language there must be some physical means by which processes can communicate.

Due to the ability of a PARLOG computation to permit the existence of more than one concurrent process, the language may be used in a manner which broadly conforms to Principle 2 (every event is an interaction among processes), although PARLOG computational style is not limited to the process view alone. Because of the semantics of logic programs, a PARLOG query which is a conjunction of calls can only succeed when all the calls succeed, and if programming by side-effects is excluded then Principle 3 (every process constructor f must be such that the behaviour of $f(P_1,...,P_n)$ depends only on the behaviours of $(P_1,...,P_n)$ is satisfied.

Milner's Principle 4 (Conjunction) is partly satisfied by PARLOG's stream-AND operator when considering the synchronous communication expressible in the language. However the language only supports a 1 to N form of communication (one producer and many consumers). Moreover, multi-way synchronisation is not the natural paradigm in the language, and it must be enforced by programming techniques, hence making it difficult to incorporate Milner's Principle 9 (Simultaneity) into the analysis of PARLOG. However, Principle 9 is supported to the extent that a PARLOG process may synchronise with more than one other process by the use of several communication channels. Encapsulation (Principle 5) is achieved to the extent that the scope rules of Logic Programming which ensure that calls made within the body of one clause are not "visible" to a procedure which calls that clause. Disjunction (Principle 6) is reflected by the parallel-OR operator in PARLOG, which coupled with the use of guards and input matching can be used to ensure either exclusive clause choice or to allow the system to arbitrarily select one clause for committal if the choice is non- exclusive. Principle 7 (Renaming) is not realisable in PARLOG. Naturally, as a language orientated towards system programming, PARLOG incorporates explicit sequencing operators to allow for such a control where needed, for example in enforcing synchronousity or in i/o operations (Principle 8).

3. Format of the examples

The example 'programs' in this paper will generally be given in the following forms:

(i)　Standard PARLOG (including modes).

(ii)   Calls in the form of $?goal_1,....,goal_n$ to the above PARLOG programs.

(iii) CCS with value passing.

4. Synchronous Communication.

We restrict ourselves to the synchronous form of communication possible in PARLOG using the method of "Incomplete messages", rather than mode reversal. We represent the encapsulated message by the tuple +/2 [1], the first argument being the message item itself, and the second argument the synchronisation variable:

$$Item + Reply$$

4.1. Synchronous 'primitives' in PARLOG

We can implement synchronous send and receive in PARLOG by the following programs:

Example 2.


```
mode synch_send(item?,tuple^).
synch_send(Item,Item+Reply)← data(Reply).


mode synch_receive(tuple?,item^).
synch_receive(Item+Reply,Item)← Reply=reply.
```

Note that *data/1* suspends until its argument is ground (ie the top level functor of a data-structure is instantiated). Thus the query *?data(X)* suspends, and *?data([H/T])* succeeds. The synchronous PARLOG programs can be composed as follows:

?synch_send(foo , Msg) , synch_receive(Msg , Item).

We model the call to synch_send/2 as the CCS form $\bar{\alpha}$(foo), synch_receive/2 as α(x), and their composition as $\bar{\alpha}$(foo) | α(x).

4.2. Producers and Consumers.

Let us consider the situation of a chained producer and consumer in PARLOG. The synchronisation primitives may be used in the formulation of a simple recursive non-terminating communicating system, where producer/1 communicates with consumer/1 via a stream consisting of a partly instantiated list of +/2 tuples. Note that the communication is constrained by the '&' operator in producer/1, and thus the stream has a maximum length of 1, ie it may consist at the most of a head element which is a +/2 tuple, and an uninstantiated tail. The producer will suspend until the consumer acknowledges receipt of the message, when it will produce the next message. The self-recursive call in the consumer will suspend until the head of the stream is instantiated, and thus "lock-step" communication is achieved.

---

[1] +/2 is an infix operator in PARLOG

Example 3.

```
mode producer(Stream^).
producer([Msg|Stream])←
    synch_send(item,Msg)& producer(Stream).


mode consumer(Stream?).
consumer([Msg|Stream])←
    synch_receive(Msg,Item), consumer(Stream).
```

? producer(Stream) , consumer(Stream).

The situation is analogous to the CCS :

Example 4.

```
P = α̅(item) . P
C = α(x) . C
PC = P | C
```

5. Communication via buffers

5.1. One place buffers

Our first example of buffered communication is that of a simple one place buffer, which may be formulated in PARLOG as:

Example 5.

```
mode buffer1a(Ins?,Outs?).
buffer1a([In|Ins],Outs)←
    synch_receive(In,X) & buffer1a(X,Ins,Outs).


mode buffer1a(Item?,Ins?,Outs^).
buffer1a(X,Ins,[Out|Outs])←
    synch_send(X,Out) & buffer1a(Ins,Outs).
```

A CCS representation of this would be:

```
B = α(x) . B(x)
B(x) = γ̅(x) . B
```

Note that the PARLOG formulation partially evaluates to:

Example 6.

```
mode buffer1b(Ins?,Outs^).
buffer1b([In|Ins],[Out|Outs])←
    synch_receive(In,X) & synch_send(X,Out)&
    buffer1b(Ins,Outs).
```

A CCS representation of buffer1b/2 would be:

$$B = \alpha(x) \ . \ \overline{\gamma}(x) \ . \ B$$

These PARLOG buffers may be composed with producer/1 and consumer/1 above in the following manner:

/? - producer(A) , buffer(A,B) , consumer(B).

and the CCS buffer may likewise be composed, assuming suitable renaming of event labels:

```
P | B | C
```

In general, we would like to reason about the equivalence of the behaviour of two programs regarding both dynamic and completion behaviour. In this case, we note that there is no completion behaviour for the producer-buffer-consumer system, which is non-terminating. Partial evaluation can be used to demonstrate that buffer1a/2 and buffer1a/3 together are 'equivalent' in operation to buffer1b/2.

Another way to specify a one place buffer in PARLOG is to reformulate buffer1a to explicitly use a list rather than an extra argument as a data store:

Example 7.

```
mode buffer1c(Ins?,Outs^).
buffer1c(Ins,Outs)←
    buffer1c([],Ins,Outs).
```

```
mode buffer1c(Store?,In?,Out^).
```

```
buffer1c([],[In|Ins],Outs)←
    synch_receive(In,Item) & buffer1c([Item],Ins,Outs).
```

```
buffer1c([Item],Ins,[Out|Outs])←
    synch_send(Item,Out) & buffer1c([],Ins,Outs).
```

The CCS formulation for this is (disregarding the rules for ADTs[2]):

```
B = B(nil)
B(nil) = α(x) . B([x])
B([x]) = γ̄(x) . B(nil)
```

Ignoring the possible overheads of constructing a list, the programs for buffer1a, buffer1b and buffer1c are operationally equivalent.

5.2. Two place buffers

Building on the formulation of buffer1a, a possible specification of a two place buffer would be:

Example 8.

```
mode buffer2a(Ins?,Outs^).
buffer2a([In|Ins],Outs)←
    synch_receive(In,X) & buffer2a(X,Ins,Outs).


mode buffer2a(Item?,Ins?,Outs^).
buffer2a(X,[In|Ins],Outs)←
    synch_receive(In,Y) & buffer2a(X,Y,Ins,Outs).

buffer2a(X,Ins,[Out|Outs])←
    synch_send(X,Out) & buffer2a(Ins,Outs).


mode buffer2a(Item1?,Item2?,Ins?,Outs^).
buffer2a(X,Y,Ins,[Out|Outs])←
    synch_send(X,Out) & buffer2a(Y,Ins,Outs).
```

The buffer2a may be run in conjunction with a producer and consumer:

/? - producer(S) , buffer(S,T), consumer(T).

A CCS representation of buffer2a is:

```
B2 = α(x) . B2(x)
B2(x) = τ . α(y) . B2(x,y) + τ . γ̄(x) . B2
B2(x,y) = γ̄(x) . B2(y)
```

Note that the use of τ on both sides the choice operator in the CCS definition for B2(x) reflects the absence of guards in buffer2a/3.

In fact, the PARLOG buffer as formulated above does not respond to the demands of its environment. By this we mean that the buffer itself will decide whether to input or output in the case of buffer2a/2. If, for

---

[2] We let nil represent the empty list, and [X] represent a list containing one item.

example, the second clause of the relation is selected, and the consumer is not ready to receive, the buffer will not at that point attempt to receive another item from the producer and will suspend. What we really need is a *guarded choice*, which in the CCS would be indicated by dropping the τ's, converting the choice in B2(x) to

$$\alpha(y) \, . \, B2(x,y) + \overline{\gamma}(x) \, . \, B2$$

The 'equivalent' PARLOG code should be:

```
mode buffer2a(Item?,Ins?,Outs^).
buffer2a(X,[In|Ins],Outs)←
        synch_receive(In,Y) : buffer2a(X,Y,Ins,Outs).


buffer2a(X,Ins,[Out|Outs])←
        synch_send(X,Out) : buffer2a(Ins,Outs).
```

Unfortunately, a property of PARLOG programs is that the guard is unsafe if it binds variables in the call, and also that no output can be made in a guard. A call to the above program for buffer2a/2 will never commit to the second clause for the second reason, and the first clause is unsafe due to the second reason. We can program around this deficiency either by using external monitor processes [Gilbert87b] or reversing the modes on the buffer-consumer stream (see the program for bufferinf_f, below).

An alternative formulation for a two-place buffer is to chain two one-place buffers. In this case, it does not matter which program for the one-place buffers is used.

Example 9. An alternative 2-place buffer


```
mode buffer2b(Ins?,Outs^) .
buffer2b(Ins,Outs) ←
    buffer1(Ins,Mid) , buffer1(Mid,Outs) .
```

In CCS, the chaining operation would be represented by:

```
B2b = B ∩ B
```
where
```
B ∩ B = ( B[δ/γ̄] | B[δ/α] ) \ δ
```

We note that the buffer implemented by buffer2b/2 above has a dynamic size, due to which processes are holding items. If we represent the cells of the buffer by either 0, a or b where a comes before b in the input stream, the buffer could be in either of the following states: <0,0> <0,a> <a,0> <b,a> <0,b>.

Can we say that buffer2a and buffer2b are equivalent in some sense? One transformation we can perform before answering this question is to reformulate buffer2a to use an explicit data structure as the data store rather than mutual recursion and arguments:

Example 10.

```
mode buffer2c(Ins?,Outs^).
buffer2c(Ins,Outs)←
    buffer2c([],Ins,Outs).


mode buffer2c(Store?,Ins?,Outs^).


buffer2c([],[In|Ins],Outs)←
    synch_receive(In,Item) & buffer2c([Item],Ins,Outs).


buffer2c([Item],Ins,[Out|Outs])←
    synch_send(Item,Out) & buffer2c([],Ins,Outs).


buffer2c([Item1],[In|Ins],Outs)←
    synch_receive(In,Item2) & buffer2c([Item1,Item2],Ins,Outs).


buffer2c([Item1,Item2],Ins,[Out|Outs])←
    synch_send(Item1,Out) & buffer2c([Item2],Ins,Outs).
```

The CCS formulation for this is (again disregarding the rules for ADTs[3]):

```
B = B(nil)
B(nil) = α(x) . B([x])
B([x]) = τ . γ̄(x) . B(nil) + τ . α(y) . B([x]<>[y])
B([x]<>[y]) = γ̄(x) . B([y])
```

As with buffer2a, the choice points should be guarded (remove the τ's in the CCS, put the calls to synch_send/2 and synch_receive/2 into guards in buffer2c/3 as appropriate). We can use the same programming techniques as before to overcome this problem.

There is a fundamental difference between the dynamic behaviour of the different *producer-buffer2n-consumer* systems. This difference is due to the number of processes in existence at any one time, but is not reflected in the CCS forms. Consider the overall behaviour of

?producer(A) , buffer2a(A,B) , consumer(B).

We note that only one of producer or consumer may access the buffer at any one time because buffer2a can only respond to either the input stream or the output stream, due to the structure of the program for buffer2a/3. The potential for parallelism in this case is reduced. On the other hand, the overall behaviour of

---

[3] We use ◇ as the concatenation operator on lists.

?producer(A) , buffer2b(A,B) , consumer(B).

is quite different: the buffer may communicate with both producer and consumer simultaneously. In both cases a situation of bounded asynchronousity exists: if the consumer is slower than the producer, the latter will be constrained by the rate of the former only after the buffer has been filled up for the first time. During the input of the first two data items, the producer will only be constrained by the rate of execution of the buffer cells.

The CCS buffers can be conceived of behaving quite differently when composed with a producer and a consumer: due to the interleaving semantics of the CCS parallel operator, neither buffer can communicate with more than one other entity at a time.

6. Unbounded buffers and queues

We now consider unbounded buffers. The formulation of an unbounded buffer using the technique of arguments to store the data items is sketched out below. This method is obviously impracticable due to the inability to specify an infinite number of relations with an increasing number of arguments:

Example 11. Unbounded buffer using arguments

```
mode bufferinf_a(Ins?,Outs^).
bufferinf_a([In|Ins],Outs)←
    synch_receive(In,X1)& bufferinf_a(X1,Ins,Outs).


mode bufferinf_a(X1?,Ins?,Outs^).
bufferinf_a(X1,[In|Ins],Outs)←
    synch_receive(In,X2)& bufferinf_a(X1,X2,Ins,Outs).


bufferinf_a(X1,Ins,[Out|Outs])←
    synch_send(X1,Out)& bufferinf_a(Ins,Outs).


mode bufferinf_a(X1?,...,Xn?,Ins?,Outs^).
bufferinf_a(X1,...,Xn,[In|Ins],Out)←
    synch_receive(In,Xn+1)& bufferinf_a(X1,...,Xn,Xn+1,Ins,Outs).


bufferinf_a(X1,...,,Xn-1,Xn,[In|Ins],Out)←
    synch_send(In,Xn)& bufferinf_a(X1,...,Xn-1,Ins,Outs).
```

The way to avoid the problem of specifying relations with the number of arguments ranging up to infinity is to use a list to represent the stored data items:

Example 12. 'Equivalent' infinite buffer

```
mode bufferinf_b(Ins?,Outs^) .
bufferinf_b(In,Out) ←
        bufferinf_b([],In,Out) .


mode bufferinf_b(Store?,Ins?,Outs^) .
bufferinf_b([],[In|Ins],Outs) ←
    synch_receive(In,Item) & bufferinf_b([Item],Ins,Outs) .


bufferinf_b([Msg|Msgs],[In|Ins],Outs) ←
    synch_receive(In,Item) &
    append([Msg|Msgs],[Item],Msgs1) ,
    bufferinf_b(Msgs1,Ins,Outs) .


bufferinf_b([Msg|Msgs],Ins,[Out|Outs]) ←
    synch_send(Msg,Out) & bufferinf_b(Msgs,Ins,Outs) .
```

The CCS formulation of the above would be (again, ignoring the ADT):

Example 13.

$$B\infty = B\infty(nil)$$
$$B\infty(nil) = \alpha(x) \cdot B\infty([x])$$
$$B\infty([x] <> 1) = \tau. \ \alpha(y) \cdot B\infty \ (\ [x] <> 1 <> [y] \ ) + \tau. \ \overline{\gamma}(x) \cdot B\infty(1)$$

Again, we note that the PARLOG's inability to permit output in a guard means that the choices are in effect unguarded. In the case that the data store of messages is empty (ie, is not of the form [Msg|Msgs]), the first clause of bufferinf_c/3 would be chosen by the PARLOG evaluator. However, in the case that there was a message on the input stream, and the store was not empty, the third clause might be selected, and an attempt made to send a message. If the consumer was never ready to receive, deadlock would occur earlier than if the buffer responded fairly to the requirements of both consumer and producer.

We give below the PARLOG program for the guarded choice formulation indicated by the CCS:

$$B\infty([x] <> 1) = \alpha(y) \cdot B\infty \ (\ [x] <> 1 <> [y] \ ) + \overline{\gamma}(x) \cdot B\infty(1)$$

In the following PARLOG program, we use a consumer which *sends* requests to remove items from the buffer, and hence both consumer/1 and producer/1 have output modes. The consumer produces a stream of -/2 tuples which have as the first argument the instruction to remove an item and the second argument as the item removed from the buffer. The buffer in this case has input modes on both the input and the output stream, and employs suspension coupled with pattern matching to repond to requests to add or remove items from the data store. Note that *fairness* is not guaranteed by this version, and we would need to use a technique similar to a *fair merge* [Shapiro84] to ensure fair attention by the buffer to each stream of requests.

Example 14.

```
mode bufferinf_b(Ins?,Outs?) .
bufferinf_b(In,Out) ←
    bufferinf_b([],In,Out) .


mode bufferinf_b(Store?,Ins?,Outs?) .
bufferinf_b([],[In|Ins],Outs) ←
    synch_receive(In,Item) &
    bufferinf_b([Item],Ins,Outs) .

bufferinf_b([Msg|Msgs],[Item + R|Ins],Outs) ←
    R=reply &
    append([Msg|Msgs],[Item],Msgs1) ,
    bufferinf_b(Msgs1,Ins,Outs) .

bufferinf_b([Msg|Msgs],Ins,[remove - X |Outs]) ←
    X=Msg &
    bufferinf_b(Msgs,Ins,Outs) .


mode consumer(Requests^) .
consumer([H|T]) ←
    request(H,X)&
    consumer(T) .


mode request(Request^,Item^) .
request( remove - Item , Item)  ← data(Item) .
```

Reformulating the unbounded buffer using individual processes to store the data items, we face the problem of creating buffers which never output anything. One such program is presented below, which inputs items, but can never output them due to infinite tail recursion creating an unbounded buffer ahead of the most recently created buffer1 cell:

Example 15.

```
mode bufferinf_d(Ins?,Outs^) .
bufferinf_d(Ins,Outs)←
    buffer1(Ins,Mids), bufferinf_d(Mids,Outs) .
```

The CCS for this buffer is:

B∞ = B ∩ B∞

Alternatively, we can create an unbounded buffer which never outputs since it can never input. In this case, there is an unbounded buffer between the most recently created buffer1 cell and the input stream.

Example 16.

```
mode bufferinf_d(Ins?,Outs^) .
bufferinf_d(Ins,Outs) ←
    bufferinf_d(Ins,Mids)  , buffer1(Mids,Outs) .
```

The CCS for the above is:

B∞ = B∞ ∩ B

We attempt to rectify the problem of no output on an infinite buffer by ensuring that items are input, and then are output whilst at the same time the buffer grows in length.

Example 17.

```
mode bufferinf_e(Ins?,Outs^) .
bufferinf_e([In|Ins],Outs) ←
    synch_receive(In,Item) &
    bufferinf_e(Ins,Mids) , buffer1a(Item,Mids,Outs) .
```

The CCS for this program is:

B1∞ = α(x) . (B1∞ ∩ B(x))

The infinite buffer may be rewritten as:

Example 18.

```
mode bufferinf_f(Ins?,Outs^) .
bufferinf_f([In|Ins],Outs) ←
    bufferinf_f(Ins,Mids) , buffer1([In|Mids],Outs) .
```

This program is not readily expressible in CCS since we are explicitly manipulating the communication streams in the PARLOG program. We note that the buffers bufferinf_e/2 and bufferinf_f/2 always increase in length with each data item processed; operationally this may swamp the system with processes. Of more importance is the fact that in an implementation, communication between buffer cells takes a finite time, and the overall delay associated with the buffer will increase with the number of items that it has ever input[4]. CCS ignores such implementation oriented problems.

---

[4] The buffer may be implemented with individual cells on different processors, so that introducing more cells may not necessarily introduce greater inefficiency due to the processing time required by each cell regardless of intercell communication.

The difference in the number of items stored in the two types of buffers (ie using a list or processes) is not apparent operationally here. Both buffers have a store in the range of $0 \leq$ size $< \infty$; however the size of the data-structure store can range dynamically within these limits, whilst the size of the process buffer monotonically increases during its execution.

Another way of formulating the infinite buffer is to use a variation of the one slot buffer, but allowing unconstrained recursion so that the producer can run ahead of the consumer. Note that the difference with buffer1b/2 is the use of the parallel-AND operator between the receive_send calls (which we bracket for clarity) and the self-recursive call to the buffer. The size of such a buffer may grow or shrink according to the rates of the producer and consumer. Messages are held in the suspended calls to synch_send/2 rather than in a system buffer as in PARLOG's naive asynchronous form of communication. The clausal form is the same as the regular one-slot buffer. In this case there is no easy inductive reasoning path towards the formulation of an n-place buffer from the one-place buffer.

Example 19.

```
mode bufferinf_g(Ins?,Outs^).
bufferinf_g([In|Ins],[Out|Outs])←
    ( synch_receive(In,X) & synch_send(X,Out) ) ,
    bufferinf_g(Ins,Outs).
```

This program is harder to model in CCS than the previous ones. If we naively translate bufferinf_g/2 as follows, we get an *infinite bag*, ie the order of output of items is not preserved:

$$B\infty = (\alpha(x) . \bar{\gamma}(x)) \mid B\infty$$

We note that one way of programming an infinite bag in PARLOG is:

Example 20.

```
mode bag(Ins?,Outs^).
bag([In|Ins],Out_all)←
    synch_receive(In,X) &
    synch_send(X,Out),
    bufferinf_g(Ins,Outs),
    merge([Out],Outs,Out_all).
```

where *merge/3* is non-deterministic.

The problem is that the streams in PARLOG impose an ordering which is lost in the CCS version. Thus in the CCS formulation above, we cannot guarantee in which order the items input into the bag will be output. We can impose an order, however, by using synchronisation signals:

Example 21.

```
B∞ = Start ∩ B'∞
Start = δ̄ . δ̄
B'∞ = Cell ∩ B'∞
Cell = ε . α(x) δ̄ . ε . γ̄(x) . δ̄ .
```

where def

$$P \cap Q = (P[\gamma/\beta] \mid Q[\gamma/\alpha]) \setminus \gamma$$

In this case, the ε and $\bar\delta$ signals are used to sequence the input and output of items; each $\alpha(x).\bar\gamma(x)$ pair is guarded by an ε which is not activated until a previous $\alpha(x).\bar\gamma(x)$ pair has input a value. Likewise, the output of the pair cannot take place until a synchronisation signal has been received from the previous pair on successful output of its value. The renaming in the definition of the chaining operation '$\cap$' is the equivalent of an infinite number of different synchronisation flags.

However, this formulation allows the 'eager' creation of buffer cells, a condition which is not strictly necessary, since more buffer cells may be created than are needed. This is not the case in the PARLOG program above, where the spawning of new buffer cells is restricted by the rate at which the input stream is being partially instantiated. Thus in the PARLOG version, no more buffer cells are spawned than are required by the producer.

To force the spawning of buffer cells to be in step with the production of messages from the producer, we formulate the following CCS 'program'. The main point to note is that the spawning of new buffer cells is guarded by the initial ε action, which waits until it receives a synchronisation signal from a previous buffer pair.

Example 22.

```
B∞ = Start ∩ B'∞
Start = δ̄ . δ̄
B'∞ = ε . (Cell ∩ B'∞ )
Cell = α(x) δ̄ . ε . γ̄(x) . δ̄
```

7. Bounded buffers

We may formulate a bounded buffer using either the data-structure technique or an explicit number of one place buffer processes. The buffer formulated using processes as stores is:

Example 23. Bounded Buffer - processes

```
mode bounded(Bound?,Ins?,Outs^).
bounded(Size,Ins,Outs) <-
    Size > 0 :
    Size1 is Size-1 ,
    buffer1(Ins,Mids) , bounded(Size1,Mids,Outs).


bounded(0,Outs,Outs).
```

Note that this program is quite different to that proposed by [Takeuchi87], which is closer to a program which uses data-structures as stores. The following example uses a **difference list** as a store, making the operation of adding to the end of the list more efficient (the second clause of *bounded/5*), and doing away with the need for a call to *append*.

Example 24. Bounded buffer - data structures

```
mode bounded(Bound?,Ins?,Outs^) .
bounded(Bound,In,Out) ←
    bounded(Bound,0,Var/Var,In,Out) .


mode bounded(Bound?,Size?,Store?,Ins?,Outs^) .

bounded(Bound,Size,Msgs/Tail,[In|Ins],Outs) ←
    Size =< Bound :
    synch_receive(In,Item) ,
    Tail=[Item|NewTail] , Size1 is Size + 1 ,
    bounded(Bound, Size1, Msgs/NewTail,Ins,Outs) .

bounded(Bound, Size, [Msg|Msgs]/Tail,Ins,[Out|Outs]) ←
    Size > Bound :
    synch_send(Msg,Out) ,
    Size1 is Size - 1 ,
    bounded(Bound, Size1, Msgs/Tail,Ins,Outs) .
```

8. Conclusions

We have discussed a class of communication based algorithms, restricted to a producer-buffer-consumer scenario. PARLOG programs have been formulated, and informal equivalences between them noted. In order to enable such comparisons, formulations in CCS have been given for the same algorithms. We note that PARLOG lacks a formal denotational semantics based on logic, and that the most recent work on semantics has been to model PARLOG in CCS [Ellis86 , Hussey87]. The inability of a PARLOG clause to produce output before committal means that in some cases choice is unguarded, and thus the behaviour of resulting programs may not be desirable. There are ways to avoid this, based on the use of external stream monitoring processes [Gilbert87b], or by metalevel programming. The above inability is a deficiency that PARLOG needs to rectify before it could be used effectively as a specification language. Advantages that PARLOG has over CCS as a 'specification-implementation' language include its 'truly parallel' model of execution whereby any number of processes may be in existence at any instant. We also note that the stream based model of communication employed by PARLOG permits easier specification of a process-based infinite buffer with non-monotonically increasing size than does the event gate model of CCS.

Acknowledgements

This paper was written while the author was funded by Alvey on "Implementation and Applications of PARLOG", Project number 043/098.

I would like to thank Reem Bahgat, Chris Hogger, Iain Phillips and Steve Vickers of Imperial College for all the time that they gave me, and the helpful comments that they made during the preparation of this paper.

Tables

The PARLOG syntax used in this article is given in the Table below.

Symbol	Meaning
←	logical implication
&	sequential-AND
,	parallel-AND
;	sequential-OR
.	parallel-OR
:	guard operator
?	input mode annotation
^	output mode annotation

Table 1 PARLOG syntax

References

Ellis86.

Mark R. Ellis, "A Relational Language into ECCS," M.Sc, Imperial College of Science & Technology, London. UK, 12th September 1986.

Gilbert87.

David Gilbert, "PARLOG: a tutorial introduction.," *Proceedings of Parallel Processing and Supercomputing*, Begian Institute for Automatic Control, Antwerp, Belgium, November 19-20, 1987.

Gilbert87b.

David Gilbert, "Executable LOTOS: Using PARLOG to implement an FDT," *Proceedings of IFIP Protocol Specification, Testing and Verification: VII, Zurich, Switzerland, 5-8 May 1987*, Elsevier Science, North-Holland, Amsterdam, Netherlands, 1987.

Giles78.

David A. Giles, "The Theory of LISTS in LCF," CSR-31-78, Department of Computer Science, University of Edinburgh, Edinburgh, 1978.

Gregory87.

Steve Gregory, *Parallel Logic Programming in PARLOG: The Language and its Implementation*, Addison-Wesely, London, UK, 1987.

Hussey87.

Charlie Hussey, "Interpreting PARLOG Programs as CCS agents," Report for MSc degree in Engineering, Imperial College of Science and Technology, London, September 1987.

Milner86.

Robin Milner, "Process Constructors and Interpretations," *Proceedings of IFIP 10th International World Computer Congress*, vol. 10, pp. 507-514, North Holland, Dublin, Ireland, September 1-5, 1986.

Shapiro83.

Ehud Shapiro and Akikazu Takeuchi, "Object Oriented Programming in Concurrent Prolog," CS83-08, Department of Applied Mathematics, Weizmann Institute of Sciences, Rehovot, Israel, June 1983.

Shapiro84.

E. Shapiro and C. Mierowsky, "Fair, Biased, and Self-Balancing Merge Operators: Their Specification and Implementation in Concurrent Prolog," CS84-07, Weizmann Institute, Rehovot, Israel, 1984.

Takeuchi87.

Akikazu Takeuchi and Koichi Furukawa, "Bounded Buffer Communication in Concurrent Prolog," in *Concurrent Prolog*, ed. Ehud Shapiro, vol. 1, pp. 464-475, MIT Press, 1987.

Specification and Verification in Communications Standards

David Freestone
Formal Methods Group
British Telecom/Information Services Standards Division
St Vincent House
Ipswich
UK

May 1988

Abstract

Formal techniques for specification and verification have been researched and developed in academic institutions for several years, but their industrial exploitation has been more recent and more cautious. This paper presents two areas of work where the technology is being transferred. The first area is the application of formal techniques to international standards for communications protocols and services (in OSI). The second area concerns a component–engineering approach to protocols and services, and results from recent research. In both cases the formal treatment of concurrency is significant.

1 Formal Specification and Verification in Industry

Formal specification languages and mechanical verification have been active research topics for several years in academic institutions. The transfer of these technologies into industrial organisations is more recent and has largely been cautious. However, the benefits of formal specifications are increasingly appreciated, especially in applications where correctness is critical. Many companies are now publishing the results of projects using it (eg [1][2][3][4]).

Distributed systems is an application area where the benefits of formal techniques are recognised: appropriate abstraction for concurrency, no unintentional ambiguity and the possibility of verification. In particular, there is growing recognition among practising engineers (both software and hardware) that informal techniques are inadequate for specifying concurrency. However, there is also widespread resistance to adopting formal techniques. Part of this resistance stems from lack of familiarity and part from lack of suitable training, but some stems from indigestible presentation of techniques. There is a real need for work bridging the gap between academic development and industrial exploitation.

The following two sections of this paper address that gap. Both concern communications protocols, an aspect of distributed systems which necessarily involves concurrency. Section 2 discusses the application of formal techniques to the international standards of Open Systems Interconnection (OSI). Section 3 presents some exploratory work on how formal techniques can facilitate a component engineering approach to protocols and services.

2   Formal Methods in Standards

International standards are non–proprietory conventions which define certain technical properties of products, but exist separately from them.   They enable compatibility between products from different sources.   Standards are important commercially because they can expand existing markets and create new ones (eg facsimile [5]).   They are particularly important for defining communcations mechanisms for use between different manufacturer's computing equipment.   Several different families of such standards have been developed and are being developed (eg [6][7][8][9]).

Communications standards are a natural candidate for the early industrial application of formal techniques.   Their independent existence leads to a high motivation for accuracy and absence of unintentional ambiguity.   Mistakes in standards can be extremely costly. For example, two products correctly implementing a mistaken standard may not interwork. Another reason why they are a natural candidate concerns their development costs. Formal techniques typically increase costs in the early stages of a product life cycle.   Such a profile already exists for communications standards, and is accepted.

One family of communications standards, Open Systems Interconnection (OSI), has begun to employ formal techniques.   OSI is a hierarchy of protocols and services, which has been developed within the International Standards Organisation (ISO) and the International Consultative Committee for Telegraphs and Telephones (CCITT) for over a decade.

The programme of work in ISO (and latterly in CCITT) has developed Formal Description Techniques (FDTs) specifically for OSI – although they are applicable more widely – and is now producing formal specifications using them.   Future work will include their use in other aspects of protocol engineering.   Significantly for this conference, the representation of concurrency is an important element in the FDTs for OSI.   In fact, some of the papers presented here cover OSI work.

ISO has declared the purposes of the FDTs to be (in summary):

- unambiguous, clear and concise specification;

- a basis for determining the completeness of protocols and services;

- a foundation for analysing protocols and services for correctness, efficiency etc;

- a basis for verifying that protocols and services meet the requirements of OSI;

- a basis for determining the conformance of implementations to the standards;

- a basis for determining the consistency of OSI standards with each other;

- a basis for implementation support.

Three FDTs are now recognised by ISO and CCITT: SDL [10], Estelle [11] and LOTOS [12].   None meets all the purposes completely, and they display different characteristics. Some of the papers in this conference demonstrate the FDTs, and a conference later this year [13] will be devoted to them.   LOTOS is used in section 3 of this paper.

Recently, a three–phase plan for introducing formal specifications (otherwse called formal descriptions) into OSI standards has been agreed within ISO [14].   The following outlines the three phases.

1   Formal descriptions are separate technical reports which can later be incorporated into standards.   During this phase it is expected that detailed knowledge and experience of FDTs will be scarce among standards–makers and standards–users.

2   Formal descriptions are non–binding annexes to standards.   During this phase there should be increasing knowledge and experience of FDTs, but still not sufficient for formal descriptions to be fairly assessed for integral inclusion in standards.

3 Formal descriptions are binding annexes to standards, with precedence over informal text in cases of conflict or ambiguity. By this phase, there should be sufficient knowledge and experience of FDTs among standards makers for fair assessment of formal descriptions.

An aspect of protocol engineering referred to in the list of purposes of FDTs is *conformance testing*. This is worth highlighting, because it refers to a distinctive feature of distributed communications systems. Products based on protocols usually have to operate within a distributed information service which is run commercially. Consequently, in addition to designers and implementers, managers of information services are also concerned with communications standards. One of their functions is to ensure that products connected to their services really do satisfy the relevant standards, in order to safeguard the interests of other users. For commercial reasons, there is usually no access to product designs, nor to the source code of implementations. By observing external behaviour alone, the manager must judge whether a product meets its obligations. This is *conformance testing*. Notice that its purpose is different from testing by implementers during product development. Notice also that it cannot be replaced, even in principle, by verification. Conformance testing should provide evidence to increase the manager's confidence that the implementation conforms.

At present, there is no systematic way of deriving useful conformance tests from OSI standards, and no way of quantifying how much confidence a customer can have in a product which a particular conformance test suite. These problems are not unique to OSI products, but their effects are particularly wide-ranging because of the need for distributed interworking. The subject is covered in more detail in [15]. The application of formal techniques to improve the situation is currently under study in ISO, and is an area where technology transfer could prove fruitful.

3 Component Engineering for Formal Specifications

The work reported in this section results from a study [16] of the scope for defining protocols from smaller, reusable specification components. One of its outcomes was a library of possible components. The idea was for each module to be fully specified and its properties understood. The ways components could be combined would also be given. These components could be used as the foundation for designing new protocols, which could be completed by customised elements. Several protocols and services, at various levels of communication, were studied to produce the library.

The format for each entry in the library is in three parts. First a trace predicate specification of the component's properties; second a LOTOS process which satisfies the trace predicate specification; and third a proof of that satisfaction. Unfortunately, lack of time prevented all entries from being completed. In particular, several proofs are absent.

Trace predicates were chosen as a fairly simple and well documented specification framework [17][18]. They have a natural relationship to LOTOS, which models sequences of atomic events. The presentation here uses standard LOTOS, but the original work used a variation called (tautologously) *Temporal LOTOS* [16][19]. The choice of LOTOS allowed the work to follow international developments in OSI. The proof calculus used is based on [18].

An interesting discovery was that two or more different traditional components of protocols could sometimes be combined into one generalised component. For example, a *plexor* was created as a generalisation of *multiplexor* and *splitter* – taking input on an arbitrary number of channels and outputting the data, redistributed, on an arbitrary number of output channels. The notion of *expedited data* from OSI protocols was also easily generalised, into *priority data* where arbitrarily many grades of priority can be used, and any protocol data unit (pdu) may overtake the preceding pdu if it has higher priority. Generalised components like these help to partition the specification into similar and separate concerns.

The *plexor* in fact proved an instructive component. It was an entry from the earliest versions of the library. However, the proof of satisfaction was not attempted until fairly recently. Our experience was that even small proofs are very time–consuming and this component seemed "obviously correct" to all the project team. Unfortunately, it turned out that the process did not satisfy the specification! The moral is that specifiers cannot always trust their intuition.

Figure 1 illustrates a *plexor*. Also sketched are the first trace predicate specification (figure 2) and the first LOTOS process (figure 3). Figures 2 and 3 use abbreviated syntax to highlight the important aspects.

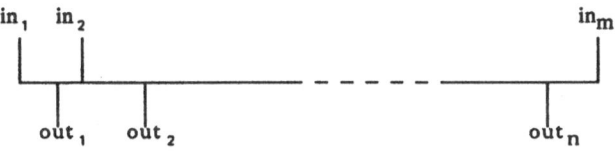

Figure 1   Illustration of *plexor*

$S(tr) \triangleq$ **there exist** globin, globout

.   globin **interleaves** $in_1, \dots, in_m$

and   globout **interleaves** $out_1, \dots, out_n$

.   globout **prefixes** globin

Figure 2   First trace predicate specification

A *trace* is a "snapshot history" of a process – the sequence of events that has occurred up to now. A *trace predicate* is a predicate that holds for all possible traces of the process.

In S(tr), *globin, globout, $in_1, \dots, in_m$, $out_1, \dots, out_n$* are all sequences of data values, passed on in_i gates or out_j gates. They are all subsequences of the (arbitrary) trace *tr* – projected onto the data value component. Each in_i is the snapshot history (trace) of data passed on gate in_i and each out_j is the snapshot history of data passed on gate out_j. So, S(tr) says that we can construct a sequence *globin* which interleaves the in_i sequences, and a sequence *globout* which interleaves the out_j sequences; and we can construct them in such a way that *globout* *prefixes* (ie is an initial subsequence of) *globin*. In other words, what comes out of the plexor must previously have gone in – but the data packets can be re–ordered to some extent.

In figure 3, the LOTOS process P comprises a sub–process for each in_i gate and a sub–process for each out_j gate. The IN_is run independently (|||) of each other and the OUT_js run independently of each other, but the two collections of sub–processes synchronise and communicate via gate *plex*, which acts like a kind of internal buffer. The operational details of *P* are relatively easy to follow from the process definition.

P ≙   **hide** plex **in**

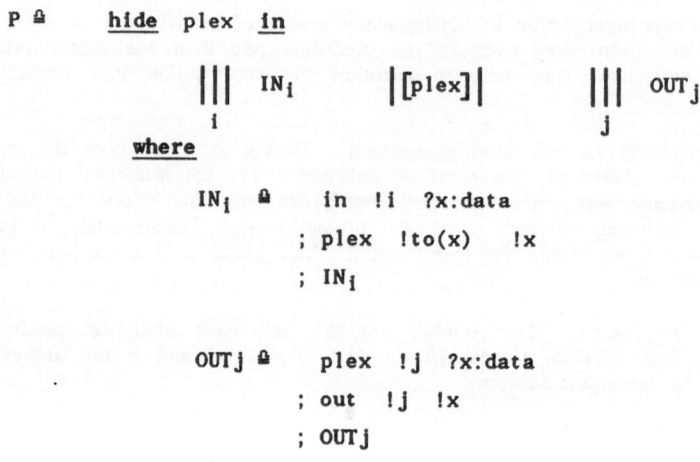

$$|||_i \; IN_i \qquad |[plex]| \qquad |||_j \; OUT_j$$

where

IN$_i$ ≙   **in** !i ?x:data
            ; plex !to(x) !x
            ; IN$_i$

OUTj ≙   plex !j ?x:data
            ; out !j !x
            ; OUTj

Figure 3   LOTOS process

The proof obligation is set out below as "Theorem", and a counter example follows which disproves it.

<u>"Theorem"</u>

<u>forall</u> tr∈traces(P) . P <u>satisfies</u> S(tr)

<u>Counter-example</u>

Take m=1, n=2

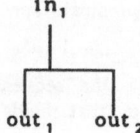

A possible snapshot history of P is:

    tr = <in !1 !x,   i,   in !1 !y,   i,   out !2 !y>

where the i's represent the hidden events on gate *plex*. In this case, the second data value input on *in$_1$* has been output (on *out$_2$*), but the first data value (due to be output on *out$_1$*) has not. This is a valid trace of P. The corresponding sequences in S(tr) are:

    in$_1$ = <x,y>   out$_1$ = <>   out$_2$ = <y>

There is only one possible *globin* and only one possible *globout*:

    globin   = <x,y>   globout   = <y>

and clearly *globout* does not <u>prefix</u> *globin*.

□

The problem with the component is in the specification. S(tr) does not handle all cases where data has been input but not yet output. We probably do want to allow the behaviour given in the counter–example, and so the specification must be changed. The following, revised specification S'(tr) does allow this behaviour.

$$S'(tr) \triangleq \textbf{there exist } globin, globout$$
$$\qquad globin \underline{\textbf{interleaves}} \quad out(in_1), \ldots, out(in_m)$$
$$\qquad \textbf{and} \quad globout \underline{\textbf{interleaves}} \quad out_1, \ldots, out_n$$
$$\qquad globout = globin$$

where out(in_i) is the unique subsequence of in_i which has selected precisely those data items which appear in some out_j.

This time, P <u>satisfies</u> S'(tr), and the following is a sketch of the proof.

Theorem

<u>forall</u> trϵtraces(P) . P <u>satisfies</u> S'(tr)

Before entering the main proof an internal lemma must be presented. Its proof is given in the appendix.

Lemma 1

t ϵ traces(P) <u>and</u> t$\neq$<>

$\Rightarrow$

$\exists$ U : $\wp$\{u(i,j,x) | u(i,j,x) <u>prefixes</u> <in !i !x, i, out !j !x> \}
$$\qquad t \underline{\text{interleaves}} \ U$$

This means that any trace of P is an interleaving of histories for individual data packets (each one symbolized by a u(i,j,x)). However, individual data history may not have been completed in the "current" trace (= snapshot history). Therefore each u(i,j,x) is some <u>prefix</u> of the eventual lifetime history.

Corollary

Each event at gate *out* corresponds to a unique event at gate *in*

Sketch of Proof of Theorem

By induction on traces(P)

$tr = \diamondsuit$

$tr = \diamondsuit \Rightarrow$ globin = globout = $\diamondsuit$, so   P <u>satisfies</u> $S'(\diamondsuit)$

$tr = t$

Inductive hypothesis: P <u>satisfies</u> $S'(t)$

write $out(in_i) = out(in_i)_t$ for all i
          $globin = globin_t$
          $globout = globout_t$

$tr = t^\frown <e>$

e   takes a value in one of three cases (by inspection of the alphabet of P):

    case 1: e = in !i !x   for some i, x
    case 2: e = <u>i</u>   representing a hidden plex !j !x
    case 3: e = out !j !x   for some j, x

case 1

$out(in_i) = out(in_i)_t$ for all i, by definition of out, so can put
$globin   = globin_t$
$globout   = globout_t$

all the clauses of $S'(tr)$ hold by hypothesis.

case 2

$out(in_i) = out(in_i)_t$ for all i, by definition of out, so can put
$globin   = globin_t$
$globout   = globout_t$

all the clauses of $S'(tr)$ hold by hypothesis.

case 3

This is the case where all the work has to be done.

By the corollary to the lemma there is a unique *in* event corresponding to e. This event is in some unique in_i. The corresponding $out(in_i)$ now includes x.   All the other $out(in_i)$s remain the same.

The idea of the proof is to define 3 positions in $globout_t$ (= $globin_t$) which determine where x is allowed to be slotted in.   It has to go after the *previous (=last) element of out_j* if there is such an element; and it has to go between the *previous* and *next elements of in_i which are also in out(in_i)*, if such elements exist. If any of these elements do not exist, then appropriate default values are taken.

The following algorithm sketches the definitions of the 3 positions.

A: if out_{jt}=<> then A=beginning else A=last(out_{jt})

B: if out(in_i)$_t$=<> then B=beginning

   else if previous(x,out(in_i)$_t$) exists then B=previous(x,out(in_i)$_t$)

   else B=beginning

C: if out(in_i)$_t$=<> then C=end

   else if next(x,out(in_i)$_t$) exists then C=next(x,out(in_i)$_t$)

   else C=end

previous(x,O) selects the position in *globout*$_t$ of the data item which is previous to *x* in *O*. *next(x,O)* selects the position in *globout*$_t$ of the data item which is next after *x* in *O*. A, B and C all exist by definition of *globout* and the inductive hypothesis.

Notice that B precedes C by definition. Lemma 2 (in the appendix) shows that A also precedes C. Diagrammatically, therefore, there are two possible orders for A B and C.

		Can put x anywhere in here	

A	B	C
B	A	C

x can always be inserted between greater(A,B) and C. Therefore, in the proof, put:

   globout = globout$_t$ with x inserted immediately after the greater of A, B.
   globin = globout

globout interleaves out$_1$ ... out$_m$ because x inserted after A
globin interleaves out$_{(in_1)}$, ... ,out$_{(in_n)}$ because x inserted between B and C.

Hence P satisfies S'(tr).

All three cases hold, so the inductive step is proven.   □

4   Conclusions

The OSI work discussed in section 2 is pioneering the application of formal techniques on large scale projects. In consequence, it has had to address many issues for the first time. Some have been technical, eg how well do techniques scale up? Others have been issues of policy, eg should formal descriptions be compulsory for every component standard? The results will be helpful for many future projects using formal techniques.

The work on specification components, presented in section 3, is on a much smaller scale. However, the proof example highlights several points that are relevant generally. First, it is easy to miss mistakes in formal specifications just as it is in processes or programs – many people had inspected the original *plexor* specification without spotting the problem. Second, trying the proof revealed the mistake, and provided hints on how to correct it. Third, the final proof gives increased confidence that the specification and process really are consistent.

Effective technology transfer requires pilot projects to demonstrate how the technology can be exploited successfully. This paper has discussed two areas where that transfer is taking place with formal techniques for concurrency.

Acknowledgements

Acknowledgement is made to the Director of Research and Technology of British Telecom for permission to publish this paper.

The work reported in section 3 was performed under FORMAP, a collaborative project now between British Telecommunications plc and The General Electric Company plc, within the UK programme (SE/051).

References

[1]   M Bailey: Formal specification using Z; Proceedings of SHARE European Association; 1987;

[2]   Paul Johnson: Using Z to specify CICS; Proceedings of SHARE European Association; 1987;

[3]   David Shepherd: Using formal methods in VLSI design; Proceedings of Refinement Workshop; University of York; January 1988;

[4]   Tom Farr: Experience of implementation from a formal specification; Proceedings of Refinement Workshop; University of York; January 1988;

[5]   A T Bence: Teleservice Developments in CCITT – Migration to the ISDN, Computer Networks and ISDN Systems; Vol 9; 1985; pp329–337;

[6]   K G Fretton, C G Davies: CCITT Signalling System No. 7; Overview; British Telecommunications Engineering; Vol 7, Part 1; April 1988;

[7]   P J Davidson: Review of the CCITT recommendations for integrated services digital network (ISDN); British Telecom Technology Journal; Vol 3, No 4; October 1985;

[8]   K D Foggarty: ISDN services and network recommendations; British Telecom Technology Journal; Vol 4, No 1; January 1986;

[9]   P A Jenkins, K G Knightson: Open Systems Interconnection – An Introductory Guide; British Telecommunications Engineering; Vol 3; July 1984;

[10]   CCITT Recommendations Z.100 – Z.104 (1984): Specification and Description Language;

[11]   ISO DIS 9074: Information Processing Systems – Open Systems Interconnection – Estelle – A Formal Description Technique based on an Extended State Transition Model;

[12]   ISO DIS 8807: Information Processing Systems – Open Systems Interconnection – LOTOS – A Formal Description Technique Based on the Temporal Ordering of Observational Behaviour;

[13]   FORTE 88: An International Conference on Formal Description Techniques; 6 – 9 September 1988; University of Stirling, Scotland;

[14]   ISO/IEC/JTC1/N87: Report of the ISO/TC97 SWG meeting on Formal Description Techniques; New York City; October 1987;

[15]   J R Halliwell, T J Holland, D Freestone: Conformance Testing of Communications Protocols in Information Systems; British Telecom Technology Journal; Vol 4 No.3; July 1986;

[16]   A Library of Specification Modules and Their Application; Report No 5 to the Alvey Directorate; UK Alvey FORMAP Project (SE/051);

[17]   C A R Hoare: Programs are Predicates; in C A R Hoare and J C Shepherdson, eds: Matheamtical Logic and Programming Languages; Prentice Hall; 1985;

[18]   C A R Hoare: Communicating Sequential Processes; Prentice–Hall International Series in Computer Science; 1985;

[19]   Chris Smith, Steve Rudkin: Time Guards and ASN.1 in LOTOS; submitted for publication.

Appendix

Lemma 1

> t ε traces(P) <u>and</u> t≠<>

➡

> ∃ U : ₽{u(i,j,x) ⏐ u(i,j,x) <u>prefixes</u> <in !i !x, i, out !j !x> }
>   .   t <u>interleaves</u> U

Proof

By induction on traces(P).

<u>tr = < ></u>   trivially true

<u>tr = t</u>   inductive hypothesis

                 write U = U_t

<u>tr = t ∧ <e></u>

e takes a value in one of three cases (by inspection of the alphabet of P):

       case 1: e ─ in   !i !x   for some i, x˙
       case 2: e ─ i   representing a hidden plex !j !x
       case 3: e ─ out !j !x   for some j, x

   <u>case 1</u>

   put   U ─ U_t U {e}   e is a fresh u(i,j,x) by construction

   <u>case 2</u>

   e ─ i   ➡   ∃i . e ─ last (tr <u>restricted to</u> αIN_i),
                                        by definition of P

           ➡   ∃i . <in !i !x, e> ─ last2 (tr <u>restricted to</u> αIN_i)
                                        from properties of ;
                                        and definition of P

           ➡   ∃i . <in !i !x> ─ last (t <u>restricted to</u> αIN_i)
                                        by definition of tr

           ➡   <in !i !x> ε U_t   by hypothesis

Therefore, can put

$$U - U_t \cup (\langle in \ !i \ !x, \ e\rangle) \ - \ (\langle in \ !i \ !x\rangle)$$

case 3

$$e - out \ !j \ !x \ \Rightarrow \ \exists j. \ e - last \ (t \ \underline{restricted \ to} \ \alpha OUT_j),$$
$$by \ definition \ of \ P$$

$$\Rightarrow \ \exists j. \ \langle \underline{i}, \ e\rangle - last2 \ (tr \ \underline{restricted \ to} \ \alpha OUT_j)$$
$$from \ properties \ of \ ; \ and$$
$$hide \ and \ definition \ of \ P$$

$$\Rightarrow \ \exists i,v,w \ . \ \wedge \langle in \ !i \ !x, \ \underline{i}\rangle \wedge w - (tr \ \underline{restricted \ to} \ \alpha IN_i)$$
$$from \ properties \ of \ ;, \ I[\ ]I$$
$$and \ definition \ of \ P$$

$$\Rightarrow \ \langle in \ !i \ !x, \ \underline{i}\rangle \ \epsilon \ U_t \qquad by \ hypothesis$$

Therefore, can put

$$U - U_t \ \cup \ (\langle in \ !i \ !x, \ plex \ !j \ !x, \ e\rangle)$$
$$- \ (\langle in \ !i \ !x, \ plex \ !j \ !x\rangle)$$

The property holds for all three cases of e. Therefore the inductive case is proven.

□

Lemma 2 A precedes C

Sketch Proof

$$in_{it} = \langle...,x,...,next,...\rangle \qquad\qquad by \ definition \ of \ next$$

$\Rightarrow$ t $\underline{restricted \ to}$ $\alpha IN_i = \langle..., in \ !i \ !x, \ \underline{i} \ , \ in \ !i \ !next \ , \ ...\rangle$
      where $\underline{i}$ represents the hidden plex !j !x,
      by definition of P and the properties of ;

$\Rightarrow$ t $\underline{restricted \ to}$ $\alpha OUT_j = \langle....,\underline{i}\rangle$
      where $\underline{i}$ represents the hidden plex !j !x,
      by definition of P, properties of ;, and value of e

$\Rightarrow$ last(out$_{jt}$) precedes the $\underline{i}$ representing the hidden plex !j !x in t

$\Rightarrow$ last(out$_{jt}$) precedes in !i !next in t
      because of relative positions of $\underline{i}$ and in!i!next from above

$\Rightarrow$ last(out$_{jt}$) precedes next in globout
      by definition of globout

□

Experience with LOTOS and environment LOTTE on an ISDN protocol

Paul A.J. Tilanus   Yan Yang

PTT - Dr. Neher Laboratories
The Netherlands

May 1988

Abstract

In November 1987 the operational services of the Netherlands PTT asked the Formal Specification group and the D-channel experts of the Communications Research department whether it was possible to produce a simulatable formal specification of layer 3 of the ISDN D-channel protocol. In this paper the experiences with LOTOS in the production of this specification, and the experiences with the LOTOS environment LOTTE are discussed.

1 Background

The experience reported on in the following sections was gained in the specification of the ISDN D-channel layer 3 protocol (*D3 protocol*) [1]. This protocol describes the signaling between the user and the local exchange. This section provides some background information on the project.

1.1 The request

Although the Netherlands PTT is involved in the definition of this protocol, more confidence that the selected options would nicely cooperate could be obtained, if some kind of simulation could be performed beforehand. For this reason DNL was asked to write a simulatable specification of the D3 protocol. Writing the specification itself was expected to be one source of 'prompts for clarification'; what the experiments with the simulator would add was an open question.

The specification had to be produced at short notice, and therefore the members of the project team all worked part-time on the project in time freed by delaying work with a lower priority. As a result the project had nine members with different background (FDT, ISDN/D-channel, Conformance Testing) who provided a total of 1.2 man years within three and a half months. Thus the project got many of the characteristics of a large project: many people with different background and numerous occasions of passing of responsibilities.

Our laboratories support two internationally standardised formal specification languages: SDL [2] and LOTOS [3]. (SDL is felt to be superior to the third international standard, Estelle [4], mainly because it has a graphical syntax and is less implementation oriented.) The reason to select LOTOS for the specification of the D3 protocol is a pragmatic one: a simulator could be bought on short notice.

1.2 Training

Not all members of the project team did know LOTOS before the project. They took a three day course on LOTOS and LOTTE (see section 3) at the beginning of the project. This course differs significantly from the at that moment available tutorials and courses in one aspect: the concept of *observational equivalence* is not treated. The idea behind this treatment is that the part that can easily be explained hardly needs explanation at all. E.g. choice between alternatives is commutative and associative; if not the operator should not be called *'choice'*!

On the other hand, the explanation of the part that is harder to explain, such as interactions on hidden gates that become internal events, becomes rather technical. Knowledge of observational equivalence in these cases does not help the specifier to construct better specifications.

2 The importance of a good architecture

2.1 The starting point

The basis for the D3 specification was a natural language document [1], augmented by a set of semi-formal SDL-diagrams [5]. These documents describe how **ONE** call is processed in a state oriented way. The description in [1] describes how the network side and the user side proceed through the call states, and indicates how the network/user distinguishes simultaneously calls/terminals. The SDL-diagrams strictly separate the network side and the user side, but they are rather informal; they even use non-SDL constructs.

The D3 protocol uses the services provided by the D-channel layer 2. The layer 2 primitives to be used are clearly described. Also the messages exchanged by the layer 3 protocol entities (one on the user side, one on the network side), and their parameters, are well defined in [1].

However, the service provided by the D3 protocol is ill defined. Not even the names of the primitives exchanged with the higher layers are given in [1]. The developers of the D3 protocol seem to rely on some *'common understanding'* of the cooperation of the D3 protocol with the higher layers.

2.2 Construction of the architecture

The system that had to be specified could naturally be divided in four parts:

- the network side protocol entity,

- the user side protocol entity,

- the layer 2 service, and

- the 'connection management'.

From these the connection management served mainly simulation purposes by providing the functions the exchange normally performs. The layer 2 service is rather simple, and could be derived from the protocol entities once these were fully specified. The protocol entities are similar, but on the network side there is a complication that several terminals (phones, facsimile, telex, etc.) of the user can respond simultaneously to one incoming call.

The specification of the four parts was done in the order of the list above. That means that we started with the most complicated part. This approach was suggested by people from the Twente University of Technology, The Netherlands, based on their experience in e.g. [6]. They also suggest to specify 'breadth first', and within the four parts these suggestions were followed as well. We endorse their suggestions.

Due to time pressure the state oriented description would be adopted to stay close to the informal description. In other places, e.g. [7], this style is called 'construction oriented'. The most straightforward approach, one process per call state, was expected to provide easy to handle building blocks. At the same time three persons worked on the architecture of the specification, gluing the building blocks together.

2.2.1 Architecture

In the informal specification the description focuses on one call. Within this call the protocol entities (subscriber, exchange) are treated simultaneously.

As a starting point for the structuring the LOTOS specification protocol entities are made independent (interleaving) processes. This gives, for example, a process per terminal of the subscriber (user) on the user side, and a process per subscriber (line) on the network side.

In ISDN a subscriber can have up to two simultaneous calls. To distinguish between them on the D-channel different call reference values are associated to different calls in the layer 3 protocol.

The second step applied in structuring is separation of different instantiations of the protocol, i.e. separation of different calls. This requires some overhead. Application of this step gave rise to processes called call_handler and one call_manager. The call manager takes care of the rule 'No two simultaneous calls of one subscriber shall use the same call reference value', by maintaining a set of call reference values in use. A call_manager handles all interactions related to its call reference value. A call_handler is 'created' by the call_manager as the latter finds out that a new call started.

The structure obtained after application of the first two steps was never changed.

Further structuring followed the good-programming-rules of

- minimise communication,

- data administration at one place only,

- don't use a process if an abstract data type operator can do the work,

- don't enforce structure in data.

The last rule is more a specification rule than a programming rule. Its application gives e.g. that a list of items is used only if the order is relevant, otherwise a set should be used.

The result of the third step leads to many independent processes connected to a manager that takes care of the common bookkeeping. This third step leaves a lot of freedom, and twice during the specification of the network side the architecture was changed significantly, in both cases directed towards more independence of processes (interleaving as much as possible). To obtain this maximum independence sometimes a rather artificial process (the create process in section 2.2.2) is required for technical reasons.

Throughout the specification the principle known as 'subgating' was used in combination with interleaving processes. This principle is one that makes LOTOS a very powerful language. We believe that it is this principle that kept the amount of changes required after the two major reshuffles to a minimum.

2.2.2 Example of a create process

Suppose one has a manager process that wants private communication with dynamically created child processes (using subgating). If the manager creates the child processes himself, as in figure 1, then the private communication becomes impossible.

```
...
Manager [ gate ] ( first_id )
where
    process Manager [ gate ] ( last_id : Id ) : noexit
    :=
        i; ( Child [ gate ] ( new( last_id ) )
            |[ gate ]|
                Manager [ gate ] ( new( last_id ) )
            )
    []
            Communicate_with_one_of_the_Children [ gate ] ( last_id )
    where
        ...
    endproc (* Manager *)
```

Figure 1: A Manager process creating Child processes directly

Note: the intention is that the manager and child use the
       Id-value of the child to force private communication.

In this case there is a trick using ignore processes. However, the communication of the manager with a child can hardly be called 'private', all children participate in *every* interaction.

Using the trick, a child process consists of two parts:

- the original child participating in all interactions with its own identification as subgate, and

- the Ignore part: a process interleaving with the original child that participates in every interaction on the gate where the subgate value is different from its own, but does nothing with these interactions.

The alternative requires a separate process to take care of the actual creation, see figure 2.

```
    . . .
  ( Manager [ gate ] (first_id )
  | [ gate ] |
    Create [ gate ]
  )
  where
      process Manager [ gate ] ( last_id : Id ) : noexit
      :=
          i;
          gate ! create ! new( last_id );
          Manager [ gate ] ( new( last_id ) )
          [ ]
          Communicate_with_one_of_the_Children [ gate ] ( last_id )
      where
            . . .
      endproc (* Manager *)
      process Create [ gate ] : noexit
      :=
        gate ! create ? new_id : Id ;
        (        Child [ gate ] ( new_id )
          |||
                Create [ gate ]
        )
      endproc (* Create *)
```

Figure 2: Child creation via a Create process

This is a real solution: the manager has private communication with a child. The advantages of this solution over the trick are:

maintainability – if a new interaction between manager and child is required, the create process will not change whereas the ignore process would require an update;

size – the create process is a modest process, whereas the size of the ignore process is proportional to the number of different interactions between manager and child;

simulatability – the number of processes in case of the create process increases by one if a child is created, whereas this number increases by two if the trick is used.

A disadvantage of the solution is that the creation requires an extra interaction.

2.2.3   The state-processes

The SDL diagrams served as a basis for the construction of the state-processes. In the (informal) diagrams signal names are given, the values they convey are *not* mentioned. For that reason it was decided to construct the state-processes in two steps:

- message (primitive) names only,

- parameters.

This approach (that works fine when specifying in SDL) was not a success. This is caused by the fact that LOTOS has no global variables, and values that have to be available in successor states should be passed as parameters of the state-processes. It required numerous iterations to get all values received in previous states at all places where they were used.

This suggests that for a state-oriented specification it is better to

- construct a transition diagram for the state-processes,

- list the values required by each state-process for direct usage,

- construct from these lists and the diagram the list of values required by each state-process for direct and/or indirect usage, and

- start construction of the state-processes based on the combined lists.

An alternative approach would be the creation of a 'global memory process' in parallel with each call process. The values of global variables can be stored in and retrieved from this memory process via interactions over a hidden gate. The advantage of this architecture is that the number of parameters of each state process remains low (and constant). The drawback, however, is that each expression must be preceded by a retrieval interaction for every variable in that expression, resulting in too many interactions. This drawback is even more severe in a specification that is intended to be simulated with a single step simulator.

2.2.4   Timers

In the D-channel layer 3 protocol timers   abound, even if all optional timers are omitted. It is sometimes regarded a drawback that LOTOS does not have a timer construct in the language. However, the usage of timers in the D3 protocol is, with one exception, restricted to 'deadlock preventing' timers, i.e. timers that are set when entering a next state, and are stopped immediately after the first interaction (reception of a signal in SDL terminology) in that next state. The exception is a timer that guards the transition to the state after the next state, but that is set again in certain circumstances in such a way that it can expire in the state after the next state only.

The timer process we use is given in figure 3. The process Timers is started in parallel with a call handling process, communicating on a hidden gate (to prevent that other processes corrupt the timers of this call).

```
process Timers [ timer ] : noexit
:=
    timers_with_timerset [ timer ] ( empty_timer_set )
where
    process timers_with_timerset [ timer ]
                                ( timerset : set_of_timer )
                                : noexit
    :=
        timer ? T : Timer Id ! expires [ T IsIn timerset ] ;
        (* timer T expires *)
        timers_with_timerset [ timer ] (Remove ( T, timerset ) )
    [ ]
        timer ? T : Timer_Id ! kill [ T IsIn Timerset ] ;
        (* timer T stopped *)
        timers_with_timerset [ timer ] ( Remove ( T, timerset ) )
    [ ]
        timer ? T : Timer_Id ! restart [ T IsIn timerset ] ;
        (* stop and start again, timerset does not change *)
        timers_with_timerset [ timer ] ( timerset )
    [ ]
        timer ? T : Timer_Id ! start [ T NotIn timerset ] ;
        timers_with_timerset [ timer ] ( Insert (T, timerset ) )
    [ ]
        ( choice T : Timer_Id [ ]
            timer ! T ! T IsIn timerset ;
            timers_with_timerset [ timer ] ( timerset )
        )
    [ ]
        timer ! kill_all ;
        stop
    endproc (* timers_with_timerset *)
endproc (* Timers *)
```

Figure 3: The Timers process

The process timers_with_timerset maintains a set of active timers. The process has six alternatives:

1. a timer expires,

2. a timer is stopped ('kill' is used because 'stop' is a keyword),

3. a timer is restarted,

4. a timer is started,

5. a timer can be inspected (after 'timer ! T1 ? active:Bool;', where T1 is a value of sort Timer_id, active is true if T1 is active, otherwise active is false),

6. the whole process can be stopped, regardless any active timers.

The first three alternatives are only possible if the timer is active, the fourth alternative only if the timer is not active.

If one wishes, a value for the duration can be added when starting or restarting a timer; this value can not be used and would only serve documentation purposes.

Nevertheless, the process in figure 3 is all one needs for deadlock preventing timers. For systems where time has to be measured (e.g. modems with echo cancellation) a completely different kind of timer is required. Those timers should have access to a clock process (internal or external) that can be specified easily, but will make simulations much harder.

2.2.5 Abstract data types

The specification of abstract data types is not so easy, so we had a small group of specialists for this task. All members of the project team were able to provide the signature (sorts and operators) of a data type, and, to be able to provide operators that did what they had to do, an informal description had to be provided for more complex operators.

Sometimes extra operators are required that are not needed in the expressions in the process part of the specification. As an example, in the D3 protocol there was a need for a set of messages. This requires the definition of the operators for equality and inequality for messages. The equations for inequality and those for equality giving true do not cause a problem, but the equations for equality giving false do. Assuming there are 30 messages, roughly 435 equations for equality giving false are required.

The usage of an extra operator, successor, reduces this number to slightly more than thirty equations. This is a trick, there is no order of messages, and the successor operator can be regarded overspecification. Therefore, the additional operators and the rewrite rules were maintained in separate types and in such a way that the project members could not see them.

We ran into two, already known, shortcomings:

- abstract data type definitions are 'flat', little structure can be given to them,

- generic types have to be renamed before they can be actualised.

We decided not to use generic types; editing a copy is in most cases easier than renaming and actualising.

3   The LOTOS environment LOTTE

The LOTOS Tool Environment LOTTE, built at DNL (with the exception of the simulator), has been used successfully in smaller projects, but never in a project with the characteristics of the one at hand. In the first subsection an overview of the functions provided by LOTTE will be given. It appeared that *specification-in-the-many* and *specification-in-the-large* demand extensions to the environment. These will be discussed in the second subsection.

3.1   An overview of LOTTE

The LOTOS tool environment LOTTE provides an environment for the integration of LOTOS tools. It aims at giving easy and uniform access to existing and future tools.

LOTTE consists of a framework in which tools can be inserted when they become available. Each additional tool will be accessible via an additional command in LOTTE, and can be added in the menu. For all commands LOTTE uses a default specification name if appropriate, being the specification name used in the previous command.

At present the main functions in LOTTE are:

sss   a syntax and static semantics checker,

gsr   an interactive report generator for gate sort lists analysis. Reports that can be asked for are:

- the lists of sorts offered at a gate of a process,
- the lists of sorts offered at a gate in a parallel expression,
- the lists of sorts if two processes communicate on indicated gates.

(gsr has not yet been updated for the latest version of LOTOS.)

hippo   an interactive single step simulator.

For certain commands successful completion can depend on successful completion of other commands. In LOTTE this is the case for the report generator and the simulator, for which the specification should not contain static semantical errors. Execution of these tools therefore depends on the successful completion of the static semantic check.

A derived mechanism is used to suppress repetition of actions. For instance, a user can ask for a check on static semantics. If, however, the specification has not been modified since a previous check, the results of the check are already available. LOTTE recognises this and will give these results instead of doing the check again.

Some minor functions present in LOTTE are:

- UNIX derived functions such as the editor vi, the change directory command cd, the list files command ls.

- Functions to show the status of a specification such as **order** and **show**.
  Order lists the files associated with a specification ordered by time (oldest first) to determine the recent history of the specification. Show gives the latest file associated with a specification.

- Functions to help in the development cycle (edit, check, correct, edit).
  The main tool for this is **recover**, which reconstructs the specification source using the information in a listing file. This enables the user to correct errors in the listing file, avoiding the switching between listing and source files.

3.2 Discussion of problems encountered and extensions of LOTTE

Although extensions to LOTTE are still for further study, several ideas emerged during the specification of the D3 protocol that are related to specification of large systems.

A LOTOS specification is structured as a tree, and the order of presentation of the objects (processes and types) in this specification is the left to right scan of this tree. A human reader of the specification seldomly reads a specification in this order. On a machine this can be overcome by a browser or environment directives (analogous to compiler directives), but on paper another solution has to be provided.

One alternative is to allow for out of order object definitions, provided the order can be restored by giving them path-names instead of identifiers. Such path-names might become part of the language when the graphical syntax for LOTOS, which is currently being developed as a joint effort of ISO and CCITT [8], becomes an addendum to the LOTOS standard.

Building a specification breadth first, not yet elaborated processes are introduced (i.e. their heading and a dummy behaviour expression are given). When the elaboration of such a process has to be checked qua syntax and static semantics, one does not need to (re)check previously checked processes and the equations of types (the operators are needed, of course). As a variation on this theme, consider a minor change in a process not affecting its arguments or functionality. At present a subsequent static semantics check requires an analysis of the complete specification where a partial check would suffice.

Going one step further, the checker might insert the headings and dummy behaviour of not yet defined processes, assuming they are local processes, giving a warning that this part of the specification was added automatically.

The monolithic aspect of LOTOS specifications poses other major problems in the development of large specifications, especially if that is team work. Some way to handle these specifications should be developed (probably borrowing from software development techniques and environments, e.g. as described in [9]) and the project environment should provide facilities for version control and piecewise editing and analysis. A tool equivalent with the **recover** tool in LOTTE, which was found very useful, should also work piecewise.

Avoiding repetition of checking would speed up the edit-check-correct-cycle considerably, especially for large specifications. One way of avoiding unnecessary repetition of checks is by a folding-function that eliminates equations and reduces processes to their functionality. If also some include-file mechanism is provided, even direct insertion of the types and/or processes to be checked is not needed.

A specification, aimed at checking process line_mngr in a file with the same name, might then look like:

```
specification D3_protocol [ sub ] : noexit
 .INCLUDE FOLD($USER/D3spec/general/data_types )
 .INCLUDE $USER/D3spec/general/behaviour
where
 .INCLUDE FOLD( $USER/D3spec/network ) WITH
                    FULL( $USER/D3spec/network/line_mngr )
 .INCLUDE FOLD( $USER/D3spec/userside )
 .INCLUDE FOLD( $USER/D3spec/l2_service )
endspec (* D3_protocol *)
```

The fold/include solution is a partial solution only, but it is relatively easy to implement and as such for the short term an attractive solution.

Some data base structure to store specifications in some internal representation (instead of using a simple ASCII file), maintaining versions and status of parts of the specification should be studied as one of the alternatives. Such a data base could also store results of functions for these parts of specification, associated with the version for which it was performed.

The usage of an internal representation would be beneficial for the performance of the tools in the environment. Currently all tools start from the ASCII file that contains the specification, and have to perform lexical and syntactical analysis. This is annoying, especially for a tool like the simulator that has to process the full specification; the include/fold solution does not help in this case, piecewise programming techniques in the more advanced solution might help. When the environment has to support both a textual and a graphical syntax, semantics oriented tools like the report generator and simulator become simpler if they take an internal representation as their starting point.

Another point of discussion is ACT ONE, the abstract data type part of LOTOS. There is an urgent need to specify structures, hide operators, and perform actualisation and renaming in a more user-friendly way. This should not be a problem; SDL [2] shows a way it can be done without loss of formality or the need to extend the mathematical basis. Also the extension of the standard library with (amongst others) integer, characters and strings of characters, all with suitable syntax (not writing 1(2) or 1 + 2 for twelve) is badly needed. Inclusion of processes, such as Timers (see section 2.2.4), in the library should also be considered.

3.3 Some experiences with the simulator HIPPO

The simulator incorporated in LOTTE is made in the SEDOS project under the EC supported ESPRIT program. This single step simulator, called HIPPO, is based on the theory developed in [10].

HIPPO offers the possibility to define and evaluate expressions for the abstract data type part. The equations are interpreted as rewrite rules. The rewrite system for the operators in the D3 specification were debugged using these features of HIPPO. This part of HIPPO is fast, and proved to be very helpful to eliminate errors (mostly oversights).

When HIPPO has digested the specification, it asks what part of the specification should be simulated. Since HIPPO gives a list of all possible interactions (menu) of the simulated part of the specification, it is nice that one is not forced to simulate the full specification. Even when only a part of the specification is simulated, the menu tends to be long (over 100 entries).

Two comments can be made about the menu.

1. The menu also contains a lot of interactions with a guard that can not be made true, e.g.

   - gate ? element : Element_sort [element IsIn empty_element_set];

   Since there is an equation

   - forall e : Element_sort (e IsIn empty_element_set = false)

   the interaction can be eliminated from the menu.

2. Simulation of large systems is generally done on a less ad hoc basis than the simulation of small systems. The person simulating a large specification can provide test sequences in LOTOS and the parallel composition of these test specifications and the part of the system to be tested limits the menu.

The latter approach is related to the constraint oriented specification style: a test sequence constrains the order of interactions to one the person performing the simulations is interested in.

The complete specification of the D3 protocol could not be simulated as a whole. The problem was that the computer ran out of memory (22 Mbyte) during the preparation of the specification by HIPPO. Therefore the specification was divided in parts (the parts mentioned in section 2.2), and each part was simulated separately. During the simulation of the most complex part of the specification, the network side, computation of a menu with about 180 entries required more than one minute (elapsed time), and only a limited number of steps could be made (memory problems, again).

It should be recognised, however, that HIPPO is still in the prototype phase, and optimisations in memory usage and response time have to be made. Also some changes in HIPPO should be considered. For example, HIPPO maintains the complete history of the simulation and allows undoing of simulation steps or jump to a previous step. If the length of the history could be limited (as an option) the problem of running out of memory after a limited number of simulation steps might be solved.

During the development of the D3 specification the '*normal behaviour*' of the parts of the specification were checked with HIPPO. ('*Normal behaviour*' means the es-

tablishment and release of a single connection without line failures, collisions, etc.)
These checks revealed mostly two kinds of errors:

- incompleteness of data types resulting in guards and predicates that did not
evaluate to true or false;

- offers that are out of order resulting in mismatches or no match at all.

The second kind of errors can be found statically by the report generator gsr, described in section 3.1. The fact that we did find so many of these errors that could
have been found earlier (and repaired at less costs) shows the need for a tool like
gsr.

4 Long term prospects

The experiences presented in this paper will also be forwarded to the RACE project
SPECS and a planned ESPRIT II project DNL is/will be involved in. These projects
will end in late 1992. In this section we express our expectations of the state of the
art at that point in time.

The RACE project SPECS aims at, amongst others, a multi-language prototype
environment that supports the specification languages SDL and LOTOS. The architecture of this SPECS environment is described in [11]. In this SPECS environment
a clear distinction is made between the syntax oriented parts of the environment
(called language towers), and the common, semantics oriented part of the environment (called the basis).

With the envisioned SPECS environment it will be possible to specify different parts
of a system in different languages. Nevertheless, simulation, deadlock detection, and
other semantics oriented tools can work on the complete system. For that purpose
a labelled transition system based mathematical representation will be used in the
basis.

The planned ESPRIT II project aims at an industrial LOTOS environment. It is
expected that part of this environment could serve as upgraded LOTOS tower. The
other part will overlap with the SPECS basis and the combination might serve as a
starting point for a set of tools on the basis with the qualities of the tools from the
ESPRIT II project.

By that time, SDL and LOTOS will both have two syntaxes: one textual and one
graphical. The tower for each language supports both syntaxes. However, being a
prototype, the towers will need some harmonisation.

Both projects will work in the area of methodology: how to obtain a high level
specification, and how to refine, add design into formal descriptions until the im-
plementable level is reached. These methodologies are expected to be supported by
test theory and tools so that lower level formal descriptions can be compared with
their higher level specifications.

We expect that in 1992 time is ripe to proceed the generation of tests from a formal description to test a lower level description in the same formalism, towards the
generation of tests for lower level descriptions in another formalism. In this way

the current gap between formal description techniques and conformance testing is expected to be bridged.

References

[1] CEPT
     Recommendation T/CS 46-30 (T/GSI 04-02/3)
     Copenhagen, September 1987

[2] CCITT
     Z.100 – SDL – Specification and Description Language
     Geneva, January 1987

[3] ISO DIS 8807
     Information processing systems – Open systems interconnection –
     LOTOS – A Formal Description Technique Based on Temporal Ordering
     of Observational Behaviour
     ISO, July 1987

[4] ISO DIS 9074
     Information processing systems – Open systems interconnection –
     The definition of the specification language Estelle
     ISO, 1986

[5] CEPT Sub-working Group SPS/PAR, Temp. Doc. 10 revised
     Issue D of layer 3 SDL diagrams
     Oslo, October 1987

[6] M. van Sinderen
     Draft formal specification of the OSI connection-oriented session
     service in LOTOS (version 5)
     ESPRIT/SEDOS/C1/WP/35/T, November 1986

[7] *Guidelines on the application of Estelle, LOTOS and Estelle*
     CCITT/ISO, to be published by the end of 1988.

[8] ISO/CCITT
     Draft answer to question Q48.4/ Working Document for Q.2/X –
     Graphical Representations for LOTOS (GLOTOS)
     ISO/IEC JTC 1/SC 21/WG 1 McL 67

[9] Gail E. Kaiser, Simon M. Kaplan and Josephine Micallef
     Multiuser, Distributed Language-Based Environments
     IEEE Software, November 1987

[10] Peter H.J. van Eijk
     Software tools for the specification language LOTOS
     Ph.D thesis, Enschede, January 1988

[11] The SPECS Consortium
     The SPECS Architecture: Towards an Integrated Specification Environment
     in 'SDL '87 – State of the Art and Future Trends'
     R. Saracco, P.A.J. Tilanus (editors)

The Specification and Design of a Nondeterministic Data Structure Using CCS

Stuart R. Matthews
Division of Computer Science
Hatfield Polytechnic
College Lane
Hatfield, Herts AL10 9AB

June 1988

Abstract

We present the development of a nondeterministic data structure representing a set. The document concentrates on the pragmatic aspects of specifying and designing this particular object in CCS. The results of the exercise will be of interest to anybody wishing to use CCS.

Our initial description of the required object is given as an ADT with a nondeterministic operation. We perform an informal translation of this ADT into a CCS agent, using the ADT to provide the types on which the CCS agent operates. Through a series of examples we introduce the ideas necessary for producing the final design. We introduce a technique for obtaining a nondeterministic choice between a variable number of elements. We define a new family of operators called *link* which is a generalisation of the chaining operators used by Milner.

One of the example designs is found not to satisfy its specification. We show how this inequivalence can be demonstrated by using the expansion principle.

We discuss the informal translation from ADT to CCS specification and the possibilities for generalising this process. We conclude with observations about the general use of CCS as a specification and design tool.

1 Introduction

Certain problems and systems find very concise and clear expression through nondeterministic specifications. This is certainly true for the *unification problem* as it is elegantly expressed by Martelli and Montanari using a nondeterministic specification in [MM82]. Martelli and Montanari's presentation of unification takes the form of a number of transformations which are applied nondeterministically to a set of equations. Thus there are two aspects to the nondeterminism in this specification: nondeterminism in the order of application of the transformations and nondeterminism in the choice of equation to be transformed.

Martelli and Montanari give their specification in a mixture of set algebra and natural language. It was required for the purposes of recent work to transcribe the specification into a uniform notation, whilst keeping as close as possible to the structure of the original presentation. CCS ([Mil80]) [1] is a suitable language for this transcription since it allows the expression of nondeterminism. Furthermore, any CCS specification can be implemented in PFL ([Mit86]). We therefore gain the bonus of an executable nondeterministic unification algorithm.

---

[1]See [Wal87] for a good introduction to CCS.

The transcription process consisted of two main areas corresponding to the two aspects of the nondeterminism described above. In this document we present the specification and design of that part of the transcribed specification which replaces the rôle of the set in Martelli and Montanari's original presentation. This component, a nondeterministic data structure, forms a system with its own particular interests. Also of interest is the specification and design process itself. We start the process by expressing the required object as an ADT with a nondeterministic operation, using the language presented by Subrahmanyam in [Sub81]. Via CCS, we arrive finally at a PFL program.

In Section 2, we present the most abstract notion of the required object as an ADT with a nondeterministic operation. We then translate this ADT into a CCS expression which forms the specification from which design proceeds. We then progress towards a design that can be implemented in PFL, introducing some necessary ideas in Sections 3 and 4. Section 3 introduces the fundamental technique used to obtain nondeterministic choice in the design. Section 4 introduces a family of operators which allow CCS agents to evolve into linked structures of agents. Also in this section, we show how a semi-formal technique can be used to prove the incorrectness of a design with respect to a specification. Section 5 presents the final design with an extensive example. In Section 6 we discuss issues arising from the specification and design exercise. In particular, we discuss how we achieve a translation of the ADT in Section 2 into CCS. The PFL implementation of the final design is included as an appendix.

2 The Required Object

First, we need to specify the object that we want to build. Figure 1 describes an abstract data type (ADT) *Set*, with constructor and accessor operations suitably consistent with the object in mind. We assume that the sorts *elem* and *bool* have already been defined in the ADTs *Element* and *Boolean* respectively. The language we use for the definition is based on that given by Subrahmanyam [Sub81], but we have assumed some minor syntactic changes.

Figure 1 gives a fairly typical ADT definition; the reader not familiar with the basic notions of ADTs is referred to, for example, [Gut77],[GHM78] and [BG81]. However, the operation *CHOOSE* requires some explanation. The way *CHOOSE* is written denotes that it is a nondeterministic operation. Intuitively, the result of such a nondeterministic operation is chosen at random from a set of possible outcomes. This set of possible outcomes is described by a predicate over the arguments and result of the particular nondeterministic operation, called the *characteristic predicate* of the operation. The characteristic predicate of *CHOOSE* is $P_{CHOOSE}(s, x)$. The definition of this predicate

$$P_{CHOOSE}(s, x) = ISMEMBER(x, s)$$

states that the result of *CHOOSE(s)* is any value e in *elem* such that *ISMEMBER(e,s) = TRUE*. Note that the result of *CHOOSE(EMPTYSET)* is left undefined.

Apart from differences of notation, *Set* is almost exactly the example given by Subrahmanyam and the reader is referred to [Sub81] for a complete explanation of the nondeterministic operation *CHOOSE*.

What we require is an equivalent of the ADT *Set* in CCS. Although it is not immediately clear that *Set* brings us any closer to being able to specify the desired object in CCS, there are two good reasons for presenting the ADT *Set*:

- ADTs are a valuable tool which, particularly since the object we intend to construct is a data structure, can only help in understanding the problem and the process of design.

ADT Set using Boolean and Element

SORTS
    set

OPERATIONS
    EMPTYSET :   → set
    INSERT   : elem set → set
    DELETE   : elem set → set
    ISEMPTY : set   → bool
    ISMEMBER : elem set → bool
    <u>CHOOSE</u>   : set   → elem ∪ { UNDEFINED }

VARIABLES
    x,y: elem
    s : set

EQUATIONS
    DELETE(x,EMPTYSET) = EMPTYSET
    DELETE(x,INSERT(y,s)) = if x = y then DELETE(x,s)
                                   else INSERT(y,DELETE(x,s))

    ISEMPTY(EMPTYSET) = TRUE
    ISEMPTY(INSERT(x,s)) = FALSE

    ISMEMBER(x,EMPTYSET) = FALSE
    ISMEMBER(x,INSERT(y,s)) = if x = y then TRUE
                                  else ISMEMBER(x,s)

    CHOOSE(EMPTYSET) = UNDEFINED
    $P_{\underline{CHOOSE}}(s, x)$ = ISMEMBER(x,s)

Figure 1: Abstract Data Type *Set*.

- *Set*, *Element* and *Boolean* will provide the types for a well-founded specification in CCS.

Allowing ourselves the liberty of mixing set algebra with CCS, we define a machine *SET* which is intended to be in some sense *equivalent* to the ADT *Set*:

$$SET \quad : \quad \{insert, delete, \overline{isempty}, ismember\_in, \overline{ismember\_out}, \overline{choose}\}$$

$$SET \stackrel{def}{=} SET'(\emptyset) \tag{1}$$

$$
\begin{aligned}
SET'(\emptyset) \stackrel{def}{=} \; & insert(e).SET'(\{e\}) \\
& + delete(e).SET'(\emptyset) \\
& + \overline{isempty}(true).SET'(\emptyset) \\
& + ismember\_in(e).\overline{ismember\_out}(false).SET'(\emptyset)
\end{aligned}
\tag{2}
$$

$$
\begin{aligned}
SET'(S) \stackrel{def}{=} \; & insert(e).SET'(S \cup \{e\}) \\
& + delete(e).SET'(S - \{e\}) \\
& + \overline{isempty}(false).SET'(S) \\
& + \overline{choose}(e \text{ where } e \in S).SET'(S) \\
& + ismember\_in(e).\overline{ismember\_out}(e \in S).SET'(S)
\end{aligned}
\tag{3}
$$

where $\emptyset$ is the empty set and S is a variable ranging over all non-empty sets. We suggest that *SET* is the closest we can get to reproducing the ADT *Set* using CCS (although it is not the only approximation). We will consider further the relationship between ADTs and CCS and the informal process by which we have moved from one to the other in Section 6. For now we simply note the following points. Operations *INSERT*, *DELETE*, *ISEMPTY* and *CHOOSE* correspond to the single *SET* ports *insert*, *delete*, *isempty* and *choose* respectively. The *ISMEMBER* operation, however, corresponds to the two ports *ismember_in* and *ismember_out*. The *EMPTYSET* operation has no counterpart port in *SET*.

We now return to the second of our reasons for presenting the ADT *Set*. Rather than using set algebra (and assuming this to be defined and understood) we could instead use the ADTs *Element*, *Boolean* and *Set* to provide the types on which the machine *SET* operates, thus:

$$
SET\_spec \quad : \quad \{insert, delete, \overline{isempty}, ismember\_in, \overline{ismember\_out}, \overline{choose}\}
$$

$$SET\_spec \stackrel{def}{=} SET\_spec'(EMPTYSET) \tag{4}$$

$$
\begin{aligned}
SET\_spec'(EMPTYSET) \stackrel{def}{=} \; & insert(e).SET\_spec'(INSERT(e, EMPTYSET)) \\
& + delete(e).SET\_spec'(EMPTYSET) \\
& + \overline{isempty}(TRUE).SET\_spec'(EMPTYSET) \\
& + ismember\_in(e).\overline{ismember\_out}(FALSE). \\
& \quad SET\_spec'(EMPTYSET)
\end{aligned}
\tag{5}
$$

$$
\begin{aligned}
SET\_spec'(INSERT(x, s)) \stackrel{def}{=} \; & insert(e).SET\_spec'(INSERT(e, INSERT(x, s))) \\
& + delete(e).SET\_spec'(DELETE(e, INSERT(x, s))) \\
& + \overline{isempty}(FALSE).SET\_spec'(INSERT(x, s)) \\
& + \overline{choose}(\underline{CHOOSE}(INSERT(x, s))). \\
& \quad SET\_spec'(INSERT(x, s))
\end{aligned}
\tag{6}
$$

$$+ \ ismember\_in(e).\overline{ismember\_out}(ISMEMBER(e,$$
$$INSERT(x,s))).SET\_spec'(INSERT(x,s))$$

Using the ADTs in this way provides us with a more rigorously formal specification than that given by using set algebra (remember that we assume the definition of ADTs *Element* and *Boolean*). We will use *SET_spec* as the specification from which to design our *SET* machine.

There is an important point to note about the specification *SET_spec* using the ADTs. The *INSERT* operation of *Set* is just a *cons*-like operation ie. the result of an operation *INSERT(x,s)* does not depend on whether *ISMEMBER(x,s) = TRUE*. But what we wish to consider is not the term resulting from an *INSERT* operation, but the behaviour of the ADT as a whole. In other words, we should not take the definition of *INSERT* to imply that a design for the *insert(e)* operation of *SET_spec* should necessarily add *e* to the underlying data structure. Rather, we should concentrate on the behavioural aspect of the ADT which means that it is possible to distinguish between an instance of *Set* s_1 where $ISMEMBER(x, s_1) = TRUE$ and an instance s_2 where $ISMEMBER(x, s_2) = FALSE$, but not between instances of *Set* which differ only in the number of instances n, $n > 0$, of an element e that they contain. Consider

$$t \ = \ INSERT(apple, EMPTYSET)$$
$$u \ = \ INSERT(apple, INSERT(apple, EMPTYSET))$$

Now *t* is not distinguishable from *u*, but both are distinguishable from *EMPTYSET*. By *not distinguishable* we mean that *t* and *u* are *extraction equivalent* ([Sub81]). Briefly, an *extractor* on a particular type *T*, is an operation which returns instances of types other than *T*. The extractors defined on *Set* are *ISEMPTY*, *ISMEMBER* and *CHOOSE*. Two instances of a type, t_1 and t_2, are *extraction equivalent* if every sequence of operations ending with an extractor yields the same result when applied to t_1 and t_2. Thus *t* and *u* are extraction equivalent because, for example,

$$ISEMPTY(DELETE(apple, t)) \ = \ ISEMPTY(DELETE(apple, u))$$
$$\underline{CHOOSE}(t) \ = \ \underline{CHOOSE}(u)$$
$$ISMEMBER(apple, DELETE(apple, t)) \ = \ ISMEMBER(apple, DELETE(apple, u))$$

and so on. Let us therefore state a requirement of our design by noting the following property of *SET_spec*:

Property 1 *If two instances of Set s_1 and s_2 are extraction equivalent then SET_spec'(s_1) and SET_spec'(s_2) are observationally congruent* [2] .

We will see the importance of these remarks in the next Section.

SET_spec cannot be implemented directly in PFL because the language does not afford ADTs with nondeterministic operations. We must therefore take into account which of the required types *are* available in PFL as we progress towards a design. In effect, we have the types *Element* and *Boolean* available, but not the type *Set*. What we must therefore achieve in moving from the specification to a design, is to replace the *Set* operations with CCS mechanisms. We shall introduce these mechanisms through a series of examples in the following sections, arriving at the completed design for *SET_spec* in Section 5.

---

[2]In other words, they have precisely the same behaviour. See [Wal87] for a detailed explanation of *observational congruence* and *observational equivalence*.

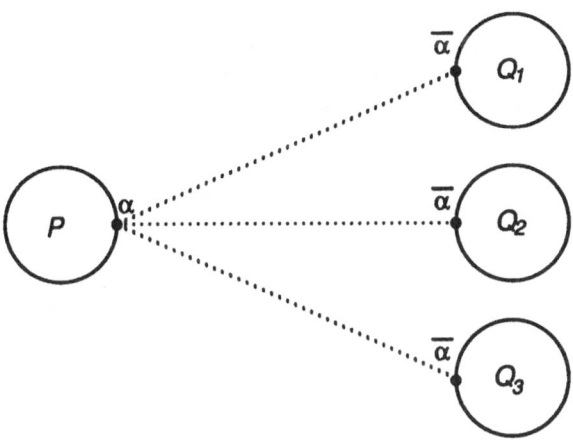

Figure 2: The machine of definition 7.

3 The Basic Mechanism Used for Choice

The first question we address is how to use CCS to make a nondeterministic choice ie. what mechanism can we use to replace the _CHOOSE_ operation used in the specification _SET_spec_?

The most obvious candidate is the + (or) operator. But with a little reflection we see that this would only be of use if the set from which the choice is to be made is fixed. For example, the agent

$$\overline{\alpha}(5) + \overline{\alpha}(9) + \overline{\alpha}(11)$$

outputs $e \in \{5, 9, 11\}$ on $\overline{\alpha}$. However, our specification requires that the cardinality of the set be variable. The technique we shall use is described below.

We define an agent, say P, which accepts a value on a port α and define machines $Q_1, \ldots, Q_n$ each of which output a value on $\overline{\alpha}$. If we compose P with $Q_1, \ldots, Q_n$ then the order in which $Q_1, \ldots, Q_n$ succeed in communicating with P is nondeterministic. To provide a concrete example, suppose that we have three Q-agents each trying to output one of the values 5,9 or 11:

$$P \stackrel{def}{=} \alpha(x).P$$
$$Q_1 \stackrel{def}{=} \overline{\alpha}(5).NIL$$
$$Q_2 \stackrel{def}{=} \overline{\alpha}(9).NIL$$
$$Q_3 \stackrel{def}{=} \overline{\alpha}(11).NIL$$

and compose these agents together thus:

$$(P \mid Q_1 \mid Q_2 \mid Q_3) \backslash \alpha \tag{7}$$

This machine is depicted in Figure 2.

Now suppose that the values output by $Q_1, \ldots, Q_n$ were exactly those of some set set S, $\mid S \mid = n$, that we wish to represent. We see that the machine P would receive a succession of nondeterministic choices from S. Using the technique of having one agent to hold each member of S will allow us to vary the contents of S, provided that we can create and kill-off the Q-agents. We will see how this is possible in the next example.

Let us move on to building a machine, say BAG, which approaches the desired SET machine. BAG is a simpler machine than SET, providing only the operations $insert$, $\overline{isempty}$ and $\overline{choose}$. We require the machine to satisfy the following specification:

$$BAG\_spec \quad : \quad \{insert, \overline{isempty}, \overline{choose}\}$$

$$BAG\_spec \stackrel{def}{=} BAG\_spec'(EMPTYSET) \tag{8}$$

$$BAG\_spec'(EMPTYSET) \stackrel{def}{=} insert(e).BAG\_spec'(INSERT(e, EMPTYSET)) \tag{9}$$
$$+ \overline{isempty}(TRUE).BAG\_spec'(EMPTYSET)$$

$$BAG\_spec'(INSERT(x, s)) \stackrel{def}{=} insert(e).BAG\_spec'(INSERT(e, INSERT(x, s)))$$
$$+ \overline{isempty}(FALSE).BAG\_spec'(INSERT(x, s)) \tag{10}$$
$$+ \overline{choose}(\underline{CHOOSE}(INSERT(x, s))).$$
$$BAG\_spec'(INSERT(x, s))$$

We can design BAG to be a collection of *cells* (like the Q-agents in the previous example). Each cell will contain one element of the bag which BAG represents and will be prepared to output this element on $\overline{choose}$. The other operations, $insert$ and $\overline{isempty}$, will be supported by a monitoring agent. This *monitor* will accept new values on $insert$ and spawn new cells containing these values. Because new values must pass through the monitor (and because elements cannot be deleted from BAG), it is easy to arrange for the correct output from $\overline{isempty}$. In CCS, the design for BAG looks like this:

$$BAG\_des \quad : \quad \{insert, \overline{isempty}, \overline{choose}\}$$

$$BAG\_des \stackrel{def}{=} Bag\_empty \tag{11}$$

$$Bag\_empty \stackrel{def}{=} insert(e).(Bag\_non\_empty \mid Bag\_element(e)) \tag{12}$$
$$+ \overline{isempty}(TRUE).Bag\_empty$$

$$Bag\_non\_empty \stackrel{def}{=} insert(e).(Bag\_non\_empty \mid Bag\_element(e)) \tag{13}$$
$$+ \overline{isempty}(FALSE).Bag\_non\_empty$$

$$Bag\_element(e) \stackrel{def}{=} \overline{choose}(e).Bag\_element(e) \tag{14}$$

Expressions 12 and 13 correspond to the idea of a *monitor* and expression 14 describes the behaviour of the cells holding each element. Note that although BAG may consist of many machines ($\mid B \mid +1$ where B is the bag represented), the user is given the impression of a monolithic machine whose behaviour is exactly that described by the specification BAG_spec. For example, suppose that an agent using BAG_des performs the sequence of actions $\overline{insert}(mango)$, $\overline{insert}(peach)$, $\overline{insert}(guava)$ and finally $choose(x)$, defined formally as follows:

$$BAG\_User \stackrel{def}{=} \overline{insert}(mango).\overline{insert}(peach).\overline{insert}(guava).choose(x).NIL$$

$$BAG\_Usage \stackrel{def}{=} (BAG\_User \mid BAG\_des)\backslash\{insert, isempty, choose\}$$

Figure 3 depicts the situation where agent BAG_Usage has reached the $choose(x)$ action.

We now return to the point raised in the previous section about the $INSERT$ operation, which has serious implications for any design for SET based on the design for BAG. Suppose that an

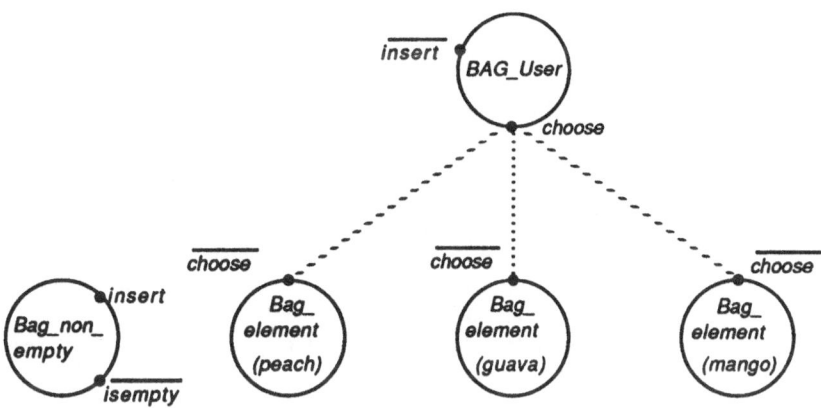

Figure 3: *BAG_User* may succeed in communicating with any one of the *Bag_element* agents.

agent derived from *BAG_des* currently represents a bag B, which is the union of a non-empty bag of identical elements $\mathcal{E}$, $\mid \mathcal{E} \mid = m$ and a bag of other elements $\mathcal{F}$, $\mid \mathcal{F} \mid = n$. Writing *BAG_des(B)* to denote this derived agent, *BAG_des(B)* consists of $n + m$ value cells (and one monitor machine). Any machine which chooses an element $x \in B$ from *BAG_des(B)*, has a probability p_e of receiving $e \in \mathcal{E}$ given by $p_e = m/(m+n)$. Therefore, the greater m, the greater the probability of receiving $e \in \mathcal{E}$ (unless $n = 0$ in which case $p_e = 1$ always). However, this type of behaviour would not contradict the requirement of property 1. If two bags B_1 and B_2 differ only in the number of one particular element that they both contain, then the agent *BAG_des(B_1)* is observationally congruent to the agent *BAG_des(B_2)* — they can perform the same actions.

   If we want to ensure that the probability of choosing each element from *SET* is the same, then we must state another requirement of the design:

Property 2 *For a machine SET_spec'(s) each e such that ISMEMBER(e,s) = TRUE must have equal probability of being returned by the $\overline{choose}$ communication.*

We are effectively expanding our notion of what it means for two agents to have *the same behaviour*, beyond observational congruence. It is necessary to include statements about the probability of actions so that we obtain a more practical form of equivalence when the design is implemented in PFL.

4   Using *link* Operators to Build Structures

In this section we move closer to being able to design a machine to satisfy the specification *SET_spec*. The machine we design, *BOX*, supports the communications *insert*, *delete*, $\overline{isempty}$ and $\overline{choose}$. We introduce a technique which allows the design for *BOX* to satisfy the requirement of property 2. However, *the design at which we arrive, BOX_des, does not satisfy the specification BOX_spec*. We present the design for two reasons.

- To introduce the concept of *linking-together* agents.

- Because it is informative to see a design which appears at first sight to satisfy a given specification, but which can be shown to be incorrect.

If the same basic technique used in *BAG_des* were to be used in a design for *SET_spec*, then how could we satisfy the requirement of property 2? The answer is, to arrange that each value that is an element of the set represented by the machine is held by only one cell. To achieve this we must ensure that when a value e is received by $SET\_des(S)$ on its *insert* port, a new value cell is spawned if and only if $e \notin S$. We therefore need to introduce into the design some means by which a newly inserted element can be deterministically compared to the values contained in all existing cells. The solution is to organise the value cells into a structure such that they can communicate amongst themselves. The links in the structure will be channels restricted to allow communication only between two joined cells. This technique is used by Milner to build pushdowns and queues ([Mil80, Chapter 8]). We will give a generalisation of the linking operators used by Milner and demonstrate the mechanism by examples.

Let us put the idea of *linking* into context by designing a new machine, *BOX*, which has property 2. The specification is given below:

$$BOX\_spec \quad : \quad \{insert, delete, \overline{isempty}, \overline{choose}\}$$

$$BOX\_spec \overset{def}{=} BOXspec'(EMPTYSET) \tag{15}$$

$$BOX\_spec'(EMPTYSET) \overset{def}{=} insert(e).BOX\_spec'(INSERT(e, EMPTYSET))$$
$$+ \, delete(e).BOX\_spec'(EMPTYSET) \tag{16}$$
$$+ \, \overline{isempty}(TRUE).BOX\_spec'(EMPTYSET)$$

$$BOX\_spec'(INSERT(x, s)) \overset{def}{=} insert(e).BOX\_spec'(INSERT(e, INSERT(x, s)))$$
$$+ \, delete(e).BOX\_spec'(DELETE(e, INSERT(x, s))) \tag{17}$$
$$+ \, \overline{isempty}(FALSE).BOX\_spec'(INSERT(x, s))$$
$$+ \, \overline{choose}(\underline{CHOOSE}(INSERT(x, s))).$$
$$BOX\_spec'(INSERT(x, s))$$

As we have discussed, the value cells in the design for *BOX* are to be linked together. To express this *linking-together* we define a new family of operators which we will call *link*. Milner defines two diadic chaining operators, $\frown$ and $\backsim$, for forming links between specific machines (single and double links respectively). The disadvantage of this scheme is that a new operator must be defined wherever the number of links or the names of the ports to be linked differs. What we would like is a definition of a class of operators that can be instantiated by the number of links and the names of the linked ports. We define the generic diadic operator $\frown$ (called *link*) below to satisfy this requirement. But first we need a preliminary definition:

Definition 1 *We will say that two ports α and β are io-complementary if and only if*

$$\overline{[\lambda/\alpha]} = [\lambda/\beta]$$

where $\lambda \notin names\{\alpha, \beta\}$.

In other words, two ports are io-complementary if and only if one is an output port and the other is an input port. Now, assuming the function sos : $seq \rightarrow set$ which converts sequences to sets in a natural way, we can give the definition of *link*:

Definition 2 *Let M_1, M_2 be the machines to be linked together, $M_1 : L_1, M_2 : L_2$, where L_1 and L_2 each contain no pairs of labels which are complementary, $\overline{L_1} \cap L_2 = \emptyset$. Let J_1 and J_2 be the sequences of ports from L_1 and L_2 that we wish to join, $sos(J_1) \subseteq L_1$, $sos(J_2) \subseteq L_2$*

Figure 4: The agent $P \overset{join}{\frown} Q$.

and $length(J_1) = length(J_2) = n > 0$, where the labels of J_1 and J_2 must be distinct, $sos(J_1) \cap sos(J_2) = \emptyset$. We also require that corresponding pairs in J_1 and J_2 are io-complementary: $\forall i.i = 1 \ldots n$ $J_1(i)$ and $J_2(i)$ are io-complementary. Let $J_1 \frown J_2$ be the operator which joins the ports J_1 to J_2, then

$$M_1 \ J_1 \frown J_2 \ M_2 =$$
$$(M_1[\lambda_1/name(J_1(1)), \ldots, \lambda_n/name(J_1(n))] \mid$$
$$M_2[\lambda_1/name(J_2(1)), \ldots, \lambda_n/name(J_2(n))]) \backslash \lambda_1, \ldots, \lambda_n$$

where $\lambda_1, \ldots, \lambda_n \notin names(L_1 \cup L_2)$ and $\lambda_1, \ldots, \lambda_n$ are distinct.

The definition of $M_1 \ J_1 \frown J_2 \ M_2$ says that corresponding ports $J_1(i)$ and $J_2(i)$ are renamed to the same label λ_i. The machines M_1 and M_2, with all ports to be linked thus renamed, are composed together and the resulting machine is restricted by the new labels $\lambda_1, \ldots, \lambda_n$. Because pairs of ports have been given the same names and these names made inaccessible from outside, M_1 and M_2 have n channels of communication which are hidden from any other machines with which $M_1 \ J_1 \frown J_2 \ M_2$ might be composed. M_1 and M_2 are effectively *linked together*. Note that $J_1 \frown J_2$ joins $J_1(1)$ to $J_2(1)$, $\ldots$, $J_1(n)$ to $J_2(n)$ ie. it is the position in which the ports appear in the sequences J_1 and J_2 that establishes their correspondence.

For example, suppose that we have two machines P and Q, $P : \{\bar{\alpha}, \beta, \rho\}$ $Q : \{\gamma, \bar{\delta}, \sigma\}$, and suppose that we wish to compose these machines together, joining $\bar{\alpha}$ to γ and β to $\bar{\delta}$. We define for this purpose an operator $\overset{join}{\frown}$ as an instantiation of the *link* operator:

$$\overset{join}{\frown} \overset{def}{=} \langle \bar{\alpha}, \beta \rangle \frown \langle \gamma, \bar{\delta} \rangle$$

using the angle brackets $\langle \rangle$ to denote a sequence. The machine composed of P and Q with the ports linked as desired can now be written

$$P \overset{join}{\frown} Q$$

This machine is depicted in Figure 4. Because the renamed versions of $\bar{\alpha}, \beta, \gamma$ and $\bar{\delta}$ are restricted (by definition of *link*) these ports are not available to outside machines. P and Q thus have two channels of communication guaranteed free from interference. P and Q can, however, continue to communicate freely with other machines on ports ρ and σ respectively.

With the *link* operator at our disposal, let us consider how we might design BOX. Suppose that we define an agent to act as a cell for holding values and aim to construct BOX from just this one type of agent. It follows that this *cell* machine must offer at least the ports *insert, delete, $\overline{isempty}$* and *$\overline{choose}$* (in order to satisfy the specification). The $\overline{choose}$ communication can clearly work as it did in the design for BAG (definition 13). To see how the other communications might be handled, let us take an example.

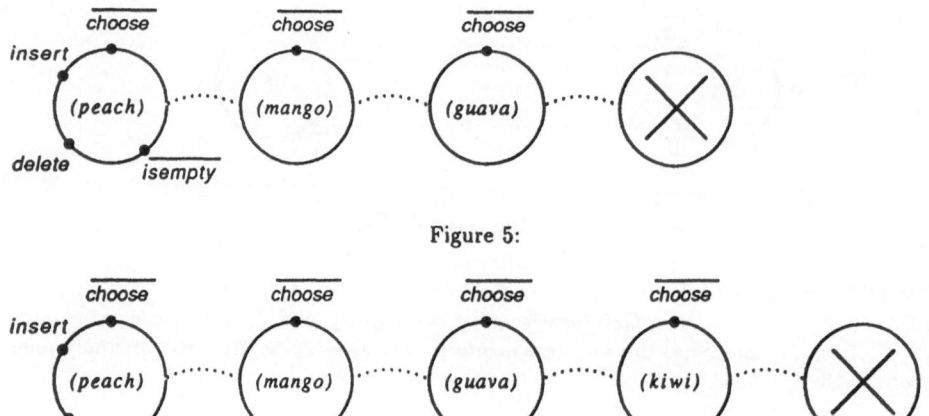

Figure 5:

Figure 6:

Figure 5 shows a *BOX* machine into which have already been inserted the values *peach*, *mango* and *guava*. There is one cell holding each of these values and the cells are linked together as denoted by the broken line (this does not denote a single link—we have not yet decided how many ports must be linked). It is the leftmost cell, that containing *peach*, which communicates with the outside world. Suppose that this cell receives a value on its *insert* port, say *kiwi*, representing a value to be inserted into *BOX*. Since we require to have each value held by only one cell, the appropriate action would be to compare the new value against that already held in this cell. If the values are found to be equal then the new value can be ignored and the cell just return to its previous state; if the values are different then the new value could be passed to the cell on its right. In this particular case the values *are* different and *kiwi* is therefore passed to the right.

The second cell behaves in exactly the same way as the first and thus *kiwi* is passed to the third cell and finally to the end cell. This end cell is currently empty (which we have chosen to denote by marking it *X*) and it accepts *kiwi* to itself become a value-containing cell. In addition, it spawns a new *X* cell which it links onto its right hand side (and thus onto the right hand end of the chain). The resulting machine is shown in Figure 6.

Suppose now that the leftmost cell receives the value *mango* on its *delete* port. It compares this value to that which it already contains and, finding them unequal, passes the value for deletion, *mango*, to its right. The second cell, however, finds that it contains the value that it receives on its *delete* port. At this point it could simply become an empty cell which forwarded messages, but that seems to be an untidy solution. It would be neater to keep the chain packed to the left. Instead then, this now empty cell (it has had its value deleted) *sucks-back* the value in the cell to its right so that it now contains *guava*. The third cell, now itself empty does the same so that it now contains *kiwi*. The fourth cell, however, finds that the cell to its right is the end-marker cell and cannot suck-back a value. In this case the fourth cell becomes the *X* cell and the fifth cell, previously the *X* cell, dies. The chain is now packed-left, as shown in Figure 7.

Finally, the $\overline{isempty}$ communication. Clearly this can be handled by the leftmost cell since if it contains a value then *BOX* is not empty. If the leftmost cell is currently the *X* cell (which would be the case if all values in *BOX* were deleted) then *BOX* is known to be empty. This also

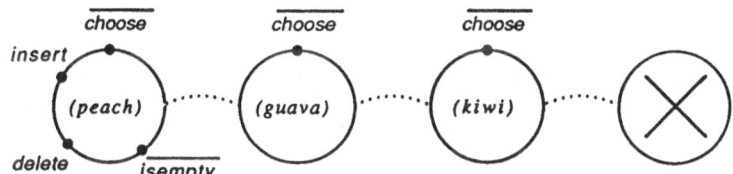

Figure 7:

leads us to conclude that in its initial state BOX should consist of just an X cell.

Let us now put these ideas into a CCS design for BOX, defining first an appropriate $\frown$ operator:

$$\overset{box}{\frown} \;=\; \langle \overline{blow}, suck, \overline{rremove}, rempty \rangle \frown \langle receive, \overline{passback}, lremove, \overline{lempty} \rangle$$

$$BOX\_des \;\overset{def}{=}\; Front\_elem \tag{18}$$

$$Front\_elem \;:\; \{insert, delete, \overline{isempty}, choose, \overline{blow}, suck, \overline{rremove}, rempty\}$$

$$Front\_elem \;\overset{def}{=}\; Front\_elem\_x \tag{19}$$

$$Front\_elem\_x \;\overset{def}{=}\; insert(e).(Front\_elem\_1(e) \overset{box}{\frown} Elem)$$
$$+\; delete(e).Front\_elem\_x \tag{20}$$
$$+\; \overline{isempty}(TRUE).Front\_elem\_x$$

$$Front\_elem\_1(v) \;\overset{def}{=}\; insert(e).(if\ e = v\ then\ Front\_elem\_1(v)\ else\ \overline{blow}(e).Front\_elem\_1(v))$$
$$+\; delete(e).(if\ e = v\ then\ Front\_elem\_0\ else\ \overline{rremove}(e).$$
$$Front\_elem\_1(v)) \tag{21}$$
$$+\; \overline{choose}(v).Front\_elem\_1(v)$$
$$+\; \overline{isempty}(FALSE).Front\_elem\_1(v)$$

$$Front\_elem\_0 \;\overset{def}{=}\; suck(e).Front\_elem\_1(e) \tag{22}$$
$$+\; rempty.Front\_elem\_x$$

$$Elem \;:\; \{receive, \overline{passback}, lremove, \overline{lempty}, \overline{blow}, suck, \overline{rremove}, rempty, \overline{choose}\}$$

$$Elem \;\overset{def}{=}\; Elem\_x \tag{23}$$

$$Elem\_x \;\overset{def}{=}\; receive(e).(Elem\_1(e) \overset{box}{\frown} Elem\_x)$$
$$+\; lremove(e).Elem\_x \tag{24}$$
$$+\; \overline{lempty}.NIL$$

$$Elem\_1(v) \;\overset{def}{=}\; receive(e).(if\ e = v\ then\ Elem\_1(v)\ else\ \overline{blow}(e).Elem\_1(v))$$
$$+\; lremove(e).(if\ e = v\ then\ Elem\_0\ else\ \overline{rremove}(e).Elem\_1(v)) \tag{25}$$
$$+\; \overline{choose}(v).Elem\_1(v)$$
$$+\; \overline{passback}(v).Elem\_0$$

$$Elem\_0 \;\overset{def}{=}\; suck(e).Elem\_1(e) \tag{26}$$
$$+\; rempty.Elem\_x$$

Note that we have not achieved our aim of building *BOX_des* from a single type of agent. Instead we have used two types of agent *Front_elem* and *Elem*. This is because the behaviour of the leftmost or *front* cell, which communicates with the outside world, needs to be slightly different from that of the other cells (it never needs to pass values back to its left and we do not want it to die having once communicated that it is currently the X cell) [3] . Note also that, as with *BAG_des*, although a derivation of *BOX_des* may consist of many machines, the user has the impression of a monolithic machine.

As we noted at the beginning of this section, *the design BOX_des does not satisfy the specification BOX_spec*. The problem is not immediately apparent from the example we have given. Indeed, the incorrectness of the design was only found by the author when the PFL implementation of *BOX_des* was tested. The incorrectness of the design manifests itself through the following type of behaviour (of the implementation): we insert into an empty *BOX_des* some values, say *mango* and *ugly*, at which point *BOX_des* represents the set $\{mango, ugly\}$; we then delete from the *BOX_des* the value *ugly*. Clearly, at this point we would like *BOX_des* to represent the set $\{mango\}$, as demanded by the specification. However, choosing a value from *BOX_des* may still yield the value *ugly*! The reason for this is simply because there is nothing in the definition of *BOX_des* which requires a delete operation to be completed before *BOX_des* performs another communication such as $\overline{choose}$. Similarly, an *insert* operation need not be completed before other communications are performed. Let us see how we might have detected this problem by analysing the specification and design for *BOX*.

If *BOX_spec* and *BOX_des* are equivalent, then we would expect the agents *BOX_spec** and *BOX_des** obtained after applying a particular sequence of actions to also be equivalent. Consider the following sequence of actions:

$$BOX\_User \stackrel{def}{=} \overline{insert}(pawpaw).\overline{insert}(melon).\overline{delete}(melon).choose(x).\overline{output}(x).NIL \qquad (27)$$

First, we apply *BOX_User* to *BOX_spec*. We use the *expansion principle* ([Wal87]) to obtain the resulting machine, writing L for $\{insert, delete, \overline{isempty}, ismember\_in, \overline{ismember\_out}, \overline{choose}\}$, as follows:

$(BOX\_User \mid BOX\_spec)\backslash L$

$= \overline{(insert}(pawpaw).\overline{insert}(melon).\overline{delete}(melon).choose(x).\overline{output}(x).NIL$
$\qquad \mid BOX\_spec'(EMPTYSET))\backslash L \qquad\qquad\qquad\qquad$ by 15 and 27

$= \tau.(\overline{insert}(melon).\overline{delete}(melon).choose(x).\overline{output}(x).NIL$
$\qquad \mid BOX\_spec'(INSERT(pawpaw, EMPTYSET)))\backslash L \qquad\qquad$ by 16

$= \tau.\tau.(\overline{delete}(melon).choose(x).\overline{output}(x).NIL$
$\qquad \mid BOX\_spec'(INSERT(melon, INSERT(pawpaw, EMPTYSET))))\backslash L \qquad$ by 17

$= \tau.\tau.\tau.(choose(x).\overline{output}(x).NIL \mid BOX\_spec'(DELETE(melon,$
$\qquad INSERT(melon, INSERT(pawpaw, EMPTYSET)))))\backslash L \qquad\qquad$ by 17

$= \tau.\tau.\tau.(choose(x).\overline{output}(x).NIL \mid BOX\_spec'(INSERT(pawpaw, EMPTYSET)))\backslash L$ by Set

$= \tau.\tau.\tau.\tau.(\overline{output}(pawpaw).NIL$
$\qquad \mid BOX\_spec'(INSERT(pawpaw, EMPTYSET)))\backslash L \qquad\qquad$ by 17 and Set

$= \tau.\tau.\tau.\tau.\overline{output}(pawpaw).BOX\_spec'(INSERT(pawpaw, EMPTYSET)))\backslash L$

The agent we have derived is stable ie. it is not capable of any internal τ actions. We have shown that the only visible action which $(BOX\_User \mid BOX\_spec)$ can perform is $\overline{output}(pawpaw)$. This is because at the point where *BOX_spec* has accepted the sequence of actions $\overline{insert}(pawpaw)$.

---

[3]We could define *BOX* in terms of one type of agent, but in practice this would be slightly odd since some ports in the leftmost cell would not be linked. It seems better to capture in the design the fact that the behaviour of the front cell is different to the behaviour of all non-front cells

$\overline{insert}(melon).\overline{delete}(melon)$, the only value it can yield to the $choose(x)$ communication is $pawpaw$. We now apply the same derivation sequence to BOX_des.

$$
\begin{aligned}
&(BOX\_User \mid BOX\_des)\\
=\ &(\overline{insert}(pawpaw).\overline{insert}(melon).\overline{delete}(melon).choose(x).\overline{output}(x).NIL\\
&\quad \mid Front\_elem\_x)\backslash L \hspace{4cm} 27,18,19\\
=\ &\tau.(\overline{insert}(melon).\overline{delete}(melon).choose(x).\overline{output}(x).NIL\\
&\quad \mid (Front\_elem\_1(pawpaw)\overset{box}{\asymp} Elem\_x))\backslash L \hspace{2.6cm} \text{by } 20\\
=\ &\tau.\tau.(\overline{delete}(melon).choose(x).\overline{output}(x).NIL\\
&\quad \mid ((blow(melon).Front\_elem\_1(pawpaw))\overset{box}{\asymp} Elem\_x))\backslash L \hspace{1.5cm} \text{by } 21\\
=\ &\tau.\tau.(\overline{delete}(melon).choose(x).\overline{output}(x).NIL\\
&\quad \mid (Front\_elem\_1(pawpaw)\overset{box}{\asymp}(Elem\_1(melon)\overset{box}{\asymp} Elem\_x)))\backslash L \hspace{0.8cm} \text{by } 24\\
=\ &\tau.\tau.\tau.(choose(x).\overline{output}(x).NIL\\
&\quad \mid ((\overline{rremove}(melon).Front\_elem\_1(pawpaw))\overset{box}{\asymp}(Elem\_1(melon)\overset{box}{\asymp} Elem\_x)))\backslash L \quad \text{by } 21\\
=\ &\tau.\tau.\tau.(\tau.(\overline{output}(melon).NIL \mid ((\overline{rremove}(melon).Front\_elem\_1(pawpaw))\overset{box}{\asymp}\\
&\quad (Elem\_1(melon)\overset{box}{\asymp} Elem\_x)))\\
&\quad + \tau.(choose(x).\overline{output}(x).NIL \mid (Front\_elem\_1(pawpaw)\overset{box}{\asymp}(Elem\_0\overset{box}{\asymp} Elem\_x))))\backslash L \quad \text{by } 25,25
\end{aligned}
$$

Although we have not yet derived a stable agent, we have done enough to prove that BOX_spec and BOX_des are not equivalent. We have found that after accepting the sequence of actions $\overline{insert}(pawpaw).\overline{insert}(melon).\overline{delete}(melon)$ BOX_des is capable of performing a $\overline{choose}(melon)$. But we have already shown that after the same sequence of actions, the only $\overline{choose}$ action of which BOX_spec is capable is a $\overline{choose}(pawpaw)$. Note that if we were to continue applying the expansion rule, we would find that after accepting $\overline{insert}(pawpaw).\overline{insert}(melon).\overline{delete}(melon)$ BOX_des is capable of performing $\overline{choose}(melon)$ or $\overline{choose}(pawpaw)$.

Having shown that BOX_des and BOX_spec have different behaviours after accepting the same sequence of actions proved that the two machines are not observationally congruent. We will consider in the next section how the problems giving rise to this inequality can be resolved. We do not give a corrected version of BOX_des. It would be almost identical to SET_des, given in the next section, but without the expressions associated with the $ismember\_in$ and $ismember\_out$ communications.

5   The design for SET

In this section we present the design for SET. Firstly, we must consider the cause of the inequality between BOX_spec and BOX_des, to ensure that the mistake is not repeated in SET_des.

Looking at the definition of $BOX\_spec'(S)$ in the previous section (definition 17), we see that whenever the agent performs one of the communications which change S (ie. $insert$ or $delete$) the agent immediately accommodates the change [4] . However, the same is not necessarily true when BOX_des performs one of these communications. Whereas derivations from BOX_spec consist of only one agent, the agent derived from BOX_des consists of $n + 1$ agents, where n is the number of values contained in BOX_des. Now, the $operation$ of deleting a value from BOX_des or inserting a value into BOX_des extends beyond the communication $delete$ or $insert$ offered by $Front\_elem$. These operations may take many internal communications between the agents constituting BOX_des and we can make no assumptions about the order in which these actions will happen. Thus the user may find that a value passed to the $delete$ port will still be yielded

---

[4]Not immediately in a temporal sense, but in the sense that the very next agent derived is $BOX\_spec'(S')$ with S' changed as desired

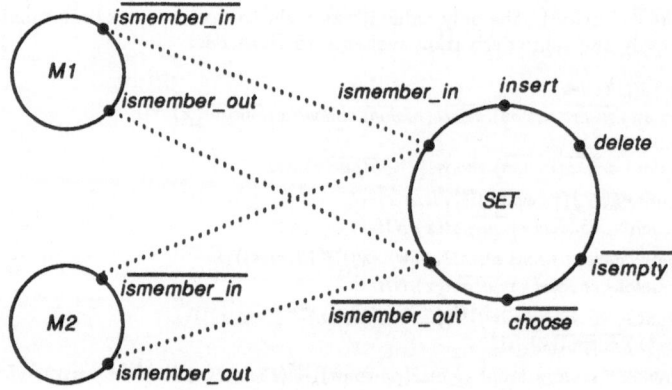

Figure 8:

by *choose(x)*, as was demonstrated in the previous section. How can we ensure that a design for *SET*, based on the design principles of *BAG_des* and *BOX_des*, is equivalent to *SET_spec*? Whatever mechanism we use must, of course, be invisible from outside.

If we view the problems of *BOX_des* as being that it accepts communications before having properly re-organised its internal structure, then the answer is to prevent the machine from performing these actions before it is ready to do so. We can see that for the communications *insert,delete* and *isempty* this might be arranged by *Front_elem* waiting for a *continue* signal from the other cells, indicating that an *insert* or *delete* operation had been completed. But the *choose* action presents a greater problem since all value-containing cells can perform *choose* but must be prevented from doing so for the duration of a delete or insert operation.

We need also to consider how the ismember operation (actions *ismember_in* and *ismember_out*) is to be handled. We can arrange that when a cell receives a value *e* on its *ismember_in* port, it outputs *TRUE* on its own *ismember_out* port if *e* is equal to the value which the cell contains. Otherwise, the cell passes the value *e* to its right-hand neighbour. If the end cell receives a value on its *ismember_in* port then it outputs *FALSE* on *ismember_out*. But there is a problem with this plan. Suppose that two agents M_1 and M_2 are using *SET*, a situation depicted in Figure 8. Now, if M_1 performs *ismember_in* followed by M_2 performing *ismember_in* and both machines wait for *ismember_out* then, under the scheme described above, we cannot be sure that each machine will accept the correct *ismember_out* communication. Clearly, these communications also need to be scheduled: having performed *ismember_in*, *SET_des* should not perform the same action again until after it has performed *ismember_out*.

With these considerations in mind, here is the design for *SET*:

$$\overset{sch}{\frown} = \{\overline{rinsert}, \overline{rdelete}, rempty, \overline{rismember}\} \frown \{linsert, ldelete, \overline{lempty},$$
$$lismember\}$$

$$\overset{set}{\frown} = \{\overline{rinsert}, suck, \overline{rdelete}, rempty, \overline{rismember}\} \frown \{linsert, \overline{passback},$$
$$ldelete, \overline{lempty}, lismember\}$$

$$SET\_des \overset{def}{=} (Scheduler \overset{sch}{\frown} Front\_elem) \backslash \{amember, continue\} \tag{28}$$

$$Scheduler : \{insert, delete, \overline{choose}, \overline{isempty}, ismember\_in, \overline{ismember\_out},$$

$$\overline{rinsert}, \overline{rdelete}, \overline{rismember}, \overline{remply}, amember, continue\}$$

$$Scheduler \overset{def}{=} Sched\_start \tag{29}$$

$$Sched\_start \overset{def}{=} remply(b).(if\ b\ then\ Sched\_empty\ else\ (amember(e). \\ Sched\_non\_empty(e))) \tag{30}$$

$$Sched\_empty \overset{def}{=} insert(e).\overline{rinsert}(e).continue.Sched\_start \\ + delete(e).Sched\_empty \tag{31} \\ + \overline{isempty}(TRUE).Sched\_empty \\ + ismember\_in(e).\overline{ismember\_out}(FALSE).Sched\_empty$$

$$Sched\_non\_empty(v) \overset{def}{=} insert(e).\overline{rinsert}(e).continue.Sched\_start \\ + delete(e).\overline{rdelete}(e).continue.Sched\_start \\ + \overline{isempty}(FALSE).Sched\_start \tag{32} \\ + ismember\_in(e).\overline{rismember}(e).continue.Sched\_start \\ + \overline{choose}(v).Sched\_start$$

$$Front\_elem : \{linsert, ldelete, lismember, \overline{lempty}, \overline{rinsert}, suck, \overline{rdelete}, \\ \overline{rismember}, remply, \overline{ismember\_out}, amember, continue\}$$

$$Front\_elem \overset{def}{=} Front\_elem\_x \tag{33}$$

$$Front\_elem\_x \overset{def}{=} linsert(e).\overline{continue}.(Front\_elem\_1(e) \overset{set}{\frown} Elem) \\ + ldelete(e).\overline{continue}.Front\_elem\_x \tag{34} \\ + \overline{lempty}(TRUE).Front\_elem\_x \\ + lismember(e).\overline{ismember\_out}(FALSE).\overline{continue}.Front\_elem\_x$$

$$Front\_elem\_1(v) \overset{def}{=} linsert(e).(if\ e = v\ then\ \overline{continue}.Front\_elem\_1(v)\ else\ \overline{rinsert}(e). \\ Front\_elem\_1(v) \\ + ldelete(e).(if\ e = v\ then\ \overline{continue}.Front\_elem\_0\ else\ \overline{rdelete}(e). \\ Front\_elem\_1(v)) \tag{35} \\ + \overline{lempty}(FALSE).Front\_elem\_1(v) \\ + \overline{amember}(v).Front\_elem\_1(v) \\ + lismember(e).(if\ e = v\ then\ \overline{ismember\_out}(TRUE).\overline{continue}. \\ Front\_elem\_1(v)\ else\ \overline{rismember}(e).Front\_elem1(v))$$

$$Front\_elem\_0 \overset{def}{=} suck(e).Front\_elem\_1(e) \tag{36} \\ + remply.\overline{continue}.Front\_elem\_x$$

$$Elem : \{linsert, \overline{passback}, ldelete, lismember, \overline{lempty}, \overline{rinsert}, suck, \\ \overline{rdelete}, \overline{rismember}, remply, \overline{ismember\_out}, amember, continue\}$$

$$Elem \overset{def}{=} Elem\_x \tag{37}$$

$$Elem\_x \overset{def}{=} linsert(e).\overline{continue}.(Elem\_1(e) \overset{set}{\frown} Elem) \\ + ldelete(e).\overline{continue}.Elem\_x \tag{38} \\ + \overline{lempty}.NIL \\ + lismember(e).\overline{ismember\_out}(FALSE).\overline{continue}.Elem\_x$$

$$Elem\_1(v) \stackrel{def}{=} linsert(e).(if\ e = v\ then\ \overline{continue}.Elem\_1(v)\ else$$
$$\overline{rinsert}(e).Elem\_1(v))$$
$$+\ ldelete(e).(if\ e = v\ then\ Elem\_0\ else\ \overline{rdelete}(e).Elem\_1(v))$$
$$+\ \overline{amember}(v).Elem\_1(v) \tag{39}$$
$$+\ lismember(e).(if\ e = v\ then\ \overline{ismember\_out}(TRUE).\overline{continue}.$$
$$Elem1(v)\ else\ \overline{rismember}(e).Elem1(v))$$
$$+\ \overline{passback}(v).Elem\_0$$

$$Elem\_0 \stackrel{def}{=} suck(e).Elem\_1(e) \tag{40}$$
$$+\ rempty.\overline{continue}.Elem\_x$$

The body of any agent derived from *SET_des* is similar to agents derived from *BOX_des* ie. linked chains of cells. But *SET_des* agents consist additionally of a *Scheduler* agent composed in parallel with the chain of cells. *Scheduler*'s purpose is to ensure that communications are not accepted from an outside user until the cell-structure is properly organised. *Scheduler* handles all communications with outside users except $\overline{ismember\_out}$. More precisely, having accepted any communication from outside, *Scheduler* will not accept further communications from outside until it receives the "all-clear" (action $\overline{continue}$) from the cell structure. So, for example, if *Scheduler* accepts a *delete* from outside (and passes it on into the cell structure) it must wait for the $\overline{continue}$ action before returning to a state in which further actions from outside can be accepted. The cell structure will not offer the $\overline{continue}$ action until it has completely re-organised itself; after a *delete* this is once the structure has been packed left (up to the cell which contained the deleted value). Thus, if the next action from outside is a *choose* (after *delete*), there is no possibility of *choose* returning the previously deleted value. After an *insert*, the $\overline{continue}$ action is offered once a new cell had been spawned. From outside, this scheduling mechanism is invisible.

The cells no longer communicate directly with a user on the $\overline{choose}$ port. Rather, it is *Scheduler* that supplies this communication. This modification is necessary to be able to prevent a *choose* being performed whilst another operation is in progress. Whenever *Scheduler* passes through its initial state, it obtains a value e from the cells via *amember* (unless the set is empty). This value e is then held by *Sched_non_empty* ready to surrender to a *choose* from outside. Because e is re-evaluated after every communication from outside, we are guaranteed a fair and correct choice from amongst the cells. For example, after an *insert(v)* *Scheduler* re-chooses from the cells, one of which now contains v. After a *delete(v)* *Scheduler* re-chooses from the cells, none of which now contain v.

For an example, we depict in Figure 9 the evolution of *SET_des* during the sequence of actions

$$insert(apple).insert(pear).\overline{choose}(pear).delete(pear)$$

The following points should be noted when studying this example:

- For simplicity we have not shown all the ports, only those which are relevant at each stage.

- A bracketed value in a cell indicates the value stored by that cell.

- The joining of two ports by a solid line indicates the occurrence of a communication.

- There are other possible derivations at the point where the *amember* communications occur.

517

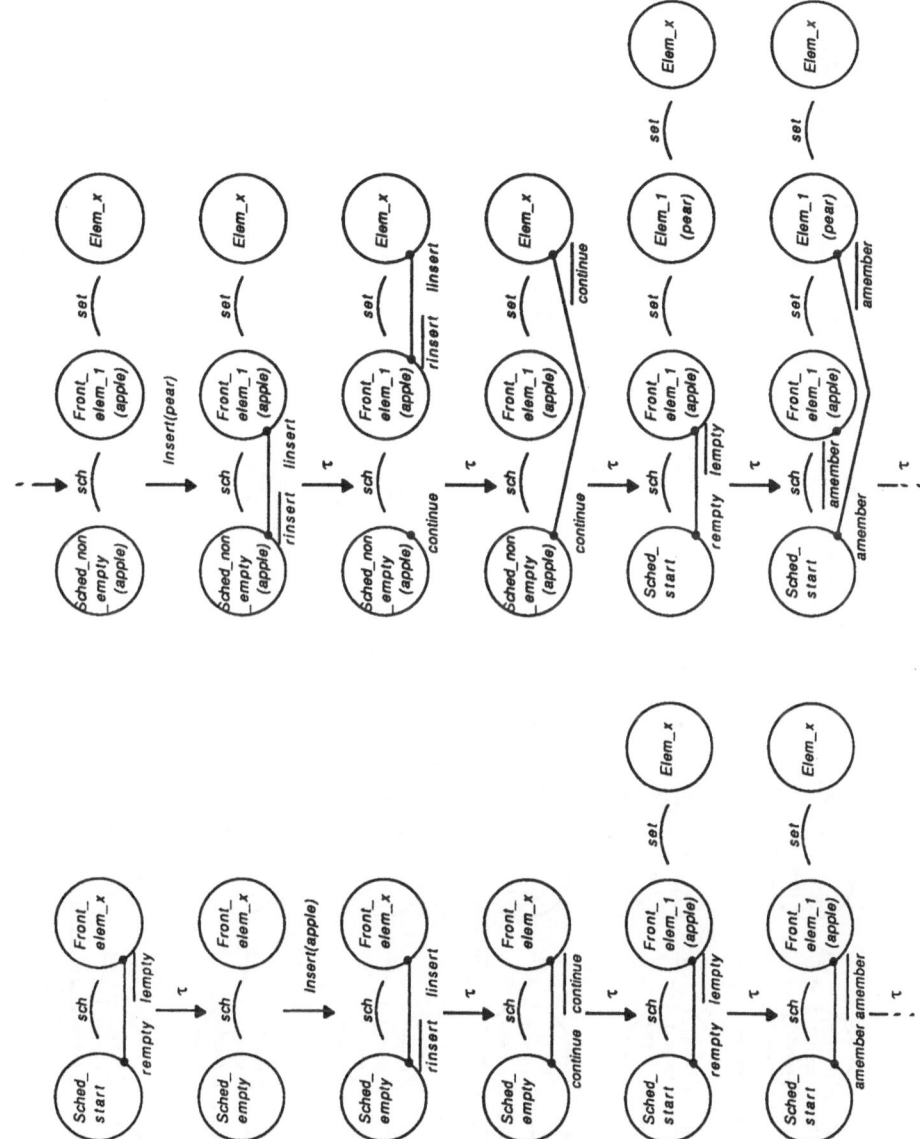

Figure 9: The evolution of *SET_des* during the sequence *insert*(apple). *insert*(pear). $\overline{choose}$(pear). *delete*(pear).

518

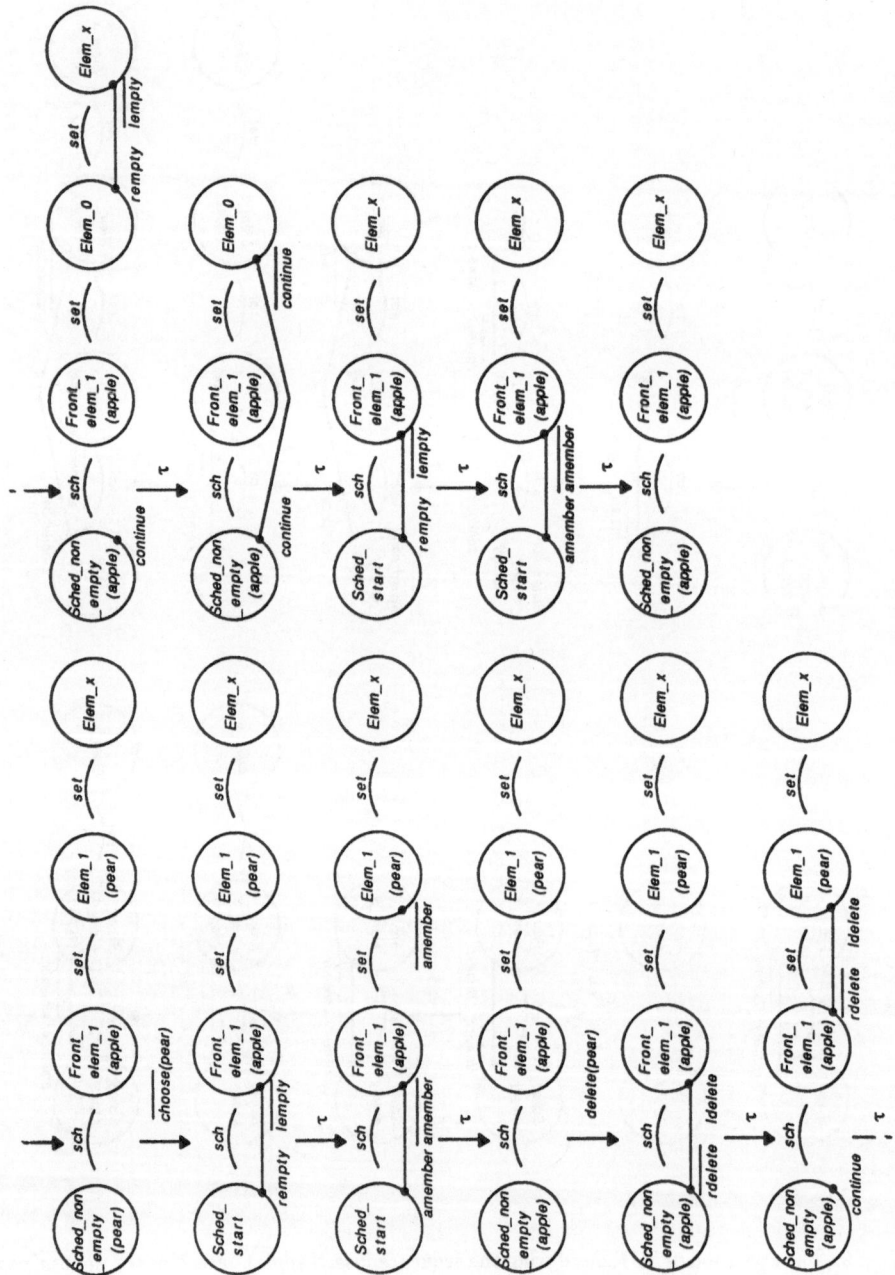

Figure 9 continued.

- It is not the same *Elem_x* which moves right and left as values are added and deleted. This effect is created by a cell being created or dying respectively (and another cell changing its state).

A PFL implementation of *SET_des* is given in Appendix A.

6 Discussion

Firstly, let us return to the issue of translating the ADT *Set* into the CCS definition of *SET* (definitions 1 to 3). What makes us choose a particular expression as the CCS counterpart of a nondeterministic ADT? Ideally we would like to be able to say that we choose the CCS definition which is the most accurate reproduction of the ADT in terms of semantics. But such a statement clearly requires a formal definition of *most accurate reproduction*. We shall not attempt such a definition since it would require a mapping between the semantics of the languages that is beyond the scope of this document. Instead, we suggest that *SET* is the most accurate reproduction of *Set* in an intuitive sense and present justification for this claim.

Let us consider briefly the nature of the two languages. The semantics of an ADT is algebraic, so that *all* the objects in the ADT world are values of various sorts. But in CCS there are agents and there are actions, which are very different classes of object from each other. If we then allow parameterised ports, we can also regard the agents as operating on data — values of various sorts. Since this is typically how CCS is viewed then we can say that this language has the classes of object: agent, action and value. These classes of object do not all share the same privileges. For example, agents cannot be passed as values in communications between ports. As a result of these and other differences, specification in the two languages has a completely different flavour.

When we translate a data type into CCS, we get a data object ie. an agent whose state we can manipulate by performing various actions upon it. The possible states of the agent correspond to the values of the data type, the actions correspond to the operations and the changes in state correspond to the results of the operations.

The translation from *Set* to *SET* can be thought of as having two stages: translation of syntax and translation of semantics. Firstly, we want to decide what actions are to be offered by *SET* in correspondence to the operations of the ADT — translation of syntax [5] . We can do this by considering the signatures of the operations in *Set*. We notice that the operations can be divided into categories based upon their signatures:

1. The category of basic constructors: those operations with no arguments and which return a value of sort *set*. The only basic constructor is *EMPTYSET*.

2. Non-basic constructors: those operations with signature *elem set* $\rightarrow$ *set*. These operations are *INSERT* and *DELETE*.

3. Basic accessors: those operations which take an argument of sort *set* and produce a result which is of a different sort viz *ISEMPTY* and *CHOOSE*.

4. Non-basic accessors: those operations which take an argument of sort *set* *and* arguments of other sorts and return a value of a sort other than *set*. There is only one such operation viz *ISMEMBER*.

---

[5] But as we will see, the division of the translation process into syntactic and semantic stages is very loose.

Taking these four categories of operations in turn, we can decide what actions should be offered by *SET*.

EMPTYSET presents a problem because its rôle as the basic constructor of type *Set* is not needed in the CCS world where we are defining a data object. The data object *exists*, it does not have to be constructed. Anyway, it would not be possible to return an empty *SET* agent from some *emptyset* port since agents cannot be passed as values in CCS. So we conclude that *SET* will have no port corresponding to the *EMPTYSET* operation. We must therefore assume that in its initial state *SET* will be empty [6] .

INSERT and *DELETE* both have the signature *elem set → set*. Because we are manipulating a data object when communicating with *SET*, there will be no visible result from an action such as inserting or deleting. The result is a change in the internal state of the machine *SET*. Therefore, we can translate these operations into the single actions *insert(e:elem)* and *delete(e:elem)* respectively.

ISEMPTY and __CHOOSE__ both take a single argument of sort *set*. Because *SET* is an object, information about the current contents of *SET* is held in the state of the machine. Hence these operations can be translated into single actions which output values of the same sort as *ISEMPTY* and __CHOOSE__. Thus we have the actions $\overline{isempty}(b{:}bool)$ and $\overline{choose}(e{:}elem)$ respectively.

Finally, *ISMEMBER: elem set → bool*. What we really want from *SET* is an action $\overline{ismember}(e \in S)$, where *S* is the set corresponding to the current state of *SET* and *e* is the element that we wish to test for membership. But this action in isolation would be invalid because *e* needs to be bound to a value. We therefore need to prefix this action with an input action which accepts *e*. Thus *ISMEMBER* translates into the pair of actions *ismember_in(e:elem)*. $\overline{ismember\_out}(b{:}bool)$.

We conjecture that this translation of operations into ports produces a machine as close as possible to *Set* (in a syntactic sense) because the translation is a minimal one. By *minimal* we mean that the sort of *SET* consists of the least possible number of actions sufficient for *SET* to be considered a counterpart to *Set*. More precisely, the least possible number of actions sufficient to allow actions and state changes to represent basic and non-basic constructor operations and to allow actions to represent basic and non-basic accessors. The only room for disagreement is that *EMPTYSET* has no counterpart action. It might be suggested that what we require as a counterpart to *EMPTYSET* is a *clearset* port, which changes the state of *SET* to be empty. But this would correspond more closely to a *CLEARSET: set → set* operation in the ADT, defined as *CLEARSET(s) = EMPTYSET*. Furthermore, adding a *clearset* port would mean that the translation was no longer minimal in the sense described above.

There is obviously more to translating *Set* to *SET* than choosing the ports. But in making this choice we have given ourselves a framework for the specification. Equally importantly, in choosing the ports we have been considering semantic aspects of the translation process. For instance, we have already decided that in its initial state *SET* must be empty. The rest of the process of defining *SET*, ie. deciding the semantics for the machine, is based on an understanding of the meaning of *Set* and intuition about the way in which these semantics must be translated to produce a useful data object. Note that we translate the fact that __CHOOSE__*(EMPTYSET)* is undefined in *Set* by specifying that *SET'(∅)* does not offer a $\overline{choose}$ communication.

Secondly, it is worth discussing briefly the context of the specification and design exercise and the scope of usefulness for the product. As we mentioned in the introduction, *SET* forms part of a larger specification. The entire exercise was intended to produce an implementation of Martelli and Montanari's presentation of unification. We desired that the implementation should reflect

---

[6] We are already considering semantics here!

the structure of their presentation as closely as possible. Since the presentation uses a set, this was one of the components that had to be specified and designed. As we have seen, the design exercise was not straightforward and the final definition of *SET_des* could hardly be said to be simple. It is therefore important to note that we are not presenting *SET_des* for its utility as a re-usable software component. Neither are we suggesting that everyone who is designing from a specification which uses sets should automatically assume that *SET_des* will form part of the final design. Using CCS as a design tool can produce some very elegant solutions to certain problems (eg. the Weavesort and Palindrome Recogniser machines in [Hen84]), but assuming that *SET_des* would be included might preclude the discovery of such elegant solutions. Thus it should be a general rule when designing in CCS not to assume that suitable components of previous designs will be re-used. However, this is not to say that there are no situations in which *SET_des* will be re-usable: whenever we require a set to exist explicitly in the design we can use *SET_des* for that purpose.

7 Conclusions

We have presented the process of specifying and designing a data object representing a set, introducing the concepts of non-deterministic choice from a varying number of elements and of *link* operators used to build chains of communicating agents.

The most abstract notion of the desired object was given as an ADT with non-deterministic operations. We translated this ADT by an ad-hoc process into a CCS definition, *SET*, specifying the desired machine. Justification was given for the choice of CCS definition by categorising the ADT operations and showing that *SET* is a minimal translation of *Set*. It is tempting to think that we could produce general rules for translating ADTs into CCS using a complete categorisation of operations based on signatures. Rules for each category could specify the ports needed to correspond to an operation. Using the ADT to provide the types on which the CCS translation operates, as we have done to provide the specification *SET_spec*, may provide a basis for translating the semantics of the ADT into semantics for the CCS agent. We could then define formally what it meant for a CCS definition to be the *most accurate reproduction* of an ADT. Those interested in the matter of linking CCS and ADTs are referred to Udo Pletat's work in [Ple87] and to LOTOS [ISO87].

The example specification and design used to introduce *link* operators illustrates one of the advantages of being able to both specify and design in the same language. By a semi-formal proof technique we were able to show that the design did not satisfy the specification (they were not behaviourally equivalent). This was a relatively straightforward exercise because both the specification and design were in the same language. To prove the two definitions inequivalent we needed only show that one could exhibit a behaviour that was impossible for the other.

Ideally we would also like to show that our final design, *SET_des*, is equivalent to the initial specification *SET_spec*. If we were to restrict the values of the sort *elem* to a *reasonable* number then we could, in theory, use the bisimulation proof technique [Wal87] for this purpose. However, in practice such a proof would be a very extensive exercise. Furthermore, how would we know how many values in *elem* would be sufficient for a complete proof when *SET_des* might potentially be used with *elem* as an infinite sort? Instead, we satisfy ourselves with the fact that we have not managed to prove *SET_des* inequivalent to *SET_spec* and with sensible behaviour from the PFL implementation of *SET_des*.

Acknowledgements

I would like to thank Wilf Nichols and Dave Smith for many useful and interesting discussions during the preparation of this document.

The author is supported by an SERC Research Studentship.

References

[BG81]   R. M. Burstall and J. A. Goguen. Algebras, theories and freeness: An introduction for computer scientists. In *Proceedings of Marktoberdorf Summer School on Theoretical Foundations of Programming Methodology*, August 1981.

[GHM78] John V. Guttag, Ellis Horowitz, and David R. Musser. Abstract data types and software validation. *Communications of the ACM*, 21(12):1048–1064, December 1978.

[Gut77]   John Guttag. Abstract data types and the development of data structures. *Communications of the ACM*, 20(6):396–404, June 1977.

[Hen84]   Matthew Hennessy. Proving systolic systems correct. Technical Report CSR-162-84, Department of Computer Science, University of Edinburgh, The King's Buildings, Edinburgh EH9 3JZ, June 1984.

[Mil80]   Robin Milner. *A Calculus of Communicating Systems*. Springer-Verlag, 1980. LNCS 92.

[Mit86]   Kevin Mitchell. *Implementations of Process Synchronisation and their Analysis*. PhD thesis, Department of Computer Science, University of Edinburgh, The King's Buildings, Edinburgh EH9 3JZ, July 1986. Published as CST-38-86.

[MM82]   Alberto Martelli and Ugo Montanari. An efficient unification algorithm. *ACM Transactions on Programming Languages and Systems*, 4(2):258–282, 1982.

[ISO87]   International Standardization Organization. Lotos (formal description technique based on the temporal ordering of observational behaviour), September 1987. Draft International Standard ISO/DIS 8807.

[Ple87]   Udo Pletat. Algebraic specifications of abstract data types and ccs: An operational junction. In *Proceedings of the 6th IFIP Conference on Protocol Specification, Testing and Verification*, pages 361–372. Elsevier Science Publishers, 1987.

[Sub81]   P. A. Subrahmanyam. Nondeterminism in abstract data types. In S. Evan and O. Kariv, editors, *Automata, Languages and Programming*, pages 148–164. Springer-Verlag, 1981. LNCS 115.

[Wal87]   David Walker. Introduction to a calculus of communicating systems. Technical report, Laboratory for Foundations of Computer Science, Department of Computer Science, University of Edinburgh, The King's Buildings, Edinburgh EH9 3JZ, 1987.

A   PFL Implementation of *SET_des*

```
fun SET(insert,delete,choose,isempty,ismember_in,ismember_out) =

    let

        fun Scheduler(insert,delete,choose,isempty,ismember_in,ismember_out,
                      rinsert,rdelete,rempty,rismember,amember,continue) =
            let

                fun Sched_start() =
                    rempty inp b => (
                        if b
                        then Sched_empty()
                        else amember inp e => Sched_non_empty(e))

                and Sched_empty() =
                        (insert inp e => rinsert out e =>
                            continue inp _ => Sched_start())
                    + (delete inp e => Sched_empty())
                    + (isempty out true => Sched_empty())
                    + (ismember_in inp e =>
                            ismember_out out false => Sched_empty())

                and Sched_non_empty(v) =
                        (insert inp e => rinsert out e =>
                            continue inp _ => Sched_start())
                    + (delete inp e => rdelete out e =>
                            continue inp _ => Sched_start())
                    + (isempty out false => Sched_non_empty(v))
                    + (ismember_in inp e => rismember out e =>
                            continue inp _ => Sched_start())
                    + (choose out v => Sched_start())

            in
                Sched_start()
            end

        and Front_elem(linsert,ldelete,lempty,lismember,rinsert,suck,rdelete,
                       rempty,rismember,amember,ismember_out,continue) =
            let

                fun Front_elem_x() =
                        (let
                            port newrinsert and newsuck and newrdelete and
                                 newrempty and newrismember
                        in
                            linsert inp e => continue out () =>
```

```
                              (Front_elem_1(e) & Elem(rinsert,suck,
                              rdelete,rempty,rismember,newrinsert,
                              newsuck,newrdelete,newrempty,newrismember,
                              amember,ismember_out,continue))
                end)
            + (ldelete inp e => continue out () => Front_elem_x())
            + (lempty out true => Front_elem_x())
            + (lismember inp e => ismember_out out false =>
                  continue out () => Front_elem_x())


        and Front_elem_1(v) =
                (linsert inp e => (
                    if e = v
                    then (continue out () => Front_elem_1(v))
                    else (rinsert out e => Front_elem_1(v))))
            + (ldelete inp e => (
                    if e = v
                    then Front_elem_0()
                    else (rdelete out e => Front_elem_1(v))))
            + (lismember inp e => (
                    if e = v
                    then (ismember_out out true =>
                        continue out () => Front_elem_1(v))
                    else (rismember out e => Front_elem_1(v))))
            + (amember out v => Front_elem_1(v))
            + (lempty out false => Front_elem_1(v))


        and Front_elem_0() =
                (suck inp e => Front_elem_1(e))
            + (rempty inp _ => continue out () => Front_elem_x())

    in
        Front_elem_x()
    end

and Elem(linsert,passback,ldelete,lempty,lismember,rinsert,suck,
         rdelete,rempty,rismember,amember,ismember_out,continue) =
    let

        fun Elem_x() =
                (let
                    port newrinsert and newsuck and newrdelete and
                        newrempty and newrismember
                in
                    linsert inp e => continue out () =>
                        (Elem_1(e) & Elem(rinsert,suck,
                            rdelete,rempty,rismember,newrinsert,
                            newsuck,newrdelete,newrempty,newrismember,
```

```
                                        amember,ismember_out,continue))
                    end)
            + (ldelete inp e => continue out () => Elem_x())
            + (lempty out true => NIL)
            + (lismember inp e => ismember_out out false =>
                    continue out () => Elem_x())

        and Elem_1(v) =
                (linsert inp e => (
                    if e = v
                    then (continue out () => Elem_1(v))
                    else (rinsert out e => Elem_1(v))))
            + (ldelete inp e => (
                    if e = v
                    then Elem_0()
                    else (rdelete out e => Elem_1(v))))
            + (lismember inp e => (
                    if e = v
                    then (ismember_out out true =>
                            continue out () => Elem_1(v))
                    else (rismember out e => Elem_1(v))))
            + (amember out v => Elem_1(v))
            + (passback out v => Elem_0())

        and Elem_0() =
                (suck inp e => Elem_1(e))
            + (rempty inp _ => continue out () => Elem_x())

    in
        Elem_x()
    end

in

    let
        port linsert and ldelete and lempty and lismember and
                rinsert and suck and rdelete and rempty and
                rismember and amember and continue
    in
        Front_elem(linsert,ldelete,lempty,lismember,rinsert,suck,
                rdelete,rempty,rismember,amember,ismember_out,continue)
            &
        Scheduler(insert,delete,choose,isempty,ismember_in,ismember_out,
                linsert,ldelete,lempty,lismember,amember,continue)
    end

end;
```

A High-Level Petri Net Specification of the Cambridge Fast Ring M-Access Service

Jonathan Billington[*]
University of Cambridge
Computer Laboratory
New Museums Site
Pembroke Street
Cambridge CB2 3QG

June 8, 1988

Abstract

Numerical Petri Nets (NPNs), a high level inhibitor net, are used to characterise the Cambridge Fast Ring Hardware at a high level of abstraction. The NPN model describes the service provided to *users* of the hardware (stations, monitors, bridges and ring transmission plant), known as the M-Access Service. The model has been developed to formalise the M-Access service definition in order to remove ambiguities and as a basis for the development and verification of the protocols using the M-Access service.

Keywords

Formal specification, high-level Petri nets, local area networks, service specification, protocols.

---

[*]The author is supported by a Telecom Australia Postgraduate Scholarship and an ORS Award.

1 Introduction

The Cambridge Fast Ring (CFR) Networking System [1] consists of a cluster of CFRs interconnected by bridges. The CFR is a slotted ring designed during the early 1980s to provide a raw 100 MBs transmission speed and to substantially increase the bandwidth between point-to-point users. Hardware for the stations, the monitor and bridges for the Cambridge Fast Ring has recently been fabricated in VLSI. The hardware implements the low level protocols between the various distributed components. The task of designing a set of protocols above the basic hardware to provide application services is underway.

An initial draft of the protocol architectures for the CFR has been compiled in [2], where it is shown that different architectures can co-exist above the basic service provided by the CFR hardware. This service is known as the M-Access Service and has been defined in [3].

The following benefits would accrue from providing an appropriate formal description of the service.

- Ambiguities which arise in narrative descriptions would be removed.

- The formal model could be executed to investigate properties of the service, such as global sequences of service primitives.

- The model can provide the basis for the development or synthesis of protocols to be implemented above the M-Access Service, such as the Unison Data Link Protocol [4].

- The model provides the basis for the verification of such protocols once specified.

The purpose of this paper is to describe in some detail the way in which the M-Access Service may be specified, using Numerical Petri Nets (NPNs) [5] as the formal description technique. NPNs have been chosen because of their visual appeal and the availability of computer aided tools to analyse them [6]. The appendix gives a brief description of NPNs.

The paper is structured as follows. Section 2 describes the M-Access Service based on [3] and [1] and discusses some of the assumptions made about the operation of the CFR. Section 3 presents the NPN specifications and various specification issues are discussed in section 4. The final section provides some conclusions.

2 CFR M-Access Service

2.1 Features of M-Access

A draft description of the M-Access Service is given in [3]. The service provides an abstract view of the features of a ring cluster: a set of rings interconnected by gateways. From a *user's* perspective the ring cluster provides the following facilities:

- Error protected communications paths;

- Fixed slots of 32 octets in which to transmit packets;

- A routing mechanism by which slots can be routed to their destinations, given that a 16-Bit address is provided by the user.

- Two types of communication paths:

    1. Point-to-point, where the address indicates the particular destination; and

    2. Broadcast, where a special address (hex FFFF) indicates that the message is to be broadcast to all other stations on the ring cluster.

- Some buffering

- The communications path has the following characteristics

    1. Packets may be lost

    2. Packets may be duplicated (this is a rare event, but possible)

    3. Packets may not be sequenced (this can only happen in an interconnected ring cluster where there is the possibility of two (or more) paths from source to destination).

2.2 Service Primitives

In the style of Open Systems Interconnection, service primitives define the communication between the users and the provider of a communications service in terms of

- what is to be transferred across the interface - this is defined by a set of parameters associated with the service primitive and the service primitive type.

- the allowable sequences of service primitives both at a local interface and globally.

- the relationships between the service primitives at each of the interfaces.

The service specification is provided at an abstract level in order to avoid over specification of the interface between the user and provider.

In the M-Access Service, two service primitives are defined:

- M-DATA request

- M-DATA indication

2.2.1 M-DATA request

This primitive is invoked to initiate the sending of data from one service user to another service user (for point-to-point operation) or to all other service users (broadcast). The primitive therefore has 3 essential information types: the source address, the destination address and the data to be transferred, which are defined as associated parameters, using the following syntax:

$M-DATA\ request(source-CFR-address, destination-CFR-address, M-data)$

In [3], two further parameters are defined: *retry-control* and *transmit-status*. We shall model the retry-control parameter at a higher level of abstraction (perhaps more appropriate for a service specification as retry-control does not involve both users (or all users in broadcast mode)), where we will allow the service provider to choose any number of retries non-deterministically. Hence any number of retries (including zero) would be allowable in a realisation of the service, and user control of this number on a packet basis would also be a possibility.

We argue here that the *transmit status* parameter should be removed from the M-DATA request primitive, as the return of its value cannot be considered atomic with the transfer of data from user to provider. This is important from the point of view of defining sequences of service primitives. If a value of *transmit-status* needs to be determined before the M-DATA request can occur, then the corresponding M-DATA indication may have occurred before it! The present time-sequence diagrams quite rightly deny this possibility - so we have a problem. This may be solved by creating a separate TRANSMIT-STATUS indication primitive. This primitive will only occur at a local interface (no global significance) and is normally excluded from service definitions, however, it appears useful to include it in a simplified form as discussed below.

2.2.2 M-DATA indication

This primitive is invoked to receive data sent by a sending user. It complements the M-DATA request primitive. The receiving user has completed the receipt of all the data when the primitive occurs.

The primitive has the same set of parameters as the M-DATA request primitive.

$$M-DATA\ indication(source-CFR-address, destination-CFR-address, M-data)$$

These parameters have the same values as those in the corresponding M-DATA request primitive. (Note: we are not considering address mapping required in multi-cluster configurations in this report. They have also not been considered in [3].)

2.2.3 M-TOG indication

The CFR hardware has the capability of telling a user of the M-Access Service that a transmission of a packet has not succeeded (ie the number of retries has been exhausted). This signal is known as 'Thrown-on-Ground' or TOG for short. A TOG will normally indicate that the receiver is busy. The effect of the TOG is to discard the packet. The user then has the option of retrying the same packet, accepting the loss, or trying another packet to a different destination before retrying sometime later. Hence the TOG signal has important consequences for the way in which the user behaves. It appears to be useful at the service level to explicitly define a primitive to express this characteristic of the service, as a form of notification service. A possible parameterless primitive would be: M-TOG indication, indicating that the current packet is considered lost by the service provider.

2.3 Sequences of Service Primitives

2.3.1 Point-to-Point

In [3], the sequences are partially determined by the time-sequence diagrams of figures 3a) and 3c). The M-DATA indication either follows the corresponding M-DATA request, or it does not occur at all. What is not covered in [3] is a statement as to how much buffering will be allowed - ie how many M-DATA requests can occur, before a M-DATA indication must occur in the case where there is no loss? This is obviously implementation-dependent and an implementation-independent specification must allow for the choice to be arbitrary.

Other important points are that duplicates are possible (although rare) and that the medium does not preserve sequence (but only in CFR clusters where there are multiple paths between source and destination).

Now that we have introduced the M-TOG indication, its affect on the allowable sequences of primitives will need to be defined. This will be done in the NPN specification.

2.3.2 Broadcast

The sequences of primitives for a successful broadcast are partially given in figures 3b) and 3c) of [3]. There are a number of questions that need to be discussed here. The set of M-DATA indications that may arise from a broadcast must occur (if at all) after the originating M-DATA request. However, nothing is stated regarding the sequences of occurrences of the resulting M-DATA indications. Obviously this depends on the ring topology, the delays through the receiver and whether or not retries are allowed when broadcasting. Retries (and hence duplicates) can occur if the source station does not receive its own broadcast because it is busy receiving another packet for example. If no retries are allowed, then the medium cannot duplicate packets. Out of sequence messages are still a possibility. A reasonable abstraction may be to assume that the occurrences of M-DATA indications are not ordered across the different receiving stations. Although it is possible that M-DATA indications will occur in the order that stations are encountered on the ring, this cannot be guaranteed, due to packets being flow controlled across the M-Access interface.

Figure 3c) suggests that if any packet is lost, all are lost. It is possible however, that a number of stations could receive a packet, while others do not. A M-DATA indication need not occur due to the receiver being busy or because of a transmission error (or other reasons). It appears reasonable to abstract from the ring topology and assume that loss by a particular station is independent of its position on the ring, even though loss due to a corrupted packet would imply that all further destinations on the ring concerned would discard the packet (except in the unlikely event of a packet's CRC being made good due to noise). Because packets can be lost due to the receiver being busy, this position-independence assumption for loss is required as the most general case.

The use of M-TOG indication in broadcast mode is rather problematic. The broadcast protocol appears to work in the following way. A broadcast is initiated by a source by setting the destination address to all ones. If other stations on the CFR are willing to accept packets from the source of the broadcast, they will do so if they are not busy. On detecting that it is a broadcast packet, the CRC is not changed. If the CRC is bad, the packet is not received, but continues on its way around the ring. If the CRC is good the packet is received and sent on its way, again with the same CRC (good). If the sender receives back the broadcast packet with a good CRC, it notes that it is a broadcast packet and no retransmissions are initiated. Hence no TOG will occur. Of course, the broadcast may well have failed to be received by many stations that were busy at the time. That is hard luck, they only get one chance in this scenario and no M-TOG indication will occur.

It is possible, however, on a multislot ring, that the sender of the broadcast packet is busy receiving another packet when its broadcast packet returns. In this case, no signal is sent from the receiver to indicate that it was a broadcast packet from itself, and it is treated like a normal packet. The good CRC will indicate that the packet has not been received and should be retransmitted. This could cause a string of duplicate broadcast messages to be received until the retry limit is exceeded. At this stage a TOG signal occurs. The user will find it difficult to use the TOG in any sensible way - as its interpretation can be that all or none of the destinations (or anything in between) have received the packet and any number of duplicates. It appears to be advisable for the TOG signal to be ignored when broadcast mode is used. Hence, at the service level, the M-TOG indication will not occur.

2.4 List of Assumptions

This section summarises the assumptions that have been discussed above.

2.4.1 Point-to-point

1. A M-DATA request must have preceded the occurrence of a corresponding M-DATA indication.

2. The parameters associated with the M-DATA indication are identical to those of the corresponding M-DATA request.

3. Duplication is possible, but only if retries are allowed.

4. In ring clusters, sequencing is not maintained in general.

5. Single packet buffering occurs in the transmitter and receiver in the current CFR implementation. There will also be buffering in any gateways.

6. Loss of M-SDUs is possible and handled in two ways:

   - reported to a user in an M-TOG indication if a retry limit is exceeded; but
   - otherwise not reported to the user.

2.4.2 Broadcast

1. A M-DATA request must have preceded the occurrence of a corresponding M-DATA indication.

2. The parameters associated with the broadcast set of M-DATA indications are identical to those of the originating M-DATA request.

3. Loss of M-SDUs is possible. In general it does not depend on the position of the station on the ring. It is not reported to the user.

4. Duplication is probable on multislotted rings, if retries are allowed.

5. Occurrences of M-DATA indications are not ordered across destination stations.

6. In general misordering is possible.

7. Same buffering as for point-to-point.

8. M-TOG indications do not occur.

3   NPN Specification

In this section we will specify various characteristics of the M-Access Service using Numerical Petri Nets. The following aspects will be investigated for both Point-to-Point and Broadcast operation.

1. An arbitrary cluster of rings, with unlimited storage.

2. A single CFR with a single packet buffer for sending and a single packet buffer for receiving in each of its stations, corresponding to the CFR implementation.

3.1   General Comments

The aim of this section is to start with the most general M-Access Service that may be envisaged and then to refine it towards the actual Cambridge Fast Ring Architecture. It will be assumed that there can be an arbitrary number of stations communicating over the service, where the number is limited by the Source Address space.

3.2   Abstracting from CFR slots

At the top level of abstraction we shall assume that any station can access the CFR simultaneously with any other station. This implies no contention over slots. In a more detailed specification the slot contention could be modelled.

3.3   Abstracting from Ring Topology

We shall also abstract from the ring topology by assuming that the service is independent of the positions of the stations around the ring. For ring clusters we shall also assume that the service is independent of the ring to which a station is attached. For example, a station on ring 5 communicating with another station on ring 1 will be treated as identical to stations communicating on the same ring. This is the general case for the CFR and CFR clusters when considering possible global sequences of service primitives as there can be arbitrary delays caused by a receiving station being busy.

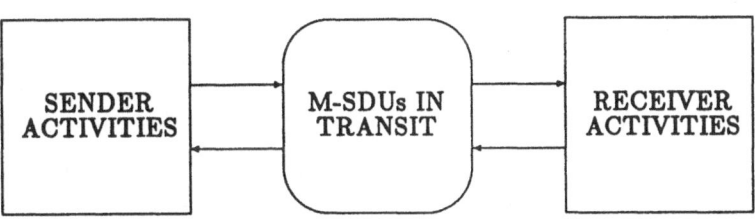

Figure 1: Means/Activity Net of CFR M-Access Service

3.4   Structure

It is assumed that the sending and receiving operations in each of the stations are identical and independent. We therefore only need to model a generic sender communicating over the ring with a generic receiver, each parameterised by the station address.

The general structure is given by Means/Activity nets. In Means/Activity nets, *activities* (actions) are represented by rectangles and *means* (resources) by ovals (or rectangles with rounded corners). An arrow from a *means* to an *activity* implies that the *means* is necessary for the *activity* to occur and arrow from an *activity* to a *means* implies that the *means* is modified by the *activity*, often by the production or consumption of a resource associated with the *means*.

The structure of the CFR M-Access Service is given in figure 1. "Sender Activities" and "Receiver Activities" involve the invocation of service primitives associated with sending and receiving data, respectively. The M-SDUs are the M-Access Service Data Units that are in transit from sender to receiver. It is assumed that M-SDUs are available for the sender and that they are forwarded on to the destination user.

3.5   NPN Specification: CFR Cluster

3.5.1   Implicit Interaction with Users

The Means/Activity Net of figure 1 may be refined into a general M-Access Service where the dynamics are specified by the Numerical Petri Net of figure 2. Implicit interaction with the M-Access Service Users is modelled by the occurrence of service primitives. Explicit interaction with M-Access Service Users is considered in the next section. Service primitives associated with sending are drawn on the left side of the diagram, those associated with describing the channel between the sender and receiver are drawn in the centre and those associated with receiving on the right.

The NPN comprises 4 transitions and one place. Three of the transitions model the three service primitives, while the forth models possible loss of M-SDUs that is not reported to a user. The place 'SP-storage' (Service Provider storage) models an unbounded number of buffers in the service provider and may be regarded as a queue with a non-deterministic service discipline. The place may store structured tokens which are triples. These tokens

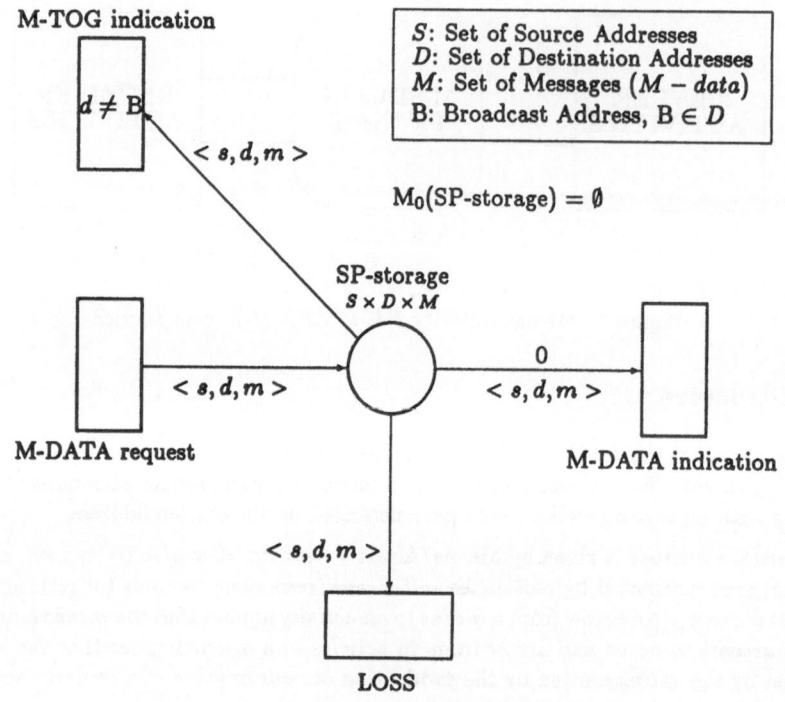

S: Set of Source Addresses
D: Set of Destination Addresses
M: Set of Messages $(M - data)$
B: Broadcast Address, $B \in D$

$M_0(\text{SP-storage}) = \emptyset$

Figure 2: Top Level NPN of CFR M-Access Service

represent the M-SDUs that are in transit from source to destination. The first variable of the triple represents the source address; the second, the destination address; and the third the data to be transferred.

The following properties of the service provider are modelled.

- Service Primitive Occurrences. The occurrences of the service primitives M-DATA request, M-DATA indication and M-TOG indication are modelled by the firing of the transitions labelled with the corresponding names. The station at which the service primitive occurs is determined by the address variables in the associated M-SDUs. The M-DATA request and M-TOG indication primitives occur at the *source* address of the associated M-SDU and similarly the M-DATA indication primitive occurs at the *destination* address of its associated M-SDU. The occurrence of the M-TOG indication indicates to the source user that the provider has discarded the M-SDU and believes that it has not been delivered to its destination.

- M-SDUs. SDUs are modelled as tokens placed in 'SP-storage'. The tokens are triples containing 3 variables: the source address, s whose domain is the source address space (integers 1 through 65534 for the CFR parameter Source-CFR-address); the destination address, d, which corresponds to the source address plus a special broadcast address, B, (the broadcast address is 65535 for the CFR) and the monitor address, M, (0 for the CFR) and; the data field, m, corresponding to the M-data parameter of the CFR, which is an arbitrary string of length 32 octets.

- Arbitrary Buffering within the service provider. In order to allow any amount of storage in the M-Access Service Provider, it is necessary to allow the place 'SP-storage' to be unbounded or to have a finite but indeterminate capacity. We have modelled the unbounded case here as it is (slightly) easier to represent. This allows a particular implementation to have any countable number of buffers and still conform to the Service Specification.

- Sequence: The place 'SP-storage', acting as a non-deterministic queue, allows arbitrary overtaking of an M-SDU by another M-SDU and hence models the "non-sequence preserving" nature of the service. Note that FIFO order is also a possibility.

- Arbitrary Loss: Loss not reported to the service user is modelled by the occurrence of the "LOSS" transition.

- Normal Transfer: M-SDUs are placed in 'SP-storage' on the occurrence of an M-DATA request. The choice of values for the parameters is chosen arbitrarily from the domains of the variables. Any number of M-DATA requests may occur, initiated by any station and destined for any station. So long as there is an M-SDU in 'SP-storage', an M-DATA indication may occur. The M-SDU is retained to allow for possible duplication (see below) or for broadcast. Normal transfer is modelled by the occurrence of the M-DATA indication transition followed by the occurrence of the LOSS transition for the same SDU, without any intervening occurrence of another M-DATA indication for the same SDU. The possibility exists for a destination not to receive M-SDUs from a source on the *hate list*. This corresponds to sequences in which M-DATA requests for a particular source-destination pair are always followed by either a LOSS event or a M-TOG indication, but not by a M-DATA indication.

- Duplication: Duplication is modelled by the occurrence of the M-DATA indication transition 2 or more times for the same M-SDU.

- Broadcast: This is indistinguishable from duplication except that the 'destination' address parameter must be the Broadcast address value, B. The individual destination address for each occurrence of the M-DATA indication for the broadcast is not known at this level of abstraction. The next section details how this information may be incorporated by a refinement of the specification in figure 2. Duplication of broadcast M-SDUs is allowed. If duplication is not intended (as is the case with the CFR) it may be removed in a further refinement as shown in a further section.

- M-TOG indication. The occurrence of this transition for a particular M-SDU prevents any further occurrences of the M-DATA indication for the same SDU (by removing it from the queue), and indicates to the source that the service provider believes (rightly or wrongly) that the M-SDU has been discarded. The main reason for this in the CFR is that the receiving station is busy and has refused to accept the M-SDU (on a number of occasions determined by the retry limit). We have deliberately forbidden the occurrence of an M-TOG indication for a broadcast M-SDU, by associating the condition $d \neq B$ with the corresponding transition. Of course, this restriction could be removed if it was felt useful for an M-TOG indication to be used for broadcast M-SDUs.

- Retry control. In this specification retry control is handled implicitly and non-deterministically. The retries are the mechanism for duplication. Duplication can only occur if retries are allowed, but it may not occur even if retries are allowed. As far as the occurrence of service primitives is concerned, the number of retries is not relevant, and is invisible to the users. One important factor to users is the number of duplicates and that they can be limited to zero by retry prevention. The present specification models arbitrary duplication. This is more general than the Cambridge Fast Ring, where the number of duplicates is bounded by the retry limit. A more detailed specification can be given to accommodate this limit by explicitly modelling a retry control parameter which passes the limit to the M-Access Service provider. At this stage the transmit side of the service specification is being refined to an interface specification.

- Quality of Service. Quality of service parameters have not been included in [3] and have thus been ignored in the current NPN specification. The Retry Control parameter may be considered as an implementation-dependent QOS parameter, as it affects a) the transfer delay, b) the probability of SDU loss and c) the probability of duplication. In a service specification it is important to abstract away from implementation choices. This is why the present service specification does not include the retry control parameter. It is considered inappropriate at the service level of specification.

- Ring Broken. It appears to be useful to include in the definition of the service a *Ring-Broken* primitive. This has not been modelled as again it does not form part of the M-Access Service definition in [3]. Given that the effect of a broken ring is to lose M-SDUs and possibly to allow for duplicates, the present specification does model this behaviour without the introduction of a specific primitive.

3.5.2 Explicit Interaction with Users

The specification of figure 3 shows how the service interacts with its users and also specifically indicates the destinations which receive M-SDUs as a result of a broadcast. Five places (and associated arcs) have been added, 3 for the source user and two for the destination user. It is quite arbitrary whether or not any, all or none of the destinations receive a broadcast M-SDU.

Each station's source has a set of messages stored in place 'Messages' which it wishes to transmit to any one of a set of destinations. (In the CFR, each message is restricted to an arbitrary string of 32 octets.) The source may also wish to broadcast the message. The Broadcast and destination addresses are stored in place 'Destinations'. When a M-DATA request occurs, an M-SDU is formed for the particular source-destination pair and the associated message and stored in the M-Access service provider. This M-SDU may now be lost (transition 'LOSS' occurs); discarded by the service provider while informing the source ('M-TOG indication' occurs); or it may be delivered to an allowed destination ('M-DATA indication' occurs). The M-TOG indication may not occur for a broadcast M-SDU. When it does occur, the discarded M-SDU is saved in the place 'Lost-SDUs'. Any number of M-DATA requests may occur concurrently from any number of stations.

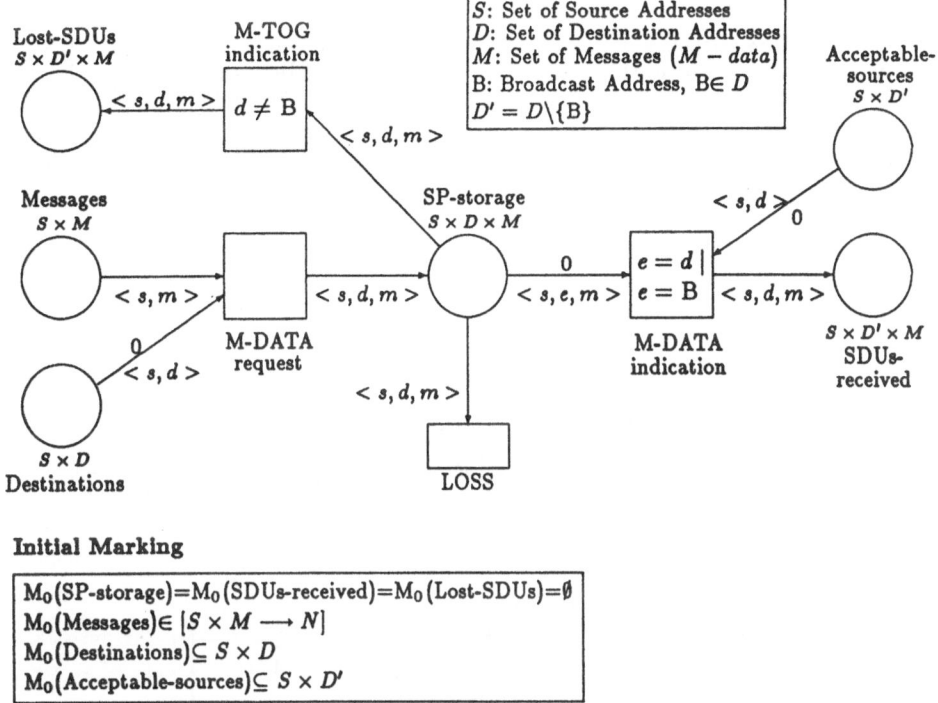

Figure 3: NPN of CFR M-Access Service: Explicit interaction with users

Each destination is prepared to receive messages from a set of sources (cf the 'hate list' and select register of the CFR). These are stored in the place 'Acceptable-sources'. If the source address of an M-SDU in 'SP-storage' is on the list of acceptable sources, the M-DATA indication may occur and the M-SDU is passed to the destination and stored in 'SDUs-received'. Duplication is allowed by the M-SDU remaining in 'SP-storage'. If it is a broadcast M-SDU, then the source address must still be acceptable to the destinations that receive it. Two points should be made regarding broadcast.

1. A destination station which receives the broadcast M-SDU is now identified.

2. Arbitrary duplication of broadcast M-SDUs is allowed to each destination that finds the source acceptable. If the specification is to be restricted to disallowing duplicates when broadcasting then a more complicated specification results. The details are presented in the next section.

3.5.3   No Duplication when Broadcasting

In this section a specification of the CFR M-Access service is developed where no duplication occurs in broadcast mode. We will not include interaction with the users explicitly, as this can be done in exactly the same manner as in the previous section.

We assume that the broadcast to each receiving station is not ordered and that any number

Initial Marking

M_0(M-SDU-reception)$=\emptyset$
M_0(In-transit-M-SDUs)$=\emptyset$

Definitions

S: Set of Source Addresses
D: Set of Destination Addresses
M: Set of Messages $(M - data)$
B: Broadcast Address, $B \in D$
$D' = D \backslash \{B\}$

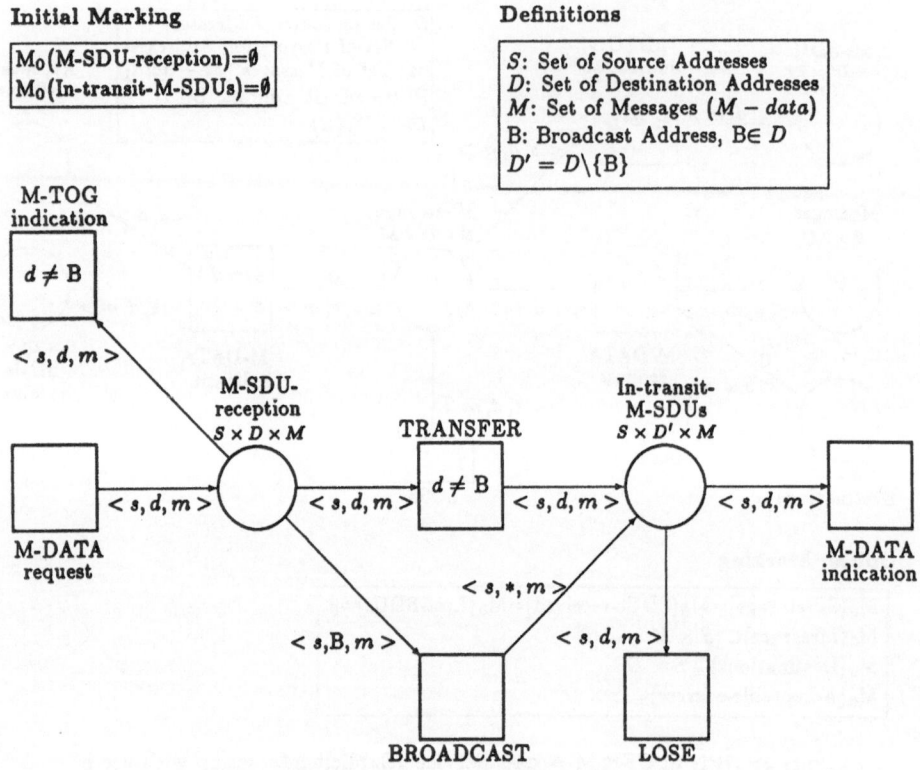

Figure 4: M-Access Service: No Duplication

Initial Marking

M_0(M-SDU-reception)=∅
M_0(In-transit-M-SDUs)=∅

Definitions

S: Set of Source Addresses
D: Set of Destination Addresses
M: Set of Messages ($M - data$)
B: Broadcast Address, B∈ D
$D' = D\backslash\{B\}$

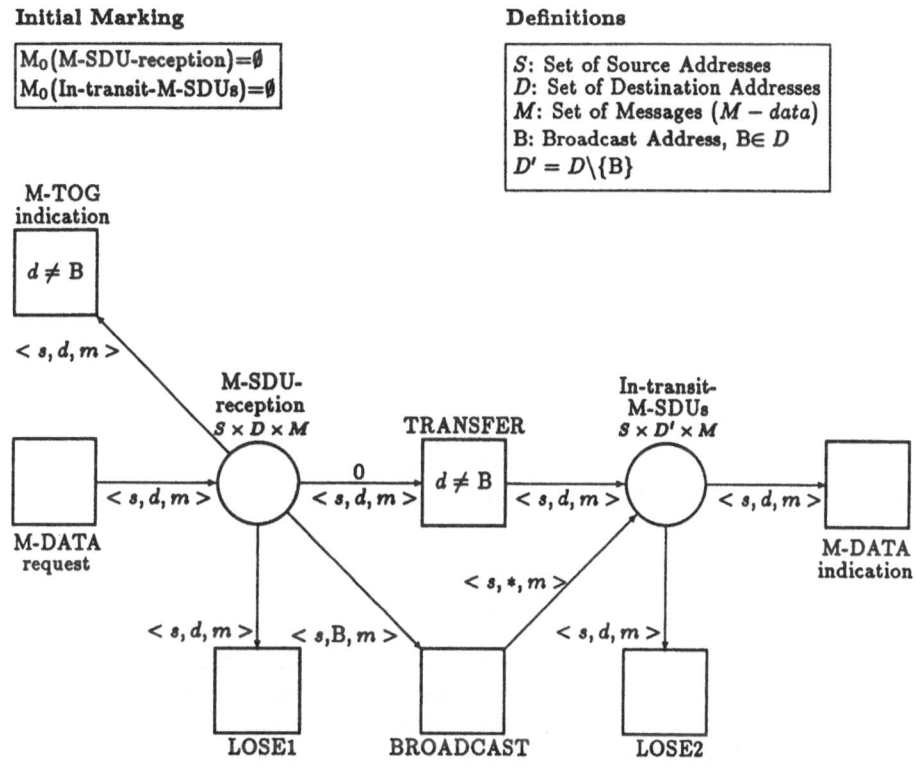

Figure 5: M-Access Service: No Duplication for Broadcast

of stations may not receive the broadcast. We firstly consider the simplest situation where there is no duplication for point-to-point. (This may be regarded as close to the initial expectation of the service to be provided by the CFR.)

The NPN specification is given in figure 4. It has been necessary to refine the service provider storage into two places: 'M-SDU-reception' which stores M-SDUs of the initial M-DATA request; and 'In-transit-M-SDUs' which stores all possible broadcast and point-to-point M-SDUs. (This has been done to ensure that M-TOG indications may only occur for point-to-point M-SDUs.) Point-to-point M-SDUs are simply transferred from 'M-SDU-reception' to 'In-transit-M-SDUs' by transition 'TRANSFER'. The transition 'BROADCAST' converts a broadcast M-SDU into a set of M-SDUs, one for each possible destination. Any M-SDU may be lost (transition 'LOSE') or successfully delivered to its destination ('M-DATA indication').

In order to allow for duplication of point-to-point M-SDUs, we retain a copy of the M-SDU in 'M-SDU-reception' and allow any number of duplicates by successive firing of 'TRANSFER'. An extra transition is included to allow the M-SDU to be removed from the provider. The situation is depicted in figure 5.

3.5.4 CFR M-TOG indications

The above specifications allow a M-TOG indication to occur any time after a M-DATA request has occurred, so long as the M-SDU remains in 'SP-storage' (figure 2) or 'M-SDU-reception' (figures 4 and 5). This allows any number of M-DATA requests to have occurred at a particular station, before an M-TOG indication occurs which relates to any one of the previous M-DATA requests.

This is more general than the situation which exists in the CFR implementation, where only single buffering is provided in each station for the transmission of M-SDUs. Thus after a M-DATA request, a M-TOG indication must occur before the next M-DATA request, if it occurs at all. In other words, for a particular CFR station, the M-TOG indication relates to the M-DATA request which immediately preceded it. Thus a strict order is imposed.

This may be specified in the NPNs of figures 4 and 5 by introducing a place (and associated arcs) which restricts the capacity of 'M-SDU-reception' to one M-SDU per source station. This construction is illustrated in the next section and is not repeated here.

In the original design of the CFR, the idea of double buffering (ie allowing two M-SDUs to be stored in the transmit and receive FIFO buffers) was considered. This would allow two M-DATA requests to have occurred before the M-TOG indication occurred for the first M-DATA request. This can also be modelled very simply by a change in the initial marking of the control place which determines the capacity of 'M-SDU-reception'. The initial marking represents the number of buffers available for each station's transmit FIFO. We can therefore allow a mixture of single and double buffering in different stations (or in general a mix of any number of buffers) by altering the initial marking of this control place.

3.6 NPN Specification: Single CFR

Each station in the CFR has two buffers: one for sending M-SDUs and the other for receiving M-SDUs. Each buffer has the capacity for just one M-SDU.

The more general case of unlimited storage was specified in the previous section. We may now refine the NPN of figure 2 to the specific case of single buffering for the CFR. In this section we shall only consider the case of implicit interaction with the users. Explicit interaction can be added trivially, in a similar way to that shown in figure 3.

We shall consider the following characteristics of the CFR

- Arbitrary number of stations

- Point-to-point and broadcast modes

- Single transmit buffer and single receive buffer for each station

- Sequence of M-SDUs preserved per source-destination flow

- Single broadcast by each station (only one broadcast per station is allowed at any one time due to the single transmit buffer)

- Arbitrary loss of M-SDUs

- Three modes of duplication:

  1. Arbitrary duplication in both point-to-point and broadcast mode;
  2. No duplication in broadcast mode, but arbitrary duplication for point-to-point operation; and
  3. No duplication

The duplication case 2 is close to the operation of the CFR, although duplication for point-to-point is rare and limited. A limit to the amount of duplication can be incorporated into the specification in a straightforward way if desired. (It requires an extra place to store the duplication limit for each station.)

We shall consider the three modes of duplication in separate NPN specifications. As usual, the left side of each diagram represents the transmitter and the right side the receiver. The transitions in the centre represent various ways in which the CFR can operate. We represent a set of transmit buffers, one for each station, by the single place 'Transmit-buffers' and we record the stations which have buffers that are empty in place 'Empty-transmit-buffers'. (This is the same as the control place for determining the capacity of M-SDU-reception mentioned above.) A similar situation exists for the receive buffers. We also include explicitly which stations are acceptable sources of M-SDUs for each of the destinations, by storing them in place 'Acceptable-sources' as we did in figure 3.

3.6.1 Arbitrary Duplication

The single CFR M-Access service with arbitrary duplication in both broadcast and point-to-point modes is specified by the NPN in figure 6. The initial state of the service is specified by the initial marking of the net. Each station connected to the CFR will have an empty buffer for transmitting and one for receiving and these are stored as tokens with a single variable that takes the value of the station address, in the places 'Empty-transmit-buffers' and 'Empty-receive-buffers' respectively. The addresses of the source stations acceptable to each destination are stored in place 'Acceptable-sources' as tokens which are pairs. The first variable stores the source address and the second the destination address. Initially all the transmit and receive buffers are empty (places 'Transmit-buffers' and 'Receive-buffers' are empty).

With this initial state, any number of stations may request the sending of an M-SDU. This is achieved by firing transition 'M-DATA request'. A token representing an M-SDU, (a triple consisting of the source address, destination address and the message contents) is placed in 'Transmit-buffers' and the token representing that the buffer was empty for that station is removed from 'Empty-transmit-buffers'. If the M-SDU is not broadcast, then one of three events may occur:

  1. The M-SDU is successfully transferred to the chosen destination. This may only occur if the source is acceptable to the destination. This is achieved in the NPN by firing transition 'TRANSFER'. A copy of the M-SDU is maintained in the transmit buffer while it is transferred to the destination's receive buffer which is removed from the list of empty buffers. The M-SDU may then be removed from the transmit-buffer which would then be marked free by the occurrence of transition 'LOSE'.

542

Definitions

S: Set of Source Addresses
D: Set of Destination Addresses
M: Set of Messages ($M - data$)
B: Broadcast Address, B$\in D$
M: Monitor Address, M$\in D$
$D' = D\backslash\{B\}$

Initial Marking

M_0(Transmit-buffers)=M_0(Receive-buffers)=$\emptyset$
M_0(Empty-transmit-buffers)$\subseteq S$
M_0(Empty-receive-buffers)=M_0(Empty-transmit-buffers)$\cup\{M\}$
M_0(Acceptable-sources)$\subseteq S \times D'$

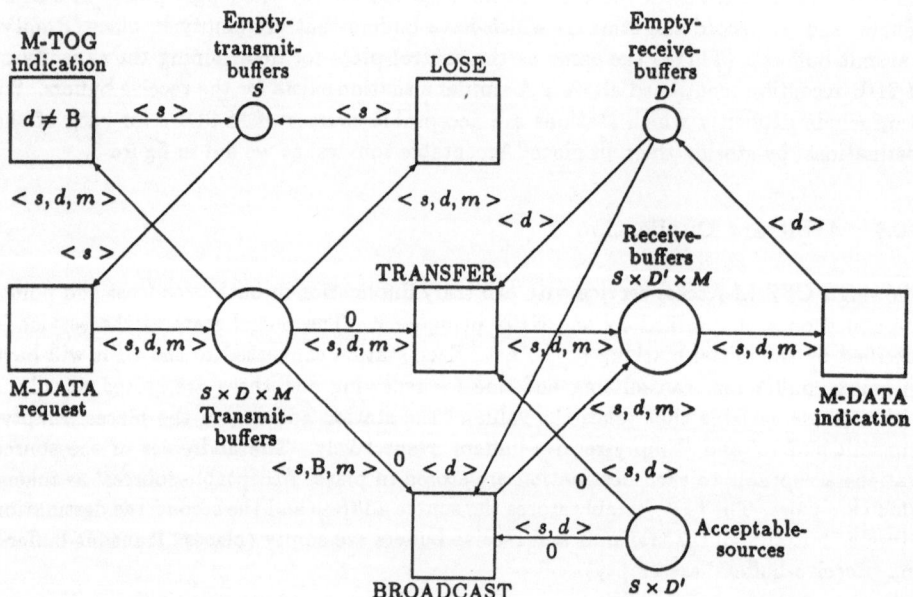

Figure 6: Single CFR M-Access Service: Duplication

Concurrently, an M-DATA indication may occur at the destination, with the M-SDU being removed from the receive-buffer which is marked free. This may be considered as the normal operation of the service. Duplication may occur by firing 'TRANSFER' twice (or more) before the occurrence of the 'LOSE' transition.

2. The M-SDU is refused by the destination and this is reported to the source user. This is achieved by firing 'M-TOG indication', which removes the M-SDU from the transmit buffer and marks it free.

3. The M-SDU is lost. The CFR transmitter hardware falsely believes that the M-SDU has been accepted by the destination, due to a CRC error in the return path. This is represented in the NPN by the firing of the 'LOSE' transition. The M-SDU is discarded and the transmit buffer marked free.

For broadcast M-SDUs, there are two possibilities.

1. The M-SDU is lost by firing transition 'LOSE'.

2. The M-SDU is broadcast one at a time to any of the allowable destinations by repetitively firing transition 'BROADCAST'. When this transition occurs, a copy of the M-SDU is retained in the transmit buffer, the M-SDU is transferred to an accepting destination and its buffer is removed from the empty list. An M-DATA indication may then occur with the consequent release of the receive buffer. This then allows duplication of the broadcast M-SDU, as the 'BROADCAST' may occur again for the same destination. It may also occur again for any other destination. The broadcast ends with the occurrence of the 'LOSE' transition, which empties the transmit buffer.

Before finishing this section, a comment is in order on transition folding. The specification of figure 6 could be made more compact by folding transitions 'TRANSFER' and 'BROADCAST' using the Transition Condition '$e = d \mid e = B$' and changing the Input Condition associated with the arc from place 'Transmit-buffers' to $< s, e, m >$ for the new transition. Exactly the same procedure has been followed in figure 3 (see transition 'M-DATA-indication'). We have chosen not to do so, in order that point-to-point and broadcast modes are clearly separated as this helps with the development of the specifications in the next two subsections.

3.6.2   No Duplication in Broadcast mode

In order to avoid duplication in broadcast mode we must keep a record of the stations to which we have broadcast. In a single CFR this is relatively easy as no simultaneous transmissions by a particular station are allowed due to single buffering. For each station, only a single point-to-point or broadcast transmission is possible and this must have completed (successfully or not) before the next transmission can occur. This allows us to use the list of allowed source-destination pairs stored in 'Acceptable-sources' to determine which station has received a broadcast M-SDU.

The NPN specification is shown in figure 7. It is the same as figure 6, except that an extra place, 'Broadcast-destinations', transition, 'LOSE2', and associated arcs have been added. Two further changes have been made. Firstly, the Transition Condition, $d \neq B$, has been

Definitions

S: Set of Source Addresses
D: Set of Destination Addresses
M: Set of Messages $(M - data)$
B: Broadcast Address, B$\in D$
M: Monitor Address, M$\in D$
$D' = D\backslash\{B\}$

Initial Marking

M_0(Transmit-buffers)=M_0(Receive-buffers) =M_0(Broadcast-destinations)=$\emptyset$
M_0(Empty-transmit-buffers)$\subseteq S$
M_0(Empty-receive-buffers)=M_0(Empty-transmit-buffers)$\cup\{M\}$
M_0(Acceptable-sources)$\subseteq S \times D'$

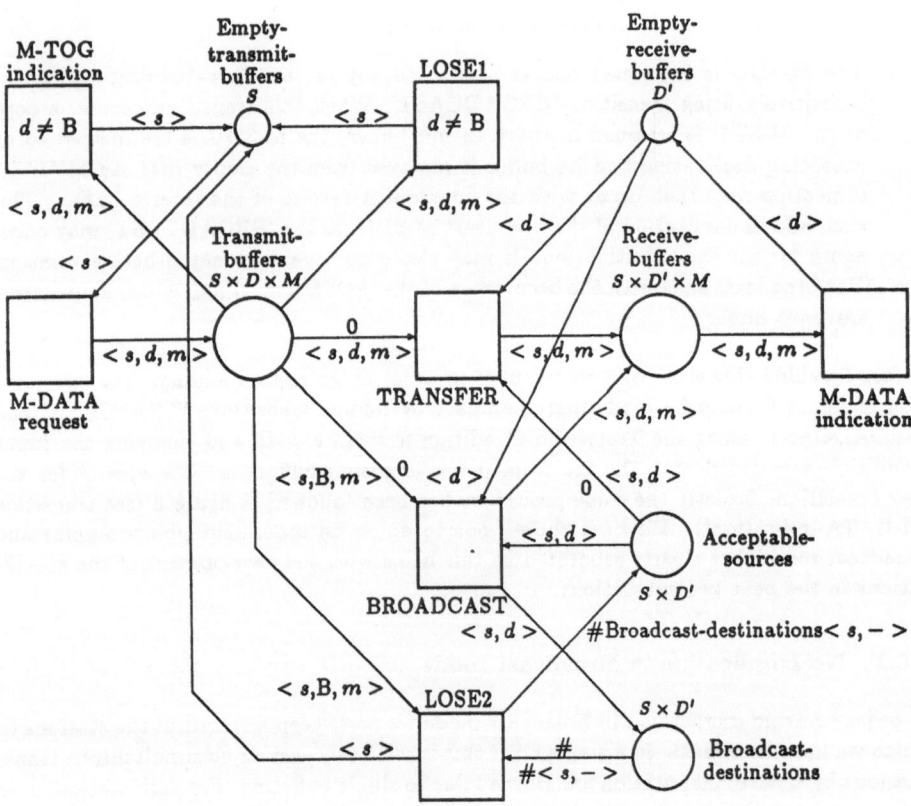

Figure 7: Single CFR M-Access Service: No Duplication for Broadcast M-SDUs

associated with the 'LOSE' transition and it has been renamed 'LOSE1'. 'LOSE1' may only lose point-to-point M-SDUs and 'LOSE2' may only lose broadcast M-SDUs. Secondly, the Destroyed Tokens inscription, '0', has been removed from the input arc from 'Acceptable-sources' to transition 'BROADCAST'. Thus when 'BROADCAST' fires, a destination which will accept M-SDUs from the broadcasting source, is removed from the list of accepting destinations stored in 'Acceptable-sources', and is placed in a list of destinations which have received a broadcast M-SDU. The list is stored in place 'Broadcast-destinations'. The broadcast will continue until the list of accepting destinations is exhausted (there will no longer be a source-destination pair token in 'Acceptable-sources' with the broadcast source address - hence 'BROADCAST' will not be enabled (for this source address) and the only remaining possibility for the broadcast M-SDU is that it is removed from the transmit buffer by firing 'LOSE2') or the M-SDU is lost by firing 'LOSE2'.

'LOSE2' is enabled by a broadcast M-SDU being in a transmit buffer. When it fires, the following actions occur atomically:

1. A particular source's broadcast M-SDU is removed from the transmit buffer;

2. The transmit buffer is marked empty (returned to 'Empty-transmit-buffers'); and

3. All tokens representing broadcast destinations that have successfully received the source's broadcast have been stored in 'Broadcast-destinations'. These tokens are removed and returned to 'Acceptable-sources'.

Of course, any number of stations could be active at the same time.

3.6.3 No Duplication

The single CFR M-Access service with no duplication at all is shown in figure 8. This differs from figure 7 in only two respects. Firstly, the Destroyed Tokens inscription, '0', on the input arc from 'Transmit-buffers' to 'TRANSFER' has been removed. Secondly, an arc from 'TRANSFER' to 'Empty-transmit-buffers' has been added, with a single variable Created Tokens inscription, which is of source address type. Now, when 'TRANSFER' fires, the M-SDU is removed from the transmit buffer and the buffer is marked empty. Hence no duplication can occur.

With the above changes, we have made the assumption that as far as users of the M-Access Service are concerned, the operation of delivery of an M-SDU to a receiving station and the freeing of the transmit buffer can be considered atomic for point-to-point operation.

3.7 Notification Service

The above specifications have included the M-TOG indication primitive as a form of notification service. Of course it is possible not to provide this service by not informing users when the transmit hardware believes that an M-SDU has been lost. This can easily be modelled by deleting the 'M-TOG indication' transition and its associated arcs from the above NPN specifications.

Definitions

S: Set of Source Addresses
D: Set of Destination Addresses
M: Set of Messages $(M - data)$
B: Broadcast Address, $B \in D$
M: Monitor Address, $M \in D$
$D' = D \backslash \{B\}$

Initial Marking

M_0(Transmit-buffers)$=M_0$(Receive-buffers) $=M_0$(Broadcast-destinations)$=\emptyset$
M_0(Empty-transmit-buffers)$\subseteq S$
M_0(Empty-receive-buffers)$=M_0$(Empty-transmit-buffers)$\cup\{M\}$
M_0(Acceptable-sources)$\subseteq S \times D'$

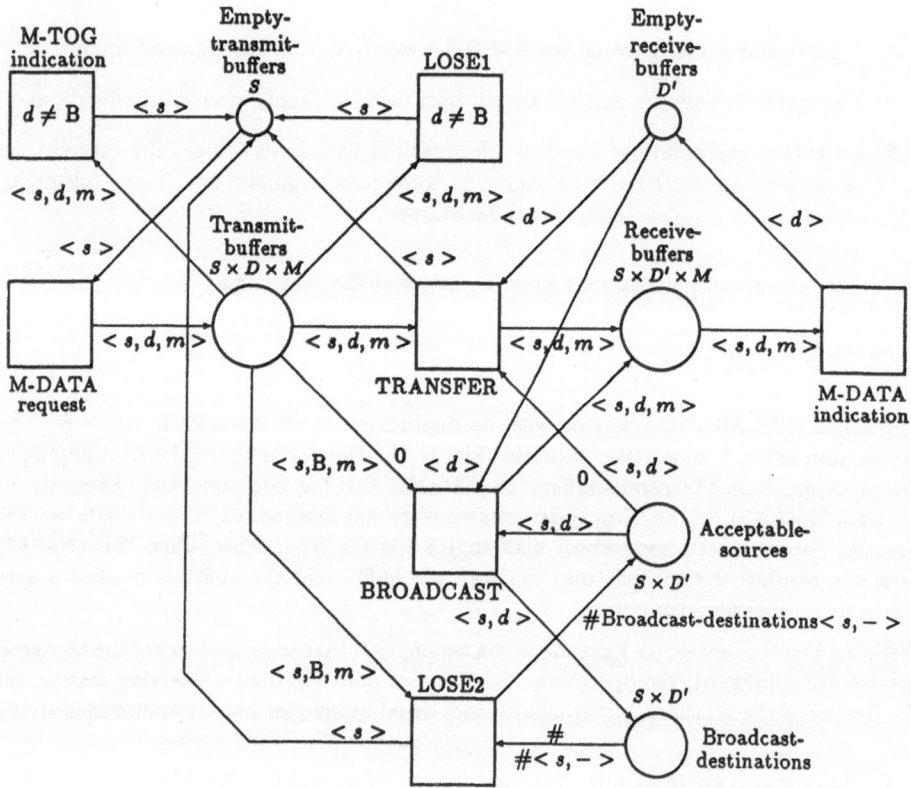

Figure 8: Single CFR M-Access Service: No Duplication

3.8 Comments on Ring Slots

The above NPN specifications model of the M-Access service for a single CFR is more general than that provided by the CFR hardware. It allows there to be simultaneous transmissions (point-to-point or broadcast) by all stations and arbitrary interleaving of all received M-SDUs. The CFR's hardware only permits there to be n simultaneous transmissions, where n is the number of slots on the ring. In most practical CFR installations, the number of stations will be greater than the number of slots.

There would be some point to a further refinement of the NPN specifications to include the slot structure if this was going to be part of a top down design using formal methods. This, however, is not the purpose of this work, which is to provide a formal basis for the development or verification of protocols being designed to operate over CFR systems. We claim that the present level of detail of the M-Access service is sufficient for this purpose.

4 Discussion of NPN Specification

There are a number of matters that require discussion regarding the NPN specification. These include: fairness; finite delay; mandatory sequences versus optional sequences and these are detailed below.

4.1 Finite Delay

The NPN specifications presented above say nothing about the time it takes before a transition fires after it is enabled - the enabling time. This is because nets have abstracted away from time. Hence the enabling time could be anything from zero to infinity. An important property that we would like to preserve in our models is that given a M-DATA request at some point in time we would like to make the temporal assertion that either an associated M-DATA indication or M-TOG indication or a LOSS event occurs some bounded time later.

For this to be the case in the net, we must ensure that a transition cannot be enabled indefinitely without firing. This is known as the finite delay property. Another way of looking at this is to consider only those sequences generated by the net where the stop state corresponds to all storage places (eg buffers for M-SDUs) are empty. For example in figures 2 and 3, the stop state is defined by $M_0(\text{SP-storage}) = \emptyset$. Hence for a particular M-SDU, the singleton sequence M-DATA request is excluded - it must be followed (at some stage) by one of the three other possible events mentioned above.

4.2 Progress Properties

Another desirable property of the service is that infinite sequences of events must include an M-DATA indication. On the other hand we are quite happy for infinite sequences to exclude the occurrence of either an M-TOG indication or a LOSS event or both. Thus we do not wish the service to be fair to events we would not consider useful.

We would like to guarantee some form of progress property. For example that there exists in every possible sequence, the subsequence "M-DATA request(u,s), M-DATA indication(u,d)" where u=(s,d,m) is an M-SDU comprising the source address, s, the destination address, d, and the M-data parameter, m; and the second parameter defines the station address at which the primitive occurs.

It appears that the CFR does not support such a progress property. For example, it is possible that every station switches its select register to "receive from nobody". In this case, it is not possible for a M-DATA indication to occur.

We may define a quasi-progress property as follows. 'No infinite sequence will contain an infinite subsequence of LOSS events'. This rules out the possibility of loss of M-SDUs occurring infinitely often.

This is probably true in a single CFR, as loss depends on the probability of a transmission error which is much less than one and hence the probability of an infinite repetition of loss events is zero. However, in ring clusters, M-SDUs may be lost for a number of other reasons. Consider the case when a station on one ring wishes to send M-SDUs to a station on another ring. The receiver may not accept an M-SDU for a number of reasons (eg the source not being selected) and if this is not recovered by retries (which is impossible if the source is not selected) then the M-SDU will be lost as no signal is passed back to the source for a M-TOG indication to occur. Thus an infinite loss sequence is possible.

The above NPN specifications allow the infinite loss case to occur. This appears to be an accurate description in the case of CFR clusters. In the case of a single CFR it may provide too general a model. To overcome this we could do one of two things:

- constrain the model to exclude the offending infinite sequences. This may be done by introducing an extra place to limit the number of LOSS transition occurrences to some finite number. Unfortunately this will increase the state space.

- eliminate the offending sequences when analysing the model.

4.3 Fairness

In the above section we have mentioned that we are quite happy for the service not to be fair to 'LOSS' events and 'M-TOG indications'. We would be delighted if these events never occurred. Another form of fairness that would probably be desirable is that the service should be fair to each of the stations. By this we mean that we want to disallow the behaviour where a set of stations can be locked out of communication with another station indefinitely by yet another station constantly gaining access to it.

The CFR allows a receiver to select a set of stations (the 'hate list') from which it will not accept M-SDUs, so in general it is not fair. However, given that a source station is not on the hate list, we would like to guarantee that eventually it will succeed. The problem is identical to that described in the previous section. We wish to guarantee that the subsequence 'M-DATA request((s,d,m),s), M-DATA indication((s,d,m),d)' occurs in a infinite sequence containing an infinite subsequence of 'M-DATA request((s,d,m),s)'s. Although allowing for this possibility, the NPN specification does not guarantee this behaviour.

4.4 Mandatory Sequences

When defining the service it is important to be able to state which sequences of service primitives are essential and which others are optional. More generally one needs to specify that one or more of a set of sequences is mandatory. Thus in order to conform to the M-Access Service, it is necessary that, there exists a sequence in which "M-DATA request((s,d,m),s), M-DATA indication((s,d,m),d)" is a subsequence. It is now debatable whether or not this should be universally quantified over all source-destination pairs. This is probably too strong, as there will be some destinations which do not want every other source to be able to send them data (cf the "hate list" in the CFR). However, it does seem reasonable to quantify over source addresses. Thus at the very least, each source must be able to send one M-SDU to one other station. On the other hand, it is obviously not mandatory for the service to include sequences which contain LOSS events. It is also necessary that the language of service primitives of the realisation of the service is a sublanguage of that defined in the service specification.

We could therefore consider figures of merit of conformance to a service specification. For example factors in a figure of merit would be the number of sequences that contained LOSS events, and the proportion of LOSS events in the sequence.

5 Conclusions

The service provided by clusters of Cambridge Fast Rings, known as the M-Access Service, has been characterised using a high-level Petri Net. The specification has been divided into a set of 'senders' (one for each station) and 'receivers' (one for each station and the monitor), communicating via a queue in the service provider.

An attempt has been made to clarify the present M-Access service definition and care has been taken to itemise the modelling assumptions.

The specification is presented at various levels of detail. In its most general form, the M-Access Service provider can re-order, duplicate or lose M-SDUs which can be transmitted either to a single destination or broadcast to all stations. This allows a very simple model of the behaviour using a high-level net consisting of just one place (representing a queue of arbitrary size and service discipline) and four transitions (3 representing service primitive occurrences and the forth representing loss of M-SDUs). This specification does not indicate which destinations receive broadcast M-SDUs, only that some broadcast M-SDUs may have been received. In this sense it is incomplete.

At the next level of detail, interaction with users is made explicit. In particular, the list of sources, with which each destination is prepared to communicate (realised in the 'hate list' and 'select register' of the CFR), is specified and this allows the destinations to which broadcasts are received to be detailed. This further detail comes at the expense of 5 extra places and associated arcs.

If the broadcast service is restricted to being duplicate free, then this can be modelled with the addition of one place and 2 transitions and associated arcs. The addition of a further transition allows the complete service to be duplicate free. This is also at the expense of a relatively complex inscription to describe sums of tokens.

A further refinement is presented where the service provided by just a single Cambridge

Fast Ring is modelled. In this specification, the sequence of M-SDUs is preserved and single buffers are modelled for transmitting and receiving M-SDUs (for each station). Duplication and loss are still possible. The list of acceptable sources is included in the specification. The service provider is conveniently modularised into service primitive actions and those associated with its internal operation on M-SDUs: loss; transference (originals or duplicates); and broadcast. Further refinements placing restrictions on the amount of duplication are also presented.

It is claimed that the models are relatively simple (half to one A4 sheet) and allow flow of data to be visualised by executing the net. Sequences of service primitives may also be generated. This allows considerable confidence to be gained in the veracity of the specification. The specification is also very general. The particular model developed here could easily be modified to represent an electronic mail service or connectionless network service for example. It also provides an adequate model for many local area networks of varying topologies (rings, buses, broadcast star networks (eg Hubnet)).

High-level nets allow very general specifications to be modelled quite simply. As these are restricted the models become more complex. The greater the degree of non-determinism and concurrency the simpler the net representation. This facilitates stepwise refinement from general specifications to more specific situations.

Limitations of the approach have been indicated. These concern the need to exclude unwanted infinite sequences and involve notions of fairness. This issue is a subject of research within the net community and elsewhere.

It has been stressed that conformance to a service specification needs to be specified in order to allow systems to be verified and the concept of a figure of merit for the conformance of protocols to service specifications has been canvassed.

Although only a particular application has been modelled in this paper, other experience indicates that the technique can be generally applied to the specification of protocol services, including connection-oriented, connectionless and N-way data transmission. For complex applications structuring specifications becomes important, a matter only briefly addressed in this paper. Hierarchical design using means/activity nets [7,8] and ideas from Harel's state charts [9] show promise for providing appropriate techniques.

6 Acknowledgements

David Tennenhouse suggested to me that it would be interesting to verify the CFR UDL protocol. This has led to the investigation of the M-Access Service described in this paper. Detailed discussions on the operation of the CFR with David and John Porter have been most useful.

The permission of the Director Research, Telecom Australia, to publish this paper is hereby acknowledged.

References

[1] Andy Hopper and Roger M. Needham. *The Cambridge Fast Ring Networking System (CFR)*. Technical Report 90, University of Cambridge Computer Laboratory, New

Museums Site, Pembroke Street, Cambridge CB2 3QG, England, June 1986.

[2] A. M. Chambers and D. L. Tennenhouse. Communications architectures for the Cambridge Fast Ring. October 1986. Draft Unison Project Document, Ref: UA004.

[3] A. M. Chambers. CFR M-Access service definition. November 1986. Draft Unison Project Document, Ref: UA008.

[4] D. L. Tennenhouse. The Unison Data Link protocol specification. September 1986. Draft Unison Project Document, Ref: UC022.

[5] G. R. Wheeler. *Numerical Petri Nets - A Definition*. Research Laboratories Report 7780, Telecom Australia, May 1985.

[6] Jonathan Billington, Geoffrey Wheeler, and Michael Wilbur-Ham. PROTEAN: a high-level Petri net tool for the specification and verification of communication protocols. *IEEE Transactions on Software Engineering, Special Issue on Tools for Computer Communication Systems*, SE-14(3):301–316, March 1988.

[7] W. Reisig. Petri nets in software engineering. In W. Brauer, W. Reisig, and G. Rozenberg, editors, *Petri Nets: Applications and Relationships to Other Models of Concurrency*, pages 63 – 96, Springer-Verlag, Berlin, 1987. Lecture Notes in Computer Science, Vol. 255.

[8] Australia. Reply to questionnaire on question 2/X. CCITT Study Group X, Question 2/X, Com X - No. 16, October 1985.

[9] David Harel. On visual formalisms. *Communications of the ACM*, 31(5):514–530, May 1988.

[10] James L. Peterson. *Petri Net Theory and the Modeling of Systems*. Prentice Hall, Englewood Cliffs, N.J., 1981.

[11] Wolfgang Reisig. *Petri Nets, An Introduction*. Volume 4 of *EATCS Monographs on Theoretical Computer Science*, Springer-Verlag, Berlin, 1985.

[12] E. Best and C. Fernandez. *Notations and Terminology on Petri Net Theory*. Arbeitspapiere 195, GMD, January 1986.

[13] W. Reisig. Place/Transition Systems. In W. Brauer, W. Reisig, and G. Rozenberg, editors, *Petri Nets: Central Models and Their Properties*, pages 117 – 141, Springer-Verlag, Berlin, February 1987. Lecture Notes in Computer Science, Vol. 254.

[14] Hartmann J. Genrich and Kurt Lautenbach. System modelling with high-level Petri nets. *Theoretical Computer Science*, 13:109–136, 1981.

[15] G.R. Wheeler. July 1987. Private Communication.

A Appendix: Numerical Petri Nets

This appendix provides an introduction to NPNs, and a subset of notation sufficient for understanding the specification of the CFR. The complete definition of NPNs can be found in [5]. The reader is assumed to have a knowledge of Petri Nets [10,11] or Place/Transition Systems as defined in [12,13].

A.1 Extensions

Numerical Petri Nets are Place/Transition (P/T) Systems with the following extensions.

- Tokens have been generalised to tuples of variables (cf Predicate/Transition (PrT) Nets [14]). This allows the convenient modelling of the parameters associated with service primitives, and in particular M-SDUs.

- A set of data variables is associated with the net. Only very simple types (integer, modulo, boolean, enumerated, strings) are presently considered. A data variable can always be represented by a place, with appropriate input and output arcs and a token carrying its present value. It is introduced purely for modelling convenience for objects such as counters. (This feature is not used in the CFR specification).

- There are two different types of place capacity. The first, K, sets a bound on the number of tokens of a particular value that can be resident in a place (this is the same as for PrT systems). The second, K*, sets a bound on the total number of tokens allowed in a place. (This feature is not used and needs to be generalised to be more useful).

- Three inscriptions are associated with the arcs of the underlying net.

    1. An Input Condition (IC) is inscribed to the left of a transition's input arc, as seen by an observer at the transition. It defines a condition which may be satisfied by a collection of tokens in the associated input place.

    2. The Destroyed Tokens (DT) are inscribed to the right of each input arc, as seen by our observer. It defines a bag (multiset) of tokens, which is removed from the associated input place (by bag subtraction), when the transition fires.

    3. The Created Tokens (CT) are inscribed next to each output arc of a transition. It defines a bag of tokens, which is deposited into the associated output place, when the transition fires.

- There are two optional inscriptions associated with each transition.

    1. A Transition Condition (TC), written next to or inside the associated transition. It defines a condition on net data variables and the variables associated with tokens residing in the transition's input places.

    2. A Transition Operation (TO), written next to or inside its transition. It defines an operation on the data variables. (This feature is not used in the CFR specification).

- An initial Marking (Mo), defining the initial allocation of tokens to each of the places in the net, and the initial value of each of the data variables.

A.2 A Generic Example

An example which illustrates the extensions is shown in figure 9. Places and transitions are given names which are strings of alphanumeric characters commencing with a letter.

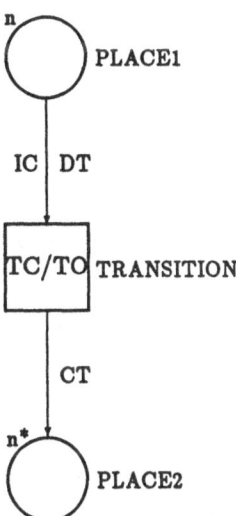

Figure 9: NPN illustrating graphical net elements and generic inscriptions

Places are represented by ellipses (usually circles) and transitions by rectangles or bars. Place capacities are represented by integers written next to the place (eg n or n^*). Note that the underlying transitions, places and arcs constitute a directed net.

A.3 Marking

An NPN marking is the net marking (the bags of tokens associated with each place) together with a vector consisting of the value of each of the data variables. The net marking is restricted by the capacities of the places, as is the vector by the type of each variable. The NPN marking may be thought of as the global state of a distributed system.

A.4 Enabling

A transition is enabled when

- the Input Condition is true for each of its input places

- its Transition Condition is true

- if fired, the capacities of its output places will not be exceeded.

A.5 Transition Firing Rule

When a transition fires (occurs), the net marking is changed by the following actions which occur indivisibly and concurrently.

- for each input place, its Destroyed Tokens are removed (bag subtraction)

- for each output place, its Created Tokens are added (bag addition)

- the Transition Operation is performed if it exists.

A.6   Net Execution

An NPN is defined with an initial marking. For this initial marking a set (possibly empty) of transitions will be enabled. (Note that a single transition may be enabled a number of times, if a number of different values of variables satisfy the ICs and TC.) An arbitrary choice is made as to which transition occurs (and which set of allowed variable values are bound). This occurrence generates a new marking, with a new set of enabled transitions. The net can be executed further, generating a set of reachable markings. The complete set of markings that can be generated this way is known as the Reachability Set. These markings are related to one another by a set of transition firing (occurrence) sequences. A directed graph which relates the set of markings (nodes of the graph) to transition occurrences (edges of the graph) is called a Reachability Graph.

A.6.1   Binding Variables

Free variables (those associated with tokens) may be part of the ICs, DTs, CTs, TC and TO specifications associated with a transition. The scope of these variables is restricted to the transition concerned. When the transition fires, the variables are bound to a particular value via consistent substitution.

A.7   Notation

The following presents the subset of NPN notation used in specifying the CFR M-Access Service. An extension to the notation is defined for bag sums.

In the following: tok is a token; $M(p)$ is the bag of tokens in place p; and $mult(x, M(p))$ is the multiplicity of token x in the bag of tokens $M(p)$ and $\emptyset$ denotes the empty bag.

A.7.1   Tokens

In general tokens are tuples of variables/constants separated by commas and enclosed in angular brackets. Variable names are strings of alphanumeric characters commencing with a letter. Some examples are $< 7 >$, $< red, yellow >$ and $< x, y, z >$. Alphanumeric strings may be variables or constants, the context making it clear.

A.7.2   Input Conditions

$IC(p, t)$	CONDITION ON INPUT PLACE MARKING, $M(p)$
tok	$tok \in M(p)$
0	$M(p) = \emptyset$
#	$\top$ (always true)

A.7.3 Destroyed Tokens

DT(p,t)	DESTROYED TOKENS BAG $D(p)$
tok	$\{tok\}$
0	$\emptyset$
#	$M(p)$
'blank'	the "enabling" tokens, $E(p)$

The "enabling" tokens are defined as follows for each input condition.

IC(p,t)	ENABLING TOKENS BAG $E(p)$
tok	$\{tok\}$
0	$\emptyset$
#	$M(p)$

A.7.4 Created Tokens

CT(t,p)	CREATED TOKENS BAG $C(p)$
tok	$\{tok\}$

A.7.5 Transition Condition

Transition Conditions comprise formulae with the usual logical connectives ('&' for and; '|' for inclusive or), negation operator ('~' for not) and two place predicates ($<, \leq, =, \neq, \geq, >$). They may involve variables, natural number expressions and string expressions.

A.7.6 Additional Notation

Bag Sums

With broadcast protocols and services we need a convenient notation to describe bag sums over the domains of the token variables. We have two cases of interest:

- A partition of the domain of the marking of a place, which is a set of sets; and

- A partition of the place marking (in general a set of bags).

In the first case a notation has been defined in [5]. We modify it to use * (instead of *underscore*) to indicate variables over which sums are made.

A notation for the second case above is not described in [5]. The need for the notation was raised by the author and the present form of the syntax is a modified and generalised version of that suggested by Geoff Wheeler [15].

Consider a place, p, with arity n_p, a set of domains of variables $\{V_1, \ldots, V_{n_p}\}$ and Marking $M(p)$ given by the multirelation

$$M(p) : V_1(p) \times \cdots \times V_{n_p}(p) \longrightarrow N$$

Let $v_i \in V_i(p)\ \forall i \in \{1, 2, \ldots, n_p\}$, then the notation N$(p)$ and its meaning (the bag $B(p)$) is defined in the following table.

$N(p)$	BAG $B(p)$
$< v_1, \ldots, v_{i-1}, *, v_{i+1}, \ldots, v_{n_p} >$	$\bigcup_{v_i \in V_i(p)} \{< v_1, \ldots, v_i, \ldots, v_{n_p} >\}$
$\#p < v_1, \ldots, v_{i-1}, -, v_{i+1}, \ldots, v_{n_p} >$	$\sum_{v_i \in V_i(p)} \{m_{v_i} < v_1, \ldots, v_i, \ldots, v_{n_p} >\}$

where $m_{v_i} = mult(< v_1 \ldots, v_i, \ldots, v_{n_p} >, M(p))$. When $M(p)$ is a relation rather than a multirelation, the bag sum reduces to set union.

These may be used in DT and CT inscriptions, for example $DT(s,t) = N(p)$, but when $s \neq p$, we must ensure that the domains of $M(p)$ and $M(s)$ are the same. An abbreviated form of the second notation can be used when the bag sum desired is associated with the same place to which the DT or CT refers, ie when $s = p$. In this case we may drop the p from the start of the notation.

$$\# < v_1, \ldots, v_{i-1}, -, v_{i+1}, \ldots, v_{n_p} > \equiv \#p < v_1, \ldots, v_{i-1}, -, v_{i+1}, \ldots, v_{n_p} >$$

Examples of the notation can be found in the CFR specification. An example of the abbreviated notation and its meaning is as follows.

$DT(p,t)$	DESTROYED TOKENS BAG $D(p)$
$\# < s, - >$	$\bigcup_{i \in D'} \{< s, i >\} \cap M(Broadcast - destinations)$, with $s \in S$

where $M(Broadcast - destinations) \subseteq S \times D'$.

Optional Place Inscription

In this report we have included for the first time a place inscription which explicitly states the domains of the token variables ie the domain of the marking of the place. The appropriate cartesian product is written next to the place.

MODELLING OF DISTRIBUTED PROBLEM SOLVING USING LOGIC MODIFIED PETRI NETS

Antonella Di Stefano, Fabio Gibilisco and Orazio Mirabella

Instituto di Informatica e Telecomunicazioni,

Facoltà di Ingegneria, Università di Catania,

Viale A. Doria 6, 95125 Catania, ITALY

Abstract

Distributed Problem Solving (DPS) combines aspects of distributed processing networks and logical inferential strategies. The design of DPS systems requires a formal model to represent distributed knowledge and inherent communication protocols. This paper presents a proposal that logic modified Petri Nets be used to represent distributed knowledge evolution. Some simple examples will be used to discuss the proposed model and to indicate the features which render it suitable for DPS representation.

1. Introduction

Distributed Problem Solving (DPS) combines aspects of distributed processing networks and logical inferential strategies. A DPS network consists of several intelligent nodes cooperating to solve a single problem. These nodes share:

- knowledge

- inferential resources.

Generally a DPS system can be regarded as a network of expert machines each having a significant expertise for solving partial problems. More precisely, the expert entities which cooperate in problem solving can receive not only the solution request but also added knowledge temporarily increasing the inferential capabilities of the DPS system.

A fundamental requirement for DPS is the development of communication protocols able to handle knowledge [SMIT] [DURF] [HALP] [LESS]. Current cooperation strategies for DPS are based upon three concepts:

1)   Inferential task decomposition:

each node of the network is viewed as expert in a subgoal of the main problem. Therefore the problem to be solved is distributed over different sites, each with partial competence.

2)   Decentralized global coherence:

a decentralized coordination policy of network knowledge provides effective cooperation between nodes.

3)   Functionally accurate cooperation:

each node has expertise to solve an incomplete and inconsistent portion of each problem. The global solution is thus obtained by iteration.

The design of concrete DPS strategies according to these (or similar) principles requires a formal model, derivation of which is best approached by analysis and evaluation of existing systems.

A model for DPS must be able to represent not only knowledge evolution, but also concurrency and nondeterminism which characterise distributed systems.

Among models developed for specifying cooperating processes, Petri Nets (PNs) are suitable for the description of concurrency and nondeterminism. This paper proposes a modification to PNs to extend their application to knowledge evolution. It will be shown how both knowledge and its inference are well suited to representation by PNs. This modelling is based on fitting the PN operation to the inferential environment.

Section 2 of the paper presents logic modified PNs (LPNs), while in section 3 DPS networks are modelled by them. Finally, in section 4, some simple examples are given which illustrate features of the proposed model.

2.   PNs for Knowledge Representation

2.1 PNs

A simple PN is a 4-tuple:A = <P,T,pre,post>

where

P is the finite set of places,

T is the set of transitions,

pre: PxT -> {0,1} is the input function,

post: PxT -> {0,1} is the output function.

A marking can be defined for a PN, as a function M: P -> {0,1}. A PN can be represented by a graph having two types of nodes, places (circles) and transitions (rectangles), and directed arcs, pre and post; the markers are black points inside places.

Usually, the places represent the system state variables, while the transitions allow switching from one state, characterized by a place marking, to a second state.

A transition can fire when its input places are marked and more than one transition can fire simultaneously.

The set R of reachable markings starting from an initial marking M is defined as:

$R(A,M_o) = \{M_i \in N^m : s\ M(s)\ \text{->}\ M_i\ \}$

where

A is a PN,

m is the cardinality of P,

M_o is the initial m-dimensional marking,

M_i is the final marking reached after the transition sequence s.

To derive reachability conditions an incidence matrix, C, can be defined starting from the pre and post functions.

Each element of C is given by

$C(i,k) = post(i,k) - pre(i,k)$ $i \in P,\ k \in T$

From C, the fundamental equation can be defined (in matrix form)

$M = M_i + C \cdot \bar{s}$

where

M_i is the final marking reachable starting from the marking M for the firing sequence s,

$\bar{s}$ is the commutative image, i.e. the array indexed by T whose components $\bar{s}(t)$ are the number of occurrences of each t in the sequence s, $t \in T$.

This fundamental equation allows us to demonstrate the following necessary condition for a reachable marking.

Theorem 1

M_1 is reachable starting from M   if $(M - M_1)$ is orthogonal to the solution set for $C^T y = O$.

Alternative theorems for marking reachability have been derived, for example that of minimal flow [BRAM].

A PN reachability analysis also allows one to prove when a certain system configuration is not reachable starting from defined initial conditions.

These characteristics render PNs well suited to transfer, with appropriate modifications, to problem solving strategies, as will be shown below.

2.2 Representation of Knowledge by LPNs

Several methods of knowledge representation are currently available, logic, semantic networks, frame and rule based systems [KOWA] [JOHN] [MINS] being the most common. We choose to use formal logic notation, and more precisely we will refer to first order logic.

Such an instrument, based on the known connectives $\frown$ (not), $\wedge$ (and), $\vee$ (or), <- (implication) and <-> (equivalence), and the two qualifiers $\exists$ (exists) and $\forall$ (for all), allows one formally to define and solve a large number of problems.

One important benefit is that any formula of the predicate calculus can always be converted into a particular form, called the clausal form, expressed generally by:

$$A \lor B \lor C \ldots \ <\text{-} \ K \land L \land M \ldots$$

where A, B, C, K, L, M are knowledge and $\lor$, $\land$, <- are clausal connectives.

Recently many tools have been developed for a particular subclass of the clauses, the Horn clauses (e.g. PROLOG [CLOC]), which are expressed as:

$$A \ <\text{-} \ K \land L \land M \ldots$$

These clauses can be modelled on PNs suitably modified as follows:

- Each place is a knowledge element.

- The transitions represent inferences.

In particular, the marking of a place represents the assertion of the corresponding knowledge. Transition firing produces a new assertion of post-knowledge (implied knowledge) starting from pre-knowledge (conditions) previously asserted.

Elimination of existing knowledge upon transition firing must be avoided; i.e. loss of marker on precondition places. This can be accomplished if, when a transition fires, the markers are not taken off. This is equivalent to saying that output arcs exist which mark the input places of the transition. These will be omitted in the graph but they must always be borne in mind.

According to the proposed definition, a problem described by the Horn clause

$$A \ <\text{-} \ K \land L \land M$$

can be represented by the graph in Fig. 1, which shows that the solution of A is derived (since reachable) from K, L and M.

Fig. 1.

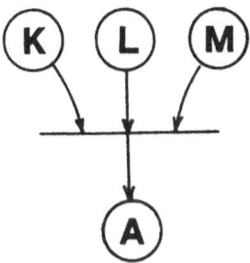

Fig. 2.

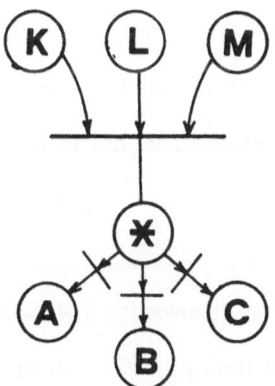

The general case of a clause

$A \lor B \lor C \; <- \; K \land L \land M$

can be represented by the graph in Fig. 2, where the place * allows resolution of

the nondeterminism typical of operator, by the structural conflict between the places

A, B and C.

Therefore the problem solution lies in reaching knowledge starting from a marker which represents acquired knowledge. In this way the solution of a problem can be viewed as the capability to reach one or more places which represent the assertion of the required fact. If the place is not reachable starting from the given initial conditions, the fact will be considered false.

According to this model, the assertion of one or more facts (which depend upon well-defined preconditions) is equivalent to demonstrating the reachability of one or more places, starting from initial markers which represent the basic knowledge and the previously stated facts.

In this way Theorem 1 allows us to determine deductively the unreachability of required knowledge when the initial hypotheses are not true. But, in converse, if such conditions are verified, the answer to the inference request is "maybe". This means that the required knowledge may be reached, but to be sure of this the PN must be explored.

Using PNs we can perform a fast "knowledge reachability" test since several PN implementations (suitable for real time process control) exist whose transition firing strategy provides a fast network evolution.

PNs inherently allow representation of different knowledge view-depths; in fact each place can represent other subnets. This means that a Knowledge Base (KB) can be represented by a PN whose complexity depends on the view-depth.

3. Modelling DPS by LPNs

Pieces of knowledge and the relationships between them can be distributed on several nodes, each one offering different inferential power about generally different knowledge domains. In this way a DPS network can be modelled by different LPNs which act in cooperation to reach the common aim.

In the proposed approach, the knowledge, represented by LPNs, is distributed on separate nodes. Each one is qualified by a competence domain representing the problem class which can be solved by its inferential capability. Thus, a node which has been requested to solve a problem, may not be able to do so due either to its limited KB or to the need for specific competences from different domains.

It is therefore reasonable to consider that several nodes must be involved in an inferential sequence and must communicate to achieve the problem solution. This cooperation can be obtained in different ways depending on the adopted strategy. In terms of LPN modelling this means that access to specific knowledge can be obtained by more than one transition sequence. These ones present transitions relevant to different intelligent nodes that can be redundant (1).

---

(1) Redundancy analysis of particular transition sequences on different nodes allows us to determine those parameters which identify the global safety of the distributed KB.

A simple example will clarify such assertions. Let us consider a network made up of three nodes X, Y, Z (each one operating in a particular knowledge domain) whose mutual relations are expressed by the LPNs shown in Fig. 3.

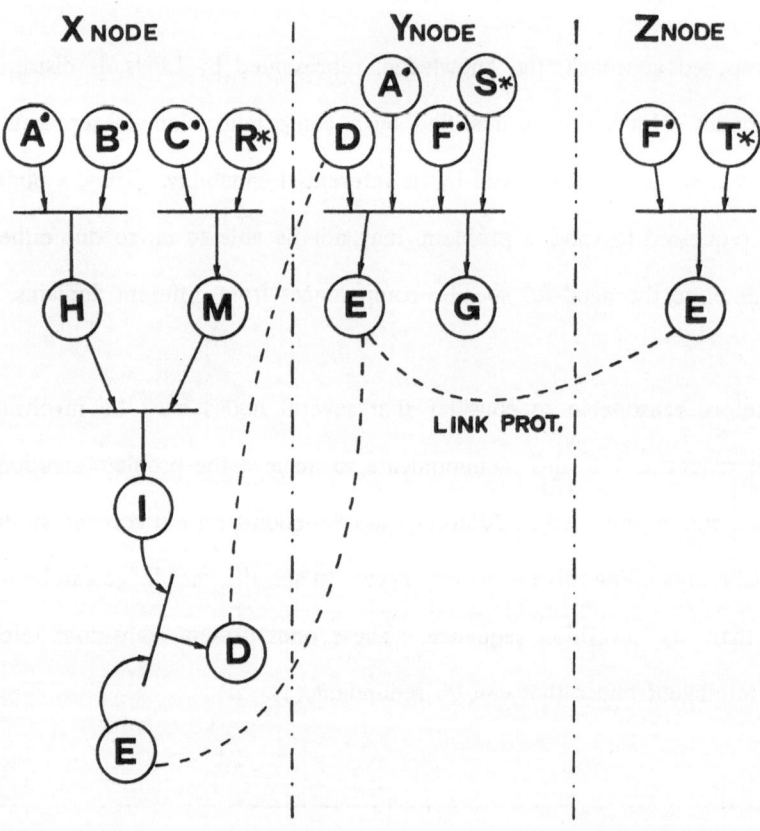

Fig. 3. A distributed LPN based upon three nodes X, Y and Z.

Let us suppose we wish to infer G, starting from actual base knowledge {R* , S* , T*} and from hypotheses {A, B, C, F}. A single node would be unable to respond to the request due to the unreachability of place G starting from the assigned pre-conditions.

In such a case the problem solution requires cooperation between the several existing nodes. Communication protocol management is then needed to link (according to a suitable policy) the knowledge at the various nodes. In this way a chaining between individual KBs is obtained, and the G assertion can be viewed as a G place reachability problem in the global LPN made up by the link. A new reachability analysis of the LPN allows us to assert G by means of a knowledge communication protocol which provides a correct and congruent link between different KBs.

A knowledge communication protocol must possess the following features:

- It must allow fast and sure knowledge exchange, thus presenting a single virtual KB to the entire network.

- It must allow questions pertinent to the particular knowledge domain of each node to be exchanged.

- It must allow (on request) the exchange of newly inferred facts and of new required facts.

- It must offer a suitable distributed inference strategy.

- It must perform a systematic analysis of the inferential process and both detect and solve (if possible) knowledge deadlocks between the different network nodes.

Distributed KB modelling by LPNs allows us to obtain an inferential progression as the possible reachability of new knowledge-places starting from the original ones. If the inferential progression stops because a necessary node cannot be reached, new nodes will be involved in the inference in order to find a new path allowing us to reach the "blocked" knowledge.

Three main DPS cooperation strategies are used today, and these will be examined in the following section together with some LPN modelling examples.

4. A Communication Protocol for DPS

As previously stated, a DPS communication protocol provides a virtual link between KBs resident in different network nodes, according to a suitable strategy for inferential task interaction and subdivision. To this end three main strategies exist: inferential task decomposition, decentralized global coherence and functionally accurate cooperation.

4.1 Inferential Task Decomposition

The global problem is split into sub-problems each of which is assigned to a single host. A typical example is the "Metaphoric Scientific Community" [KORN] where each node is an expert on a particular scientific field (see Fig. 4). In this way a problem can be formalized by several LPNs which model the knowledge domains inherent to different specializations. A coordinator M will present suitable sub-problems to each expert and will ultimately try to merge all the collected results with the aim of solving the complete problem. Each LPN allows us to infer the specialized sub-tasks if a transition sequence exists for reaching the several sub-goals.

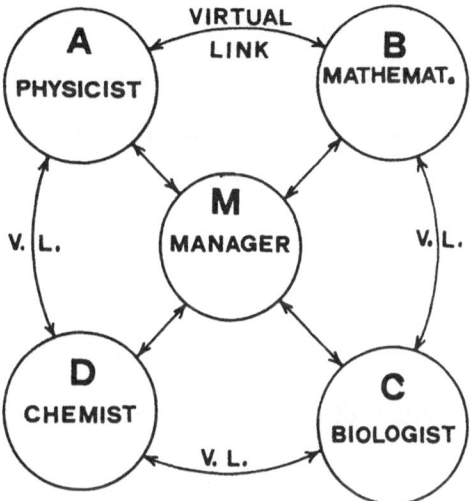

Fig. 4. Metaphoric scientific community.

The required communication protocol will be characterized by a knowledge of the expertise domain distribution and by the presence of question/answer exchange mechanisms for each expert.

4.2 Decentralized Global Coherence

The function of manager is inherent to the strategy of network cooperation. Moreover it must allow an effective global coherence.

In addition to the features of the first strategy, the protocol should permit a suitable exchange of knowledge acquired during the inference. In this case particular attention must be given to the degree of total coverage offered by all the KBs of the network in order to solve the problem. This coverage, cannot generally be assured a priori.

4.3 Functionally Accurate Cooperation

Each node represents expertise in several knowledge domains, but is generally unable to solve a single sub-problem alone. Therefore a node will be required to solve an incomplete and inconsistent portion of the global problem which contributes to an iterative solution.

In this case the communication protocol, in addition to providing rules for exchange of queries, answers and partial facts, must also manage the iterative process by which the problem will be solved.

A detailed example based on DPS, analogous to that of Fig. 3, follows. As previously stated, the assertion of G by LPN modelling is equivalent to affirmation of the reachability of the place G starting from the knowledge of the work hypotheses.

The network user presents an answer possibly relevant to all the nodes. Fig. 5 shows the sequence describing the protocol. As can be seen, the messages exchanged between nodes are: questions on inherent knowledge and facts already asserted during the inference evolution.

We now analyze the evolution of a simple example in order to define the main functions and tools that a knowledge protocol should offer.

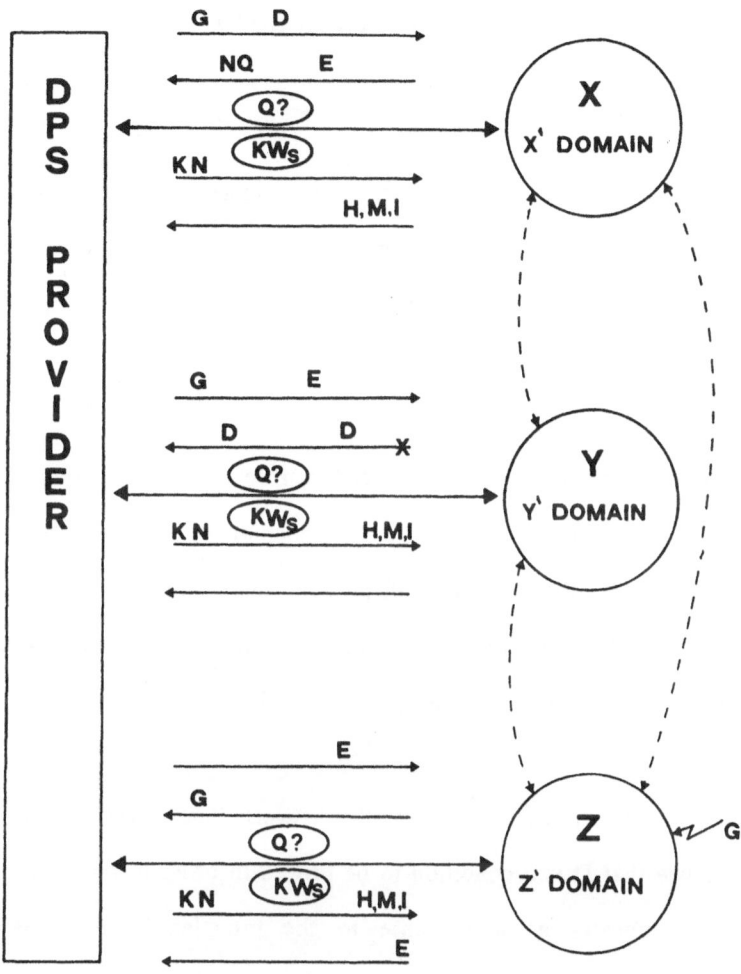

Fig. 5. An example of a functionally accurate cooperation protocol.

Under the hypothesis that the query G is presented at node Z, it, being unqualified to determine the answer, questions the network by sending a query "?G" to the DPS provider.

The DPS provider proposes the query to all the nodes which can contribute to inferring G according to their competence domain. If the network offers no competence on the property domain of G the query is rejected and the user receives a negative response to the problem.

But, in the proposed example, the query "?G" can be sent to the nodes X and Y which are competent in the domain of G. Consequently X will try to assert G and will obviously fail; a negative response will then be sent to the provider. For Y, the reachability analysis also gives negative results, since it is not possible to obtain G starting from the initial markers (A, F, S*).

However, by an automatic examination of the network (backtracking strategy), it is possible to deduce that D is a condition to be proved in order to reach G. For this reason, Y communicates as a response to the provider, a new query "?D". Recursively, the provider proposes the new query "?D" to the network. In particular "?D" is queried to X, which cannot affirm D since E is not marked.

Therefore a new query, "?E", is proposed to the entire network and the response by Y is a new request "?D". This occurrence represents a typical knowledge deadlock.

The recognition by the provider of such a condition is fundamental to the system. It initiates the alternative policy for problem solving. The block can be overcome if another node gives a response about blocked knowledge. In the above example the provider overcomes the blocking of E through the KB of Z which is able to affirm E. The assertion of E allows D to be reached, after which Y can infer G.

This example demonstrates that the provider must be able to produce a query and receive a response in a short time; moreover it must be able to transfer asserted facts during the inferential iteration. In addition, the provider must keep track of current inferential sequences in order to identify deadlock and, where possible, activate a recovery strategy.

Conclusions

The proposed approach to modelling of distributed KBs by LPNs is a first contribution to representation of DPS. It offers the advantages of transforming the problem of distributed inference into the reachability iteration of a PN. Moreover the use of LPNs as the basis for DPS development allows us, where appropriate, to exploit automatic tools already in existence for traditional PNs.

References

[BRAM]   G.W. Brams, "Reseaux de Petri: Théorie et Pratique", ESI, Paris, 1983.

[CLOC]   W.F. Clocksin and C.S. Mellish, "Programming in Prolog", Springer, Berlin, 1984.

[DURF]   E.H. Durfee, V.R. Lesser and D.D. Corkill, "Increasing Coherence in a Distributed Problem Solving Network", 9th Int. Joint Conf. on AI, Los Angeles, Aug. 1985.

574

[HALP]   J.Y. Halpern and Y. Moses, "Knowledge and Common
            Knowledge in a Distributed Environment", IBM Research
            Report, 1984.

[JOHN]   P.N. Johnson-Laird et al, "Only Connections: A Critique
            of Semantic Network", Psychological Bulletin 96, 1984.

[KORN]   W.A. Kornfeld and C.E. Hewitt, "The Scientific Community
            Metaphor", IEEE Trans. Sys. Man. Cybernet, Smc-11 1981.

[KOWA]   R. Kowalski, "Predicate Logic as a Programming
            Language", IFIP 74, North-Holland, Amsterdam, 1974.

[LESS]   E.H. Durfee, V.R. Lesser and D.D. Corkill, "Towards
            Coherent Cooperation in a Distributed Problem Solving
            Network", DAI, Pitman, London, 1987.

[MINS]   M. Minsky, "A Framework for Representing Knowledge",
            in P.H. Winston (ed.)   The Psychology of Computer
            Vision, McGraw-Hill, New York, 1975.

[SMIT]   R.G. Smith and R. Davis, "Framework for Cooperation in
            Distributed Problem Solving", IEEE Trans. Sys. Man.
            Cybernet, Jan. 1981.

An animator for CSP implemented in HOPE.

L.D. Natanson and W.B. Samson

Department of Mathematics and Computer Studies,
Dundee College of Technology,
Bell Street,
Dundee.

ABSTRACT

This paper describes the construction of an animator for
C.A.R. Hoare's Communicating Sequential Processes (CSP) in
the functional programming language HOPE. The animator
accepts declarations of CSP systems and generates HOPE
functions to represent them. An informal introduction to CSP
is given. Hoare's original LISP animations are discussed and
the evolution of the HOPE animator is traced. The development
of a HOPE language interpreter for the IBM PC is presented.
Possible uses of the animator in the areas of
verification and software metrics for concurrent systems are
discussed.

1. CSP : an introduction.

1.1 The perspective :

Hoare's Communicating Sequential Processes (CSP) [Hoare85] is
in part an attempt to throw light on and give fresh ideas to the
job of programming - in general. Concurrency and parallelism are
shown to be convenient notions to use in the decomposition and
modularisation of computer systems into subsystems.

The decomposition strategies Hoare promotes are to be seen to
have general applicability : subroutines (the oldest
modularisation technique) find a place in the CSP scheme. However,
the more novel notions of concurrency and process are useful in
the particular cases of computer systems that interact with their
environment. In these systems, conventional techniques, such as
subroutines are inadequate to reduce the complexity of subsystems
making up the whole to a manageable level.

Ideally, one should come to CSP without prejudging the meaning of terms such as concurrency, process and event. The parallelism in a system will become apparent as one attempts to prescribe the system in CSP. Clearly, in todays environment, such a naive stance is impossible to take up. In what follows the reader is asked to bear in mind that familiar terms will be given meanings useful towards building a notation in which the modularisation of complex systems can be expressed.

1.2 An example situation.

Imagine the kitchen of a large and popular restaurant. There are many, fifty say, chefs of similar skill. Tonight a particular dish is being served and each chef spends the evening repeatedly following a set of instructions (known as a recipe) to prepare a single customer serving of the dish. There are many chefs, but only one recipe.

In the kitchen there are only twenty burners, say. Depending on an individual chef's skill with knives and so on, each chef will arrive at a burner at different rates and times. Queues will form for the use of a burner. As restaurant managers we would expect the chefs to use their common sense and if there are no burners free get on with something else that is useful. We wish to maximise throughput (that is customer servings produced per hour). We would not expect to have to write in the recipe instructions to handle cases like all the burners being in use.

The chefs get annoyed. Tonight's dish depends on thickening a sauce - the recipe states that once the cornflour paste has been added the sauce must be put on a burner within thirty seconds. The chefs' argument is that they cannot possibly know when it is safe to add the paste without asking the other forty nine whether they have just added paste or not. On top of this, we have employed another thirty five chefs to prepare a marinade for tomorrow night and these chefs too are competing for the burners.

The chefs only know about cuisine. They say that we cannot allow the recipes to carry instructions that are to do with managing the fact that they are all working at once. That is our job - as restaurant managers.

In our role as managers of this system, we decide to develop a notation in which to express the strategies to be employed to handle these situations. What level of abstraction should we go to? We are only interested in what the chefs are doing at any time, not how they are doing it. And we're only interested in what they are doing where it might affect other chefs.

So we begin to think of the chefs as a sequence of events. These events are our tags for what stage they get to in the recipe. At the end of the evening we can say that chef A behaved like this

event1, event2, event7, event1, ... , STOP.

Being true managers we are more likely to make this stronger and say that chef A was event1, event2, event7, ... , STOP.

At the beginning of the evening things are not quite so simple. We are building our strategy, and looking at the recipe we find a bit like this :

```
1     get a jug
2     IF there is fresh milk THEN
3        pour 1 pint in the jug
      ELSE
4        put four tablespoons of dried milk in the jug
5        fill the jug with water to the 1 pint mark
6        whisk
7     pour jug into a pan and heat on burner
```

To make our plan in advance we have to reconcile ourselves to the best estimate of what a chef will be is

1,2,3,7   or   1,2,4,5,6,7

So at the beginning of the evening we have to say that each chef is a set containing two lists -

{ [1,2,3,7] , [1,2,4,5,6,7] }

We have arrived at a position where we can say that to talk about managing parallel situations the 'atoms' of our discourse should be events.

Events can be thought of as marker points in the lifetime of a process and our management of processes. We can't interfere with processes between events, we have no information about what happens then anyway and we're not interested. The events are rendezvous between us (the management) and the processes. The process, by being ready to meet us is making known to us what point in it's life it has reached. If we acknowledge and meet the process at the event we allow the process to go on (till the next rendezvous).

So what of processes ? They are another two levels up from events. In between there is a trace - a description of what the history of a process might be. A trace then is a (possibly infinite) list of events. A process is a (again possibly infinite) list of traces.

Our notation now has three classes of data : events, traces and processes and we are in a position to ask what operations do we need to be able to do on these items.

As the atoms are events, we must be able to put together a list of events to form a trace, and a set of traces to form a process.

Also we need to be able to express such things as the kitchen process consists of the simultaneous activity of all the chef processes :

kitchen <= chef 1 || chef 2 || chef 3 || ... || chef 4

|| is a (right associative) binary operator that takes two process and returns a process that is the parallel execution of the original two.

If we look at the above example, it would be nice to be able to modularise the process and think of it as a combination of subprocesses.

We have the process { [1,2,3,7],[1,2,4,5,6,7]}. It is clear that we could break this down into

{[1,2]} followed by either ({[3]} followed by {[7]})
            or   (({[4,5,6]}) followed by {[7]})

In fact if we use e -> P to mean a process that first engages in event e and then behaves like process P we have more compactly

1 -> (2 -> (3 -> {[7]} | 4 -> {[5,6,7]}))

1.3 The CSP notation.

For the purposes of this paper, CSP can be considered to be the following framework :

Processes, Alphabets and events

A process is a function which maps events to processes. The set of events for which a process is defined is known as its alphabet.

BLEEP is a process that raises an exception whenever it is applied to an event in its alphabet.

STOP is a process that maps every event to the process BLEEP.

The prefix, choice and parallel operators.

-> - The prefix operator.

-> is a function that maps (event,process) pairs to processes.

<u>| - the choice operator.</u>

| is a function that maps a pair of (event,process) pairs to processes.

<u>|| - the parallel operator.</u>

|| is function that maps (process,process) pairs to processes.

<u>1.3.1 Examples using ->, | and ||.</u>

These examples are taken from Hoare.

A process SVM (Simple Vending Machine) has alphabet
{ coin, choc }

SVM = coin -> (choc -> SVM)

This is interpreted as a machine that will engage in event sequences such as

coin, choc, coin, choc, ...

but not

coin, coin, choc,...

A process VM (Vending Machine) has alphabet
{ in1p, in2p, coin, choc }

VM =(in1p -> (toffee -> VM) | (in2p -> (choc -> VM))

VM will engage in sequences such as -

in2p,choc,in2p,choc,in1p,toffee,in2p,choc ...

For an example using || consider a process CUST, with alphabet { coin, curse, choc, toffee } and a process CVM with alphabet { coin,clink, toffee }

CUST = coin-> ((toffee -> CUST) | (curse-> choc -> CUST))

This customer prefers toffee to chocolate. If he is about to receive a chocolate, he accepts it but first curses.

CVM = coin ->(clink ->(choc -> CVM))

This machine produces only chocolates, and reassuring clinks as the coins fall into the money boxes.

What interpretation can we give to the parallel combination of these two processes? Clearly, the customer will be dissatisfied continuously : the engaging sequence of CVM || CUST when restricted to CUSTs alphabet will be like

coin,curse,choc,coin,curse,choc,...

When restricted to CVMs alphabet, sequences will be like

coin, clink, choc, coin, clink, ...

CVM || CUST should have an alphabet

{ coin,clink,choc,curse,toffee }

and will engage in sequences that are interleavings of the restricted sequences above : e.g.

coin,clink,curse,choc,coin,curse,clink,choc, ...

Notice, in particular, that it may engage in clinks before or after curses; since these events are not in the intersection of CUSTs and CVMs alphabets they do not provide synchronisation between the two processes.

2. Implementation of CSP in LISP

2.1 Rationale

In his book [Hoare85], Hoare presents a mathematical treatment of concurrency. We feel that a factor that contributes greatly to the book's accessibility to its intended audience, is that Hoare illustrates his theory with examples of LISP. (His intended audience is the ' aspiring programmer ').

For his purposes, the exposition of a theory, we judge this choice as admirable, since LISP is widely available and known about. It is also readily understood by computing professionals who may be mathematically naive.

2.2 Examples

For example, having introduced the notion of process and the prefix and choice operators, Hoare wishes to introduce the reader to a mathematical notation -

A process P is expressible as

$$\mu X.(x : B \to F(x,X)) \qquad \qquad \ldots\ldots E1$$

That is, a process P can be regarded as a function, F, with domain B. The set B contains the events in which a process is initially prepared to engage. F(x) defines the future behaviour of the process, if it first engages in x.

To unravel the terse expression (E1 above) further, Hoare points out that a process P can be represented as a function in LISP that takes events as arguments and returns a function (that maps events to processes) as a result. This is in general, there is however a particular process BLEEP which cannot engage in any event.

The CSP for prefix and choice can then be implemented thus :

PREFIX

e -> P is the process that first engages in event e and then behaves as process P.

e -> P can be represented as

$$\lambda x. \quad \text{if } x = e \text{ then P else "BLEEP}$$

CHOICE

e1 -> P1 | e2 -> P2 is the process that can engage in either event e1 (and then behaves as process P1) or event e2 (and then behaves as process P2).

e1 -> P1 | e2 -> P2 can be represented as

$$\lambda x. \quad \text{if } x = e1 \text{ then P1}$$
$$\text{else if } x = e2 \text{ then P2}$$
$$\text{else "BLEEP}$$

2.3 Drawbacks

Hoare's LISP extracts are intended as illustrations. We wish to build an animator which, as well as illustrating CSP, could be used for such purposes as verification, measurement and rapid prototyping of real systems.

The intention is that the animator should allow a system of processes to be declared and then build a program which can be added to and modified by an experimenter. The experimenter would be free to add in whatever analytical functions seem useful.

Any reasonable system leads to fairly intractible LISP. Modifying this is made more complicated by the fact that LISP interpreters will allow just about anything, as LISP has no types to be strongly typed about.

In addition, the underlying framework of CSP is functional; with growing complexity LISP systems start to lose their correspondence with CSP.

For example, consider the object BLEEP, which is introduced very early on in Hoare's development. Its status is unclear - at the mathematical level it is a process (the result of a function acting on an event, which in turn is a function itself). In LISP it has the same status as an event (simply an atom).

3. Animating CSP in HOPE

3.1 Rationale

The decision to use a functional programming language appeared natural, for the usual reasons. Development in a functional language is fast and easy; we wanted to experiment with our fundamental representation of the process before moving on to animation. This original experimentation helped enormously to clarify our view of CSP, as we hope to show.

3.2 First attempts

Our first attempts were in Standard ML. These are interesting to see as they betray our preoccupation with thinking of a process as a (possibly infinite) list. Since Standard ML does not support any lazy evaluation of lists, ruses were employed such as the one below. Our target was to be able to define a Vending Machine VM as closely as possible to

```
VM  =    coin -> ( choc -> VM )
```

```
 exception undef;      (* exception for undefined results *)

 datatype event = coin | choc | bleep;
 datatype process = proc of  unit -> event * process
                    | BLEEP;

 (* a simple chocolate machine  *)
 fun VM() =  ( choc, proc VM1)
    and
 fun VM1() = ( coin, proc VM);
```

Although this approach did capture the vending machine, it
wasn't clear how we could define a function prefix, say, to enable
us to construct VM along the lines of

```
      fun VM() = prefix( coin, prefix( choc, VM));
```

3.3 Higher order functions

A more natural representation of processes is :

```
      datatype event = coin | choc ;
      datatype process = BLEEP
              | proc of event -> process;
```

We can now declare a function prefix as

```
      fun prefix( e: event, p:process)  x =
          if x = e then p
                  else BLEEP;
```

Notice that prefix is a function that maps (event, process)
pairs to   processes (which   are functions   themselves). The
declaration above is declaring such a higher order function by
declaring its effect when applied. The type of element in its
domain need not be declared - the ML interpreter can decide this
by the way in which the element is used within the implementation
part of the declaration.

3.4 APPLY

The problem with representing a process is that a process has two facets : we either deal with it as a process in its own right, as when we combine it using the prefix or choice operations, or we want to treat it as a mapping, events to processes. We can use the representation we have so far and make use of pattern matching in ML to switch between these two facets :

```
fun apply ( proc(p)) = p;
fun apply(BLEEP)     = raise undef;
```

We can now do things like

```
apply(VM) coin
```

although the returned result (from the Standard ML interpreter) is just the signature of the function which is not as meaningful as we would like.

3.5 HOPE

Our attentions then turned to Imperial College's version of HOPE, which seemed to offer some advantages over ML, having a lazy function for constructing lists : lcons. We had also read of the availability of a HOPE interpreter for the PC. As it happened, ICHOPE, the PC version, didn't support lcons or quite a few other things.

The HOPE we had been running was in fact a port of a VAX-Pascal version to a VAX-750 running ULTRIX. The original interpreter was written by Victor Wu Wai Hung (Imperial College 1984). This was ported and modified by Andy Wakelin and Rubik Sadeghi (Dundee College of Technology, 1985). As we still had the source code of Wu's original, it occurred that we could actually construct an animation environment, since we could add to the HOPE whatever facilities we needed. The process of transferring HOPE to the PC is covered in section 5.1.

Progress in HOPE seemed much easier. Apart from some small detours into trying to simulate infinite lists of possibly infinite lists, the basic representation scheme we fixed on bears a strong resemblance to the later ML models.

As this model forms the basis of the animator, it is worthwhile going into it in some depth.

3.5.1 The Model

We start with a datatype for the events of interest in our discourse :

```
data event == coin ++ choc ++ toffee ++ clink ++ curse;
```

This set could be generated by an animator as the union of all the alphabets of all the processes in a system.

The fundamental definition is of process :

```
data process == proc( event -> process) ++ BLEEP;
```

This however focuses on only one aspect of a process. What is needed is a function to prise out its functional character (an inverse to the proc constructor function). This is accomplished by pattern matching :

```
dec apply : process -> ( process -> event);
--- apply( proc ( p)) <= p;
--- apply( BLEEP)     <= undefine;
```

Undefine raises an exception in HOPE.

3.5.2 The operators

It is possible to declare functions to be infix in HOPE. For the prefix operator, we chose the symbol >>, which is a legal identifier (-> would have been closer to the Hoare notation, but is illegal as an identifier in HOPE).

```
infix >> : 5;  ! allows the use of the function >>
               ! as an infix operator
               ! that is x >> y instead of >>(x,y)
dec >>    : event X process -> process;
--- e >> p <=
            proc(( lambda el =>
                    if e = el
                    then p
                    else BLEEP ));
```

Let us examine this more closely. Ideally, we would like to say that processes are mappings from events to processes. The nearest we can approximate to this is to say that processes are labelled by the constructor function, proc, applied to a mapping from events to processes.

With prefix, we are trying to define a process : something that can be used to create other processes and so on. We are not interested in the functional facet of the process, just the process part of it.

Our implementation of prefix is saying -

e >> p is proc applied to ...

This is correct - proc labels objects of type process. The
tail of the above sentence must be an object that can be in the
domain of proc; that is, it must be a function from events to
processes. So we have -

e >> p is proc applied to
    a function f : event -> process

Which of all such functions do we want ? The one that takes
the value p for the event e, and the value BLEEP for all other
members of its domain. Lambda caters for this. It means ' call el
the event that f would be applied to '. We now have -

e >> p is proc applied to
    a function f : event -> process
      which is defined by   f(e) = p and f(x) = BLEEP (e =x)

In retrospect we wonder what all the fuss was about, since
this is really very similar to the original LISP implementation in
Hoare's book. However, the task of declaring the prefix function
helped to clarify the meaning of Hoare's treatment.

Prefix can now be used, for example :

            dec VM : process;
            --- VM <= coin >> (choc >> VM);

The only minor irritation with this is that although the
original declaration corresponds almost exactly with Hoare's
notation, the HOPE interpreter expands the definition and if asked
will return

    --- VM <= proc ((lambda el =>
                    if (coin = el)
                    then proc ((lambda el =>
                                    if (choc = el)
                                    then VM
                                    else BLEEP))
                    else BLEEP));

The choice operator can also be declared as infix. We chose
to use OR :

```
infix OR : 3;
dec OR : ( event X process ) X ( event X process) -> process;
--- ( e1, p1) OR ( e2, p2) <=
          proc(( lambda e =>
                    if e = e1
                       then p1
                       else if e = e2
                               then p2
                               else BLEEP));
```

Using this we can, for instance, have -

```
dec CUST : process;
--- CUST <=
coin >> ((toffee, CUST) OR ( curse, choc >> CUST));
```

The parallel operator, ||, is treated slightly differently. Up till now we have not been too concerned with the alphabets of a process. To combine two processes which have coincident alphabets is straightforward

```
infix || : 3;
dec || : process X process -> process;
--- p || q <= proc(( lambda e =>
                   if (( apply(p)(e) = BLEEP)
                      or
                      ( apply(q)(e) = BLEEP))
                   then BLEEP
                   else apply(p)(e) || apply(q)(e))));
```

To cater for processes with distinct or disjoint alphabets, we introduce a new function

```
dec alphabet : process -> list(event);
```

The values this takes are given explicitly; for example

```
--- alphabet( CUST ) <= [coin,choc,curse,toffee];
```

We can then recast our implementation of || as

```
--- p || q <=
            proc(( lambda e =>
                    if e ismember alphabet( p)
                        and
                    e ismember alphabet( q)
                    then
                      if ( apply(p)(e) = BLEEP)
                          or
                          ( apply(q)(e) = BLEEP)
                      then BLEEP
                      else apply(p)(e) || apply(q)(e)
                    else
                      if e ismember alphabet(p)
                      then apply(p)(e) || q
                      else
                        if e ismember alphabet(q)
                        then p || apply(q)(e)
                        else p || q       ))));
```

This is quite a complicated function. The function ismember is an infix set membership operator returning true or false. What the above declaration means is that when two processes are operating in parallel, they each only engage in events in their own alphabets, other events are completely transparent to them.

4.1 The language

The CSP animator presented here is a system by which a set of processes can be defined; that is their alphabets can be declared. Also, processes may be defined to be constructed from other processes by expressions involving ->, | and ||.

Following the definition of a set of processes, an observer event stream can be defined and the animator will report on the consequences of the processes engaging in the events that make up the observer stream.

So CSP is considered in this context as a language, whose semantics have been expressed informally above and whose syntax is described by the following B.N.F.(the distinguished symbol is <systdec>).

```
<systdec>    ::= { <declaration>; }*
<declaration> ::= <streamdec> | <procdec>
<streamdec>  ::= <id> = <alphabet>
<procdec>    ::= <id> ( <alphabet> } <= <procexp> ;
<procexp>    ::= <procfact> ( <par><procfact> }*
<procfact>   ::=  ( <procexp> )| <choice>| <prefix> | <id>
<choice>     ::=    <lbra><procterm><or><procterm><rbra>
<prefix>     ::= <id> -> <procexp>
<procterm>   ::= <id> -> <procexp>
<alphabet>   ::= ( <id> (, <id> }* )
<lbra>       ::= "("
<rbra>       ::= ")"
<or>         ::= "|"
<par>        ::= "||"
<id>         ::= <letter>{ <letter> | <digit> | . | _ }*
<letter>     ::= a | b | ... | z | A | B | ... | Z
<digit>      ::= 0 | 1 | ... | 9
```

The form of extended B.N.F. used here is taken from [Milne87]. Braces are used for iteration : { a } allows zero or one iteration. { a }* allows zero or more iterations. "a" signifies that the quoted string is a literal (used to avoid confusion with metasymbols.

4.2 Using the animator

A user can (from within or without HOPE) edit a file consisting of declarations according to the syntax of section 3.1.

Example declarations are

A vending machine process, with alphabet { coin, choc} -

VM (coin, choc) <= coin -> choc -> VM;

(Notice that parentheses are not necessary in this case, since the animator treats the prefix operator, ->, as right associative.)

A more complex vending process, with alphabet { onep, twop, choc, toffee } -

CVM (onep, twop, choc, toffee) <=
  { onep -> toffee -> CVM | twop -> choc -> CVM };

(This is an example of the choice, |, operator).

Processes can also be declared using the parallel operator, ||, as in the following :

```
CUST ( coin, curse, choc, toffee )
    <= coin -> ( toffee -> CUST | curse -> choc -> CUST);

NOISYVM ( coin, clink, choc)
    <= coin -> clink -> choc -> NOISYVM;

NOISYCUST <= CUST || NOISYVM;
```

Notice that the alphabet for NOISYCUST need not be declared,
the animator will assume it to be the union of its components'
alphabets.

In addition to process declarations, lists of events can also
be declared; typically these will be used to provide a stream of
events for processes to engage with. For example -

```
RUNSTREAM = ( coin, choc, coin, coin );
```

From within HOPE, the user can invoke an interpret function
that will translate a set of declarations into HOPE declarations.
In addition interpret also declares the following functions and
datatypes :

```
data event      == ....... ;
data process    == proc( event) -> process;
type traces     == list( event);
data process    == proc( event -> process);
dec  alphabet    : process -> traces;
dec  BLEEP       : process;
dec  STOP        : process;
dec  >>          : event X process   -> process;
dec  ||          : process X process -> process;
dec  OR          : ( event X process ) X ( event X process)
                        -> process;
dec  apply       : process           -> ( event -> process );
dec  istrace     : traces X process  -> truval;
dec  menu        : process -> traces;
dec  interact    : process X traces -> list( traces);
```

These are each described in the next subsection.

4.2.1 The animator's kit of functions.

Since the intention is to allow the user to modify and add to
the HOPE implementation of a system of processes, all the system
construction functions are provided; the parallel operator, ||,
the prefix operator, >>, and the choice operator, OR, are defined
to be infix operators.

To extract the purely functional aspect of a process, the function apply is used. For instance, if VM is the coin->choc->VM then to get the result of VM engaging in event coin, we would use

apply(VM)(coin)

The result though, itself a process, will be returned by the hope interpreter but not in as convenient a form as choc -> VM !

The utility functions are istrace, menu and interact:

istrace is an infix operator that determines if the supplied list of events is a possible trace of the process it is applied to.

For instance, if we have the original declarations -

```
VM ( coin, choc )  <= coin -> choc -> VM;
RUNSTREAM1         = ( coin, choc, coin, coin );
RUNSTREAM2         <= ( coin, choc, coin);
```

we would have

```
RUNSTREAM1    istrace VM    : false
RUNSTREAM2    istrace VM    : true
(RUNSTREAM2 <> [choc,coin]) istrace VM    : true
```

(<> is the ordinary HOPE list catenation operator.)

menu returns the list of events that the process it is applied to is prepared initially to engage in.

For example, with VM as above

menu(VM)   : [coin]

A more complicated example, that may be quite useful is

menu(apply(VM)(coin))   : [choc]

interact returns the list of menus (list of events) that arise if the process engages in the sequence of events supplied.

For example, if we have declarations for CUST as in the previous section

```
interact(CUST,[coin,clink,curse,choc,coin,choc]) would return
[ [coin],
  [curse, toffee],
  [choc],
  [coin],
  [curse, toffee]
  nil ]
```

5. Implementation issues.

The case for using a functional language for this project has already been made. The authors chose HOPE in particular since it was available and modifiable.

5.1 The PC version of HOPE

One of the fruits of this work has been that the authors now have a version of HOPE that will run on a PC and has all the facilities available on the UNIX implementation that was previously used.

The VAX sources for the HOPE interpreter were transferred to a PC and compiled using Borland's Turbo Pascal 4.0. Originally it was thought that the only changes that would need to be made were in the lexical analysis sections (where strings were handled) and the filing routines.

It turned out that the original interpreter had some fairly obvious bugs. Odd characters had crept into the source code. For example, a HOPE declaration entered as

```
dec max : num X num -> num;
```

would be returned with a lower case x as in

```
dec max : num x num -> num;
```

As already mentioned, Andrew Wakelin and Rubik Sadeghi had done a similar transfer of HOPE (to UNIX on a VAX) at Dundee College of Technology in 1985. When these small bugs became apparent, their notes were examined more carefully. They had encountered the same (and more) minor irritations, and had gone on to include some useful extensions. Among these was a facility to invoke a full screen editor to modify function implementations without the need to leave HOPE.

This seemed a useful idea, and was incorporated in the PC implementation.

5.2 The animator.

The animator can be considered as a compiler, with source language a subset of CSP and object language HOPE. The compiler itself was implemented in Pascal; being a very simple language, writing a standard recursive descent parser took very little time. It appeared to the authors to be an easier task to include code generating statements in a recursive descent parser than any other. However, A.C. Milne [Milne88] presents an interpreter (implemented in HOPE) constructed directly from a formal syntactic and semantic language specification. It may well be advantageous to go over to such a system, both from an aesthetic and formal point of view.

6. Future work.

The animator presented here is only an experiment in HOPE and CSP. In terms of Hoare's work, the operators that have been implemented are developed within the first two chapters of its seven chapters.

However, even with such a small outline, it would appear feasible to extend and work on the animator so as to explore the measurement, specification and verification of concurrent systems.

6.1 System measurement.

Rudimentary measures of conventional sequential software make measurements on lines of code : so many lines are unreachable, testing has covered such and such a proportion of lines of code. The parallel analogue to these features may well be events - what is the significance of an unengageable event ? These occur in the parallel composition of processes. The NOISYCUST process of section 1.3 cannot engage in the toffee event, whereas one of its components, CUST, can.

The proportion of engageable events engaged in during an animation may turn out to be a useful measure of the system testing process. Another such 'coverage metric' may be the proportion of possible choices exercised in animation [Smith87].

The CSP specification of a process has a diagramatic form similar to finite state transition diagrams. Since the animator has a representation of such a graph, it may be used to explore complexity issues in a similar graph-theoretic manner to McCabe [McCabe76]. Paul Smith [Smith87] suggests two kinds of complexity - the static complexity of an individual process and the synchronisation complexity of communicating processes.

6.2 Other systems.

CSP is by no means the only mathematical specification notation for concurrent systems. A Calculus for Communicating Systems (CCS) [Milner80] is perhaps more widely known. It is of a very similar flavour to CSP. CCS and CSP complement each other very well; reading either gives an insight into the other. CCS concerns itself with the equivalence of systems. From a verification point of view this presents the possibility of checking a specification in CCS against the transformation of an implementation into CCS for an equivalence.

References

[Hoare85] 'Communicating Seqeuntial Processes'.
          C.A.R. Hoare, Prentice-Hall, 1985.
[McCabe76] 'A complexity measure'.
          T.J. McCabe, IEEE Trans. Software Eng. SE-2, 1976.
[Milne87] 'The analysis and manipulation of BNF definitions'.
          A.C. Milne, EUUG autumn conference proceedings,1987.
[Milne88] 'On the construction of a functional interpreter from
           a formal language specification'.
          A.C. Milne, in preparation, 1988.
[Milner80] 'A Calculus of Communicating Systems'.
          R. Milner, Springer Lecture Notes in Computer
          Science 92, Springer-Verlag, 1980.
[Smith87] 'Metrics for Communicating Sequential Processes'.
          P. Smith, A discussion note for first MUSE/QR
          workshop, 1987.

A Concurrent Approach to the Towers of Hanoi

W.D. Crowe & P.E.D. Strain-Clark
Mathematics Faculty,
The Open University,
Milton Keynes,
UK

July 25th, 1988

Abstract

The tower of Hanoi problem has long been known to have a closed (ie non-recursive) solution. In this paper we analyse two approaches to this solution which involve concurrency. This is not an end in itself, but serves to introduce the main ideas and notations of CAP (Communicating Asynchronous Processes) - a revision of CSP which the authors have used successfully to develop correct Occam programs. In turn, CAP is part of a wider development method (ODM) which is being studied with the intention of prototyping software tools to assist developers of concurrent software.

Introduction

In this paper we investigate two approaches to the "Towers of Hanoi" problem which exploit concurrency. We do so using a new notation, referred to as CAP (short for Communicating Asynchronous Processes), which is a variant of CSP [1],[4].

We shall illustrate some stages in a development method (ODM) which we have used successfully to obtain reliable Occam code for problems involving concurrency. In particular we hope

(i) to stimulate the search for concurrency in new areas ;

(ii) to describe the Process Analysis approach to concurrency which concentrates on processes and channels of communication ;

(iii) to further the use of CAP (or CSP) as a specification language for such problems, and to establish some standard notations ;

(iv) to use the compact identities available in CAP to perform the expansions necessary for formal verification without an unnecessary combinatorial explosion of possibilities .

These points will become clearer throughout the paper but before embarking on a detailed analysis of the problem we must say what we mean by CAP.

Communicating Asynchronous Processes

Communicating Asynchronous Processes (CAP) consists of just that part of CSP and just those rules which model the language Occam [5] and its laws [7], while still maintaining all the flexibility and non-determinism available in CSP.

Both notations deal with processes and communications. In CSP processes are built from atomic activities, while in CAP there is no distinction between these two ideas - activities are processes. In CSP each process is associated with an alphabet which is just an unstructured set. In CAP the alphabet has much structure - it contains elementary processes that are either internal or associated with receiving or transmitting messages, the latter being thought of as typing the channels.

We use the notations

in ? var

out ! expr

to represent input and output processes, while

ch.a

represents an internal process which at a higher level of refinement corresponds to a fulfilled communication consisting of a matched input/output pair.

The operations in CAP are mostly given by symbols familiar from CSP, but some of the interpretations are subtly different. Notable are the following:

(i) Since we deal only with processes there is only one form of sequential combination and it uses the symbol ; rather than → . This is a binary associative infix operator and so has an associated prefix operator, of arbitrary -arity, for which we use the symbol Σ .

(ii) Parallel combination is thought of as an operator over collections of processes and channels, which form unions of connected components of the process diagram. It is only well formed if there are restrictions on alphabets amounting to avoidance of shared variables and concurrent use of channels. This contrasts with CSP where parallel composition is often described as a binary, associative, idempotent, ... , operator. Nevertheless we too use the symbol ‖ for this prefix operator. We shall often abuse notation and use ‖ as a binary infix operator (as in CSP) but its true interpretation should always be clear from the corresponding process diagram.

(iii) The symbol ⫾ is only used for the operation corresponding to ALT in Occam and gives a guarded choice between input processes on a set of channels. Hence it is a true prefix operator of variable -arity, though as usual we shall abuse notation and use it as an infix operator. Associated with this is a binary prefix (or infix) operator ⫾▷ which denotes prioritised guarded choice. That this is a binary operator reflects the fact that there are currently only two levels of priority in the transputer's scheduling firmware.

(iv) We use the symbol * as WHILE in Occam and the brackets ◁ ▷ as IF in Occam.

(v) We use the symbol ⨅ for non-deterministic choice (but we usually reserve its use to non-determinism associated with external input or the timing of the operation ⫾. It is a prefix operator of arbitrary -arity and so can be used as a binary infix operator in the usual way.

Finally we need to describe the 'laws of CAP'. This will be done elsewhere in detail [2] and can be thought of as a synthesis of the 'laws of Occam'[7] and the 'laws of CSP'[4]. All that we need to mention here is how to expand large products of parallel processes. In CSP, where it is assumed that there is some external observer with some absolute time, it is necessary to write down all the sequences of activities they might observe. For us the 'principle of relativity' tells us that there is no possibility of such an external observer - different observers will disagree as to the ordering of events in parallel processes. We use this principle to justify the choice of any one sequence of events from an equivalence class of indistinguishable sequences. In this way we not only cut down on the combinatorial explosion of traces, but by judicious choice of representative we can make the process of verification more straightforward. The end product of this work will be an interactive system where expansions are generated automatically and selection is in the hands of the designer.

Notation

Throughout the paper ⊖ and ⊕ represent subtraction and addition modulo 3, respectively.

Towers of Hanoi - The Specification

The "Towers of Hanoi" is a well known problem [3] with a straightforward specification. The simplest formal specification in our notation is

$$HANOI = INIT ; MOVE \qquad (1)$$

where, on termination of the process INIT, a representation of the three towers has been initialised with all the discs correctly stacked on one tower.

fig 1

At some later time the process MOVE has led to a representation of all the discs stacked similarly on another tower. Of course, the problem restricts the type of move allowed. Hence a fuller specification demands that the traces of the process MOVE represent legal moves, with smaller discs being placed on larger discs. The problem has a unique most-efficient solution [6] which is to iterate pairs of moves of a specific nature. The first move of a pair takes the smallest disc in a direction which is fixed at the outset (we shall choose clockwise). The other half of the move-pair involves the two towers which no longer have possession of the smallest disc, and is determined by legality. We extend our specification to include a description of this solution, and examine the traces of the process MOVE in order to verify it.

Process Diagrams and Parallel Decomposition

We shall investigate concurrency at two levels of organisation - the tower and the disc. The former enjoys a significant advantage when we come to implementation, since the latter requires rather more in the way of process inter-connections than is convenient with a transputer network.

The Towers

The three towers are arranged in a circle, with arrays of channels connecting clockwise and anticlockwise neighbours as in the following Process Diagram:

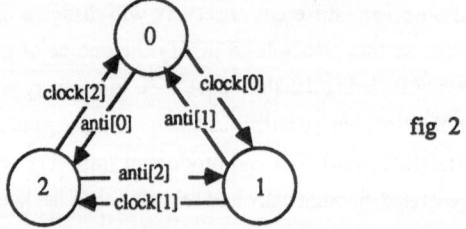

fig 2

This gives the parallel decomposition

$$HANOI = \|_{0 \leq i \leq 2} \ TOWER(i, clock[i], anti[i], anti[i \oplus 1], clock[i \ominus 1]) \qquad (2)$$

which can be written less formally as

$$\text{HANOI\_T} = \parallel_{0 \le i \le 2} \text{TOWER}(i) \tag{3}$$

or even

$$\text{HANOI\_T} = \text{TOWER}(0) \parallel \text{TOWER}(1) \parallel \text{TOWER}(2) \tag{4}$$

The Discs

The discs are best considered labelled consecutively from the smallest (=0), through the increasingly large movable discs to the three bases. If the number of movable discs is max then the bases are labelled base(n) = max + n, for n = 0,1,2. All the discs must be able to communicate with each other and so a doubly indexed linking array is required. To avoid complexity this array can be made available to the entire array of processes. Part of the Process Diagram is

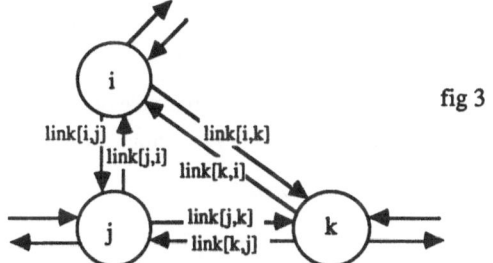

fig 3

Corresponding to this we have the parallel decomposition

$$\text{HANOI\_D} = \parallel_{0 \le i \le N-1} \text{DISC}(i, \text{link}[\,,\,]) \tag{5}$$

where N is the number of discs (including the three bases as discs).

Initialisation and Reusable Components

The original specification (1) allows for initialisation and before we describe this in detail we need to know the internal repesentation of data within the TOWER and DISC processes.

The Towers

Each of the towers is described by a <u>stack</u> of discs, with the exposed top disc and the tower index stored separately for convenience. The stack is an example of a reusable component - that is, rather than define it in detail here we assume that the existence of such a process with the usual functional properties has been defined elsewhere. Since we are dealing with communicating processes our stack takes the form of a process with three channels:

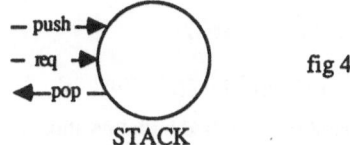

fig 4

STACK

The channel 'push' is used to push values onto the stack; the channel 'req' is used to request the top value on the stack which is returned on the channel 'pop'. The stack is used to refine the

Process Diagram for a typical tower as follows:

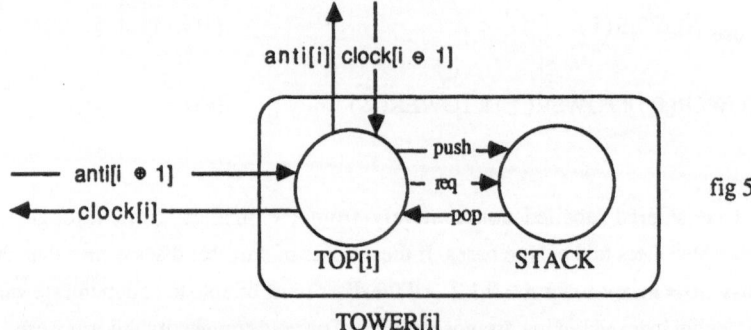

TOWER[i]

This gives the (informal) refinement of the specification

$$\text{TOWER}(i) = \text{TOP}(i) \parallel \text{STACK} \tag{6}$$

where the stack does not need to know the index i of the tower but only the local names push, req,

pop of the channels used to communicate with it. As a reusable component the process STACK will run in parallel with the process TOP indefinitely without further specification and hence

initialisation only needs to be defined for the process TOP(i).

By analogy with (1) we require

$$\text{TOP}(i) = \text{INIT}(i) ; \text{MOVE} \tag{7}$$

No index is required for the process MOVE as after initialisation all the towers behave similarly, even using the same local names for channels.

Routine manipulation using the "principle of relativity" built into CAP gives

$$
\begin{aligned}
\text{HANOI\_T} \ &= \ \parallel_i \text{TOWER}(i) \\
&= \ \parallel_i \{ \ \text{TOP}(i) \parallel \text{STACK} \ \} \\
&= \ \{ \parallel_i \text{TOP}(i) \ \} \parallel \{ \parallel_i \text{STACK} \ \} \\
&= \ \{ \parallel_i (\ \text{INIT}(i) ; \text{MOVE} \ ) \ \} \parallel \{ \parallel_i \text{STACK} \ \} \\
&= \ \{ \parallel_i \text{INIT}(i) \ \} \parallel \{ \parallel_i \text{STACK} \ \} ; \{ \parallel_i \text{MOVE} \ \} \parallel \{ \parallel_i \text{STACK} \ \}
\end{aligned}
$$

which is of the form of (1).

If we define initialisation in the obvious way

$$\text{INIT}(i) = (\text{top} := \text{base}(i) \ \triangleleft \ i \neq 0 \ \triangleright \ \text{push ! base}(0) ; .... ; \text{push ! 1} ; \text{top} := \text{smallest}),$$

where base(0) = max is the number of movable discs, expansion gives

$$\{ \parallel_i \text{INIT}(i) \ \} \parallel \{ \parallel_i \text{STACK} \ \}$$

$$= \quad \text{top}_1 := \text{base}(1) ; \text{top}_2 := \text{base}(2) ;$$

$$\text{push}_0.\text{max} ; .... ; \text{push}_0.1 ; \text{top}_0 := \text{smallest}$$

which is a satisfactory representation of the specified initial position of the stacks.

The Discs

With the disc model there is no need to introduce complicated datatypes. Each disc can be satifactorily described in terms of the tower on which it is currently placed, whether it is exposed at the top of a tower, and which disc is beneath it. It is possible to describe the smallest disc without saying that it is exposed, and similarly it is possible to describe the bases without mentioning that there is nothing beneath them.

Hence we will use distinct processes

SMALLEST = DISC(smallest, link[,])

BASE(j) = DISC(base(j), link[,]) ,   j = 0,1,2.   (8)

Initialisation and subsequent movement can be treated as before, with

SMALLEST = SETUP_S ; RUN_S(link[,])

DISC(i, link[,]) = SETUP(i) ; RUN(i, link[,]),   i = 1, ... , max-1

BASE(j) = SETUP_B(j) ; RUN_ B(j, link[,]) ,   j = 0,1,2   (9)

and with

HANOI_D = {SETUP_S || (||ᵢSETUP(i)) || (||ⱼSETUP_B(j)) } ;

$\qquad$ { RUN_S(link[,]) || (||ᵢRUN(i, link[,])) || (||ⱼRUN_B(j, link[,])) } ,

Setting

SETUP_S = tower := 0 ; next := 1

SETUP(i) = tower := 0 ; exposed := FALSE ; next := i+1 $(1 \leq i \leq max-1)$ (10)

SETUP_B(j) = tower := j ; (exposed := FALSE ◁ j = 0 ▷ exposed := TRUE   (j = 0,1,2)

we can agree that this is a reasonable representation of the initial condition of the stacks of discs.

Moving the Discs - The Tower Model

We have informally described the algorithm by which the discs are moved from stack to stack. We now intend to formalise this in terms of communications between processes. There is clearly a difference between the behaviour of the tower on which the smallest disc is stacked and the other two. Suppose this tower has index a and the others b, c in clockwise order. The tower a is clearly in control. It passes the smallest disc clockwise to tower b which must submit and accept it, and change its state accordingly. Now the second disc on tower a is exposed and this must be compared with the top disc on tower c to see which is to move. To make the comparison a copy must be sent to tower c which either accepts the disc as smaller, and acknowledges this fact, or it returns its own top disc. In either case the state of the tower must be updated and the cycle can begin again. Formally the above description gives

MOVE =   (CONTROL ◁ top = smallest ▷ SUBMIT)   (11)

emphasising the different role of the tower that is control, and then

```
CONTROL                                             (12)
  =   out_clock ! smallest ;          -- Pass smallest clockwise
      req ! popitem ; pop ? top ;     -- pop the stack and pass
      out_anti ! top ;                -- copy anticlockwise .
      in_clock ? reply;              -- Await reply,
      ( req ! popitem ; pop ? top ;  -- pop the stack and
        SUBMIT                        -- submit again
          ◁ reply = accepted ▷       -- if accepted or
        push ! top ;                  -- push back on stack and
        top := reply ;                -- replace top with reply
        SUBMIT )                      -- before submitting again
and
  SUBMIT                                              (13)
  =   in_clock ? any                 -- If contacted clockwise
        push ! top ; top := smallest ;  -- add smallest at top
        CONTROL                       -- and take control
  []                                  -- Alternatively if contacted
      in_anti ? query               -- anticlockwise with query
      ( push ! top ; top := query    -- add to top
        out_clock ! accepted         -- and acknowledge

          ◁ query < top ▷           -- if query smaller or
        out_clock ! top              -- otherwise return top
        req ! popitem ; pop ? top ) ;  -- and pop the stack ,
      SUBMIT                          -- before continuing
```

Deadlock Analysis for Towers

Before contrasting the above apparently straightforward specification with that for the disc model it is worthwhile performing some deadlock analysis. Deadlock analysis for us consists of expanding a parallel combination to obtain a trace of successful communications (or otherwise!). It is here that CAP has an advantage over CSP. In CSP though parallel processes are considered to run asynchronously it is assumed that there is some external observer who could record all the possible traces. This is not physically tenable and prompts us to take a slightly different approach. We regard two traces as <u>observationally equivalent</u> if their restrictions to each sequential sub-process of the parallel composition are the same. (A trace may in fact be observationally equivalent to one which no physical observer could possibly record - if only for relativistic reasons!) By choosing an appropriate representative of an equivalence class we can reduce to manageable proportions the number of traces that need be considered. Expanding from the initial position from (11), (12) and (13) gives

$\{ \parallel_i \text{MOVE} \} \parallel \{ \parallel_i \text{STACK} \} = \text{CONTROL}_0 \parallel \text{SUBMIT}_1 \parallel \text{SUBMIT}_2 \parallel \{ \parallel_i \text{STACK} \} =$

    clock[0].smallest ;

    $\text{push}_1.\text{base}(1)$; $\text{top}_1 := \text{smallest}$;

    $\text{req}_0.\text{popitem}$; $\text{pop}_0.1$;

    anti[0].1 ;

    $\text{push}_2.\text{base}(2)$; $\text{top}_2 := 1$;

    clock[2].accept ;

    $\text{req}_0.\text{popitem}$; $\text{pop}_0.2$;

    $\text{SUBMIT}_0 \parallel \text{CONTROL}_1 \parallel \text{SUBMIT}_2 \{ \parallel_i \text{STACK} \}$

$\sqcap$

    clock[0].smallest ;

    $\text{push}_1.\text{base}(1)$; $\text{top}_1 := \text{smallest}$;

    clock[1].smallest ;

    $\text{push}_2.\text{base}(1)$; $\text{top}_2 := \text{smallest}$;

    $\text{req}_0.\text{popitem}$; $\text{pop}_0.1$;

    $\text{req}_1.\text{popitem}$; $\text{pop}_0.\text{base}(1)$;

    $(\text{anti}[0] ! 1 ; ... \parallel \text{anti}[1] ! \text{base}(1) ; ... \parallel \text{clock}[2]. \text{smallest} ; ...) \parallel \{ \parallel_i \text{STACK} \}.$

Unfortunately the bracketed expression

    $(\text{anti}[0] ! 1 ; ... \parallel \text{anti}[1] ! \text{base}(1) ; ... \parallel \text{clock}[2]. \text{smallest} ; ...)$

is clearly equal to STOP (ie deadlock!).

Hence our expansion has given rise to a nondeterministic choice between two options, one of which leads to deadlock. The system is not deadlock-free and some changes need to be made in the specification. This is typical of specifications involving parallelism. Great care must be taken to avoid the <u>repetitive</u> behaviour of parallel components becoming out of phase.

Fortunately there are both general and specific ways round the problem. Implicit in the original algorithm is the idea that the towers are in 'lock-step', performing a sequence of pairs of moves. Unfortunately in our present specification the smallest disc can move <u>twice</u> before the second half of the first move is completed. This is easily remedied by passing a message telling the anticlockise tower to wait and hence alerting it to the direction of communication in this current cycle. The first lines of (12) and (13) become

CONTROL   (12')

    =   out_anti ! wait   -- Alert isolated tower before

        out_clock ! smallest ;   -- passing smallest clockwise

        ....

and

SUBMIT   (13')

   =   in_clock ? any ;   -- If contacted clockwise

             push ! top ; top := smallest ;   -- add smallest at top

             CONTROL   -- and take control

     []   — Alternatively if contacted

        in_anti ? any   -- anticlockwise then

           in_anti ? query   -- wait for the

          ....   -- anticlockwise query

With these alterations the first option of the expansion begins

$$CONTROL_0 \parallel SUBMIT_1 \parallel SUBMIT_2 \parallel \{ \parallel_i STACK \} =$$

anti[0].wait;

clock[0].smallest ;

....

and the second option leading to deadlock is eliminated. This expansion shows that the specification is initially correct.

It leads from state

$$CONTROL_0 \parallel SUBMIT_1 \parallel SUBMIT_2 \parallel \{ \parallel_i STACK \}$$

to state

$$SUBMIT_0 \parallel CONTROL_1 \parallel SUBMIT_2 \parallel \{ \parallel_i STACK \}$$

via a sequence of deadlock-free communications and with the representation of the stacks being correctly updated. This leads to an inductive proof that our specification is correct in its entirety. Rather than do this here we wish to stress the strategy that we adopted to eliminate deadlock. If a specification has been given for a repetitive algorithm, which can be shown to be deadlock-free provided that processes are limited to a single cycle, then the specification is deadlock-free if all processes are committed to a particular cycle before any one process has completed that cycle. Forcing processes to operate in lock-step may seem contrary to the spirit of parallelism, especially if it can only be done at the cost of extensive communication and resulting loss of efficiency. However we have seen that the above specification can be rendered safe by the addition of a single communication per cycle of moves. In fact another variant of the specification maintains lock-step with the same number of communications as the original! We simply modify the algorithm to send a copy of the disc beneath the smallest disc before we send the smallest disc itself. That is (12) becomes

CONTROL   (12'')

     =   req ! popitem ; pop ? top ;   -- Pop the stack and pass

           out_anti ! top ;   -- copy anticlockwise before

           out_clock ! smallest ;   -- passing smallest clockwise.

           in_clock ? reply;   -- Await reply,

         ....

In the above analysis there is much detail which can be ignored. Here we stress the importance of properly specified reusable components, for if the stack processes are known to be correct, and if they are used in accordance with their specifications, then they cannot precipitate a deadlock. With this in mind we could conceal all the stack-related activities and consider only the inter-tower communications as a possible source of deadlock. This leads to significantly less writing!

Termination

It is clear from the deadlock analysis that the algorithm we have supplied does not terminate as long as the stacks are nonempty. Hence it will meet our specification of moving all the discs from one tower to another, but if we were to continue the analysis we would find a disorderly ending depending on what happens when we make a pop request of an empty stack. There is an ad hoc strategy for avoiding this problem based on the use of a more elaborate stack process. We permit a stack to report its size if requested, and also to terminate if instructed.The channel req can carry messages 'popitem' , 'stacksize', and 'terminate', the former two requesting a reply on channel 'pop'.The termination strategy that we will adopt is that after the first move if any tower (necessarily in control) discovers that it has all the discs on its stack then it knows to terminate after first alerting the the other two towers to the situation. This can be accomplished by modification of the process SUBMIT as follows

```
SUBMIT =
    in_clock ? query                              --
        ( req ! terminate                         -- Terminate stack and SKIP
            ◁ query = terminate ▷                 -- if requested.
        push ! top ; top := smallest ;            -- Otherwise,
        req ! stacksize ; pop ? size ;            -- before taking control,
        ( out_clock ! terminate ;                 -- terminate all
            out_anti ! terminate; req ! terminate; -- processes if
                ◁ size = max ▷                    -- end position reached
            CONTROL ) )                           --
    []                                            --
    in_anti ? query                               --
        ( req ! terminate                         -- Terminate if requested,
            ◁ query = terminate ▷
        ( push ! top ; top := query              -- otherwise
        out_clock ! accepted                      -- continue
            ◁ query < top ▷                       -- as
        out_clock ! top                           -- before
        req ! popitem ; pop ? top ) ;
        SUBMIT )
```

We have now completed the analysis of the tower model for the "Towers of Hanoi" problem. This specification has been refined and used to produce Occam code which has run successfully on three Transputers.

Moving the Discs - The Disc Model

We use the same algorithm, with the smallest disc always moving in a clockwise direction, and making alternate moves. However, we can no longer think of this smallest disc as being in control. Though it knows its position (and hence that to which it must move) it does not know the index of the disc in this position which will subsequently become the disc beneath it. On the other hand the exposed discs don't know which is about to be covered but they are able to let the smallest disc know their respective positions. The smallest disc can then make the decision as to which disc it will cover and can inform it accordingly. Before updating its state, the smallest disc must pass on to the other exposed disc the index of the disc that was beneath it as this cycle began.

This remaining exposed disc now knows the index of the disc with which it must be compared. If its own index is smaller then it must move, updating its state, but not before informing the disc next beneath that it is now to be exposed.

On the other hand, if its own index is larger it knows that it is about to be covered. First it must inform the disc that is to do the covering, and before moving this disc must in turn inform the disc beneath it that it is now to be exposed.During these communications any disc that is not exposed may be about to receive a communication and so must be prepared. Bases behave exactly the same way as ordinary discs except near the end of the algorithm. If we demand that there is orderly termination, as for our last tower model, then if any base is contacted by any other base it knows they have arrived at the final position and must act accordingly. A little thought shows that all other discs are awaiting some communication and so can be informed immediately of the situation and can all terminate independently.

This verbal specification translates directly into CAP as follows. From the preceding work (8,9,10), we only need to define RUN_S, RUN, RUN_B

```
RUN_S =
    running := TRUE                                    -- Start
    running *                                          -- running.

       ( ▯ i=1,...base(2)
            (link[i,smallest] ? loc1                   -- Take input
                ( running := FALSE                     -- and terminate
                ◁ loc1 = terminate ▷                   -- if requested or

           ▯ j=1,...base(2)
                (link[j,smallest] ? loc2               -- take next input
                    ( clock := i ; anti := j           -- and decide
                        ◁ loc1 = position ⊕ 1 ▷        -- where to
                    clock := j; anti := i );           -- move.
                    ( link[smallest, clock] ! cover    -- Inform discs
                    || link[smallest, anti] ! compare, next ) ;  -- concerned
                    position := position ⊕ 1           -- and make
                    next := clock )))) 		           -- move.
```

```
RUN(n) =
    running := TRUE                                    -- Start
    running *                                          -- running by
       ( ( TALK                                        -- talking
              ◁ exposed ▷                              -- or
           LISTEN ) )                                  -- listening.
where
  TALK =
    link[n,smallest] ! position ;                      -- Alert smallest and
    link[smallest,n] ? cmd ;                           -- await command.
    ( exposed := FALSE                                 -- Cover if commanded
           ◁ cmd = cover ▷                             -- or learn
      link[smallest,n] ? second ;                      -- of comparison.
      ( link[n,second] ! move ;                        -- Alert disc to move
        exposed := FALSE                               -- and cover self
              ◁ second < n ▷                           -- if disc smaller or
        link[n,next] ! expose ;                        -- expose next
        position := position ⊕ 1 ;                     -- and move
        next := second )                               -- self.
           ◁ cmd = cover ▷
       SKIP )
  LISTEN =
    ▯ i=1,...base(2)
       ( link[i,n] ? cmd ;                             -- Await any command
          (running := FALSE                            -- and terminate
              ◁ cmd = terminate ▷                      -- if requested.
           link[n,next] ! expose ;                     -- Expose next
           position := position ⊖ 1 ;                  -- and move
           next := i                                   -- to exposed
           exposed := TRUE                             -- position
              ◁ cmd = move ▷                           -- if requested or
           exposed := TRUE                             -- simply expose
              ◁ cmd = expose ▷
           SKIP ) )
  RUN_B(n) =   running := TRUE
               running *
               ( ( TALK
                     ◁ exposed ▷
                  LISTEN ) )
where
```

TALK =
    link[base(n), smallest] ! position ;   -- Behave as
    link[smallest, base(n)] ? cmd ;
-- ordinary disc
    (exposed := FALSE
             ◁ cmd = cover ▷
      link[smallest,base(n)] ? second ;
      (link[base(n),second] ! move ;
        exposed := FALSE
            ◁ second < max ▷
-- until final position
        ||$_{i \neq base(n)}$link[base(n),i] ! terminate ;   -- and then terminate
        running := FALSE)   -- all processes
            ◁ cmd = compare ▷
      SKIP)
LISTEN =
    ▯ $_{i=1,...base(2)}$ (link[i,base(n)] ? cmd ;   -- Await command and
      (running := FALSE   -- either terminate
           ◁ cmd = terminate ▷   -- or
        exposed := TRUE))   -- expose.

Deadlock Analysis

As before proof that this specification is deadlock-free and satisfies the original specification proceeds by induction. There are three situations which may be encountered. Suppose the two exposed discs have indices a and b, with $position_b$ clockwise from $position_{smallest}$, then either

    (i) $next_{smallest} < a$, where $next_{smallest}$ is the index of a disc process ;

    (ii) $a < next_{smallest}$, where a is the number of a disc process

or

    (iii) a and $next_{smallest}$ are the indices of base processes.

In all cases the expansion for the inductive step begins

    ((link[a,smallest].position$\ominus$1 ; link[b,smallest].position$\ominus$1)

    ⊓

    (link[b,smallest].position$\oplus$1 ; link[a,smallest].position$\ominus$1)) ;

    link[smallest,b].cover ; link[smallest,a].compare,$next_{smallest}$.

In case (i) this continues (with variables without a suffix referring to the smallest disc) link[a,next].move ; link[next,next$_{next}$].expose ;

position := position $\oplus$ 1 ; next := b ;

position$_{next-}$:= position$\ominus$1 ; next$_{next-}$:= a ; exposed$_{next-}$:= TRUE ;

exposed$_{next*}$:= TRUE ; exposed$_a$:= FALSE ; exposed$_b$:= FALSE ;

(where next- is the value of next before the last assignment and next* is next$_{next-}$).

This contains deadlock-free communications and the assignments give a representation of the required cycle of moves.

In case (ii) the expansion continues

link[a,next$_a$].expose ;

position := position $\oplus$ 1 ; next := b ;

position$_a$:= position$_a$ $\oplus$ 1 ; next$_a$:= next- ;

exposed$_{next"}$:= TRUE ; exposed$_b$:= FALSE ;

(where next" is next$_a$).

Again this is deadlock-free. Finally in case (iii) the expansion continues

position := position$\ominus$1 ; next := b ;

exposed$_{next}$:= TRUE ; exposed$_b$:= FALSE ;

$\Sigma_{i \neq a}$ link[a,i].terminate ;

Σ_i running$_i$:= FALSE

where the assignments correspond to a single move followed by termination of all processes.

This completes the verification of the inductive step since it is not hard to show that the assignments all represent legal stack moves, and so return the parallel combination to a combined state from where the next step can start (provided we have not reached termination). Combining the inductive step with initialisation gives us a proof of correctness.

Conclusions

This paper has demonstrated several features of ODM. Parallelism has been sought and exploited. The CAP notation has been used for specification and verification and it is easy to see how the resulting refined specification can be turned into OCCAM code and run on a Transputer network. These five finger exercises are part of our initiative to develop graphical software tools to be used interactively for specification and the OCCAM code generation applicable to more complex problems.

References

[1] S. D. Brookes; C. A. R. Hoare and A. W. Roscoe, *A Theory of Communicating Processes,* Journal ACM **31** (7), 560-599 (1984).

[2] W.D. Crowe and P.E.D. Strain-Clark, *ODM Technical Note 3*, Open University pre-print, (in preparation).

[3] M. Gardiner, *Mathematical Puzzles and Diversions.* Simon and Schuster, New York (1959).

[4] C.A.R.Hoare, *Communicating Sequential Processes,* Prentice-Hall International Series in Computer Science (1985).

[5] INMOS Ltd, *OCCAM Programming Manual,* Prentice-Hall International Series in Computer Science (1985).

[6] J. S. Rohl, Towers of Hanoi, *The Derivation of Some Iterative Versions,* The Computer Journal **(20)** (3), 282-285 (1977).

[7] A. W. Roscoe and C.A.R. Hoare, *The Laws of OCCAM Programming,* Oxford University Computing Laboratory, Programming Research Group.

Author Index